Climate Change and Ecotoxicology

Climate Change and Ecotoxicology

Editor: Giselle Tang

R CALLISTO REFERENCE

www.callistoreference.com

Callisto Reference,
118-35 Queens Blvd., Suite 400,
Forest Hills, NY 11375, USA

Visit us on the World Wide Web at:
www.callistoreference.com

ISBN: 978-1-63239-834-5 (Hardback)

Cataloging-in-publication Data

Climate change and ecotoxicology / edited by Giselle Tang.
    p. cm.
Includes bibliographical references and index.
ISBN 978-1-63239-834-5
1. Climatic changes. 2. Environmental toxicology. 3. Ecology. 4. Biodiversity. I. Tang, Giselle.
QC903 .C55 2017
551.6--dc23

Table of Contents

Preface

Climate change is a growing threat to the environment. This book on climate change and ecotoxicology deals with the relationship between the release of organic pollutants such as chlorofluorocarbon and aerosols and the damage caused by them through the phenomenon categorized as climate change. Research within ecotoxicology brings forward the exact nature and extent of destruction that greenhouse gases cause and suggests remedies and mitigation measures. This book, which seeks to incorporate the latest information on climate change mitigation, will be of great help to students and researchers of environmental design, environmental technology and waste management. This book explores all the important aspects of climate change in the present day scenario. The various sub-fields of ecotoxicology and climate change along with technological progress that have future implications are glanced at in this book. As this field is emerging at a rapid pace, the contents of this book will help the readers understand the modern concepts and applications of the subject.

This book unites the global concepts and researches in an organized manner for a comprehensive understanding of the subject. It is a ripe text for all researchers, students, scientists or anyone else who is interested in acquiring a better knowledge of this dynamic field.

I extend my sincere thanks to the contributors for such eloquent research chapters. Finally, I thank my family for being a source of support and help.

Editor

Phylogeny Predicts Future Habitat Shifts Due to Climate Change

Matjaž Kuntner[1,2,3*]**, Magdalena Năpăruş**[4,5]**, Daiqin Li**[2,6]**, Jonathan A. Coddington**[3]

1 Institute of Biology, Scientific Research Centre, Slovenian Academy of Sciences and Arts, Ljubljana, Slovenia, **2** Centre for Behavioural Ecology and Evolution, College of Life Sciences, Hubei University, Wuhan, Hubei, China, **3** National Museum of Natural History, Smithsonian Institution, Washington, D. C., United States of America, **4** Centre of Landscape–Territory–Information Systems - CeLTIS, University of Bucharest, Bucharest, Romania, **5** Tular Cave Laboratory, Kranj, Slovenia, **6** Department of Biological Sciences, National University of Singapore, Singapore, Singapore

Abstract

Background: Taxa may respond differently to climatic changes, depending on phylogenetic or ecological effects, but studies that discern among these alternatives are scarce. Here, we use two species pairs from globally distributed spider clades, each pair representing two lifestyles (generalist, specialist) to test the relative importance of phylogeny versus ecology in predicted responses to climate change.

Methodology: We used a recent phylogenetic hypothesis for nephilid spiders to select four species from two genera (*Nephilingis* and *Nephilengys*) that match the above criteria, are fully allopatric but combined occupy all subtropical-tropical regions. Based on their records, we modeled each species niche spaces and predicted their ecological shifts 20, 40, 60, and 80 years into the future using customized GIS tools and projected climatic changes.

Conclusions: Phylogeny better predicts the species current ecological preferences than do lifestyles. By 2080 all species face dramatic reductions in suitable habitat (54.8–77.1%) and adapt by moving towards higher altitudes and latitudes, although at different tempos. Phylogeny and life style explain simulated habitat shifts in altitude, but phylogeny is the sole best predictor of latitudinal shifts. Models incorporating phylogenetic relatedness are an important additional tool to predict accurately biotic responses to global change.

Editor: Mónica Medina, Pennsylvania State University, United States of America

Funding: This research was supported in part by a Raffles Museum for Biodiversity Research (RMBR) Short-term Fellowship and the grants P10236, BI-US/09-12-016 and MU-PROM/12-001 from the Slovenian Research Agency to M.K. and by the NSFC grant (31272324) and Singapore Ministry of Education (MOE) AcRF grant (R-154-000-476-112) to D.L. The funders had no role in study design, data collection and analysis, decision to publish, or preparation of the manuscript.

Competing Interests: The authors have declared that no competing interests exist.

* E-mail: kuntner@gmail.com

Introduction

Biotic change due to global warming is increasing dramatically. Although climatic changes are increasingly well understood and future predictions are improving, it is much more difficult to predict biotic responses to climate change [1,2]. Models predict global climates to become warmer in the following decades (Fig. S1), and precipitation patterns will change accordingly, but how organisms will respond to such changes is less clear. Naturally, different life histories and even clades of organisms may respond in various ways [3]. For example, they may respond with a shift in time, e.g. changing their phenology [2,4], or a shift in space, such as moving to different habitats, latitudes, or altitudes [2,5,6,7], insofar as habitats are available [8]. Crucial species interactions may be broken by both types of shifts [9]. Tropical species, in particular, may be severely affected [4,10]. Actual adaptive evolutionary change within a few decades may be possible for some organisms with extraordinary genotypic and phenotypic plasticity [11,12] but is unlikely for most as it would require over 10,000 times faster rates of adaptive change than those estimated

[13]. Finally, the species unable to cope with anthropogenic global changes, and those whose dispersal ability is low, are likely to become extinct either globally or locally [1,3,14].

Communities, ecosystems and biomes are also likely to be forced to shift in space [2], become increasingly destabilized due to biotic loss [15], or be subject to thermophilization [16] and invasion [17]. In addition, food web and trophic interactions may be altered [18]. While biomes, notably tropical and subtropical forests, may expand in range due to increased rainfall and atmospheric CO_2 levels [19], such trends are expected to be countered by human habitat degradation, probably resulting in net habitat loss [20]. In some cases, temperate assemblages may increase in species richness or exhibit a higher species turnover [21]. However, at the global level with all factors combined, most biomes and communities are predicted to deteriorate, and thus may simply shrink or disappear. Their space then may be taken over by exotic and invasive species and communities, whose success can sometimes be directly attributed to global change [17].

Terrestrial organisms may be most prone to global climate changes (but see freshwater and marine reviews [22,23]), yet

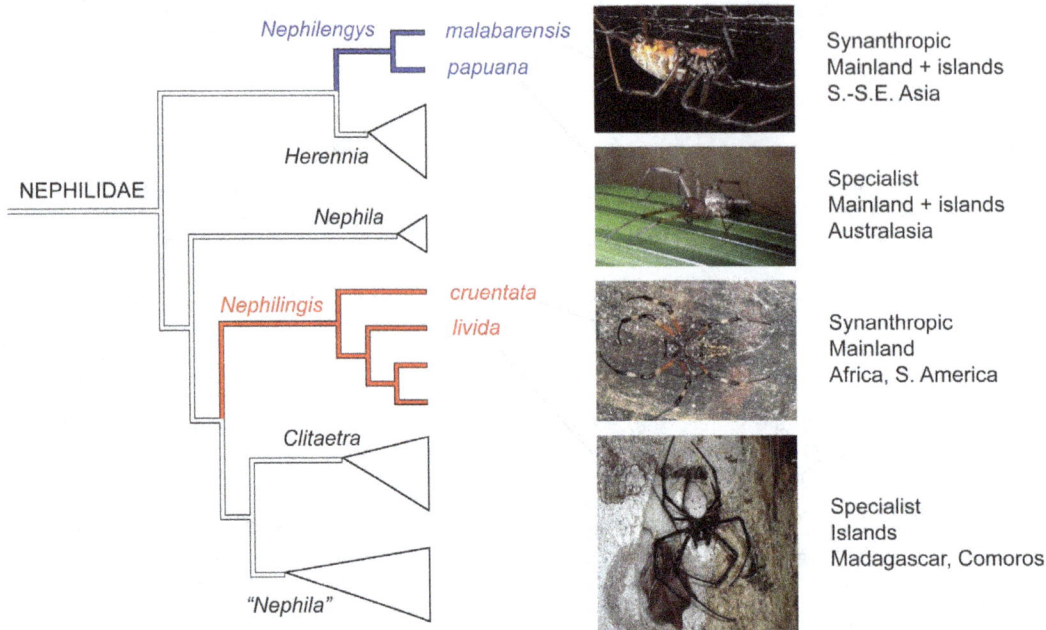

Figure 1. Two nephilid species pairs, their phylogeny, and basic ecology. One species from each clade is synanthropic and the other one a habitat specialist. This sample tests the relative importance of phylogeny versus life history on species responses to climatic changes. The phylogenetic hypothesis builds on a nephilid species level study that used 4kB of nucleotide data in addition to morphology, and an array of analytical approaches and sensitivity analyses [28].

surprisingly few studies have modeled the future of closely related terrestrial lineages—clades—in order to discern among phylogenetic and ecological effects while studies modelling single species continue to abound [24,25]. As an example of clade predictions, Diamond et al. [26] examined models for thermal tolerance in ants based on current and future climates and found that tropical ants had lower tolerances to warming. Tropical ant faunas, thus, are more susceptible to global warming, and that is precisely where the most diversity lies. However, is this necessarily true for all tropical ants? One would expect that phylogeny may affect biotic responses to climate change at least as much as ecology or natural history [27]. Fine grained studies discerning among these factors are currently lacking [3].

To test the relative importance of phylogeny versus ecology in predicted biotic responses to climate change, we selected two clades within which species have occupied two distinct ecological niches. The spider genera *Nephilengys and Nephilingis* [28] are large terrestrial invertebrate predators. The Asian *Nephilengys malabarensis* is synanthropic and hence relatively generalized, and the Australasian *Ny. papuana* is a forest species, thus relatively specialized [29]. Likewise the African-South American *Nephilingis cruentata* is synanthropic, whereas the Madagascan-Comoroan *Ni. livida* is a forest species (Fig. 1) [29,30,31]. The four species are fully allopatric, but combined they occupy all subtropical-tropical regions [29]. Ecologically the two specialists, *Ni. livida* and *Ny. papuana* should be more prone to habitat loss than the generalists *Ny. malabarensis* and *Ni. cruentata*. This prediction derives from the assertions that species with greater ecological generalization should be more likely to shift to higher elevation and latitude [3] and that niche breath positively correlates with species geographic ranges [32]. Alternatively, phylogenetic relatedness may better explain response trends. We first used predicted temperature and precipitation changes (Fig. S1) and a GIS simulation tool to model habitat degradation, gain, and loss in the four species (Fig. 2,

Fig. S2) at 20 year intervals from 2000–2080. We then investigated the extent to which phylogeny versus ecology explained habitat changes. A priori predictions are difficult: if commonly derived species undergo life history adaptation to reduce competition, then ecology may be a better predictor. If not, phylogeny may be used to predict habitat shifts due to climate change.

Materials and Methods

Modeling current habitat suitability

We used all available locality data for the four species [29]: *Ni. cruentata* (specimen records N = 436), *Ni. livida* (N = 138), *Ny. malabarensis* (N = 138) and *Ny. papuana* (N = 39). Since our previous study phylogenetic evidence has emerged placing *Ni. cruentata* and *Ni. livida* in the genus *Nephilingis*, which is phylogenetically distant from *Nephilengys* containing *N. malabarensis* and *Ny. papuana* [28]. This is currently the best available phylogenetic hypothesis of nephilid species relationships, having utilized 4 kB of nucleotide data in addition to morphology, and an array of analytical approaches and sensitivity analyses [28]. That the previously congeneric species in fact belong to distinct clades, which nevertheless contain ecologically dissimilar species (one synanthropic and one habitat specialist within each genus), adds power to our testing of phylogenetic versus life history effects on habitat suitability. Our previous model selected best fit pairs of ecological parameters for each species current habitat suitability assessment based on the outcomes of backward linear regression: annual mean temperature and elevation was used for the *Ni. cruentata* model, global land cover and elevation for the *Ni. livida* model, and global land cover and annual mean precipitation for both *Ny. malabarensis* and *Ny. papuana* models [29]. Here, we used the same pairs of parameters for all four species models, which are indicative of global climate change: annual mean temperature (TMA) and annual mean precipitation (PMA). Based on these two parameters,

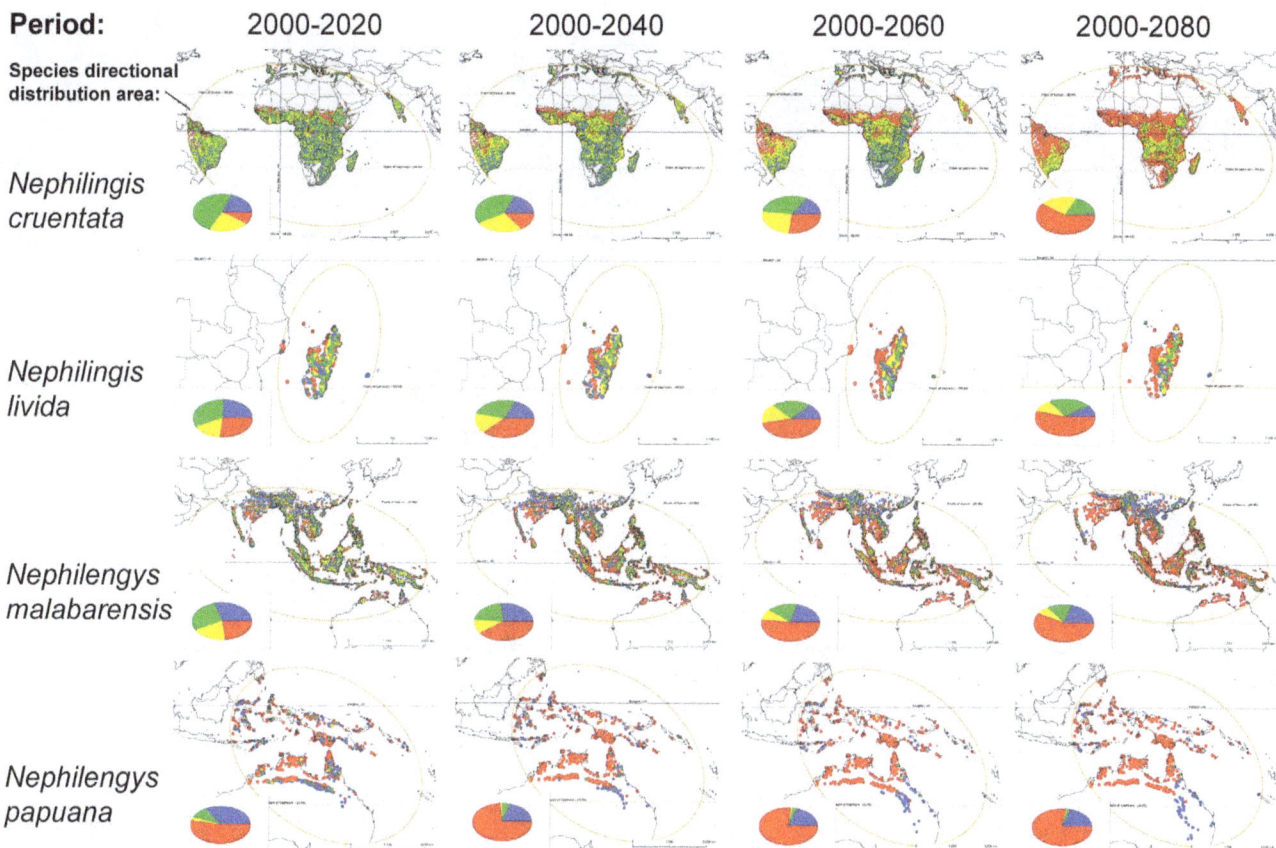

Figure 2. Predicted habitat suitability changes of two *Nephilingis* **and two** *Nephilengys* **species over increasing time ranges.** The predictions are based on the IPCC scenario A1B for temperature and precipitation (see Figs S1–S2) and are modeled for the time ranges 2000 vs. 2020, 2040, 2060, and 2080. Future habitat is classified as new, unchanged, degraded or lost.

we modeled the starting point for the year 2000, taken as each species' current habitat suitability.

Modeling future habitat suitability

We downloaded the raster data for TMA and PMA for the years 2020 to 2080 from the Downscaled Global Circulation Models data portal (http://www.ccafs-climate.org), and the current distribution (1950–2000) of climates from the WorldClim Database [33]. The raster layers are based on the IPCC Special Report on Emissions Scenarios (SRES) [34], where several scenario families of predicted climate change were proposed. The scenario A1 builds on the assumption of maximum energy requirements, and the sub-choices are based on emissions expected from the use of different fuel sources: fossil intensive (A1F1), technologically developed non-fossil sources (A1T) and a balance across sources (A1B). Among the worst case scenario (A1) we chose the balanced use of fuel sources (A1B) as the basis for our models. The remaining three scenarios build on minimum energy requirements and emissions (B1), on high energy requirements (A1F1) and on lower energy requirements; however, as the scenarios A2 and B2 only predict a limited number of years (2020, 2050 and 2080), and the A1 scenario predicts each decade, our choice of scenario (A1B) seemed justified. For global modeling use, IPCC recommends the use of A1 and B1 [35], among which we chose to model a more pessimistic version due to the availability of

predicted climate change (for B1 no associated datasets exist). The A1B scenario uses 24 Global Circulation Models [36]. All rasters have a spatial resolution of 2.5 arc-minutes, using a WGS84 datum.

The A1 group of scenarios predicts a global average surface warming until 2100 from 1.4 to 6.4°C [35]. Fig. S1 shows projected global changes in TMA and PMA for the period covered by our models (current = 2000; projected 2020, 2040, 2060, 2080). Based on these, we modeled habitat suitability of each of the four species following our GIS modeling methodology [29]. These maps define each species' directional distribution as its potential target area and thus potential dispersal range [29], and within it, we used the frequency distribution values for TMA and PMA for current specimen records [29]. We here explore the changes for each species in three habitat categories: high, moderate and low habitat suitability (Fig. S2). Several data sources exist that model global changes according to the A1B scenario. We choose the data from the Canadian Centre for Climate Modelling and Analysis - CGCM3.1 (T47), 2005 [37]. This data model version is the basis for the suite of model simulations in the IPCC Fourth Assessment Report [35]. T47 version of CGCM3.1 data model has a surface grid whose spatial resolution is roughly 3.75 degrees latitude/longitude and 31 levels vertically, and an ocean grid resolution of roughly 1.85 degrees, with 29 levels vertically [38,39].

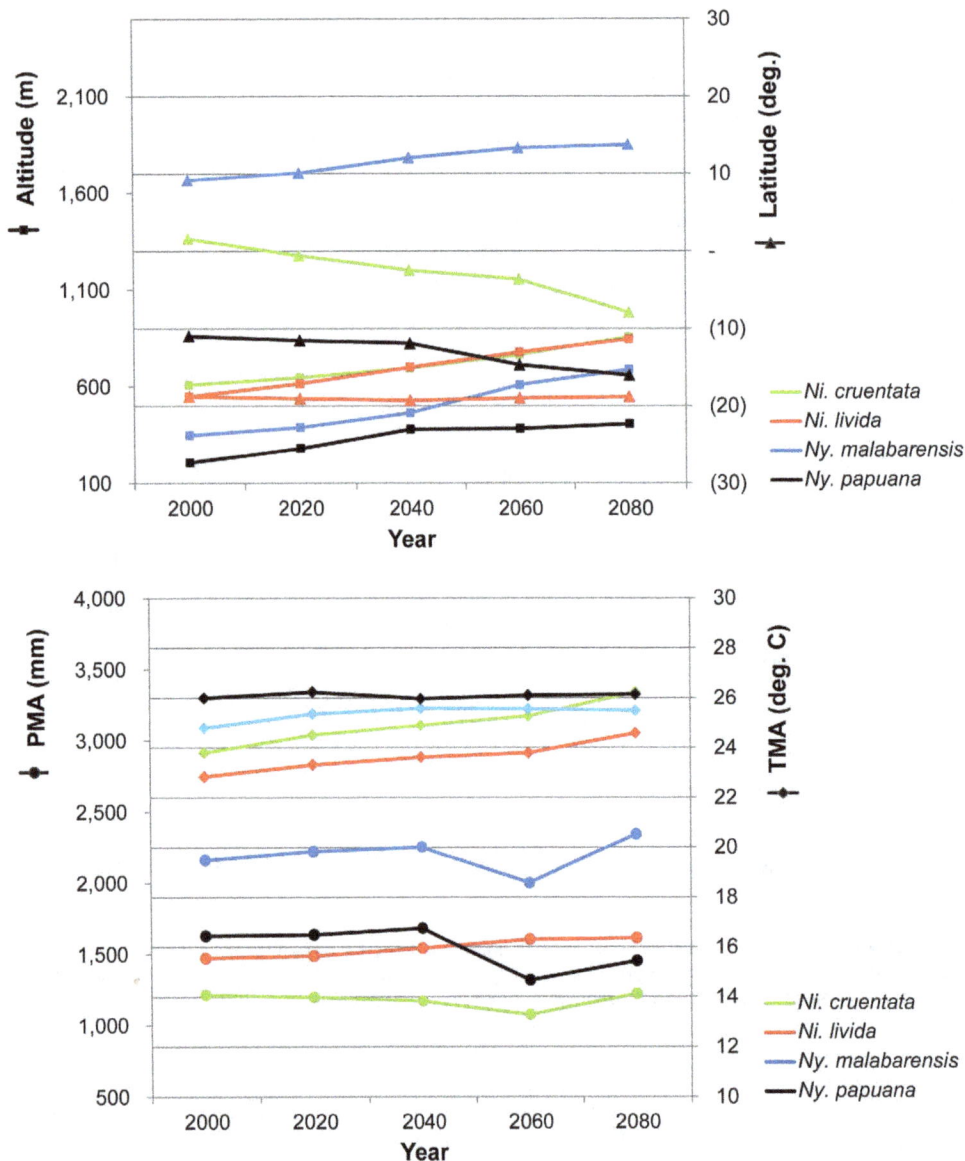

Figure 3. Changes for all species in altitude and latitude (above), and temperature and precipitation (below). Species maintain their characteristic climatic conditions by shifting towards higher altitudes and latitudes. Geographical and ecological averages for the modeled suitable habitats combined all categories (low, medium and high suitability). NI = *Nephilingis*, NG = *Nephilengys*, PMA = mean annual precipitation, TMA = mean annual temperature.

Assessing species habitat loss

To assess the degrees of projected habitat losses or gains due to climatic changes, we compared the differences in predicted habitat suitability categories for the following time periods: 2000–2020, 2000–2040, 2000–2060, and 2000–2080. We customized our previous GIS model [29] with additional tools in ArcGIS [40]: i) First, we reclassified each year's habitat suitability raster data where no habitat suitability was shown from 'NoData' to '99' using the tool 'Reclassify'; those values that showed a habitat suitability value were left as they were (high = 3; moderate = 2; low = 1); ii) We then compared using the 'Combine' tool the habitat category changes between the year 2000 and other time periods and reclassified the resulting combinations into the following nominal values: 1 = new habitat (indicating projected habitat suitability gain); 2 = unchanged habitat (indicating no change in habitat suitability); 3 = degraded habitat (indicating a loss in habitat

suitability value, but not below 1) and 4 = lost habitat (indicating a complete loss of projected habitat suitability, i.e. resulting in a change to 99).

Statistical analyses

Since the TMA and PMA data were not normally distributed, we employed standard non-parametric statistics to test the effects of phylogenetic relatedness (*Nephilingis* vs. *Nephilengys*) and life history (synanthropic vs. habitat specialist) on current species ecological preferences. To determine whether phylogenetic or ecological factors are good predictors of the simulated habitat shifts, we first devised a randomized dataset of 10,000 simulated georeferenced records per period per species (totaling 200,000 records; Table S1), calculated latitudinal and altitudinal shifts and coded them as categorical data (i.e., shift or not) based on the shifted values. We performed generalized linear model (GLM)

Table 1. Predicted percentages of future habitat change per species relative to the year 2000.

Species	Habitat categories	2020	2040	2060	2080
Nephilingis cruentata	new	19.9%	19.2%	16.1%	1.2%
	unchanged	48.5%	40.0%	31.7%	17.8%
	degraded	20.9%	25.1%	25.6%	20.4%
	lost	10.7%	15.7%	26.6%	60.6%
Nephilingis livida	new	24.5%	19.1%	14.2%	10.6%
	unchanged	33.9%	26.6%	21.6%	23.3%
	degraded	15.2%	17.8%	17.7%	11.3%
	lost	26.4%	36.5%	46.5%	54.8%
Nephilengys malabarensis	new	29.3%	26.8%	21.5%	21.0%
	unchanged	29.4%	21.6%	16.8%	13.2%
	degraded	17.8%	12.5%	9.7%	7.2%
	lost	23.5%	39.1%	52.0%	58.6%
Nephilengys papuana	new	33.3%	20.8%	20.2%	20.9%
	unchanged	10.5%	4.2%	2.5%	1.5%
	degraded	3.2%	1.0%	0.7%	0.5%
	lost	53.0%	74.0%	76.6%	77.1%

using ordinal logistical model twice, once for latitudinal shifts and once for altitudinal shifts, with phylogeny, life style, period (2020, 2040, 2060 and 2080) and habitat suitability (high = 3; moderate = 2; low = 1) as predictors. Both period and habitat suitability showed the same effects on shifts in the initial GLM tests, thus being excluded from the final models, which tested the effects of only two factors (phylogeny and life style) and the 2-way interaction on the categorical latitude and altitude shifts. The model was fitted using multinominal distribution with cumulative logit link error. Akaike information criterion (AIC: the smallest is the best) was used to select the best model. All statistical analyses were done using IBM SPSS Statistics 21 (IBM Corporation, USA). Reported p-values are two-tailed tests, with $\alpha = 0.05$.

Results

The global habitat suitability models for 2000, 2020, 2040, 2060 and 2080 predict overall shrinking or disappearance of suitable habitat for all four species (Fig. S2), although minor amounts of

newly appropriate habitat emerge over time. Categorizing habitat change as unchanged, degraded, lost, or new, the major trend across all time ranges is dramatic increase in either lost or degraded habitat for all species (Fig. 2; Table 1). Losses and gains can take different tempos. *Ni. cruentata* shows a dramatic loss by 2080, but in *Ny. papuana* the most dramatic loss occurs by 2040. Others are intermediate. In no case does new habitat tend to increase (Table 1). By 2080, habitat loss ranges from 54.8% (*Ni. livida*) to 77.1% (*Ny. papuana*). Even the widespread, generalist species *Ni. cruentata* and *Ny. malabarensis* lose 60.6% and 58.6% of their habitat. According to the model, all four species roughly maintain their known temperature and precipitation preferences (Fig. 3). However in order to do so, all species shift towards higher altitudes and latitudes (except *Ni. livida*, whose island habitats are obviously fixed) (Fig. 3).

Phylogeny significantly explains variation in current temperature preferences (Mann-Whitney U = 24,965, $df = 1$, $p < 0.0001$). Both phylogeny and lifestyle significantly explain current precipitation preferences, but phylogeny much more strongly than

Table 2. Results from a generalized linear model (GLM) testing the effects of two explorative factors (phylogeny and lifestyle) on how spiders respond to climate conditions (based on the IPCC scenario A1B for temperature and precipitation) by shifting latitude or altitude.

Shift	Exploratory factors	Wald χ^2	df	p
Latitude	Phylogeny (genus-pair)	310.711	1	<0.0001
	Life style	3.65	1	0.056
	Phylogeny × Life style	13.730	1	<0.0001
Altitude	Phylogeny (genus-pair)	30.825	1	<0.0001
	Life style	17.350	1	<0.0001
	Phylogeny × Life style	0.082	1	0.775

The model that included two factors and 2-way interaction was fitted using multinominal distribution with cumulative logit link error.
GLM test for latitude shift: Goodness of fit: AIC = 116.261; Omnibus test: $\chi^2 = 328.2$, $df = 3$, $p < 0.0001$; GLM test for altitude shift: Goodness of fit: AIC = 107.247; Omnibus test: $\chi^2 = 48.242$, $df = 3$, $p < 0.0001$.

lifestyle (Mann-Whitney U = 25,803, $df=1$, $p<0.0001$ and U = 13,503, $df=1$, $p<0.016$).

GLM analyses on simulated data show that phylogeny predicts latitudinal shifts better than life style does (Goodness of fit: AIC = 116.261; Omnibus test: $\chi^2=328.2$, $df=3$, $p<0.0001$; Table 2): *Nephilengys* is more likely to shift latitude than *Nephilingis*. Both phylogeny and life style are good predictors of the simulated altitudinal shifts (Goodness of fit: AIC = 107.247; Omnibus test: $\chi^2=48.242$, $df=3$, $p<0.0001$; Table 2): altitudinal shifts are more likely in *Nephilingis* than *Nephilengys* and in synanthropic species than in habitat specialists.

Discussion

Models that forecast biodiversity patterns have been grouped into four categories [41]: 1) those considering individual species, 2) those grouping species by niche, 3) those based on general circulation or coupled ocean-atmosphere-biosphere theories, and 4) those based on species-area curves [41]. We argue that models using phylogenetic relatedness should also be considered because our results have shown that phylogeny can strongly predict climatic and habitat preference.

Studies considering individual species fail to account for species interactions and phylogenetic factors and are therefore of limited general use. In one case, the already limited range of the golden striped salamander decreased even more [42]; in another an invasive species, the Australian redback spider, increased its already substantial distribution [43]. Outside of a phylogenetic context, these patterns make no general predictions for evolutionarily or ecologically closely related organisms.

More taxonomically inclusive studies often predict changes in large clades of organisms, notably amphibians [44,45] or ants [26], but again, fine-grained clade-based studies are too preliminary. For example, the ranges of European amphibians and reptiles are generally predicted to shrink by 2050 [44], but this study was based on two opposite, and equally unlikely assumptions of unlimited versus no dispersal. A study of plant, bird and mammal assemblages in Europe only found a weak relationship between phylogeny and climate change vulnerability [46]. However, Willis et al. [47] investigated the changes in phenology and abundance of a temperate flora over 150 years and found that different clades are quite differently affected by climate change.

Niche modelling is limited as it fails to account for differences in species natural histories [3]. Only recently have studies taken into account species trait variability in assessing predicted responses to climate change. Angert et al. [3] examined to what extent species' traits are predictive of expanding their ranges. Predicting ecologically general species with greater dispersal abilities to more easily shift to higher latitudes and altitudes in response to climatic changes, they did find the expected relationships in passeriform birds and odonates, but their models yielded low explanatory power.

Our study extended this logic on a phylogenetically fine-grained taxonomic sample, by testing the relative importance of phylogeny versus species traits in predicted biotic responses to climate change

in two species pairs from globally distributed spider clades, each pair representing two lifestyles. Despite the prediction that the unrelated specialists (forest-dwelling species) would cope with projected global changes worse than the unrelated generalists (synanthropic species), our results project habitat shrinkage patterns that may better be predicted by phylogeny than life style.

Our modeling approach assumes genotypic and phenotypic species homogeneity, perhaps an unwarranted assumption considering wide geographical ranges these species occupy. In reality, even widespread generalists show considerable genetic and phenotypic adaptations, spiders being no exception [48]. The predicted habitat shifts therefore may not be equally likely throughout the entire range of these taxa, and it may be possible that levels of adaptive change could take place at short temporal scales such as these presented in our study. The modes of habitat loss and extinction may depend on many additional factors, such as geography or basic natural history that are too complex to model accurately [3]. Phylogeny—which does broadly predict many life history traits—can simplify the prediction of biotic responses to future climate change. For accurate modelling, of course, many other factors are important as well, such as species interactions [49,50], persistence abilities [51], population dynamics and changes in genetic diversity [52]. However, more work assessing phylogeny versus ecology and life history as explanations for responses to climate change is needed.

Supporting Information

Figure S1   Predicted global changes in temperature and precipitation based on the IPCC scenario A1B. These predicted changes were used as bases for modeling species distribution 2000–2020, 2000–2040, 2000–2060 and 2000–2080.

Figure S2   Models predicting future habitat suitability for two *Nephilingis* and two *Nephilengys* species. The models for the time periods 2000, 2020, 2040, 2060 and 2080 are based on the IPCC scenario A1B for temperature and precipitation changes (see Fig. S1).

Acknowledgments

We thank Mitch Aide, Ingi Agnarsson and Gregor Aljančič for discussions and three anonymous referees for their constructive comments.

Author Contributions

Conceived and designed the experiments: MK MN. Performed the experiments: MK MN. Analyzed the data: MK MN DL JAC. Contributed reagents/materials/analysis tools: MK MN DL JAC. Wrote the paper: MK MN DL JAC.

References

1. Sandel B, Arge L, Dalsgaard B, Davies RG, Gaston KJ, et al. (2011) The Influence of Late Quaternary Climate-Change Velocity on Species Endemism. Science 334: 660–664.
2. Bellard C, Bertelsmeier C, Leadley P, Thuiller W, Courchamp F (2012) Impacts of climate change on the future of biodiversity. Ecology Letters 15: 365–377.
3. Angert AL, Crozier LG, Rissler LJ, Gilman SE, Tewksbury JJ, et al. (2011) Do species' traits predict recent shifts at expanding range edges? Ecology Letters 14: 677–689.
4. Pau S, Wolkovich EM, Cook BI, Davies TJ, Kraft NJB, et al. (2011) Predicting phenology by integrating ecology, evolution and climate science. Global Change Biology 17: 3633–3643.
5. Franco AMA, Hill JK, Kitschke C, Collingham YC, Roy DB, et al. (2006) Impacts of climate warming and habitat loss on extinctions at species' low-latitude range boundaries. Global Change Biology 12: 1545–1553.
6. Saikkonen K, Taulavuori K, Hyvonen T, Gundel PE, Hamilton CE, et al. (2012) Climate change-driven species' range shifts filtered by photoperiodism. Nature Climate Change 2: 239–242.

7. Beale CM, Baker NE, Brewer MJ, Lennon JJ (2013) Protected area networks and savannah bird biodiversity in the face of climate change and land degradation. Ecology Letters 16: 1061–1068.

8. Lawler JJ, Ruesch AS, Olden JD, McRae BH (2013) Projected climate-driven faunal movement routes. Ecology Letters 16: 1014–1022.

9. Hegland SJ, Nielsen A, Lazaro A, Bjerknes AL, Totland O (2009) How does climate warming affect plant-pollinator interactions? Ecology Letters 12: 184–195.

10. Colwell RK, Brehm G, Cardelus CL, Gilman AC, Longino JT (2008) Global warming, elevational range shifts, and lowland biotic attrition in the wet tropics. Science 322: 258–261.

11. Erwin DH (2009) Climate as a Driver of Evolutionary Change. Current Biology 19: R575–R583.

12. Munday PL, Warner RR, Monro K, Pandolfi JM, Marshall DJ (2013) Predicting evolutionary responses to climate change in the sea. Ecology Letters 16: 1488–1500.

13. Quintero I, Wiens JJ (2013) Rates of projected climate change dramatically exceed past rates of climatic niche evolution among vertebrate species. Ecology Letters 16: 1095–1103.

14. Jump AS, Penuelas J (2005) Running to stand still: adaptation and the response of plants to rapid climate change. Ecology Letters 8: 1010–1020.

15. Post E (2013) Erosion of community diversity and stability by herbivore removal under warming. Proceedings of the Royal Society B-Biological Sciences 280.

16. Gottfried M, Pauli H, Futschik A, Akhalkatsi M, Barancok P, et al. (2012) Continent-wide response of mountain vegetation to climate change. Nature Climate Change 2: 111–115.

17. Sandel B, Dangremond EM (2012) Climate change and the invasion of California by grasses. Global Change Biology 18: 277–289.

18. Barton BT, Schmitz OJ (2009) Experimental warming transforms multiple predator effects in a grassland food web. Ecology Letters 12: 1317–1325.

19. Banfai DS, Bowman DM (2006) Forty years of lowland monsoon rainforest expansion in Kakadu National Park, Northern Australia. Biological Conservation 131: 553–565.

20. Heubes J, Kuhn I, Konig K, Wittig R, Zizka G, et al. (2011) Modelling biome shifts and tree cover change for 2050 in West Africa. Journal of Biogeography 38: 2248–2258.

21. Hillebrand H, Soininen J, Snoeijs P (2010) Warming leads to higher species turnover in a coastal ecosystem. Global Change Biology 16: 1181–1193.

22. Heino J, Virkkala R, Toivonen H (2009) Climate change and freshwater biodiversity: detected patterns, future trends and adaptations in northern regions. Biological Reviews 84: 39–54.

23. Brierley AS, Kingsford MJ (2009) Impacts of Climate Change on Marine Organisms and Ecosystems. Current Biology 19: R602–R614.

24. Razgour O, Juste J, Ibanez C, Kiefer A, Rebelo H, et al. (2013) The shaping of genetic variation in edge-of-range populations under past and future climate change. Ecology Letters 16: 1258–1266.

25. Bennie J, Hodgson JA, Lawson CR, Holloway CTR, Roy DB, et al. (2013) Range expansion through fragmented landscapes under a variable climate. Ecology Letters 16: 921–929.

26. Diamond SE, Sorger DM, Hulcr J, Pelini SL, Del Toro I, et al. (2012) Who likes it hot? A global analysis of the climatic, ecological, and evolutionary determinants of warming tolerance in ants. Global Change Biology 18: 448–456.

27. Kearney MR (2013) Activity restriction and the mechanistic basis for extinctions under climate warming. Ecology Letters 16: 1470–1479.

28. Kuntner M, Arnedo MA, Trontelj P, Lokovšek T, Agnarsson I (2013) A molecular phylogeny of nephilid spiders: Evolutionary history of a model lineage. Molecular Phylogenetics and Evolution 69: 961–979.

29. Năpăruş M, Kuntner M (2012) A GIS model predicting potential distributions of a lineage: A test case on hermit spiders (Nephilidae: Nephilengys). PLoS ONE 7: e30047.

30. Kuntner M, Agnarsson I (2011) Biogeography and diversification of hermit spiders on Indian Ocean islands (Nephilidae: Nephilengys). Molecular Phylogenetics and Evolution 59: 477–488.

31. Kuntner M (2007) A monograph of Nephilengys, the pantropical 'hermit spiders' (Araneae, Nephilidae, Nephilinae). Systematic Entomology 32: 95–135.

32. Slatyer RA, Hirst M, Sexton JP (2013) Niche breadth predicts geographical range size: a general ecological pattern. Ecology Letters 16: 1104–1114.

33. Hijmans RJ, Cameron, S E., Parra, J L., Jones, P G. and Jarvis A. (2005) Very high resolution interpolated climate surfaces for global land areas. International Journal of Climatology 25: 1965–1978.

34. IPCC (2000) Special Report Emissions Scenarios - Summary for Policymakers.

35. IPCC (2007) A Report of Working Group I of the Intergovernmental Panel on Climate Change - Summary for Policymakers.

36. Ramirez-Villegas J, Jarvis A (2010) Downscaling Global Circulation Model outputs: the Delta method decision and policy analysis working paper no. 1 Cali, Colombia International Center for Tropical Agriculture, CIAT. 1–18 p.

37. Scinocca JF, McFarlane NA, Lazare M, Li J, Plummer D (2008) Technical Note: The CCCma third generation AGCM and its extension into the middle atmosphere. Atmospheric Chemistry and Physics 8: 7055–7074.

38. Randall DA, Wood RA, Bony S, Colman R, Fichefet T, et al. (2007) Cilmate Models and Their Evaluation. In:Climate Change 2007: The Physical Science Basis. Contribution of Working Group I to the Fourth Assessment Report of the Intergovernmental Panel on Climate Change [Solomon, S., D. . Qin, M. Manning, Z. Chen, M. Marquis, K.B. Averyt., M. Tignor and H.L. . Miller (eds.)]. Cambridge: Cambridge University Press.

39. Flato GM, Boer DY (2001) Warming Asymmetry in Climate Change Simulations. Geophys Res Lett 28: 195–198.

40. ESRI (2010) ArcGIS 9.3.1 Environmental Systems Research Institute, Inc., USA.

41. Botkin DB, Saxe H, Araujo MB, Betts R, Bradshaw RHW, et al. (2007) Forecasting the effects of global warming on biodiversity. Bioscience 57: 227–236.

42. Teixeira J, Arntzen JW (2002) Potential impact of climate warming on the distribution of the Golden-striped salamander, Chioglossa lusitanica, on the Iberian Peninsula. Biodiversity and Conservation 11: 2167–2176.

43. Vink CJ, Derraik JGB, Phillips CB, Sirvid PJ (2011) The invasive Australian redback spider, Latrodectus hasseltii Thorell 1870 (Araneae: Theridiidae): current and potential distributions, and likely impacts. Biological Invasions 13: 1003–1019.

44. Araujo MB, Thuiller W, Pearson RG (2006) Climate warming and the decline of amphibians and reptiles in Europe. Journal of Biogeography 33: 1712–1728.

45. Hof C, Araujo MB, Jetz W, Rahbek C (2011) Additive threats from pathogens, climate and land-use change for global amphibian diversity. Nature 480: 516–U137.

46. Thuiller W, Lavergne S, Roquet C, Boulangeat I, Lafourcade B, et al. (2011) Consequences of climate change on the tree of life in Europe. Nature 470: 531–534.

47. Willis CG, Ruhfel B, Primack RB, Miller-Rushing AJ, Davis CC (2008) Phylogenetic patterns of species loss in Thoreau's woods are driven by climate change. Proceedings of the National Academy of Sciences of the United States of America 105: 17029–17033.

48. Kuntner M, Agnarsson I (2011) Phylogeography of a successful aerial disperser: the golden orb spider Nephila on Indian Ocean islands. Bmc Evolutionary Biology 11.

49. Pigot AL, Tobias JA (2013) Species interactions constrain geographic range expansion over evolutionary time. Ecology Letters 16: 330–338.

50. Araujo MB, Luoto M (2007) The importance of biotic interactions for modelling species distributions under climate change. Global Ecology and Biogeography 16: 743–753.

51. Hof C, Levinsky I, Araujo MB, Rahbek C (2011) Rethinking species' ability to cope with rapid climate change. Global Change Biology 17: 2987–2990.

52. Habel JC, Rodder D, Schmitt T, Neve G (2011) Global warming will affect the genetic diversity and uniqueness of Lycaena helle populations. Global Change Biology 17: 194–205.

2

Growth, Development and Temporal Variation in the Onset of Six *Chironex fleckeri* Medusae Seasons: A Contribution to Understanding Jellyfish Ecology

Matthew Gordon*, Jamie Seymour

School of Marine and Tropical Biology, James Cook University, Cairns, Queensland, Australia

Abstract

Despite the worldwide distribution, toxicity and commercial, industrial and medical impacts jellyfish present, many aspects of their ecology remain poorly understood. Quantified here are important ecological parameters of *Chironex fleckeri* medusae, contributing not only to the understanding of an understudied taxon, the cubozoa, but also to the broader understanding of jellyfish ecology. *C. fleckeri* medusae were collected across seven seasons (1999, 2000, 2003, 2005–07 and 2010), with growth rates, temporal variation in the medusae season onset and differences in population structure between estuarine and coastal habitats quantified. With a mean of 2 September ±2 d (mean ±95% confidence limits), the earliest date of metamorphosis was temporally constrained between seasons, varying by only 7 d (30 August to 5 September). Juvenile medusae appeared to be added over an extended period, suggesting polyp metamorphosis was an ongoing process once it commenced. At a maximum of 3±0.2 mm d^{-1} IPD, medusae growth to an asymptotic size of ~190 mm IPD was rapid, yet, with the oldest medusae estimated to be ~78 d in age, medusae did not appear to accumulate along the coastline. Furthermore, a greater proportion of juveniles were observed along the coastline, with estuarine populations typified by larger medusae. With key aspects of *C. fleckeri*'s ecology now quantified, medusae season management protocols can be further developed.

Editor: Jack Anthony Gilbert, Argonne National Laboratory, United States of America

Funding: Mission Beach Tourism Commission (http://www.missionbeachtourism.com/), Lions Foundation of Australia (www.lionsclubs.org.au), National Geographic (www.nationalgeographic.com.au), Australian Geographic (www.australiangeographic.com.au), Cairns City Council (www.cairns.qld.gov.au), Cardwell Shire Council (www.csc.qld.gov.au), Growing the Smart State Queensland PhD grants (www.smartstate.qld.gov.au) and Comalco (http://www.riotintoalcan.com/). The funders had no role in study design, data collection and analysis, decision to publish, or preparation of the manuscript.

Competing Interests: The authors have read the journal's policy and have the following conflicts. Commercial funders: Mission Beach Tourism Commission (http://www.missionbeachtourism.com/), Lions Foundation of Australia (www.lionsclubs.org.au), National Geographic (www.nationalgeographic.com.au), Australian Geographic (www.australiangeographic.com.au), Cairns City Council (www.cairns.qld.gov.au), Cardwell Shire Council (www.csc.qld.gov.au), Growing the Smart State Queensland PhD grants (www.smartstate.qld.gov.au) and Comalco (http://www.riotintoalcan.com/).

* E-mail: matthew.gordon@jcu.edu.au

Introduction

The occurrence of jellyfish, particularly in blooms, negatively affects a range of recreational, industrial and commercial activities. For instance, while some commercially important fisheries have been unable to function when jellyfish have clogged fishing nets [1,2], in other fisheries, jellyfish have become predators of and competitors to those species being targeted [2,3]. Serious industrial issues have also been attributed to increased jellyfish abundances, with power station shut down necessary when water intake pipes have become clogged with medusae [4,5,6]. For other regions, it is the medical liability that jellyfish represent that continues to adversely affect the tourism industry which is often integral to local and regional economies [7]. Despite these significant issues, quantitative data documenting key aspects of jellyfish ecology are, in general, lacking. As a result, determining whether claims of increased season length or intensity and frequency of blooms are difficult to validate. For the Australian tropics, it is the seasonal occurrence of *Chironex fleckeri* Southcott that significantly impacts the way in which coastal areas are utilised, yet many of the currently favoured theories relating to the temporal variation in medusae occurrence, medusae growth and development as well as population structure are based on sting records or qualitative data.

The generalisation that *C. fleckeri*'s life cycle is seasonal is based on numerous reports of medusae and stings from the warmer months of the year [8,9,10,11,12,13,14,15,16,17,18] contrasted with a lack of such reports from the winter months. Although the life history of cubozoans is complex [19] in which an asexually reproducing polyp phase alternates with a sexually reproducing medusae phase [15,16,20,21,22,23,24], the timing of the shift from the polyp to medusae phase appears to vary both spatially and temporally [10,12,15,16,18]. For instance, while *C. fleckeri* medusae typically appear along the far north Queensland coastline in December, they have been reported as early as October in some seasons [12,15,18]. There is also a suggestion that the season commences earliest on the west coast, shortly after the first rains of the wet season, with the onset of the stinger season delayed if the wet season is late [10,25]. Other authors discount the relevance of the wet season [8,26], however, suggesting that medusae arrival is associated with rising water temperature [10,16,26,27]. With the polyp habitat thought to be located some distance from the coastline within estuary systems [16], the onset of the stinger season is unlikely to accurately reflect the timing of polyp metamorphosis given that the colloquial term 'stinger season' typically refers to the arrival of medusae along the coastline rather than the timing of polyp metamorphosis.

Few extrapolations can be drawn from other cubozoan species either, with studies identifying a species cue for metamorphosis limited to a handful of species. Here, increasing water temperature [28], an interaction between increasing water temperature and food [29,30] or increasing water temperature and photoperiod [31] have been identified as cues. While the development to a specific number of tentacles [32] and the presence of photo-symbiotic algae [33] have also been suggested as potential cues for metamorphosis, these links have not been validated quantitatively. Similarly, while the correlation between rainfall events and successive pulses of juvenile cubomedusae [34,35] may implicate salinity as a cue for metamorphosis, quantitative data demonstrating this link is lacking. Given this paucity in data, a need therefore exists for research distinguishing between those mechanisms driving polyp metamorphosis and those merely correlated with this process.

At metamorphosis, juvenile C. fleckeri medusae are approximately 1.2 to 1.4 mm in size [24], but increase in size rapidly [15,16,17], reaching sexual maturity late in the season [10,15]. While such growth patterns are reported for a number of scyphozoan species [36], for cubomedusae, it is only for Chiropsella bronzei Gershwin that growth parameters have been quantified [34,35]. This paucity in data is largely due to the lack of a reliable method by which cubomedusae can be aged. Size of cubomedusae, for instance, is an unreliable indicator of age, given that degrowth of the bell can occur in cases where feeding regimes are inadequate (for example, an underfed captive C. fleckeri medusa [37]). While some authors have used the number of tentacles per pedalium as an indicator of development [25], tentacles are added across a size range, and hence, may still be somewhat dependent upon feeding regime. More recently, the statoliths contained within the statocysts of cubomedusae rhopalia (eye bearing sensory structures) have been shown to contain fine growth rings that are added on a daily basis [35,38,39,40,41].

Insights into the ecology of a species can also be gained from population structure data. For example, several cohorts of C. bronzei medusae occurring within a single season, as well as a correlation between cohort appearance and significant rainfall events, were elucidated from population structure data [34,35]. Comparable data for C. fleckeri populations is, however, currently limited to the generalisation that an abundance of small individuals early in the season progresses to fewer but larger medusae late in the season [15]. Given that the shift from estuarine to coastal habitats is thought to accompany the shift from polyp to medusae phases of the life cycle [16] whereby medusae are washed from within estuaries at the onset of the season [15,42], coastal populations are likely to be dominated by larger and older medusa, while smaller and younger medusae would typify estuarine populations. Yet, it would appear that large medusae can be present even in early season samples [10,14] and "no specific rule can be given to the range of sizes encountered in a given area at a specific time" [10]. Whether the presence of only large medusae at a given location and small medusae in another indicates the presence of a nursery location [14], or perhaps the polyp habitat itself, is yet to be demonstrated. A need therefore exists to quantify aspects of population structure if the overall ecology of C. fleckeri is to be better understood.

The current paper aims to quantify some of the long held theories relating to the ecology of C. fleckeri medusae. Key aspects to be investigated include the:

○　relationship between statolith size and growth rings
○　temporal variability in the onset of medusae production (metamorphosis)
○　growth and development of C. fleckeri medusae
○　population structure of coastal and estuarine medusae populations.

Methods

Sample Sites

A total of 484 medusae were collected from seven sites at Weipa during the 1999, 2000, 2003, 2005, 2006, 2007 and 2010 stinger seasons. While not all sites were visited on each occasion or within each season, Landfall Point (12°34′53″S, 141°39′50″E), Andoomajetti Point (12°36′20″S, 141°49′21″E), Rocky Point (12°37′10″S, 141°52′38″E), Jessica Point (12°40′05″S, 141°51′42″E), Hey Point(12°44′23″S, 141°53′35″E), Wooldrum Point Beach (12°42′19″S, 141°48′03″E) and Westminster South (12°50′11″S 141°44′56″E) represented Weipa (Figure 1). A further 46 medusae were collected during the 2005, 2006 and 2007 seasons from 12 east coast sites between Cairns and Townsville, which included Mission Beach, Gin Camp, Yorkeys Knob, Port Douglas, Cardwell, Palm Cove, Buchan Point, Lugger Bay, Townsville and the Tully, Murray and Hull Rivers. Given the low numbers collected on each occasion, east coast sites were excluded from most analyses.

Weipa sites were classified as coastal or estuarine sites. Coastal sites were those that occurred along a beachfront or the embayment of Albatross Bay, such as Wooldrum Point Beach. Estuarine sites were those that occurred wholly within an estuary system or at the intersection of an estuary system and the coastal embayment of Albatross Bay, such as Andoomajetti Point.

Statolith Technique

Medusae were collected by hand and their inter pedalia distance (IPD), the distance between the mid line of alternate pedalia along the line passing through the rhopalia, was measured to the nearest mm. Each of the four rhopalial niches and a gonad sample (if gonads were evident) were removed and preserved in 98% ethanol.

For 71 medusae, the statoliths of two rhopalia were dissected from the base of their eye set. Undamaged and unshattered statoliths were cleaned of any cellular material before being embedded in resin in the profile plane (kidney shape was evident). The top 50% of each statolith was ground using 1200 gauge wet and dry sand paper, polished with Brasso, rinsed and then polished with tooth paste. The number of rings present in each statolith was counted under oil immersion using ×400 magnification on a light microscope. Each dark band with a light band either side was considered to be one growth ring. The length of each statolith was measured using a calibrated stereo dissector microscope, with length taken to be the longest distance between curved apical ends of a statolith. Both the average number of rings and average length of each statolith pair were then calculated, with the relationship between the number of growth rings and statolith length investigated using regression analysis.

Two statoliths from a further 437 medusae were dissected and their length measured under a calibrated stereo dissector microscope. The average length of each statolith pair was calculated, from which, the number of rings was estimated using the relationship between statolith length and number of rings.

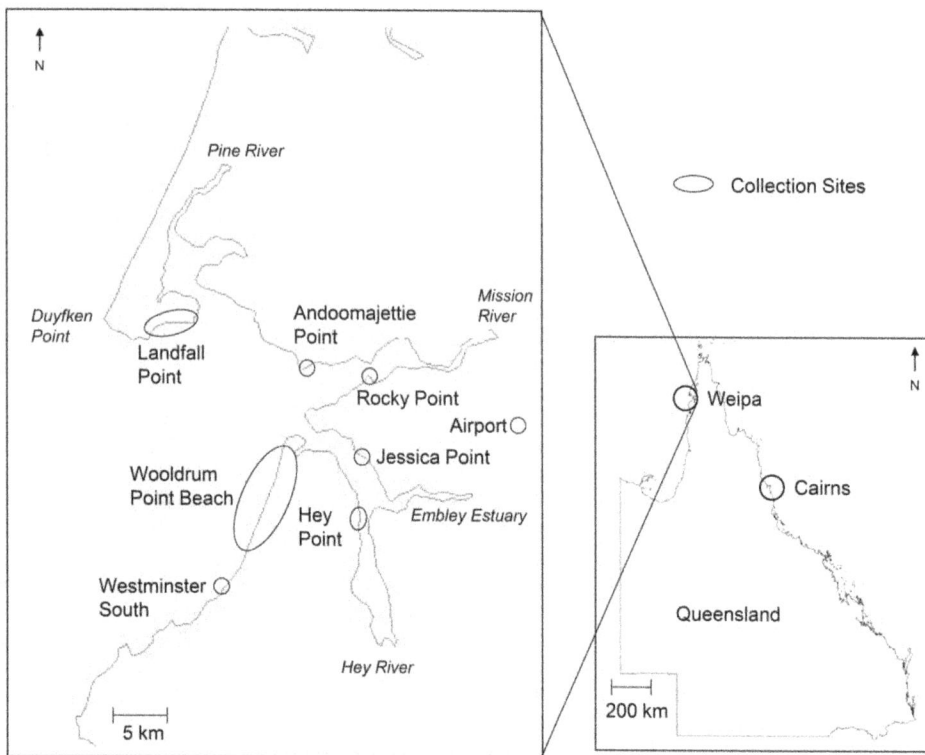

Figure 1. Geographic location of Weipa medusae collection sites, western Cape York, north Queensland, Australia.

Quantifying the Time Frame of Statolith Growth Rings

Establishing a significant relationship between medusae size (mm IPD) and age (number of rings in statolith) was a two stage process. Firstly, regression analysis was used to determine whether medusae size (IPD mm) could be predicted from tentacle number, and secondly, whether tentacle number was correlated with ring number. Given that medusae can undergo both growth and degrowth, tentacle number was considered a more reliable indicator of medusae development as tentacles are not lost once added.

To be able to age medusae, however, it was necessary to quantify the unit of time represented by successive growth rings. While tetracycline is widely used for growth ring time frame validation in fish otoliths, several attempts at applying this technique to *C. fleckeri* medusae were unsuccessful, despite various concentrations of tetracycline being trialled and medusae being housed in large, custom made cylindrical tanks. While this has limited the methods by which the interval between successive rings can be quantified, several pieces of evidence suggest that successive rings are added on a daily basis. Firstly, if growth rings were added at hourly or weekly intervals, not only are these arbitrary units of time that medusae would be unable to measure, but would also make medusae within this study less than four days (hourly) or 1.5 years (weekly) old. Monthly or annual units of time are also unrealistic given that the oldest medusae would have been 7 or 80 yrs of age respectively. Rather, if successive rings were added on a daily basis, medusae collected within this study would have ranged between one and three months of age. Not only are these realistic age estimates, given that medusae are unlikely to survive between seasons, but growth rings are added on a daily basis in three other species of cubozoa [35,38,39,40,41]. In this scenario, alternating dark and light bands would reflect the diurnal behaviours of medusae [25,43]. That is, throughout the day,

medusae expend considerable amounts of energy swimming and feeding [15], with less energy available for growth. In contrast, reduced activity levels at night [43] would allow relatively more energy to be devoted to growth. Under this scenario, alternating light and dark bands would arise from the variation in statolith density associated with differential growth between day and night. Collectively then, the most plausible unit of time between consecutive growth rings is daily, as has been shown in other cubozoans [35,38,39,40,41].

Calculating Date of Metamorphosis

Only medusae collected from Weipa sites were included in these analyses, as small sample sizes from east coast sites made analysis unreliable. The age of 461 medusae was taken to be either (a) the number of rings observed within their statoliths (64 medusae) or (b) the number of rings estimated from the relationship between statolith length and ring number (397 medusae). The metamorphosis date of each individual was calculated by subtracting an individual's age from its date of capture. The percentage of each sample that underwent metamorphosis on a given day was calculated and plotted against season number. Season number was used in preference to year number as it allowed successive samples within a season to be plotted together. That is, a sample collected in January 2007 was denoted a season number of 2006 as was a sample collected in November 2006.

Temporal variation in the onset of the 2000, 2003, 2005, 2006, 2007 and 2010 seasons was quantified by calculating the average earliest metamorphosis date of these six seasons and the associated 95% confidence limits.

A number of environmental parameters are potentially relevant to the shift from the polyp to medusae phase of the life cycle, with the following parameters quantified for the five weeks prior to the onset of metamorphosis in each season:

○   water temperature was taken to be the daily sea surface temperature at midday at 16°35′23″S, 141°33′36″E (Albatross Bay) and was obtained from the Integrated Marine Observing System at www.marine.csiro.au/remotesensing/imos

○   daily rainfall totals were obtained from the Bureau of Meteorology for Weipa Eastern Avenue (location 27042).

○   photoperiod, or the total number of hours of daylight per day, was calculated from sunrise and sunset times for Weipa obtained from Geoscience Australia at http://www.ga.gov.au/geodesy/astro/sunrise.jsp

○   tidal amplitude was calculated from hourly tide height (m) data for Humbug Wharf, obtained from Maritime Safety Qld at http://www.msq.qld.gov.au/Home/Tides.

Data were assigned to a week category (1 to 5) in which week category represented the number of weeks prior to the onset of metamorphosis for that season. The weekly variation between years in each of these parameters was investigated using a two way Analysis of Variance in which both week category and year were fixed factors. The date of the full moon within the five weeks prior to the onset of metamorphosis in each season was obtained from Geoscience Australia at http://www.ga.gov.au/earth-monitoring/geodesy.

Growth Curve Calculations

Gordon et al. [35] established that a Gompertz growth equation most accurately described the growth parameters of a closely related cubozoan, C. bronzei, using the criteria described by Kauffman [44]. Parameters for a four criteria Gompertz growth curve were estimated in Sigma Plot 10 using size at age data for 461 medusae. Size was taken to be IPD at time of capture and age was either (a) the number of rings present in an individual's statolith (64 medusae) or (b) an estimate using the relationship between statolith size and ring number (397 medusae). The maximum daily growth rate (mm d^{-1}) and 95% confidence limits were estimated by regression analysis of the linear component of the growth curve, which occurred between 40 and 70 d. The time to the onset of sexual maturity was also estimated using the resulting growth equation.

Investigation of Population Structure

Since each site was not visited on each occasion, data for medusae collected in Weipa were pooled for all seasons and sites within habitat type. East coast sites were not analysed here due to small sample sizes. Those medusae in which gonads had not developed were denoted as immature juveniles. For mature individuals, sex was determined by examining gonad samples under a stereo dissecting microscope. Individuals in which ova were visible were denoted as females, while those in which sheets of convoluted filamentous tissue were evident were denoted as males. In those cases where it was unclear whether tissue was that of a small male or an immature specimen, individuals were classified as undistinguishable (~60 individuals) and grouped with immature specimens. A chi squared homogeneity test was used to determine whether any significant difference between the proportion of males, females and juveniles existed between the estuarine and coastal habitats.

Results

A significant and positive relationship between statolith length (mm) and number of rings was established (F = 151.243,

df = 1×69, n = 71, P<0.001, R^2 = 0.687), whereby the number of rings present within a statolith increased as did statolith length. Within the size range of statoliths sampled, a linear relationship (Figure 2) provided a better fit than curvilinear equations, with this linear relationship best described by the equation:

$$R = (78.213 \times SL) - 0.088$$

where R is the number of rings within a statolith and SL is statolith length in mm.

A significant and positive relationship was established between the number of tentacles per pedalium and inter pedalia distance (mm) (F = 1201.176, df = 1×420, n = 422, P<0.001, R^2 = 0.740) whereby medusae size (IPD in mm) increased as did the number of tentacles per pedalium (Figure 3). A power curve described this relationship most appropriately, with inter pedalia distance increasing at a faster rate as more tentacles were added to each pedalium. This curve provided a minimum medusae size at the one tentacle stage (ie following metamorphosis) of ~1.8 mm and is best described by the equation:

$$S = 1.846 \times T^{1.628}$$

where S is medusa size (IPD) in mm and T is the number of tentacles per pedalium.

Medusae appeared to add tentacles in pairs, with only ~8% of the 422 medusae for which tentacle number was collected possessing an even number of tentacles per pedalium. A significant and positive relationship was identified between the number of rings within a statolith and the number of tentacles suspended from each pedalium (F = 639.733, df = 1×410, n = 412, P<0.001, R^2 = 0.609). Within the age range of medusae sampled, a linear relationship in which tentacle number increased as did ring number (Figure 4) provided a better fit than did curvilinear equations and is best described by the equation:

$$T = (0.189 \times R) - 0.899$$

where T is the number of tentacles per pedalium and R is the number of rings per statolith.

The earliest date of metamorphosis was 30 August and occurred in the 2007 season. Despite the earliest date of specimen collection varying by 33 d, the earliest date of metamorphosis did not vary by more than 7 d. That is, while medusae were first collected on the 18th October in the 2000 season in which the earliest date of metamorphosis was 31 August, in the 2010 season for which the earliest metamorphosis date was 5 September, medusae were not collected until the 20th November. The mean earliest date of metamorphosis was 2 September ±2 d (mean ±95% confidence limits).

Within the 2005, 2006 and 2007 seasons in which sample sizes were large, medusae were added to the population on an almost daily basis, with each date of metamorphosis represented by ~2%, with no one date accounting for more than 10% of a sample (Figure 5). Metamorphosis also appeared to be an ongoing process once it commenced. That is, in those seasons where the interval between successive samples (between grey and black arrowhead lines) was approximately one month, as for the 2006 and 2007 seasons, metamorphosis dates were either continuous (2006 season) or overlapped by a small amount (2007 season). In the 2005 season, however, successive samples were collected ~60 d apart (November 2005 and January 2006), with the gap in metamorphosis corresponding to the length of time between sampling occasions.

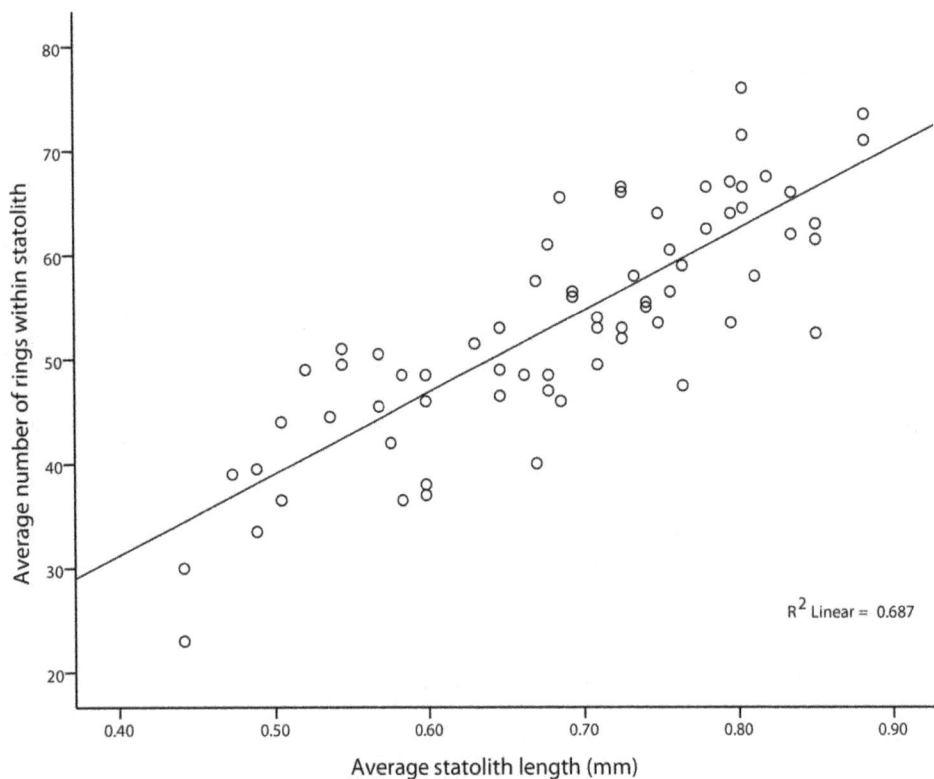

Figure 2. Positive linear relationship between statolith length and number of rings within a *Chironex fleckeri* **statolith.** Statolith length is in mm and number of rings is averaged for statolith pairs.

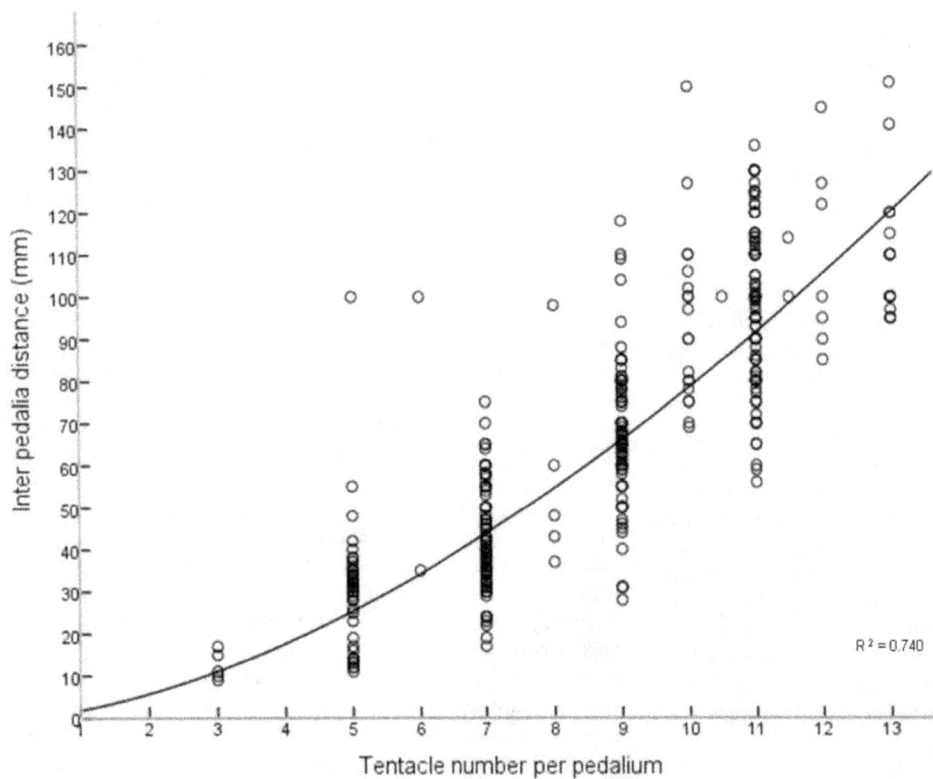

Figure 3. Positive curvelinear relationship between number of tentacles and Inter Pedalia Distance for *Chironex fleckeri* **medusae.** Inter pedalia distance is in mm and tentacle number is per pedalium.

Growth, Development and Temporal Variation in the Onset of Six Chironex fleckeri Medusae...

13

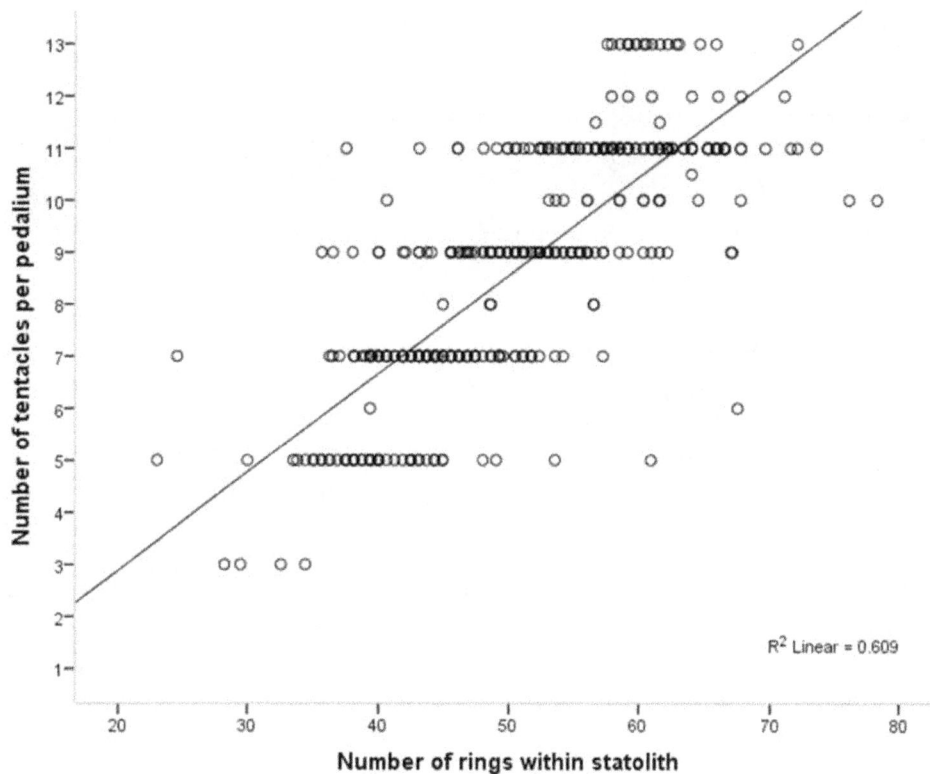

Figure 4. Positive linear relationship between number of rings and number of tentacles for *Chironex fleckeri* medusae. Number of rings is the average per statolith and tentacle number is per pedalium.

Water temperature (°C), daily rainfall (mm) and tidal amplitude (m) each showed significant weekly variation between years in the five weeks prior to the earliest date of metamorphosis within each season (Table 1). Photoperiod (h daylight d^{-1}), however, did not vary significantly by week between years (Table 1). The dates of the full moon in the five weeks preceding the onset of metamorphosis differed by 19 d across the six seasons studied here, ranging from 9 August in 2006 to 28 August in 2008.

The degree to which each environmental parameter varied in the five weeks prior to the onset of metamorphosis was parameter specific. That is, daily rainfall total (DRFT in mm) showed the greatest degree of variation, having a $\log_{10}$ transformed CV value of ~ 1, while photoperiod showed the least amount of variation with a $\log_{10}$ transformed CV value of ~ -3 (Figure 6). The variation in both water temperature (°C) and tidal amplitude (m) was also considerably greater than photoperiod, ranging from ~ 0 (tidal amplitude in m) to ~ -1.5 for water temperature (°C) (Figure 6).

A significant and positive relationship between IPD (mm) and age (days) of medusae (F = 423.3479, df = 3×457, n = 461, $P<0.0001$, $R^2 = 0.735$) whereby medusae increased in size with age towards an asymptotic size of ~ 190 mm IPD (Figure 7). The Gompertz growth equation had the following format:

$$S = 1.5 + 186.5617 \times e^{-e^{-\left(\frac{t - 51.8408}{21.9887}\right)}}$$

where S is medusa size (IPD) in mm and t is medusa age in days.

Regression analysis of the linear component of this relationship (from 40 to 70 d) revealed a maximum growth rate of ~ 3 mm d^{-1} (± 0.2 mm d^{-1}). The minimum size (IPD in mm) at which males

and females could be reliably distinguished was 46 mm IPD and 50 mm IPD respectively. According to the above growth equation, it would take ~ 45 d to reach the average size (IPD mm) of sexual differentiation. At an average 96 mm IPD for male medusae and 97 mm IPD for females, the mean size of male and female medusae did not vary significantly between the sexes (F = 0.25, df = 1×240, n = 242, $P = 0.621$).

The age structure of the overall population differed significantly between the estuarine and coastal habitats ($\chi^2 = 49.477$, df = 2, n = 281, $P<0.001$). Younger medusae were not as well represented within the estuarine habitat, with a predominance of larger and older individuals evident instead. Along the coastline, however, a greater spread of males, females and immature medusae was observed, ranging in age from 30–70 d and from five to 13 tentacles per pedalium (Figure 8). Furthermore, the oldest medusae within the estuarine habitat were older than those along the coastline (Figure 8). Although medusae added new tentacles over an age range, it was typically possible to determine the sex of an individual providing it possessed more than nine tentacles. This was consistent between the estuarine and coastal populations for all but three medusae within the coastal habitat who were classed as immature at the 11 tentacle stage.

Discussion

One area of jellyfish ecology that has received increasing attention of late is the shift from the polyp to the medusae phase, and the factors associated with, or acting as cues for this shift. Not only was the onset of medusae production temporally constrained between seasons, but it also commenced earlier than expected. That is, the earliest metamorphosis date was 30 August (2007

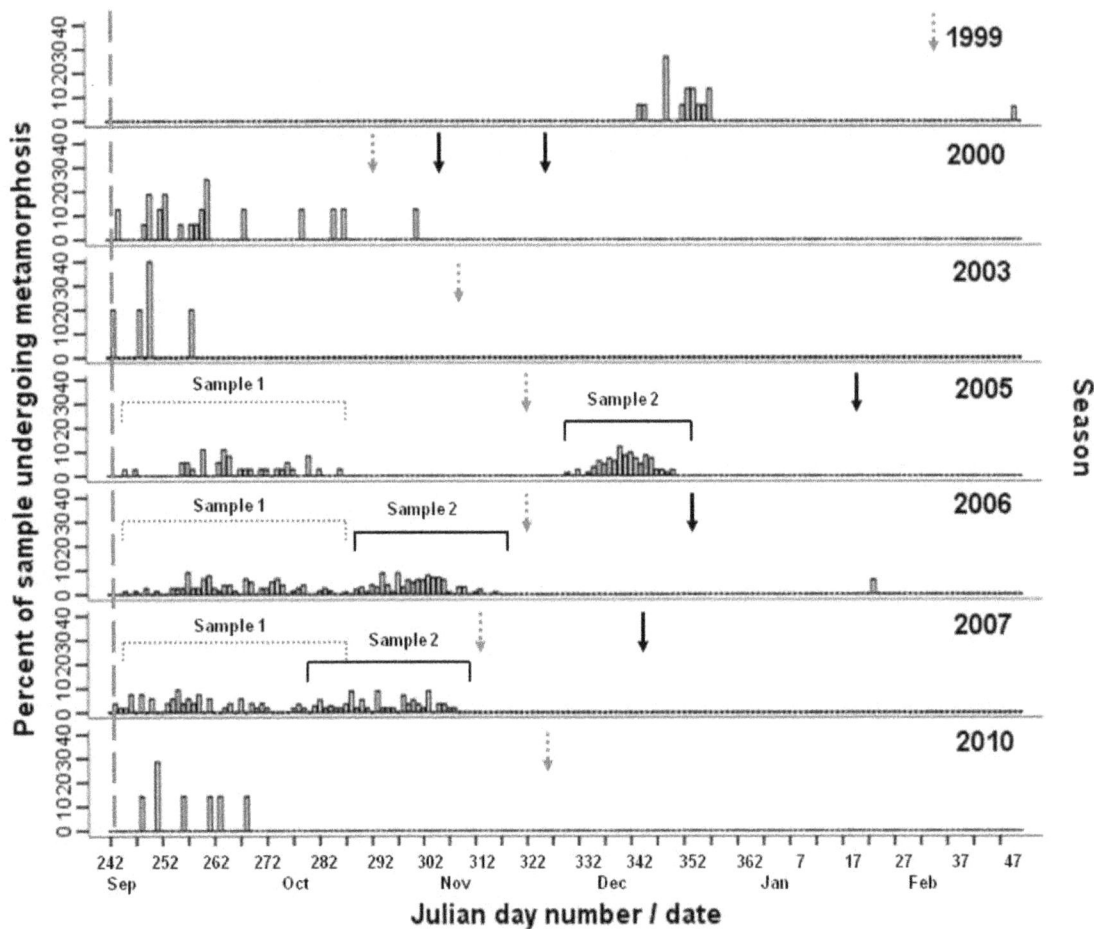

Figure 5. Percent frequency of *Chironex fleckeri* **medusae that metamorphosed each day.** Frequency is percentage within a sample and date is Julian day number or month, with 30 August referenced by a grey dash line. The first sampling occasion within a season is denoted by a grey arrowhead line, while subsequent sampling occasion(s) are denoted by a black arrowhead line.

season) and varied by only ~7 d across six seasons. Not only is September also the earliest month in which juvenile medusae are reported in plankton tows from east coast studies [15], but medusae of 120 mm collected in December on Magnetic Island [14] would be ~70 d of age (from the above growth curve), also giving them a metamorphosis date in September. This result is particularly significant in terms of modelling the overall medusae

Table 1. ANOVA results for the variation in parameters potentially associated with metamorphosis.

Parameter	Interaction	F	df	P
Water Temperature (°C)	week category×year	2.031	20×192	0.008
Total daily rainfall (mm)	week category×year	1.814	20×210	0.022
Tidal amplitude (m)	week category×year	1.696	10×697	0.030
Photoperiod (h daylight d^{-1})	week category×year	0.088	20×210	1.000

Water temperature is in °C, total daily rainfall is in mm, tidal amplitude is in m and photoperiod is the number of h of daylight d^{-1} for each week in the five weeks prior to the onset of metamorphosis in each season.

season in that the onset of each season can now be defined with greater accuracy.

Quantifying when polyp metamorphosis commenced also allows the factors associated with the shift from the polyp to medusae phase to be identified. It is these factors that polyp based studies should include when quantitatively indentifying the cue for metamorphosis. For instance, the significant between year variation in weekly mean water temperature suggests that water temperature is unlikely to have provided the temporal periodicity observed in the onset of metamorphosis within this study. A similar case exists for both tidal amplitude and rainfall. That is, while unusually high amplitude tides could result in salinity changes at the polyp habitat by pushing higher salinity waters further into estuary systems or allowing fresh waters to drain further down estuaries, the timing of any tidally driven salinity fluctuations would have varied between years. Likewise, significant between year variation in total daily rainfall (mm) suggests that rainfall (or rainfall driven salinity changes) was unlikely to have acted as a cue for the onset of polyp metamorphosis in the seasons investigated within this study. Indeed, the climate of Weipa is dominated by strong seasonal patterns [45,46,47], with low rainfall and elevated, stable salinity typical for late August/early September [46]. Salinity did not appear to be related to the metamorphosis of *C. fleckeri* polyps in laboratory based trials [15,24] either, although dilution rates may have induced encystment rather than

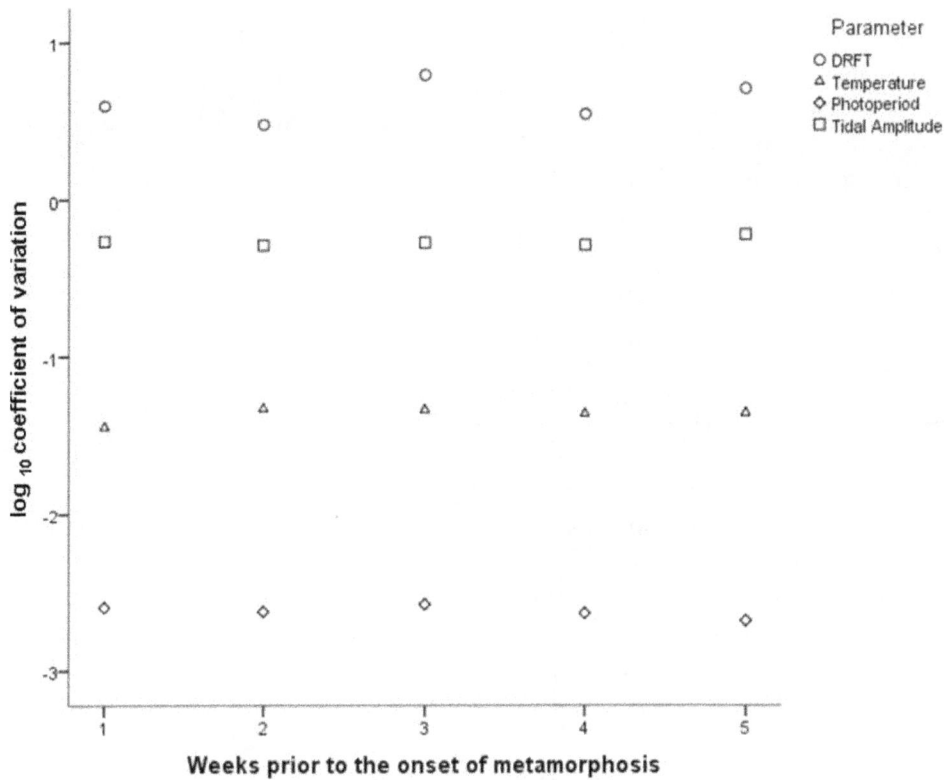

Figure 6. Coefficient of variation (log_{10} transformed) of environmental parameters. CV for daily rainfall total (mm), tidal amplitude (m), water temperature (°C) and photoperiod (h daylight d^{-1}) for the five weeks preceding the earliest date of metamorphosis in each season.

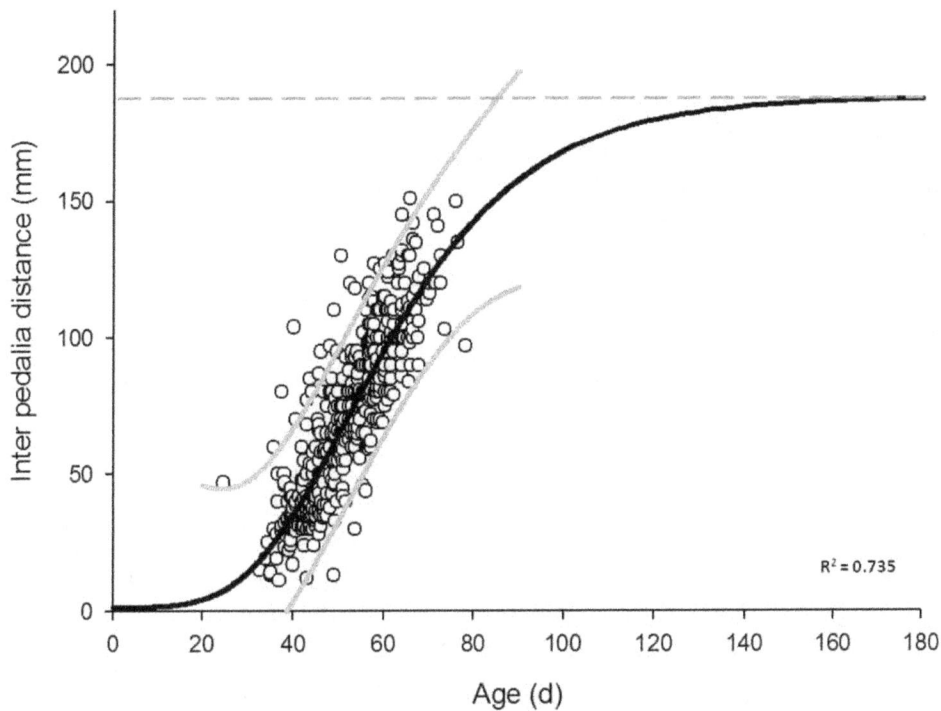

Figure 7. Gompertz growth curve for significant, positive relationship between age and IPD of _Chironex fleckeri_ medusae. Growth curve shows medusae size (IPD) in mm at age in d with 95% prediction limits (grey solid line).

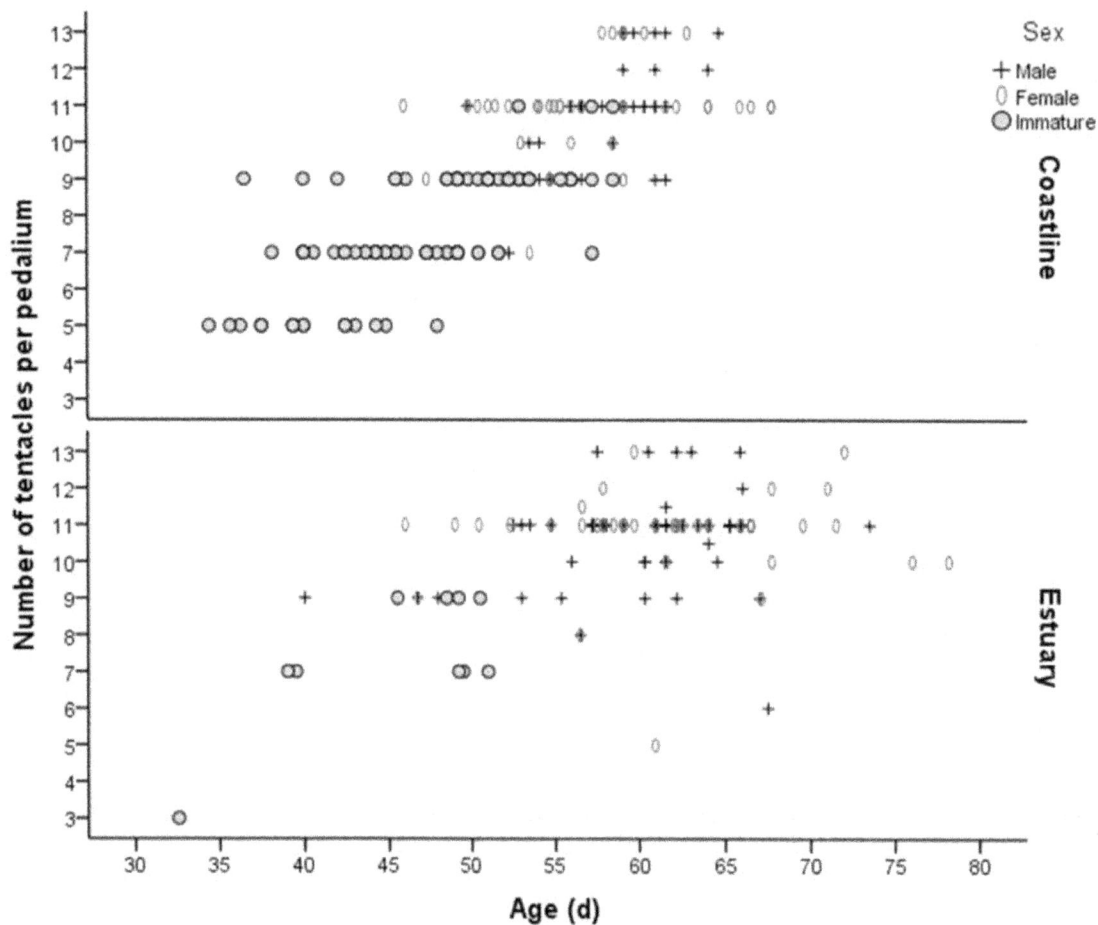

Figure 8. Age of *Chironex fleckeri* medusae with respect to tentacle number. Distinction is made between males, females and immature medusae within the coastal and estuarine habitats of Weipa, where age is in d and tentacle number is per pedalium.

metamorphosis. The influence of temperature is also unclear from polyp based studies, with all trials conducted at 28°C [15,24]. While the results of this study suggest that salinity, temperature, tidal amplitude or moon phase are unlikely to provide the temporal consistency in the onset of metamorphosis as was observed within this study, their role as interacting variables cannot, as yet, be disregarded.

One parameter that could provide a higher degree of temporal consistency in the onset of metamorphosis is photoperiod, with the average hours of daylight consistent between seasons in the five weeks prior to the earliest date of metamorphosis within each season. The influence of photoperiod on polyp metamorphosis remains largely untested, however, with *Carybdea morandinii* the only cubozoan for which a link between photoperiod and metamorphosis has been established [31]. Not only has light been positively correlated with asexual reproduction for some Scyphozoans [48,49], but Purcell [48] suggests that melatonin, a light sensitive hormone, may also play an important role in coordinating the strobilation of the Scyphozoan *Aurelia labiata*. Photoperiod, which has been shown to coordinate breeding cycles in some marine invertebrates [50], may play a similar role in coordinating polyp metamorphosis in *C. fleckeri*, and as such future polyp based research should quantify the significance of photoperiod on polyp metamorphosis.

Metamorphosis did not appear to be a single or pulse event for *C. fleckeri*, rather, an ongoing process whereby low numbers of

medusae were produced on an almost daily basis (between 2 and 10% d^{-1}). The collection of a medusa in March 2000 (metamorphosis date of February 19) and another in March 2007 (metamorphosis date of January 23) further suggests that medusae production continued over an extended time frame. Reports of 0.6 to 1.8 mm juvenile *C. fleckeri* medusae in estuarine plankton samples between September and January [15], the presence of both adult and small medusae in the first arrivals along the coastline [10,14], as well as the collection of 6 mm medusae in January and February when 120 mm medusae were collected in December further suggest that metamorphosis occurs over an extended timeframe. Reports of juvenile medusae occurring in successive waves in the only laboratory based study conducted on *C. fleckeri* polyps [24] initially appear contradictory to the results of this study, however, without the timeframe between successive waves quantified, the potential for pulses to have occurred on a daily basis, as observed within this study, cannot be disregarded.

Once within the sexual phase of the life cycle, medusae growth was rapid at up to 3±0.2 mm d^{-1}, which is up to three times that established for a closely related and often co occurring species, *C. bronzei* [34,35]. Although growth rates are likely to vary between individuals due to prey availability [8,15], medusae would typically reach their estimated asymptotic size of ~190 mm (IPD in mm) after ~140 days. That *C. bronzei* has an estimated asymptotic size of 71 mm IPD [34,35] and *Chiropsalmus quadrumanus* Agassiz is reported to reach 110 mm [51] suggests that *C. fleckeri* medusae attain a larger size than do other Chirodropid species. On an applied level, Hartwick [16] has previously suggested that medusae reach a size dangerous to humans within approximately two to three months. Indeed, *C. fleckeri* medusae undergo an ontogenetic shift in their cnidome (and diet) that potentially explains the lethality of larger medusae to humans [52] at ~60–100 mm IPD [52], a size they would reach after ~50–65 d. This is an important parameter to consider in the further development of management protocols in that the time at which medusae are likely to become lethal to humans can now be defined with greater reliability.

That the onset of sexual maturity occurred at approximately ~50 mm IPD compares favourably to estimates provided by Barnes [10] who noted that the development of a very large area of gonad material commenced at the eight tentacle stage (~60 mm IPD based on the regression equation developed here) [25]. At the rapid rate of growth quantified here, medusae would become sexually mature after ~45–50 d, which is considerably less time than the typical length of a season (~180 d). Given that the oldest medusa was ~78 d, it does not appear that medusae accumulate as the season progressed [15]. Potentially, medusae relocated from within the estuarine and coastal areas sampled here, with the infrequent collection of medusae up to several km from shore [8,37,53] suggesting that some form of emigration could take place. Alternatively, medusae that underwent metamorphosis in early September would have had several months of stable conditions and an abundant food supply prior to the onset of the wet season in which to grow and mature. That is, September to December falls within Weipa's dry/pre wet season [47,54] when salinity regimes are typically stable and elevated [47], water temperatures are typically increasing [45,46,47] and an abundance of post larval prawns occurs within the Embley Estuary [46,47]. Whether several generations of medusae occur within a single season is an aspect of *C. fleckeri*'s ecology that future research should examine.

Medusae development can also be considered with respect to tentacle number, with some authors using tentacle number rather than size when discussing the development of medusae (e.g. [25]). The maximum number of tentacles per pedalium observed within this study was 13, which compares favourably to maximums of 12 [10,11] and 13 [25], but suggests that 15 tentacles may be limited to those individuals of ~300 mm IPD which are rarely observed [8,12,55]. Given that newly metamorphosed medusae possess one tentacle per pedalium [24] and only 8.3% of medusae within this study possessed an even number of tentacles, *C. fleckeri* would appear to add tentacles in pairs. With 12 medusae possessing an odd number of tentacles and 10 possessing a even number of tentacles, samples in Kinsey [25] suggest that tentacles are added singularly, however this may be an artefact of a small sample size. *C. bronzei* also appears to add tentacles in pairs, with 92% of the 1652 medusae collected possessing an odd number of tentacles [34]. On a more applied level, tentacle number may provide a more standardised method by which groups such as Surf Life Saving can provide consistent estimates of medusae size and age, given the significant relationships that exist between these variables.

A difference in the population structure of coastal and estuarine habitats would be expected if a seasonal alternation in generations and habitats [16] is occurring for *C. fleckeri*. That is, juvenile medusae would be representative of estuarine populations while coastal populations would be typified by both a greater range in medusae size as well as an accumulation of larger individuals as the season progressed. Not only were medusae from coastal sites typically smaller (fewer tentacles) and younger than those found within the estuarine habitat, but the oldest medusae (~78 d) was also collected from within the estuarine habitat. While these results appear contradictory to expected patterns, juvenile medusae reported by Hartwick [15] were as small as 0.6 mm and collected in plankton tows [15,16], with the visually based collection techniques used within this study possibly possessing an inherent bias against such small individuals. Differences in population structure may also reflect the suitability and availability of prey within the estuarine and coastal habitats. For instance, not only have mangrove areas of the Embley Estuary been shown to be important nursery areas for many species [56], but a greater abundance of fish [54] and prawn species [57] have been reported for intertidal areas adjacent to mangrove stands. In contrast, both species diversity and overall abundance of fish was lower along the coastline [58]. Further data quantifying medusae abundance, gastrovascular cavity content and prey abundance is required to further validate these claims.

Collectively, the results of this study are relevant in both an applied and an ecological context. By quantifying growth and development rates as well as the temporal variation in the onset of polyp metamorphosis between seasons for *C. fleckeri* medusae, this paper has contributed to the understanding of the ecology of an understudied taxon, the cubozoa, as well as to the broader understanding of jellyfish ecology. On an applied level, this study has presented quantitative data upon which models predicting the seasonal occurrence of this species can be developed. For instance, estimates of when medusae are likely to present a considerable risk to humans can now be based on medusae growth and development rates, allowing significant events, such as the ontogenetic shift in cnidome, to be modelled with greater accuracy. It is in this way that the negative effects of the stinger season can be managed more effectively. Such models are of particular relevance given the way in which the seasonal occurrence of *C. fleckeri* impacts the way in which the tropical Australian coastline is utilised throughout the warmer months of the year. However, it is only when a complete understanding of the medusae phases ecological relationships are developed that the occurrence and distribution of *C. fleckeri* can be modelled with accuracy and reliability. This study represents the first attempt at quantifying such parameters, however, further long term studies are required if management practices are to be optimised and broader ecological questions regarding season length or the intensity and frequency of jellyfish blooms are to be validated.

Acknowledgments

Sincere thanks to Teresa Carrette, Robert & Julieanne Gordon, Jennie Gilbert, Glenda, Amelia & Benjamin Seymour and Ben Kelly for their assistance in field work.

Author Contributions

Conceived and designed the experiments: MG JS. Performed the experiments: MG JS. Analyzed the data: MG JS. Contributed reagents/materials/analysis tools: MG JS. Wrote the paper: MG JS.

References

1. Purcell JE, Uye S, Lo W (2007) Anthropogenic causes of jellyfish blooms and their direct consequences for humans: a review. Marine Ecology Progress Series 350: 153–174.

2. Graham WM, Martin DL, Felder DL, Asper VL, Perry HM (2003) Ecological and economic implications of a tropical jellyfish invader in the Gulf of Mexico. Biological Invasions 5: 53–69.

3. Purcell JE, Arai MN (2001) Interactions of pelagic cnidarians and ctenophores with fish: a review. Hydrobiologia 451: 27–44.

4. Rajagopal S, Nair KVK, Asariah J (1989) Some observations on the problem of jellyfish ingress in a power station cooling system at Kalpakkam, east coast, India. Mahasagar Quarterly Bulletin, National Institute of Oceanography (Goa) 22: 151–158.

5. Matsumura K, Kamiya K, Yamashita KHF, Watanabe I, Murao Y, et al. (2005) Genetic polymorphosm of the adult medusae invading an electric power station and wild polyps of *Aurelia aurita* in Wakasa Bay, Japan. Journal of the Marine Biological Association of the United Kingdom 85: 563–568.

6. Masilamoni JG, Jesudoss KS, Nandakumar K, Satpathy KK, Nair KVK, et al. (2000) Jellyfish ingress: A threat to the smooth operation of coastal power plants. Current Science 79: 567–569.

7. Burnett JW (2001) Medical aspects of jellyfish envenomation: pathogenesis, case reporting and therapy. Hydrobiologia 451: 1–9.

8. Burnett JW, Currie B, Fenner PJ, Rifkin J, Williamson JA (1996) Cubozoans (Box Jellyfish). In: Williamson JA, Fenner PJ, Burnett JW, Rifkin J, eds. Venomous and Poisonous Marine Animals - a Medical and Biological Handbook. Sydney: University of New South Wales. pp 98–118.

9. Barnes J (1965) *Chironex fleckeri* and *Chiropsalmus quadrigatus* - morphological distinctions. North Queensland Naturalist. pp 13–22.

10. Barnes J (1966) Studies on three venomous cubomedusae. The Cnidaria and their Evolution: Symposium of the Zoological Society of London. London: Academic Press. pp 307–332.

11. Southcott R (1956) Studies on Australian Cubomedusae, including a new genus and species apparently harmful to man. Australian Journal of Marine and Freshwater Ecology 7: 254–280.

12. Southcott R (1971) The Box-Jellies or Sea-Wasps. Australian Natural History. pp 123–128.

13. Keen TEB (1971) Comparison of tentacle extracts from *Chiropsalmus quadrigatus* and *Chironex fleckeri*. Toxicon 9: 249–254.

14. Brown T (1973) *Chironex fleckeri* - Distribution and movements around Magnetic Island, North Queensland. Townville: James Cook University. 30 p.

15. Hartwick R (1991a) Distributional ecology and behaviour of the early life stages of the box-jellyfish *Chironex fleckeri*. Hydrobiologia 216/217: 181–188.

16. Hartwick R (1987) The Box Jellyfish. In: Covachevich J, Davie P, Pearn J, eds. Toxic Plants and Animals. Brisbane: Queensland Museum Press. pp 99–105.

17. Hamner W, Doubilet D (1994) Australia's Box Jellyfish - A Killer Down Under. National Geographic 186: 116–130.

18. Fenner PJ, Williamson JA (1996) Worldwide deaths and severe envenomation from jellyfish stings. Medical Journal of Australia 165: 658–661.

19. Leonard JL (1980) Cubomedusae belong to the Cubozoa, not Scyphozoa. Nature 284: 377.

20. Werner B, Chapman D, Cutress C (1976) Muscular and nervous systems of the cubopolyp (Cnidaria). Experimentia 32: 1047–1049.

21. Werner B, Cutress C, Studebaker J (1971) Life cycle of *Tripedalia cystophora* Conant (Cubomedusae). Nature 232: 582–583.

22. Werner B (1973) Spermatozeugmen und Paarungsverhalten bei *Tripedalia cystophora* (Cubomedusae). Marine Biology 18: 212–217.

23. Werner B (1975) Bau und Lebensgeschichte des Polypen von *Tripedalia cystohpora* (Cubozoa, class. nov., Carybdeidae) und seine Bedeutund fur die Evolution der Cnidaria. Helgorlander wiss Meeresunters 27: 461–504.

24. Yamaguchi M, Hartwick R (1980) Early life history of the sea wasp, *Chironex fleckeri* (Class Cubozoa). In: Tardent R, Tardent R, eds. Development and Cellular Biology of Coelenterates: Elsevier/North-Holland Biomedical Press. pp 11–16.

25. Kinsey BE (1986) Barnes on box jellyfish. Unpublished folio manuscripts held in the Archives of James Cook University.

26. Jacups S (2010a) Global warming - rising sea surface temperatures - a longer box jellyfish (*Chironex fleckeri*) stinger season for the Northern Territory? The Northern Territory Disease Control Bulletin 17: 25–28.

27. Jacups S (2010b) Warmers waters in the Northern Territory-Herald an earlier onset to the annual *Chironex fleckeri* stinger season. EcoHealth 7: 14–17.

28. Laska-Mehnert G (1985) Cytologische Veranderungen wahrend der metamorphoses des cubopolypen *Tripedalia cystophora* (Cubozoa, Carybdeidae) in die medusae. Helgorlander Meeresuntersuchungen 39: 129–164.

29. Stangl K, Salvini-Plawen L, Holstein TW (2002) Staging and induction of medusae metamorphosis in *Carybdea marsupialis* (Cnidaria, Cubozoa). Life and Environment 52: 131–140.

30. Straehler-Pohl I, Jarms G (2005) Life cycle of *Carybdea marsupialis* Linnaeus, 1758 (Cubozoa, Carybdeidae) reveals metamorphosis to be a modified strobilation. Marine Biology 147: 1271.

31. Straehler-Pohl I, Jarms G (2011) Morphology and life cycle of *Carybdea morandinii*, sp. nov. (Cnidaria), a cubozoan with zooxanthellae and peculiar polyp anatomy. Zootaxa 2755: 36–56.

32. Arneson A, Cutress C (1976) Life history or *Carybdea alata* Reynaud, 1830 (Cubomedusae). In: Mackie G, ed. Coelenterate Ecology and Behaviour: Plenum Press. pp 227–236.

33. Hofmann DK, Kremer BP (1981) Carbon metabolism and strobilation in *Cassiopea andromeda* (Cnidaria: Scyphozoa): Significance of endosymbiotic dinoflagellates. Marine Biology 65: 25–33.

34. Gordon M (1998) Ecophysiology of the Tropical Australian Chirodropid *Chiropsalmus sp.* (Haeckel) [Honours]. Cairns: James Cook University. 114 p.

35. Gordon M, Seymour J, Hatcher C (2004) Growth and age determination of the tropical Australian cubozoan *Chiropsalmus* sp. Hydrobiologia 530/531: 339–345.

36. Arai M (1997) A Functional Biology of Scyphozoa. London: Chapman Hall.

37. Hamner WH, Jones MS, Hamner PP (1995) Swimming, feeding, circulation and vision in the Australian box jellyfish, *Chironex fleckeri* (Cnidaria: Cubozoa). Marine and Freshwater Research 46: 985–990.

38. Kawamura M, Ueno S, Iwanaga S, Oshiro N, Kubota S (2003) The relationship between fine growth rings in the statolith and growth of the cubomedusae *Chiropsalmus quadrigatus* (Cnidaria: Cubozoa) from Okinawa Island, Japan. Plankton Biology and Ecology 50: 37–42.

39. Ueno S, Arimoto Y, Tezuka J (2003) Growth and life history of the cubomedusae, *Carybdea rastoni*, in the southern Japan inshore waters. 7th International Conference on Coelenterate Biology.

40. Ueno S, Imai C, Mitsutani A (1995) Fine growth rings found in statolith of a cubomedusae *Carybdea rastoni*. Journal of Plankton Research 17: 1381–1384.

41. Ueno S, Imai C, Mitsutani A (1997) Statolith formation and increment in *Carybdea rastoni* Haacke, 1886 (Scyphozoa: Cubomedusae): evidence of synchronization with semilunar rhythms. In: den Hartog JC, ed. Proceedings of the sixth International Conference on Coelenterate Biology 1995. Leiden: Nationaal Natuurhistorisch Museum. pp 491–493.

42. Cropp B, Cropp L (1984) The Deadliest Creature on Earth. The Scuba Diver 3: 42–48.

43. Seymour J, Carrette T, Sutherland P (2004) Do box jellyfish sleep at night? Medical Journal of Australia 181: 706.

44. Kaufmann KW (1981) Fitting and using growth curves. Oecologia 49: 293–299.

45. Cyrus DP, Blaber SJM (1992) Turbidity and salinity in a tropical northern Australian estuary and their influence on fish distribution. Estuarine, Coastal and Shelf Science 35: 545–563.

46. Vance DJ, Haywood MDE, Heales DS, Kenyon RA, Loneragan NR (1998) Seasonal and annual variation in abundance of postlarval and juvenile banana prawns *Penaeus merguiensis* and environmental variation in two estuaries in tropical northeastern Australia: a six year study. Marine Ecology Progress Series 163: 21–36.

47. Vance DJ, Haywood MDE, Heales DS, Staples DJ (1996b) Seasonal and annual variation in abundance of postlarval and juvenile grooved tiger prawns *Penaeus semisulcatus* and environmental variation in the Embley River, Australia: a six year study. Marine Ecology Progress Series 135: 43–55.

48. Purcell JE (2007) Environmental effects on asexual reproduction rates of the scyphozoan *Aurelia labiata*. Marine Ecology Progress Series 348: 183–196.

49. Lieu WC, Lo WC, Purcell JE, Chang HH (2009) Effects of temperature and light intensity on asexual reproduction of the scyphozoan, Aurelia aurita (L.) in Taiwan. Hydrobiologia 616: 247–258.

50. Sachlikidis NG, Jones CM, Seymour JE (2005) Reproductive cues in *Panulirus ornatus*. New Zealand Journal of Marine and Freshwater Science 39: 305–310.

51. Guest W (1959) The occurrence of the jellyfish *Chiropsalmus quadrumanus* in Matagorda Bay, Texas. Bulletin of Marine Science of the Gulf and Caribbean 9: 79–83.

52. Carrette T, Alderslade P, Seymour J (2002) Nematocyst ratio and prey in two Australian cubomedusans, *Chironex fleckeri* and *Chiropsalmus* sp. Toxicon 40: 1547–1551.

53. Kinsey BE (1988) More Barnes on box jellyfish. Unpublished folio manuscripts held in the Archives of James Cook University.

54. Blaber SJM, Brewer DT, Salini JP (1989) Species composition and biomasses of fishes in different habitats of a tropical Northern Australian estuary: Their occurrence in the adjoining sea and estuarine dependence. Estuarine, Coastal and Shelf Science 29: 509–531.

55. Barnes J (1960) Observations on jellyfish stings in North Queensland. The Medical Journal of Australia 11: 993–999.

56. Vance DJ, Haywood MDE, Heales DS, Kenyon RA, Loneragan NR, et al. (1996a) How far do prawns and fish move into mangroves? Distribution of juvenile banana prawns *Penaeus merguiensis* and fish in a tropical mangrove forest in northern Australia. Marine Ecology Progress Series 131: 115–124.

57. Vance DJ, Haywood MDE, Staples DJ (1990) Use of a mangrove estuary as a nursery area by postlarval and juvenile banana prawns,*Penaeus merguiensis* de Man, in Northern Australia. Estuarine, Coastal and Shelf Science 31: 689–701.

58. Blaber SJM, Brewer DT, Salini JP (1995) Fish communities and the nursery role of the shallow inshore waters of a tropical bay in the gulf of Carpentaria, Australia. Estuarine, Coastal and Shelf Science 40: 177–193.

3

Titanium Dioxide Nanoparticles Increase Sensitivity in the Next Generation of the Water Flea *Daphnia magna*

Mirco Bundschuh, Frank Seitz, Ricki R. Rosenfeldt, Ralf Schulz*

Institute for Environmental Sciences, University of Koblenz-Landau, Landau, Germany

Abstract

The nanoparticle industry is expected to become a trillion dollar business in the near future. Therefore, the unintentional introduction of nanoparticles into the environment is increasingly likely. However, currently applied risk-assessment practices require further adaptation to accommodate the intrinsic nature of engineered nanoparticles. Combining a chronic flow-through exposure system with subsequent acute toxicity tests for the standard test organism *Daphnia magna*, we found that juvenile offspring of adults that were previously exposed to titanium dioxide nanoparticles exhibit a significantly increased sensitivity to titanium dioxide nanoparticles compared with the offspring of unexposed adults, as displayed by lower 96 h-EC_{50} values. This observation is particularly remarkable because adults exhibited no differences among treatments in terms of typically assessed endpoints, such as sensitivity, number of offspring, or energy reserves. Hence, the present study suggests that ecotoxicological research requires further development to include the assessment of the environmental risks of nanoparticles for the next and hence not directly exposed generation, which is currently not included in standard test protocols.

Editor: Elena A. Rozhkova, Argonne National Laboratory, United States of America

Funding: The Ministry of Science Rhineland-Palatinate (MBWJK) funded this study, which is linked with studies conducted within the research group INTERNANO, which is funded by the German Research Foundation (DFG). Moreover, we acknowledge the Fix-Stiftung, Landau for financial support of the research infrastructure. The funders had no role in study design, data collection and analysis, decision to publish, or preparation of the manuscript.

Competing Interests: The authors have declared that no competing interests exist.

* E-mail: R.Schulz@uni-landau.de

Introduction

More than 1,000 products, including sunscreens, textiles, and self-cleaning surfaces, either contain engineered nanoparticles or are produced by nanotechnology. As a result, the unintentional introduction of engineered nanoparticles into the environment is increasing [1]. However, potential environmental risks and effects elicited by such nanoparticles on the integrity of ecosystems remain largely unknown [2]. This situation has arisen due to the lack of information regarding current environmental concentrations of nanoparticles, which is primarily caused by the absence of suitable analytical techniques [3]. Models predicting environmental concentrations of nanoparticles are compromised for this reason. A recent study reported a median titanium dioxide nanoparticle ($nTiO_2$) concentration within surface waters in the ng to low μg/L range [4]. Moreover, the test protocols currently utilised for environmental risk assessment of nanoparticles were developed to accommodate traditional chemicals, such as pesticides [5]. As nanoparticles exhibit distinct physical and chemical properties, they often present different behaviours relative to their bulk phases and other chemical stressors [6,7]. Their rapid aggregation, which causes an accumulation at the bottom of the experimental test units, could adversely affect benthic aquatic organisms [8]. At the same time, their bioavailability in the aqueous phase in ecotoxicological experiments is reduced [9]. These unique properties and associated experimental challenges require modifications of the respective test protocols [5].

Here, we investigated the chronic adverse effects of $nTiO_2$ on the reproduction of *Daphnia magna*, whose parthenogenetic reproduction cycle is for instance described in detail by Zaffagnini [10], by applying two commercially available products with differing crystalline structure, namely P25 (Evonik, Germany; Figure 1) and A-100 (Crenox, Germany), as an additive-free, size-homogenised, stable suspension. The experimental procedure followed largely the standard test protocol designated by the Organisation for Economic Cooperation and Development (OECD) [11]. However, the experiments were performed in a flow-though system avoiding the accumulation of $nTiO_2$ aggregates at the bottom of the test vessels as a potential source of ecotoxicological effects (Figure 2) [8]. To allow for inferences on potential effects on the next (=filial) generation, the fifth brood released by the exposed adults was assessed separately for each treatment (0.00, 0.02, 2.00 mg/L) regarding its acute sensitivity to $nTiO_2$ following the respective OECD test protocol [12]. This pathway of effect may be considered as relevant in the context of the present study since similar observations were reported for other chemical stressors such as algae toxins and polyfluorinated substances [13,14] but also silver nanoparticles [15].

Materials and Methods

General Study Design

In total, five sets of experiments were performed to assess potential effects of $nTiO_2$ on the next generation of exposed adult *D. magna* (Figure 3A). During the first set of experiments

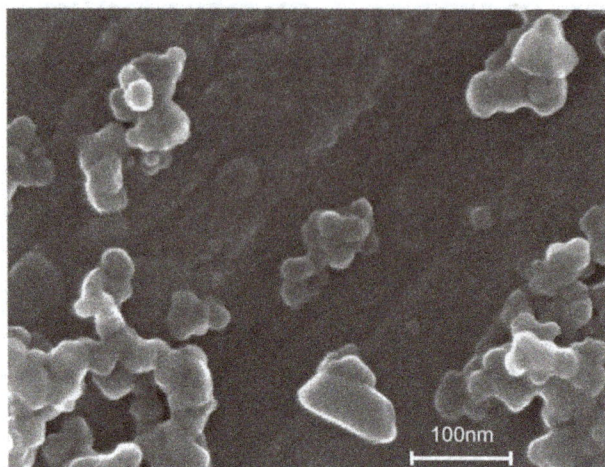

Figure 1. Scanning electron microscope image. An image of the size-homogenised, stable $nTiO_2$ suspension of the product P25 taken by an scanning electron microscope using 100,000-fold magnification (Hitachi SU8030).

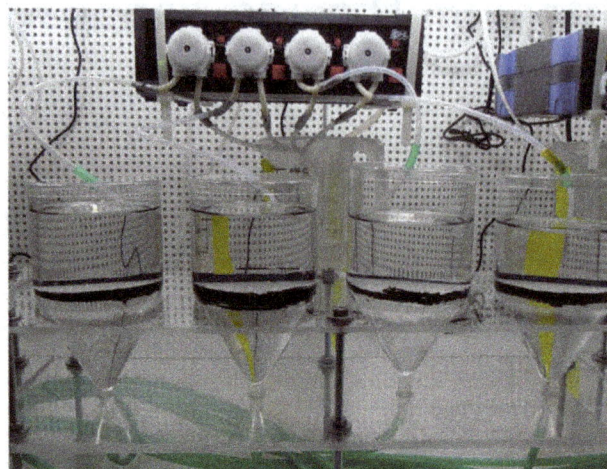

Figure 2. The experimental set-up. The flow-through testing apparatus, showing four experimental units (volume, 500 mL) with five *D. magna* each. Approximately 40 mL of the test medium was introduced every other hour slightly below the surface of the water. The old test medium was passively discharged using hydrostatic pressure at the bottom of each vessel. A fine mesh screen (0.1 mm) prevented the loss of recently hatched juvenile daphnids.

carried out once, juveniles of the fifth brood released by adult daphnids exposed to 0.00, 0.02 or 2.00 mg/L $P25-nTiO_2$ for 21 days were introduced into acute toxicity experiments. The second set of experiments was performed three times to assess the importance of an "early exposure" of juveniles towards $nTiO_2$ directly after their release (Figure 3B). Therefore, half of the adults of each treatment, including the control, were transferred to test medium not containing $P25-nTiO_2$ approximately 23 hours subsequent to the release of the fourth brood, i.e. after approximately 18 days. Thus, the exposure of newly born juveniles to nanoparticles prior to the initiation of the acute toxicity experiments was avoided. The remaining adults, in contrast, released their juveniles in the respective $nTiO_2$ treatment (as in the first set of experiments). The third set of experiments – performed once – intended to investigate implications on juveniles' sensitivity potentially driven by the exposure of daphnid's eggs towards $nTiO_2$ within the brood pouch. Therefore, adult daphnids were exposed to 0.00 and 2.00 mg/L $nTiO_2$ starting with the release of the third brood and lasting until 23 h after the release of the fourth brood resulting in an exposure period of approximately 3 days. This procedure limited any chronic implications potentially transferred from adults to juveniles and represented the maximum time period eggs may have been exposed to $nTiO_2$ during earlier experiments. Subsequently, adults were transferred to clean medium for the release of the fifth brood. To investigate whether the results obtained with P25 may be transferable to other $nTiO_2$ forms, the experimental procedure used for the first set of experiments was carried out with the product A-100 during the fourth set of experiments. The fifth set is not further described in the manuscript. It was utilized to assess the sensitivity of adult daphnids following the 18 days of exposure to 0.00 and 2.00 mg/L $nTiO_2$ (P25), which is displayed in Figure S1.

Organisms

D. magna (Clone V, Eurofins-GAB laboratories, Germany) were cultured at $20 \pm 1°C$ with a 16:8 hour (light:dark) photoperiod in ASTM reconstituted with hard freshwater that was enriched with selenium, vitamins (thiamine hydrochloride, cyanocobalamine, biotine) and seaweed extract (Marinure®, Glenside, Scotland). The

daphnids were fed with the green algae *Desmodesmus* sp. on a daily basis ($\sim$200 µg carbon per organism).

Flow-through Tests

The experimental procedure followed the standard test protocol designated by the OECD [11], with minor deviations. Briefly, each replicate consisted of 500 mL test medium and five *D. magna* that were younger than 24 hours of age at the start of the experiments. The study duration was set at 21 days, and the offspring were counted daily as a measure of the sublethal effect. Most importantly, the experiments were performed in a flow-though system. Therefore, the ASTM medium (148 mL) was mixed with the $nTiO_2$ stock solution (12 mL) immediately before delivery in equal proportions to four independent replicates, each with a volume of 500 mL. The old test medium was passively discharged through a mesh at the bottom of the vessel (Figure 1). This procedure was repeated every other hour, ensuring a complete water exchange within 24 hours. The ASTM medium was renewed daily and amended in an age-dependent manner with food (*Desmodesmus* sp.; 50–100 µg carbon/test organism). The $nTiO_2$ stock solutions were renewed at 72 h intervals. All of the pumps were equipped as completely as possible with Teflon tubes to minimise the loss of $nTiO_2$ during pumping. The flow-through apparatus avoided the accumulation of $nTiO_2$ aggregates at the bottom of the test vessels, eliminating one potential source of ecotoxicological effects [8]. Moreover, the medium was amended with seaweed extract to simulate dissolved organic matter that is normally present in natural water, which stabilised $nTiO_2$ within the aqueous phase [16]. This procedure ensured a continuous exposure to $nTiO_2$ particles of sizes <150 nm (Table S1). The average particle size was monitored daily during the chronic experiments for the 2.00 mg/L $nTiO_2$ treatment. Moreover, the actual zeta potential of both products in the test medium was analysed. Because even the highest test concentration delivered in this study was still too low to measure the zeta potential, the respective $nTiO_2$ stock suspension was diluted in test medium prior to measurement (Table S1).

A

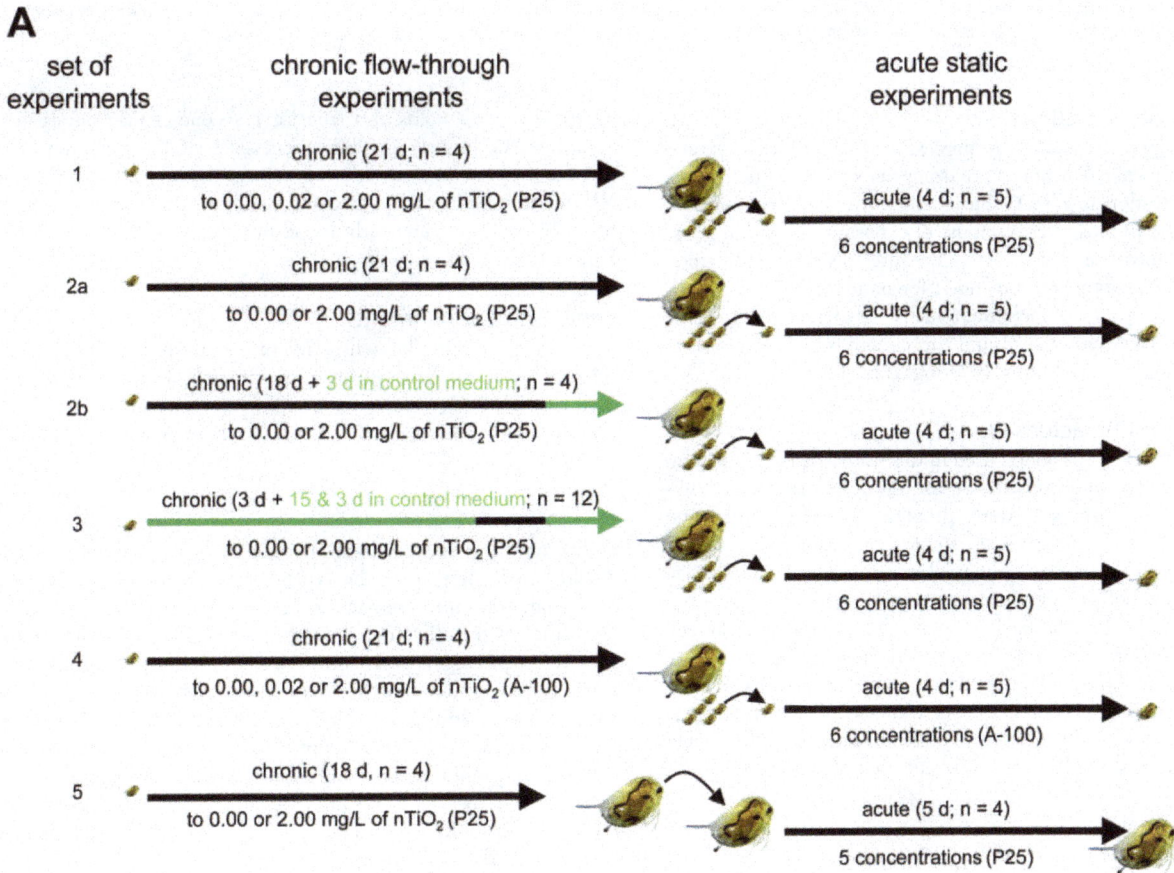

B

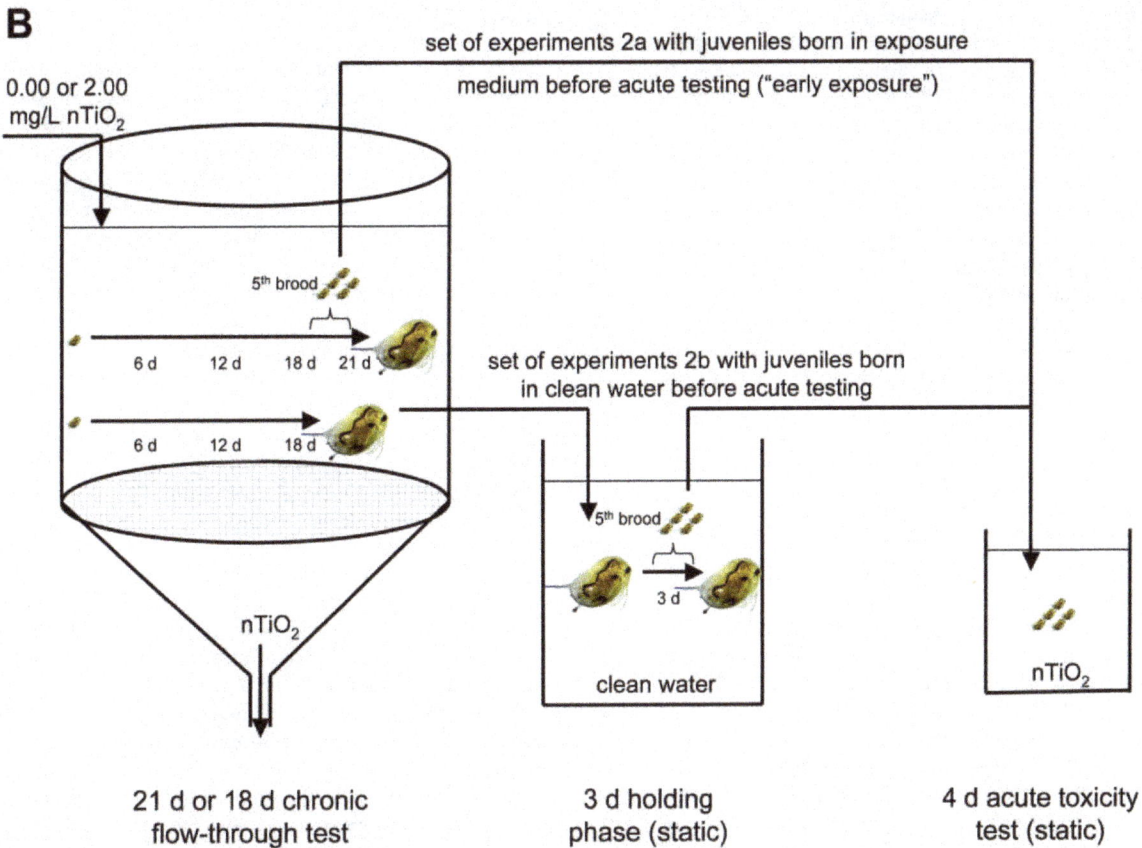

Figure 3. Experimental design. (*A*) Schematic diagram illustrating the experimental procedure of each of the five sets of experiments conducted. (*B*) Visualised experimental procedure for the assessment of the "early exposure" hypothesis (second set of experiments).

Acute Toxicity Experiments

The fifth brood released by the exposed adults was assessed separately for each treatment regarding its sensitivity to $nTiO_2$. The concentration of $nTiO_2$ that resulted in 50% immobility of the juvenile daphnids after 96 hours of exposure (96 h-EC_{50}) was used as the measure of sensitivity. Because this prolonged study duration was recently recommended for nanoparticle testing [9], the OECD test protocol for acute toxicity tests with *D. magna* [12] was adapted correspondingly, and the daphnids were exposed to 0.00, 0.50, 1.00, 2.00, 4.00 or 8.00 mg/L $nTiO_2$.

Nanoparticle Characterisation

Both of the $nTiO_2$ products used in this study, P25 (Evonik) and A-100 (Crenox), were purchased as powdered reagents. Subsequently, both products were prepared as dispersant- and additive-free, size-homogenised, stable suspensions by stirred media milling [9]. The zeta potential and the actual particle size distribution of both suspensions were determined *via* electrophoretic mobility and dynamic light scattering (Delsa^{TM} Nano C, Beckman Coulter, Germany), respectively. The concentrations of $nTiO_2$ in the 2.00 mg/L treatment were verified weekly by inductively coupled plasma mass spectrometry [9]. These analyses were supplemented by scanning electron microscope (Hitachi SU8030) imaging for P25 for the verification of particle size.

Statistical Analysis

Immobilisation data gained from the acute toxicity tests were adjusted by Abbott's formula if necessary and fitted to adequate dose-response models – based on Akaike Information Criterion and expert judgement – in order to determine 96 h-EC_{50} values using the drc extension package [17] for the statistics program R version 2.13.0. Confidence interval (CI) testing was accomplished to assess for statistically significant differences between 96 h-EC_{50} values of juveniles released by daphnids exposed to the control and those exposed to $nTiO_2$ obtained during the first, third, fourth and fifth set of experiments [18]. The 96 h-EC_{50} values of the first and second set of experiments were combined in a meta-analysis assessing for difference between juveniles of the fifth brood released by daphnids not exposed to $nTiO_2$ and those exposed to either 0.02 or 2.00 mg/L P25-$nTiO_2$. Therefore, a fixed effect model based on the standardised effect size Cohen's *d* was applied. An effect size was considered to be statistically significant when the respective CI did not include the zero value. Additionally, the respective *p*-values were computed [19].

Results

First and Second Set of Experiments

The P25-$nTiO_2$ particles were dispersed within the aqueous phase throughout the whole chronic study duration and displayed a mean size of 135.8 nm, with approximately 30% of the particles <100 nm (Table S1). The nanoparticles did not adversely affect the mean number of offspring released by exposed adults (Figure S2 left). However, subsequent acute toxicity studies (first set of experiments) showed for the fifth brood released by adults exposed to 2.00 mg/L P25-$nTiO_2$ a significantly lower 96 h-EC_{50} value (difference between 96 h-EC_{50} values: 4.39 mg/L, 95% CI of the difference 0.62 to 8.15; Figure S3). Additionally, a meta-analysis, which considered all 96 h-EC_{50} values of the first and the second set of experiments, supported these results (p = 0.0021 with n = 7;

Figure 4A). This holds also true for a second meta-analysis that accounted exclusively for the effect sizes of the offspring that originated from the adults exposed to $nTiO_2$ but that were released into nanoparticle-free test medium, which hence avoided the possibility of an early exposure (p = 0.0169 with n = 3; Figure 4B).

Third Set of Experiments

No difference in the sensitivity of the offspring (fifth brood) released by adults exposed for approximately 3 days to 0.00 and 2.00 mg/L $nTiO_2$ was observed (the difference between 96 h-EC_{50} values: 0.64 mg/L, 95% CI of the difference −1.04 to 2.32; Figure 5).

Fourth Set of Experiments

The chronic exposure to A-100 (for particle characteristics see Table S1) did not affect the reproduction of the test species *D. magna*, at concentrations up to 2.00 mg/L (Figure S2, right). Nevertheless, the offspring released by adults exposed to 0.02 mg/L A-100 were significantly more sensitive than those released by adults from the nanoparticle free control (difference between 96 h-EC_{50} values: 1.26 mg/L, 95% CI of the difference 0.62 to 1.90; Figure 6, right). The sensitivity of the offspring released by adults exposed to high $nTiO_2$ concentrations was increased by a factor of at least five. However, it was not possible to verify this deviation by statistical analysis due to the lack of suitable quantitative approaches (Figure 6, right).

Discussion

The first set of experiments showed that the fifth brood released by adults exposed to 2.00 mg/L P25-$nTiO_2$ was significantly more sensitive than the juvenile offspring released by the non-exposed control adults (Figure S3), indicating the possibility of effects passed from the parental to the filial generation of daphnids. However, these results might also be explained by the exposure of the offspring immediately after hatching and prior to their introduction into the acute toxicity tests (Figure 3A). In contrast, the exposure of the juveniles from the nanoparticle-free control commenced up to 24 hours later, with the start of the acute toxicity experiments. To test this "early exposure" possibility, we performed further *D. magna* reproduction assays. However, half of the adults for each treatment during the second set of experiments (Figure 3B), including the control adults, were transferred to test medium containing no $nTiO_2$ approximately 23 hours subsequent to the release of the fourth brood. The exposure of the newly released fifth brood to nanoparticles prior to the initiation of the acute toxicity experiments was thus avoided. In contrast, the remaining adults released their offspring into the respective treatment conditions. Three additional independent experiments were performed using this experimental set-up. All of the 96 h-EC_{50} (n = 7) values obtained for the offspring released by adults exposed to $nTiO_2$ were combined into a meta-analysis (Figure 4A), which was based on a fixed effect model using Cohen's *d* as a standardised effect size. Furthermore, a second meta-analysis was performed considering exclusively the effect sizes of the offspring that originated from the adults exposed to $nTiO_2$ but that were released into nanoparticle-free test medium (n = 3; Figure 4B). Both meta-analyses revealed a significant increase in the sensitivity of the offspring released by adults exposed to

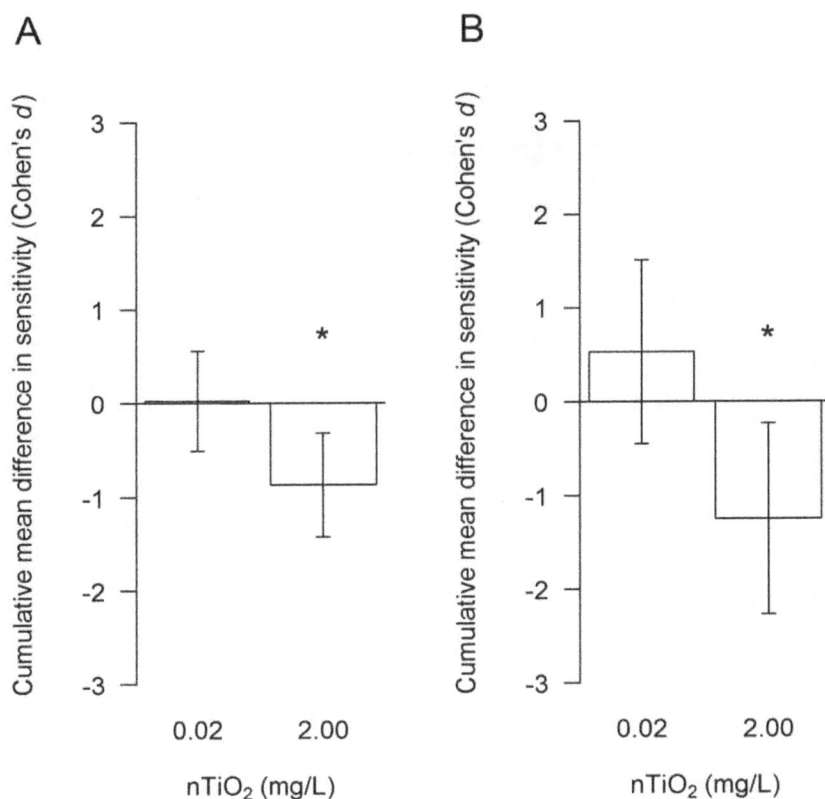

Figure 4. The sensitivity of juveniles released by nTiO$_2$-exposed adults. The cumulative mean ($\pm$95% CIs) difference in sensitivity – in terms of 96 h-EC50 values – of the offspring (fifth brood) released by adults exposed to 0.02 or 2.00 mg nTiO2/L and offspring released by control (uncontaminated) daphnids is displayed using the standardised effect size Cohen's d. (A) The cumulative effect sizes for all bioassays conducted (n = 7) with the fifth brood during the first and second set of experiments. (B) The cumulative effect sizes for acute toxicity experiments conducted with offspring released in the control medium by adults previously exposed to the above-mentioned nTiO2 concentrations during the second set of experiments (n = 3). The statistical significance of a cumulative effect is highlighted by an asterisk (*). Negative effect sizes indicate increased toxicity.

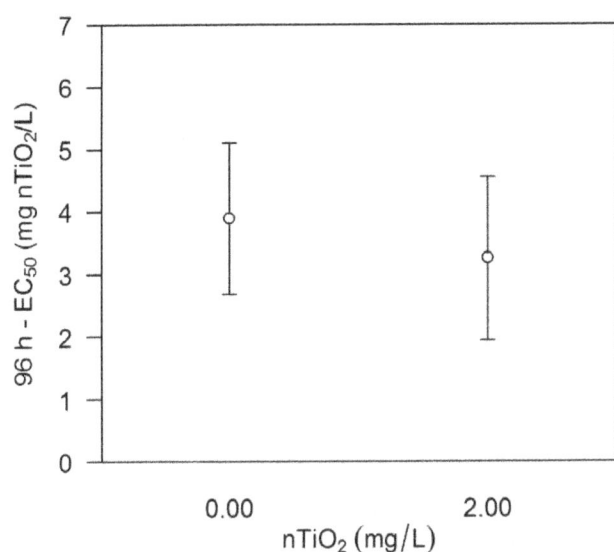

Figure 5. The sensitivity of juveniles released during the third set of experiments. 96 h- EC$_{50}$ values with respective 95% CIs of the fifth brood released by adults exposed to 0.00 and 2.00 mg/L P25-nTiO$_2$ during the third set of experiments, which considered exclusively potential implication of nTiO$_2$ exposure within the brood pouch. No statistically significant difference among treatments was observed.

2.00 mg/L nTiO$_2$ compared to offspring released from adults not suffering from nTiO$_2$ exposure. These results suggest that "early exposure" is insufficient to explain the approximately twofold increase in sensitivity (Figure 6, left) observed in the fifth brood released from the exposed adults. However, the daphnids' eggs might already be exposed to nTiO$_2$ within the brood pouch, which could potentially affect the sensitivity of the subsequent offspring [20]. To assess this issue, adult daphnids were exposed to 0.00 and 2.00 mg/L nTiO$_2$ starting with the release of the third brood and lasting until at least 23 h after the release of their fourth brood (third set of experiments). This procedure alleviated any long-term implications potentially transferred from adults to offspring and represented the maximum time period over which eggs may have been exposed to nTiO$_2$ during our earlier experiments. Subsequently, the adults were transferred to nTiO$_2$-free medium for the release of the fifth brood. Acute toxicity experiments revealed no difference in the sensitivity of the offspring released by adults exposed to 0.00 and 2.00 mg/L nTiO$_2$ (Figure 5), indicating that the exposure of the eggs during the early phases of development could not account for the observed effects. Consistent with another study [15], which uncovered adverse effects of silver nanoparticles ingested by adult fruit fly Drosophila melanogaster (Meigen) passed on to their offspring, a similar explanation might apply for this study. Comparable observations were also reported for other chemical stressors like algal toxins [13], perfluorooctane sulfonic acid, perfluorooctanoic acid [14], vinclozolin and 5-azacytidine [21]. However, in this study, we have uncovered effects induced by

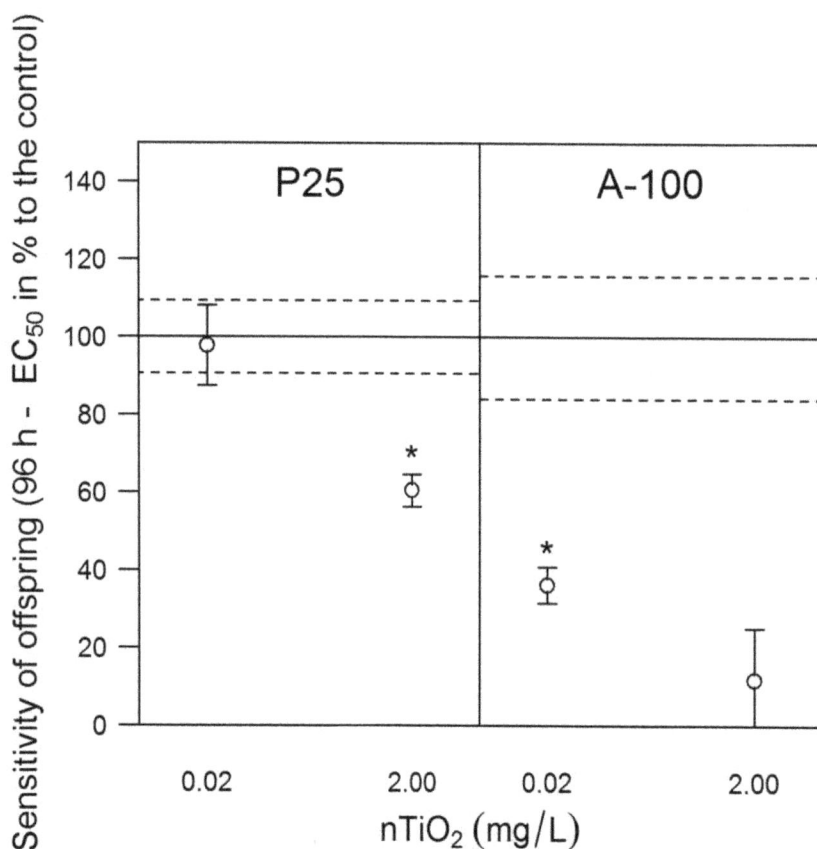

Figure 6. The sensitivity of juveniles released by adults exposed to different nTiO$_2$ products. Sensitivity, displayed as percent relative to the 96 h-EC$_{50}$ of the respective control, of the fifth brood released by adult *D. magna* exposed to different nTiO$_2$ treatments using the products P25 or A-100. The data displayed for P25 represent the weighted mean values of the seven experiments (first and second set of experiments), each with four replicates per treatment, whereas the 96 h-EC$_{50}$ for the offspring released from the control parents was 3.13 mg/L nTiO$_2$. For the product A-100, the results of one experiment with four replicates of pre-treatment are displayed (fourth set of experiments). In this situation, the 96 h-EC$_{50}$ for the offspring released from the control parents was 1.98 mg/L nTiO$_2$. The error bars and dashed lines indicate the standard error. The dashed lines are related to the control. Asterisks (*) denote significant differences between a treatment and the respective control.

nTiO$_2$ on the next generation of *D. magna* but not the parental generation, in terms of quantitative ecotoxicological endpoints: sensitivity to nTiO$_2$ after 18 days of exposure, number of offspring, and lipid content (Figure S1; S2, left; S4), while the mechanism resulting in this observation have no yet been understood.

Although this study demonstrates, for the first time, the possibility of nTiO$_2$ effects on the next generation of *D. magna* in which the parental generation exhibits no obvious effects, whether these effects would also result from exposure to other TiO$_2$ products remains unclear. Therefore, another flow-through experiment followed by acute toxicity experiments was performed with the product A-100 (fourth set of experiments; Figure 3A). Exposure to A-100 also did not affect the reproduction of the test species *D. magna*, measured as the number of offspring released at a product concentration of up to 2.00 mg/L (Figure S2, right). Dabrunz et al. [9] reported a 96 h-EC$_{50}$ of 0.74 mg/L for this product. This deviation in effects may be explained by the amendment of the test medium with seaweed extract [16] as well as by the avoidance of nTiO$_2$ deposition on the bottom of the test vessels [8]. Nonetheless, the offspring released by adults that were exposed to even 0.02 mg/L A-100 were significantly more sensitive—by an approximate factor of 3—than those released by the control parents (Figure 6, right). Moreover, the sensitivity of the offspring released by adults exposed to high nTiO$_2$ concentra-

tions was increased by a factor of at least five (Figure 6, right). These results might represent general safety implications of nTiO$_2$ products, although their intensity varies among products and the concentration applied.

Conclusion

Although the test design of this study ensured the bioavailability of the investigated nanoparticles over the whole study duration, the parental *D. magna* did not exhibit any indications of toxic stress under these standard test conditions. However, the offspring generation of daphnids released by parental *Daphnia* that were previously exposed to nTiO$_2$ were significantly more sensitive to nTiO$_2$. Finally, the present study suggests that the standardised testing protocols that comprise the foundation of the current environmental risk-assessment approaches for nanoparticles underestimate risks and require modification. Apart from the recently recommended extension of the study duration for acute toxicity testing [9], the OECD test guidelines for the assessment of chronic ecotoxicity need to be improved by considering the effects on the next generation, even on offspring that has never been directly exposed to the agents.

Supporting Information

Figure S1   120 h- EC$_{50}$ values with respective 95% CIs of adult *D. magna* following 18 days of exposure to nTiO$_2$ (P25) in the flow-through system (fifth set of experiments). No statistically significant deviations regarding the sensitivity were detected.

Figure S2   Boxplot (bold line represents the median) of the offspring per test organism (n = 20) exposed to P25 (first set of experiments) or A-100 (fourth set of experiments) nTiO$_2$ after 21 days of exposure to 0.00, 0.02 or 2.00 mg nTiO$_2$/L.

Figure S3   96 h-EC$_{50}$ values with respective 95% CIs of the fifth brood released by adults exposed to P25 nTiO$_2$ during the flow-through experiment (first set of experiments); Asterisk (*) denotes statistically significant difference between the juveniles released from adults exposed to 2.00 mg/L TiO$_2$ and the control based on confidence interval testing (difference between 96 h-EC$_{50}$ values 4.39 mg/L, 95% CI 0.62 to 8.15).

Figure S4   Lipid content per adult *D. magna* after 21 days of exposure to P25 nTiO$_2$ (first set of experiments).

References

1. Klaine SJ, Koelmans AA, Horne N, Carley S, Handy RD, et al. (2012) Paradigms to assess the environmental impact of manufactured nanomaterials. Environ Toxicol Chem 31: 3–14.
2. Behra R, Krug H (2008) Nanoecotoxicology - Nanoparticles at large. Nat Nanotechnol 3: 253–254.
3. von der Kammer F, Ferguson PL, Holden PA, Masion A, Rogers KR, et al. (2012) Analysis of engineered nanomaterials in complex matrices (environment and biota): general considerations and conceptual case studies. Environ Toxicol Chem 31: 32–49.
4. Gottschalk F, Ort C, Scholz RW, Nowack B (2011) Engineered nanomaterials in rivers - exposure scenarios for Switzerland at high spatial and temporal resolution. Environ Pollut 159: 3439–3445.
5. Handy RD, Cornelis G, Fernandes T, Tsyusko O, Decho A, et al. (2012) Ecotoxicity test methods for engineered nanomaterials: practical experiences and recommendations from the bench. Environ Toxicol Chem 31: 15–31.
6. Owen R, Handy R (2007) Formulating the problems for environmental risk assessment of nanomaterials. Environ Sci Technol 41: 5582–5588.
7. Ferry JL, Craig P, Hexel C, Sisco P, Frey R, et al. (2009) Transfer of gold nanoparticles from the water column to the estuarine food web. Nat Nanotechnol 4: 441–444.
8. Bundschuh M, Zubrod JP, Englert D, Seitz F, Rosenfeldt RR, et al. (2011) Effects of nano-TiO$_2$ in combination with ambient UV-irradiation on a leaf shredding amphipod. Chemosphere 85: 1563–1567.
9. Dabrunz A, Duester L, Prasse C, Seitz F, Rosenfeldt RR, et al. (2011) Biological surface coating and molting inhibition as mechanisms of TiO$_2$ nanoparticle toxicity in Daphnia magna. PLoS One 6: e20112.
10. Zaffagnini F (1987) Reproduction in *Daphnia*. In *Memorie dell'Istituto Italiano di Idrobiologia "Dott. Marco De Marchi"*, Peters, R. H.; de Bernardi, R., Eds. Pallanza, 1987; 245–284.
11. OECD (2008) Guideline No. 211: *Daphnia magna* reproduction test.
12. OECD (2004) Guideline No. 202: *Daphnia sp.*, acute immobilisation test.
13. Ortiz-Rodríguez R, Dao TS, Wiegand C (2012) Transgenerational effects of microcystin-LR on *Daphnia magna*. J Exp Biol 215: 2795–2805.
14. Ji K, Kim Y, Oh S, Ahn B, Jo H, et al. (2008) Toxicity of perfluorooctane sulfonic acid and perfluorooctanoic acid on freshwater macroinvertebrates (*Daphnia magna* and *Moina macrocopa*) and fish (*Oryzias latipes*). Environ Toxicol Chem 27: 2159–2168.
15. Panacek A, Prucek R, Safarova D, Dittrich M, Richtrova J, et al. (2011) Acute and chronic toxicity effects of silver nanoparticles (NPs) on *Drosophila melanogaster*. Environ Sci Technol 45: 4974–4979.
16. Hall S, Bradley T, Moore JT, Kuykindall T, Minella L (2009) Acute and chronic toxicity of nano-scale TiO$_2$, particles to freshwater fish, cladocerans, and green algae, and effects of organic and inorganic substrate on TiO$_2$ toxicity. Nanotoxicology 3: 91–97.
17. Ritz C, Streibig JC (2008) Non linear regression with R. Springer Science & Business Media: New York, USA.
18. Wheeler MW, Park RM, Bailer AJ (2006) Comparing median lethal concentration values using confidence interval overlap or ratio tests. Environ Toxicol Chem 25: 1441–1444.
19. Borenstein M, Hedges LV, Higgins JPT, Rothstein HR (2009) Introduction to Meta-Analysis John Wiley & Sons Ltd: West Sussex, UK.
20. Baird DJ, Barber I, Soares AMVM, Calow P (1991) An early life-stage test with *Daphnia magna* straus: An alternative to the 21-day chronic test? Ecotoxicol Environ Saf 22: 1–7.
21. Vandegehuchte MB, Lemière F, Vanhaecke L, Vanden Berghe W, Janssen CR (2010) Direct and transgenerational impact on *Daphnia magna* of chemicals with a known effect on DNA methylation. *Comp Biochem Physiol C* 151: 278–285.

Table S1   Particle characteristics of P25 and A-100: The table displays the 10th and the 90th percentile of the particle size distribution, the mean percentage of particles below a particle size of 100 nm as well as the mean particle size together with the polydispersity index. This index provides information on the range of the particle size distribution. A value above 0.3 indicates unreliability of the measurement, due to masking of small particles by large ones. Additionally the zeta potential of the particles and their measured concentration in the test medium is given. nTiO$_2$ concentrations were measured following Dabrunz et al.

Acknowledgments

The authors thank W. Fey for assistance during ICP-MS analysis, U. Gernert (ZELMI, Technical University Berlin) for the SEM picture, and C. Schilde (Institute of Particle Technology, Technical University Braunschweig) for the provision of stable nanoparticle dispersions. D.J. Baird, M.C. Newman and K. Schwenk are acknowledged for their valuable comments on an earlier draft of this manuscript and A.A. Kotov for advice on the early embryology od *Daphnia*.

Author Contributions

Conceived and designed the experiments: MB FS RS. Performed the experiments: FS RRR. Analyzed the data: MB FS RS. Wrote the paper: MB FS RS.

4

Tracking Signals of Change in Mediterranean Fish Diversity Based on Local Ecological Knowledge

Ernesto Azzurro[1,2]*, Paula Moschella[3], Francesc Maynou[1]

1 Institut de Ciències del Mar (ICM-CSIC), Barcelona, Spain, **2** High Institute for Environmental Protection and Research (ISPRA), Laboratory of Milazzo, Milazzo, Italy, **3** Commission Internationale pour l'Exploration Scientifique de la Mer Méditerranée (CIESM), Monaco, Monaco

Abstract

One of the expected effects of global change is increased variability in the abundance and distribution of living organisms, but information at the appropriate temporal and geographical scales is often lacking to observe these patterns. Here we use local knowledge as an alternative information source to study some emerging changes in Mediterranean fish diversity. A pilot study of thirty-two fishermen was conducted in 2009 from four Mediterranean locations along a south-north gradient. Semi-quantitative survey information on changes in species abundance was recorded by year and suggests that 59 fish species belonging to 35 families have experienced changes in their abundance. We distinguished species that increased from species that decreased or fluctuated. Multivariate analysis revealed significant differences between these three groups of species, as well as significant variation between the study locations. A trend for thermophilic taxa to increase was recorded at all the study locations. The Carangidae and the Sphyraenidae families typically were found to increase over time, while Scombridae and Clupeidae were generally identified as decreasing and Fistularidae and Scaridae appeared to fluctuate in abundance. Our initial findings strongly suggest the northward expansion of termophilic species whose occurrence in the northern Mediterranean has only been noted previously by occasional records in the scientific literature.

Editor: Adina Maya Merenlender, University of California, Berkeley, United States of America

Funding: This study has been partially supported by the CIESM/Albert II Foundation (Project: Tropical signals) and by the Euro-Mediterranean Center for Climatic Change and the Italian Ministry for the Environment and the Territory (Project: The impacts of biological invasions and climate change on the biodiversity of the Mediterranean Sea). The funders had no role in study design, data collection and analysis, decision to publish, or preparation of the manuscript.

Competing Interests: The authors have declared that no competing interests exist.

* E-mail: eazzurr@tin.it

Introduction

Global change is having an ever increasing influence on the abundance and distribution of living organisms worldwide [1] but documenting the resulting biological trends is often constrained by the lack of information from studies at the appropriate temporal and spatial scales. The Increasing success of thermophilic biota colonizing the Mediterranean Sea [2] is one clear example of rapid changes that are happening at the regional scale. This is particularly evident when we look at fish colonization. Indeed a number of native species with tropical or subtropical affinity seem to have already moved towards the northern and colder sectors of the Mediterranean [3,4]. This phenomenon, that has been named as "meridionalization" within the Mediterranean literature [5] is actually parallel to a number of poleward expansions of low latitude species that have been recorded all over the world [6,7] for a variety of species such as plants [8], butterflies [9], birds [10], insects [11] and fish [12]. We are also expecting some cold water species to decline in the near future following forecast temperature rise [13]. Changes in the distribution of Mediterranean fishes are generally revealed by casual observation of scattered individuals outside the species known geographical range [14], being these sporadic records the only source of information that has been used to study meridionalization so far [4]. Consequently, the extent of these changes may be under appreciated, because of the limited and non-continuous nature of scientific monitoring [14]. The efforts that would be needed to monitor and survey marine

habitats at scale large enough to perceive temporal and spatial trends is huge and this is clearly one of the major obstacles to researching global change [15]. There is an urgent need to fill this information gap by the use of new methodologies, suitable to deepen our capability to perceive the complex process of change.

In the last decades, "Local Ecological Knowledge" has emerged as an alternative information source on species presence or qualitative and quantitative indices of species abundance [16–22]. can be defined as the information that a group of people have about local ecosystems. We usually rely on knowledge gained by individuals over their lifetimes, and not what information has been handed down through the generations [23]. To extract data and information from individuals' memory, semi-structured or unstructured conversations between the researcher and a participant are commonly used, a practice commonly called "oral history" [24].

Here we aimed to explore of the utility of Local Ecological Knowledge to provide reliable information about some emerging changes in Mediterranean fish diversity. In the past, other studies used this approach to reconstruct trends in abundance of marine fish species, especially declines in abundance [25,26,27]. Our specific aims were threefold:

1. Identify indicators of meridionalization.
2. Reconstruct historical trends of abundance for indicator species.
3. Evaluate the potentiality of using Local Ecological Knowledge to the study large scale changes in Mediterranean fish diversity.

Methods

Study area

Interviews were carried out between August 2009 and October 2010, over 4 different locations in the Mediterranean Sea (Figure 1): Linosa and Lampedusa, belonging to the archipelago of Pelagie Islands (Sicily Strait); Milazzo (Southern Tyrrhenian Sea) and Porto San Giorgio (Central Adriatic Sea).

Sampling population

The study was directed to local and recreational fishermen with more than ten years of experience at sea. Fishermen were met during their land activities, while for example cleaning the nets in the harbors or fixing their boats. Special attention was placed to the approach, since fishermen generally mistrust fisheries researchers and managers. To counter this, formalities were avoided and conversations was focused on which species may have appeared or disappeared in the last decades. Once they showed interest in the subject through discussion then the interview was formally started. In other occasions, when we knew one local fisherman we asked him to introduce us to other fishermen.

The survey questionnaire

People were interviewed on the basis of a detailed-protocol procedure (the "interview protocol" that has been officially adopted as part of the international basin-wide monitoring program "CIESM Tropical Signals" http://www.ciesm.org/marine/programs) and a standard questionnaire was developed around the following central questions:

"Can you tell us what kind of fishes showed the greatest variation in abundance in the last decades?" Do you know species that have appeared or disappeared?

The interview-protocol was developed to guide the interviewers in their task i.e. to extract the required information from the fisherman knowledge. At first, the fisherman's age and fishing gear used by the fisherman was recorded. Photographs from a field guide [28] were used to match local fish names with taxonomic ones. Only those species that were mentioned by the fisherman were registered. Respondents were also asked to provide a qualitative ranking of the abundance of these taxa through time, on an annual basis, according to 6 different grades: 0 = ABSENT; 1 = RARE (once in a year); 2 = OCCASIONAL (sometimes in a year); 3 = COMMON (regularly in a year); 4 = ABUNDANT (regularly in captures and abundant); 5 = DOMINANT (always in captures and with great abundances). At the end of each interview, each recorded taxa was assigned to a trend factor: species that "INCREASE" (level "I"); species that "DECREASE" (level "D"); and species that "FLUCTUATE (level "F") over the respondent experience period.

Conceptual approach for experimental design and statistical analysis

Considering the subjectivity of fishermen's knowledge we used a simple design in which the variability between the different interviews was taken into account and tested against our hypotheses (i.e. H_0 of no differences between species groups; H_0 of no differences between the different locations). The species mentioned in each interview were used to build a presence-absence dataset in which the "interviews" were the samples and the "species" the variables. Each interview taken from the same location was considered as an independent replicate sample. Finally our dataset was explored by means of multivariate and univariate analysis. A two way PERMANOVA [29,30] based on Bray-Curtis resemblance matrix was used to test for the terms "Location" (with 4 levels) and "Trend" (with 2 levels: "Increase" or "decrease") that were considered as fixed crossed factors. Fluctuating species were not included in the analysis because of the

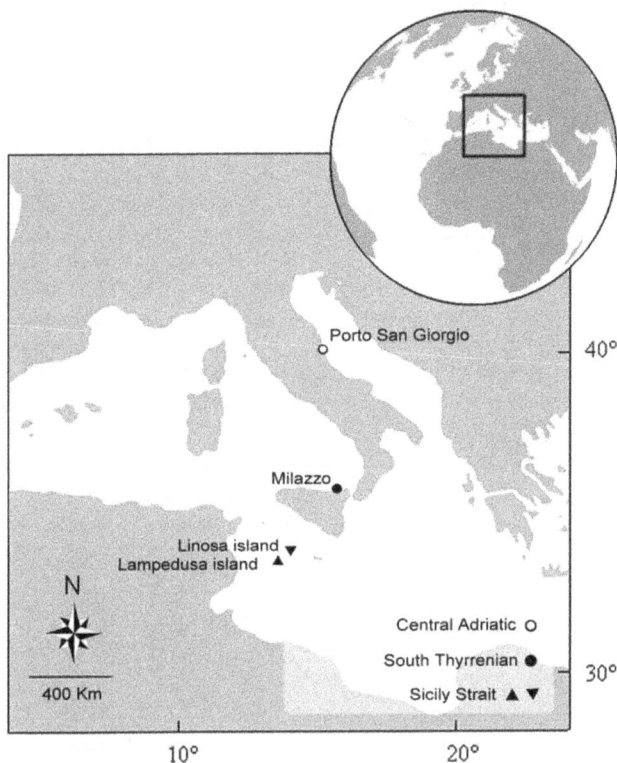

Figure 1. Study locations in the Mediterranean Sea.

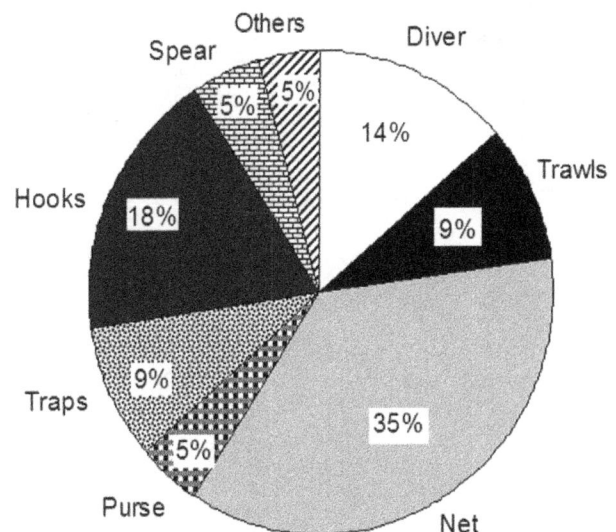

Figure 2. Percent distribution of fishing methods adopted by the respondents. Interviewed (Tot N = 32) were both recreational (N = 8, 25%) and professional (N = 24, 75%) fishermen.

Table 1. List of fish taxa cited by the respondents.

Taxa	I	D	F	Taxa	I	D	F
Ammodytidae		5		Pomatomidae	6		
Gymnammodytes cicerelus		5		Pomatomus saltatrix	6		
Atherinidae		1		Rajidae	1	1	
Atherina sp		1		Raja sp	1	1	
Balistidae		6		Scaridae	8		5
Balistes capriscus		6		Sparisoma cretense	8		5
Belonidae		2		Sciaenidae	1	2	
Belone belone		2		Sciaena umbra		2	
Carangidae	24	5		Umbrina cirrosa	1		
Caranx crysos	12			Scomberesocidae		6	
Lichia amia	4			Scomberesox saurus saurus		6	
Seriola dumerilii	4	3		Scombridae	7	24	1
Trachinotus ovatus	3			Auxis rochei		3	1
Trachurus mediterraneus	1	1		Sarda sarda	1	2	
Trachurus trachurus			1	Scomber japonicus		6	
Carcharhinidae		1		Scomber scombrus		17	
Prionace glauca		1		Thunnus thynnus		2	
Centracanthidae		2		Scophthalmidae		1	
Spicara sp.		2		Scophthalmus rhombus		1	
Clupeidae	3	21		Scorpaenidae			1
Engraulis encrasicolus		12		Scorpaena scrofa			1
Sardina pilchardus		4		Serranidae	2	3	3
Sardinella aurita	3			E. marginatus	2	3	3
Sprattus sprattus		5		Siganidae	1		1
Coryphaenidae	6			Siganus luridus	1		1
Coriphaena hippurus	6			Soleidae		4	
Dasyatidae	2	6		Solea vulgaris		4	
Dasyatis pastinaca	2	2		Sparidae	7	16	1
Exocoetidae		4		Boops boops	1	2	
Fistulariidae			7	Dentex dentex		2	
Fistularia commersonii			7	Diplodus annularis		1	
Gadidae		4		Diplodus sargus	2	1	
Merlangus merlangus		2		Diplodus vulgaris		2	
Trisopterus minutus capelanus		2		Lithognathus mormyrus		2	
Gobiidae		9		Oblada melanura		2	
Aphia minuta		2		Pagrus pagrus		2	
Gobius niger		4		Sarpa salpa	1	4	1
Gobius sp		3		Spondyliosoma chantarus		1	
Labridae		2		Sphyraenidae	14		
Coris julis		2		Sphyraena spp.	14		
Merlucciidae		3		Syngnathidae		3	1
Merluccius merluccius		3		Hippocampus sp		2	1
Mugilidae	3			Syngnathus sp		1	
Muraenidae	1	4		Torpedinidae		1	
Muraena helena	1	4		Torpedo sp.		1	
Phycidae		1		Triglidae	1		1
Phycis phycis		1		Trigla lucerna	1		1

Table 1. Cont.

The number of times in which they were assigned to the groups 'INCREASE' ('I'), 'DECREASE' ('D') and 'FLUCTUATE' ('F') is reported.

high number of empty cells. The Similarity Percentages Procedure (SIMPER) was used to identify the most important fish taxa in typifying the groups "I", "D" and "F". Cut off for low contributions was set at 90%. A Non-metric Multi Dimensional Scaling (nMDS) ordination was performed separately for the group of species "I" and "D" to visualize geographical patterns. All the multivariate analyses were performed with PRIMER 6+PERMANOVA software package from Plymouth Marine Laboratory, UK.

We applied breakpoint structural analysis [31,32] to the time series of semi-quantitative abundance data to assess the year(s) of statistically significant change in abundance. For each year in the period 1969–2010 the median value of the semiquantitative abundance index was computed for species contributing to the typification of groups "I" and "D". The technique of breakpoint analysis allows to identify statistically significant changes in the level of subsets of a time series. A time series is randomly split in 2 or more subsets ("data windows") and the mean level compared through a modified F test ("structural change" or sc test [32]). The procedure is repeated iteratively until all significant breakpoints (if any) are identified [31]. In our case, we applied the Bayesian Information Criterion (BIC) as objective criterion to determine the number of breakpoints and their associated dates [32] with the corresponding 95% confidence intervals (CI). The breakpoint analyses were performed with the R library strucchange, developed by A. Zeileis at the University of Economics, Vienna (Wirtschaft-suniversität Wien, Austria).

Results

A total of 32 artisanal fishermen were interviewed (9 from Linosa, 2 from Lampedusa, 10 from Milazzo and 11 from Porto San Giorgio). They were recreational (N = 8) and professional (N = 24) fishers with more than 10 years of experience at sea (15% N = 5 began before 1970; 28% N = 9 between 1970 and 1979; 38% N = 12 between 1980 and 1989; 19% N = 6 between 1990 and 1999) and an average age of 55 years. More than one fishing method was often adopted by the respondents, being the "nets" the most common fishing gear (35%) and diving performed by some of them (14%) (Figure 2). Out of 35 fishermen that have been

Table 2. PERMANOVA Analysis.

Source	df	SS	MS	Pseudo-F	P(perm)
Location	3	26.582	8.8607	3.9332	0.001
Trend	1	10.378	10.378	4.6067	0.001
Location × Trend	3	25.388	8.4625	3.7565	0.001
Res	59	132.91	2.2528		
Total	66	203.82			

Permutational multivariate analysis of variance based on the Euclidean dissimilarity measure for presence-absence data. The test was done using 9999 permutations under the reduced model. The group 'FLUCTUATE' was excluded from the analysis.

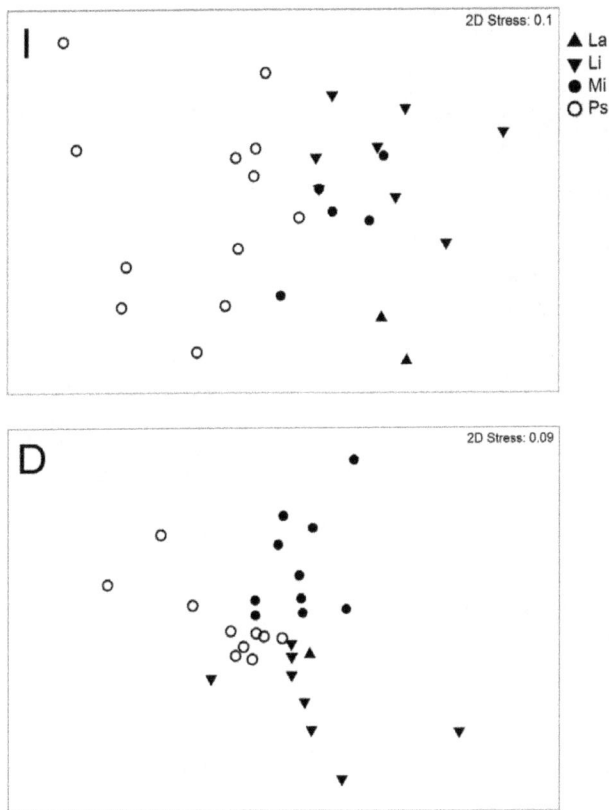

Figure 3. Non-metric Multi Dimensional Scaling (nMDS) ordination comparing interviews outputs across the different study locations. The position of each dot is defined by the assemblage of species recorded in each interview. **La** = Lampedusa; **Li** = Linosa; **Mi** = Milazzo; **Ps** = Porto San Giorgio. **I** = Group 'INCREASE'; **D** = Group 'DECREASE'.

Table 3. Most important fish taxa in typifying the groups 'I' and D by SIMPER analysis.

Taxa	Av. frequency of occurrence	Contribution (%)	Group
Sphyraena viridensis	0.42	34.92	I
Caranx crysos	0.36	34.82	
*Sparisoma cretense**	0.24	8.75	
Coriphaena hippurus	0.18	4.67	
Balistes capriscus	0.18	3.79	
Scomber scombrus	0.5	51.95	D
Engraulis encrasicolus	0.35	18.04	
Scomberesox saurus saurus	0.18	4.57	
Gymnammodytes cicerelus	0.15	3.07	
Solea vulgaris	0.12	2.45	
Sarpa salpa	0.12	2.45	
Sprattus sprattus	0.15	2.2	
Muraena helena	0.12	1.8	
Gobius niger	0.12	1.51	
Sardina pilchardus	0.12	1.51	
Fistularia commersonii	0.5	24.14	F
Sparisoma cretense§	0.36	22.17	
Epinephelus marginatus	0.21	16.26	

List of fish taxa in decreasing order of their importance in typifying the groups 'INCREASE' ('I') and 'DECREASE' ('D') by SIMPER analysis performed on presence/absence data. Cut off for low contributions: 90.00%. Group 'I' average similarity 18.05; Group 'D' average similarity 12.51; Group 'F' average similarity 15.44.
*Milazzo,
§Linosa and Lampedusa.

contacted, only 3 refused to be interviewed. The duration of interviews ranged between 35 minutes and 1 h with an average of 44 minutes. Nevertheless, the whole amount of time needed to collect data included also the search for fishermen and informal conversations before and after the interview. On average we performed 4 interviews/day.

According to our compilation of local knowledge, 59 fish species belonging to 35 families resulted to have passed through drastic changes in their abundance (Table 1). Of them, 26 species fell in the group "INCREASE" ("I"); 42 in the group "DECREASE" ("D") and 8 were indicated as "FLUCTUATE" ("F"). The total number of citations for these groups was: 91, 123 and 20 respectively.

The Carangidae family had the most of citations for the group "I" (24 citations; 6 species) followed by Sphyraenidae (14 citations; 1 species). As for the group "D", the mostly cited family was Scombridae (24 citations; 4 species) followed by Clupeidae (21 citations; 3 species), finally two families had the most of citations for the group "F": Fistularidae (7 citations; 1 species) and Scaridae (5 citations; 1 species).

PERMANOVA analysis (Table 2) revealed significant differences for the terms "Location" and "Trend" and a significant interaction between these two factors.

The Non-metric Multi Dimensional Scaling (nMDS) ordinations showed the geographical structure of the "I" and "D" datasets (Figure 3) with no real prospect of a misleading

interpretation. Looking at the "I" plot, it is quite clear the separation of the Adriatic location (Ps) from the remaining ones (Mi; La; Li) whilst in the "D" plot a separation between South Tyrrhenian (Mi) and the Pelagie islands (La and Li) seems to be apparent. According to SIMPER analysis (Table 3), 6 species contributed mostly to typify the group "I"; 11 species were important to characterize the group "D" and 3 species characterized "F". The 'I vs D' average dissimilarity was of 98.85, highlighting the almost complete separation of "I" and "D" groups of species. Among the group "I", two new fish species appeared (i.e. the species was cited from a location but was previously unknown by all the respondents) at Pelagie Islands (*Siganus luridus* and *Fistularia commersonii*). Four (*Caranx crysos*, *F. commersonii*, *Sparisoma cretense*. *Trachinotus ovatus*) were new to Milazzo and seven were new to Porto San Giorgio (*Balistes carolinensis*, *Coryphaena hippurus*, *Epinephelus marginatus*, *Lichia amia*, *Pomatomus saltatrix*, *Seriola dumerilii*, *Sphyraena viridensis*). No species were said to have completely disappeared or become locally extinct.

The time series of median abundance of the species most responsible for change, as identified in the SIMPER analysis, are shown in Figure 4, with the results of the breakpoint analysis given in Table 4. The analysis of the semiquantitative time series shows that most species undergoing an increase in abundance (and fluctuating species with recent increase), as well as many decreasing species, had significant breakpoints in the late 1990s. This figure also shows that a significant decrease in abundance happened in the 1990s for most commercial species: *Engraulis encrasicolus* from dominant to common in 1993, *Scomber scombrus* from abundant to occasional in 1999, *Solea vulgaris* from abundant

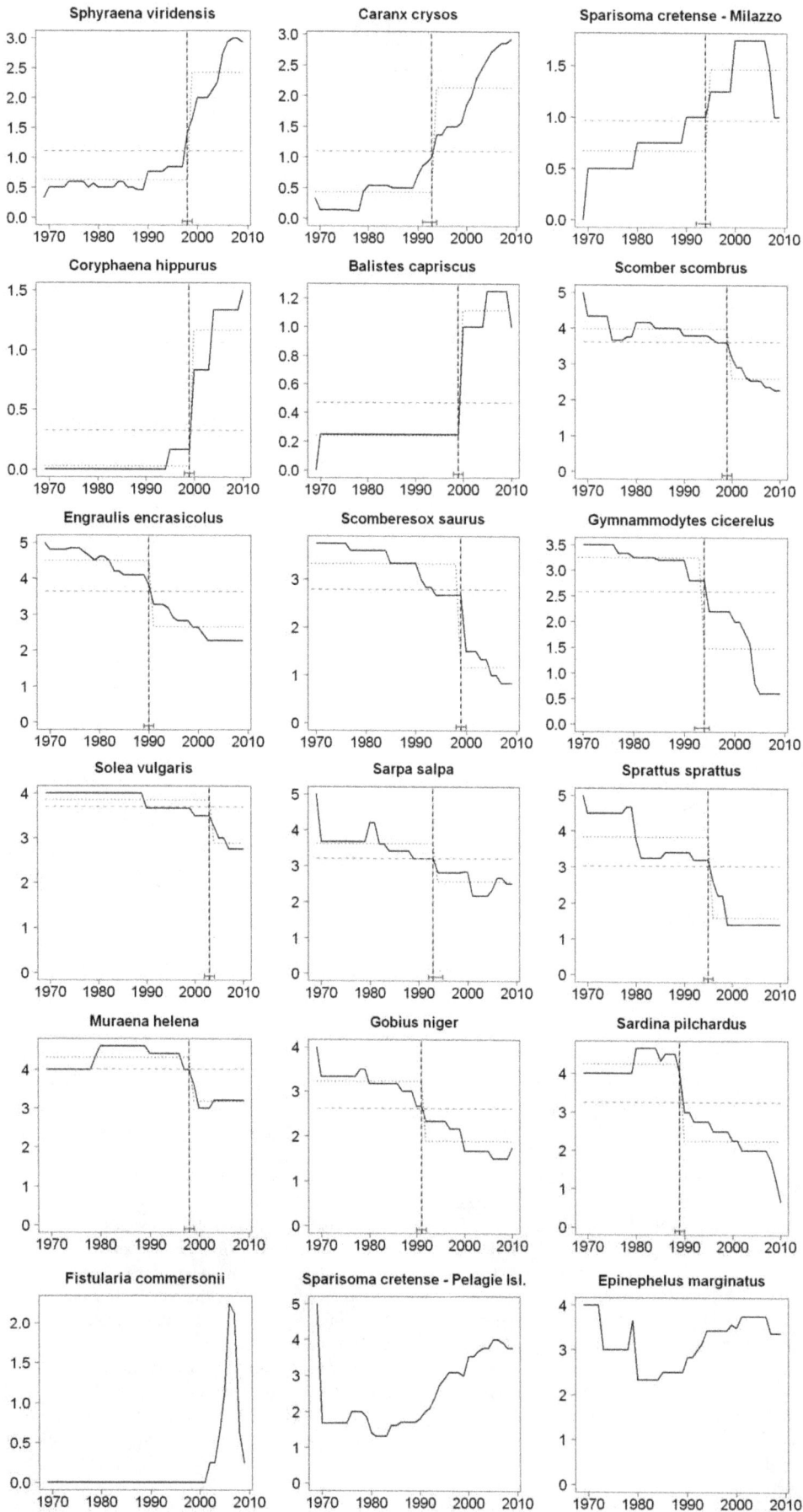

Figure 4. Dynamic of the abundance of 'SIMPER species', according to fisher's perceptions. Trends of relative abundance on a scale from 0 to 5 (see text) of the species contributing mostly to the SIMPER analysis. Bold continuous line: mean relative abundance; dashed green line: null model of no temporal change in relative abundance; dashed blue lines: best fitting local regressions before and after break point; vertical dashed line: breakpoint or year of significant change in the temporal evolution of abundance, with 95% confidence intervals in red brackets.

to common in 1999, and *Sprattus sprattus* from dominant to rare in 1995; while some pelagic species increased in abundance slightly later: *C. hippurus* from absent to rare or occasional in 2003.

Discussion

The six species that contributed the greatest to increased fish numbers over time (i.e. *Sphyraena viridensis. Caranx crysos, Sparisoma cretense, Coryphaena hippurus* and *Balistes capriscus*) are thermophilic fishes, typical from the southern sectors of the Mediterranean and their increase is consistent with what we would expect with climate warming. Remarkably the recent literature is rich in records of these species moving northward with respect to their previously known distribution. In particular, published observations are available for *S. viridensis* [33,34], *C. crysos* [35,36], *S. cretense* [37–39], *C. hyppurus* [40], *B. capriscus* [35,41]. Moreover, for *C. crysos* [42,43] and *B. capriscus* [44] northward expansions have been also documented in the Atlantic Ocean. Our data confirms that these organisms are good indicators of changes associated with warming in the marine environment [3,12]. The group "DECREASE" ("D") was represented by 42 species that are important for commercial fisheries. The most important families for this group were Scombridae and Clupeidae, two heavily fished groups all over the world. Obviously for "D" species, local negative impacts such as fishery, habitat degradation and pollution represent confounding effects to the search of global related variability.

Table 4. Results of the breakpoint structural analysis.

Species	Trend	Year	Sup Ftest
Balistes capriscus	F⁺	1999 (1998–2000)	1060.94
Caranx crysos	I	1993 (1991–1994)	150.77
Coryphaena hippurus	I	1999 (1998–2000)	503.85
Engraulis encrasicolus	D	1990 (1989–1991)	257.69
Gobius niger	D	1991 (1990–1992)	195.78
Gymnammodytes cicerelus	D	1992 (1994–1995)	125.30
Muraena helena	D	1998 (1997–1999)	170.02
Sardina pilchardus	D	1989 (1988–1990)	201.95
Sarpa salpa	D	1993 (1992–1995)	86.03
Scomber scombrus	D	1999 (1998–2000)	145.62
Scomberesox saurus	D	1999 (1998–2000)	227.22
Solea vulgaris	D	2003 (2002–2004)	138.97
Sparisoma cretense (Pelagie Isl.)	F⁺	2001 (2000–2005)	27.65
Sparisoma cretense (Milazzo)	I	1994 (1992–1995)	90.57
Sphyraena viridensis	I	1998 (1997–1999)	262.36
Sprattus sprattus	D	1995 (1994–1996)	150.59

The trend (increase 'I', decrease 'D' or fluctuate 'F'), the year of statistically significant change (with CI at the 95% level) and results of the modified F test are showed for the species that contribute significantly in the SIMPER analysis. Confidence interval not shown because outside data time interval;
⁺Increasing in last 10 years.
Probability of the Sup F <0.001 in all cases.

Disentangling the response to climate change from the effects of overfishing is particularly difficult for commercial species, nevertheless the decline observed for some boreal species such as *Sprattus sprattus* confirmed the existing concern on their resilience under a regime of climate warming [13]. To identify the species that may be in danger with the increase of temperature is a critical task [45] and traditional/local ecological knowledge can be considered valid source of such information, especially when scientific information is unavailable [25–27].

Finally in the group "FLUCTUATE" ("F"), two fishes were cited more often: *Fistularia commersonii* and *S. cretense*. The former is a non-native species that entered the Mediterranean in 2000 through the Suez Canal [46] spreading soon afterwards across the entire Mediterranean Sea [47–49]. Despite rapid geographic spread, the current status of these populations in the Central and in the Western Mediterranean is uncertain. Here, Local Ecological Knowledge provided a coherent indication of the instability of these populations which showed a rapid increase in 2003–2004 and a sharp decline soon afterwards. It is therefore possible that this species is not fully adapted to the new conditions of these Mediterranean sectors or that their dynamics are influenced by environmental fluctuations.

As far as *S. cretense* is concerned, in the 1970s this parrotfish was considered common in the Strait of Sicily [50] but absent from the northern Sicily [51]. Many respondents reported an increase of this species but others, the oldest ones with more than 30 year of experience reported a period of decline in the 70s and in the 80s followed by a clear trend of increase in the last ten years. Interestingly, the occurrence of important past fluctuations of the Mediterranean parrotfish is confirmed by the existence of a few historical observations of this species northwards with respect to its present distribution; for example, along the coast of France [52] and in the Central [53] and North Adriatic [54]. Thus, this species is considered one of clearest indicators of meridionalization because of its increasing abundance over the last two decades.

The analysis of historical trends of abundance revealed coherent species responses for the different study locations. The only exception was *Scomber japonicus* and probably more data is needed to know the real history of this species in the Mediterranean. As far as increasing species are concerned, the breakpoint analysis identified critical changes in the late 1990s and early 2000s. Actually, the first evidences of northward expansion of the range of warm-water come from the 1990s [37,55–58] and it is probably in the last decade that this phenomenon has become more apparent. A positive trend of increase of water temperature and important changes in the water circulation of the Mediterranean Sea are visible since the 1980s [13,57] and this might be at the basis of the geographical spread and success of thermophilic biota. The critical changes described by the structural analysis are indicative of the species as being fully established or even abundant in the ecosystem, while scientific records typically detect vagrant individuals or the early stages of colonization [14]. Despite the low number of interviews (only thirty-two), the perceived increase of thermophilic taxa was clear and coherent across the different Mediterranean sub-regions. Nevertheless in the future, additional surveys could be used to achieve a more precise reconstruction of historical trends, especially from earlier years that are typically difficult to sample. As usually happens in oral history surveys, the information we got was unequally distributed throughout the time. In fact, given the limited number of living people who have an early experience, pre-1970 data were based on only 5 interviews, compared to 21 respondents to the pre-1980 period and 26 respondents to the pre-1990.

Surveying Local Ecological Knowledge about changes in fish presence and abundance provided historical information that otherwise cannot be obtained. Perceived changes in species abundance can be clearly influenced by fishing methods and equipment (e.g. trawl, pelagic fishery, nets, lines and so on) and this influence could be better address with more data. Moreover, increases of some other termophilic species, such as *Thalassoma pavo* [59], could have passed unnoticed, simply because these species are not captured. Therefore it will be important in the future to broaden the number of people involved and to consider different kinds of users of natural resources such as scuba divers and long time local residents.

In local knowledge it is reasonable that not all the subjects and episodes are equally retained. In this, the capture of an "new fish" was certainly a special event that resulted to be easily remembered by the fishermen. This media property of species "never seen before" increase the potentialities of Local Ecological Knowledge as monitoring tools for these unusual occurrences that are typically difficult to monitor [14]. This possibility should be seriously taken into consideration due to our increasing need to approach large-scale patterns in the marine environment, such as species distribution shifts under climate change scenarios. Due to the preliminary nature of our results, this pilot study will hopefully serve in the future as a guide to carry out large scale investigation. In fact, more data over a broader spatial scale would allow a better definition of species temporal trends and to link these changes to environmental variables, especially along South-North gradients of the Mediterranean Sea, such as along the Italian peninsula or the Spanish coasts. Conversely, studying east-west gradients will be necessary in order to better understand the relevance of non-native species in recent changes in Mediterranean biodiversity.

Acknowledgments

The present work was developed within the framework of the project "Tropical signals" (CIESM, Commission Internationale pour l'Exploration Scientifique de la Mer Méditerranée) founded by Fondation Albert II of Monaco and with the support of the Euro-Mediterranean Center for Climatic Change and the Italian Ministry for the Environment and the Territory (project: The impacts of biological invasions and climate change on the biodiversity of the Mediterranean Sea)".

Author Contributions

Conceived and designed the experiments: EA. Performed the experiments: EA. Analyzed the data: EA FM. Contributed reagents/materials/analysis tools: PM. Wrote the paper: EA PM FM.

References

1. Parmesan C, Yohe G (2003) A globally coherent fingerprint of climate change impacts across natural systems. Nature 421: 37–42.
2. Bianchi N (2007) Biodiversity issues for the forthcoming tropical Mediterranean Sea. Hydrobiologia 580: 7–21.
3. Azzurro E (2008) The advance of thermophilic fishes in the Mediterranean sea: overview and methodological questions. In: Briand F, ed. Climate warming and related changes in Mediterranean marine biota. Monaco: CIESM Workshop Monographs 35. pp 39–46.
4. Ben Rais Lasram F, Mouillot D (2009) Increasing southern invasion enhances congruence between endemic and exotic Mediterranean fish fauna. Biol Inv 11(3): 697–711.
5. Riera F, Grau AM, Pastor E, Pou S (1995) Faunistical and demographical observations in balearic ichthyofauna. Meridionalization or subtropicalization phenomena. In: Act Coll Sci OKEANOS. La Mèditerranèe: variabilités climatiques, environnement et biodiversité, Montpellier. pp 213–220.
6. Root TL, Price JT, Hall KR, Schneider SH, Rosenzweig C, Pounds A (2002) Fingerprints of global warming on wild animals and plants. Nature 421: 57–60.
7. Walther GR, Post E, Convey P, Menzel A, Parmesan C, et al. (2002) Ecological responses to recent climate change. Nature 416: 389–395.
8. Huntley B (1991) How plants respond to climate change: migration rates, individualism and the consequences for the plant communities. J Bot 67: 15–22.
9. Parmesan C, Ryrholm N, Stefanescu C, Hillk JK, Thomas CD, et al. (1999) Poleward shifts in geographical ranges of butterfly species associated with regional warming. Nature 399: 579–583.
10. Thomas CD, Lennon JJ (1999) Birds extend their ranges northwards. Nature 399: 213.
11. Hickling R, Roy DB, Hill JK, Thomas CD (2005) A northward shift of range margins in British Odonata. Glob Change Biol 11: 502–506.
12. Perry AL, Low PJ, Ellis JR, Reynolds JD (2005) Climate change and distribution shifts in marine fishes. Science 24(308): 1912–1915.
13. CIESM (2008) Climate warming and related changes in Mediterranean marine biota. CIESM Workshop Monographs 35. Briand F, editor. Monaco. 152 p.
14. Azzurro E (2010) Unusual occurrences of fish in the Mediterranean sea: an insight on early detection. In: Golani D, Appelbaum-Golani B, eds. Fish Invasions of the Mediterranean Sea: Change and renewal Sofia-Moscow. pp 99–126.
15. Polunin NVC, Gopal B, Graham NAJ, Hall SJ, Ittekkot V, et al. (2008) Trends and global prospects of the Earth's aquatic ecosystems. In: Polunin NVC, ed. Aquatic Ecosystems. Cambridge. pp 353–365.
16. Ferguson AD, Williamson RG, Messier J (1998) Inuit knowledge of long-term changes in a population of arctic tundra Caribou. Arctic 51: 201–219.
17. Huntington HP (2000) Using traditional ecological knowledge in science: methods and applications. Ecol Appl 10: 1270–1274.
18. Moller H, Berkes F, Lyver P, Kislalioglu M (2004) Combining science and traditional ecological knowledge: monitoring populations for co-management. Ecol Soc. 9: 2. Available at http://www.ecologyandsociety.org/vol9/iss3/art2/print.pdf. accessed 2011 Jan 14.
19. Chapman PM (2007) Traditional ecological knowledge (TEK) and scientific weight of evidence determinations. Mar Poll Bull 54: 1839–1840.
20. Anadón JD, Giménez A, Ballestar R, Pérez I (2009) Evaluation of local ecological knowledge as a method for collecting extensive data on animal abundance. Conserv Biol 23: 617–625.
21. Gerhardinger LC, Hostim-Silva M, Medeiros RP, Matarezi J, Bertoncini A, et al. (2009) Fisher's resource mapping and goliath grouper *Epinephelus itajara* (Serranidae) conservation in Brazil. Neotrop Ichthyol 7: 93–102.
22. Rasalato E, Maginnity V, Brunnschweiler JM (2010) Using local ecological knowledge to identify shark river habitats in Fiji (South Pacific). Environ Conserv 37(1): 90–97.
23. Olsson P, Folke C (2001) Local ecological knowledge and institutional dynamics for ecosystem management: a study of Lake Racken watershed, Sweden. Ecosystems 4: 85–104.
24. Fogerty JE (2001) Oral history: a guide to its creation and use. In: Egan D, Howell EA, eds. The Historical Ecology Handbook: A Restorationist's Guide to Reference Ecosystems. Washington DC. pp 101–120.
25. Dulvy N, Polunin NVC (2004) Using informal knowledge to infer human-induced rarity of a conspicuous reef fish. Anim Conserv 7: 365–374.
26. Sáenz-Arroyo A, Roberts CM, Torre J, Cariño-Olvera M (2005) Using fishers' anecdotes, naturalists' observations and grey literature to reassess marine species at risk: the case of the Gulf grouper in the Gulf of California, Mexico. Fish Fish 6: 121–133.
27. Lavides MN, Polunin NVV, Stead SM, Tabaranza DG, Comeros MT, Dongallo JR (2010) Finfish disappearances around Bohol, Philippines inferred from traditional ecological knowledge. Environ Conserv 36(3): 235–244.
28. Louisy P (2005) Guide d'identification des poissons marins Europe et Méditerranée. Paris: ULMER Press. 430 p.
29. Anderson MJ (2001) A new method for non-parametric multivariate analysis of variance. Austral Ecol 26: 32–36.
30. Anderson MJ, Gorley RN, Clarke KR (2008) PERMANOVA+ for PRIMER: Guide to Software and Statistical Methods The University of Auckland Press. 214 p.
31. Bai J (1994) Least squares estimation of a shift in linear processes. J Time Ser Anal 15: 453–472.
32. Zeileis A, Kleiber C, Krämer W, Hornik K (2003) Testing and Dating of Structural Changes in Practice. Comp Stat Data Anal 44: 109–123.
33. Quignard JP, Tomasini JA (2000) Mediterranean fish biodiversity. Biol Mar Medit 3: 1–66.
34. Dulčić J, Soldo A (2004) On the occurrence of the yelowmouth barracuda *Sphyraena viridensis* Cuvier, 1829 (Pisces: Sphyraenidae) in the Adriatic Sea. Annales Ser Hist Nat 14(2): 225–230.
35. Bradai MN, Quignard JP, Bouain A, Jarboui O, Ouannes-Ghorbel A, et al. (2004) Ichtyofaune autochtone et exotique des côtes tunisiennes: recensement et biogéographie. Cybium 28(4): 315–328.
36. Psomadakis PN, Bentivegna F, Giustino S, Travaglini A, Vacchi M (2010) Northward spread of tropical affinity fishes: *Caranx crysos* (Teleostea: Carangidae), a case study from the Mediterranean Sea. Ital J Zool 1: 1–11.

37. Bianchi CN, Morri C (1994) Southern species in the Ligurian Sea (northern Mediterranean): new records and a review. Boll Muse Ist Biol Uni Ge (1992–1993) 58–59: 181–197.

38. Guidetti P, Boero F (2001) Occurrence of the Mediterranean parrotfish *Sparisoma cretense* (L.) (Perciformes, Scaridae) in south-eastern Apulia (SE Italy). J Mar Biol Ass UK 81(4): 717–719.

39. Dulčić J, Pallaoro A (2001) Some new data on *Xyrichthys novacula* (Linnaeus, 1758) and *Sparisoma* (Euscarus) *cretense* (Linnaeus, 1758) from the eastern Adriatic. Annales 23: 35–40.

40. Dulčić J (1999) First record of larval *Brama brama* (Pisces: Bramidae) and *Coryphaena hippurus* (Pisces: Coryphaenidae) in the Adriatic Sea. J Plankton Res 6: 1171–1174.

41. Dulčić J, Kršinić F, Kraljević M, Pallaoro A (1997) Occurrence of fingerlings of grey triggerfish, *Balistes carolinensis* Gmelin, 1789 (Pisces: Balistidae), in the eastern Adriatic. Annales 11: 271–276.

42. Swaby SE, Potts GW, Lees J (1996) The first records of the blue runner *Caranx crysos* (Pisces: Carangidae) in British waters. J Mar Biol Ass UK 76: 543–544.

43. Bañón R, Casas JM (1997) Primera cita de *Caranx crysos* (Mitchill, 1815) en aguas de Galicia. Bol Inst Esp Ocean 13: 79–81.

44. Bañón R, Del Rio JL, Piñeiro C, Casas M (2002) Occurrence of tropical affinity fish in Galician waters, north-west Spain. J Mar Biol Ass UK 82: 877–880.

45. Ben Rais Lasram F, Guilhaumon F, Albouy C, Somot S, Thuiller W, Mouillot D (2010) The Mediterranean Sea as a "cul-de-sac" for endemic fishes facing climate change. Global Change Biol 16: 3233–3245.

46. Golani D (2000) First record of the bluespotted cornetfish from the Mediterranean Sea. J Fish Biol 56: 1545–1547.

47. Azzurro E, Pizzicori P, Andaloro F (2004) First record of *Fistularia commersonii* (Fistularidae) from the central Mediterranean. Cybium 28: 72–4.

48. Karachle PK, Triantaphyllidis C, Stergiou KI (2004) Bluespotted cornetfish, *Fistularia commersonii* Rüppell, 1838: a Lessepsian sprinter. Acta Icht Piscat 34: 103–108.

49. Sánchez-Tocino L, Hidalgo Puertas F, Pontes M (2007) First record of *Fistularia commersonii Ruppell*, 1838 (Osteichtyes: Fistulariidae) in Mediterranean waters of the Iberian Peninsula. Zool Baetica 18: 79–84.

50. Tortonese E (1975) Fauna d'Italia: Osteichthyes Vol. XI Calderini Press. 636 p.

51. Quignard JP, Pras A (1986) Scaridae. In: Whitehead PJP, Bauchot ML, Hureau JC, Nielsen J, Tortonese E, eds. Fishes of the North-eastern Atlantic and the Mediterranean Vol II. pp 943–944.

52. Moreau E (1891) Histoire naturelle des poissons de la France Libraire de l'Academie de Médecine press. 607 p.

53. Soljan T (1975) I pesci dell'Adriatico Arnoldo Mondadori Italy. 524 p.

54. Ninni E (1924) Sulla presenza dello *Scarus cretensis* (Ald.) nell' Adriatico. Boll Ist Zool Univ Roma 24: 71–74.

55. Francour P, Boudouresque CF, Harmelin JG, Harmelin-Vivien ML, Quignard JP (1994) Are the Mediterranean waters becoming warmer? Information from biological indicators. Mar Poll Bull 28(9): 523–526.

56. Astraldi MC, Bianchi N, Gasparini GP, Morri C (1995) Climatic fluctuations, current variability and marine species distribution: a case study in the Ligurian Sea (north-west Mediterranean). Ocean Acta 18(2): 139–149.

57. Bianchi CN (1997) Climate change and biological response in the marine benthos. Proc It Ass Ocean Limn 12(1): 3–20.

58. Vacchi M, Sara G, Morri C, Modena M, La Mesa G, et al. (1999) Dynamics of marine populations and climate change: lessons from a Mediterranean fish. Porc Mar Nat Hist Soc New 3: 13–17.

59. Guidetti P (2002) Temporal changes in density and recruitment of the Mediterranean ornate wrasse *Thalassoma pavo* (Pisces, Labridae). Arch Fish Mar Res 49: 259–267.

5

Evaluating Social and Ecological Vulnerability of Coral Reef Fisheries to Climate Change

Joshua E. Cinner[1]*, Cindy Huchery[1], Emily S. Darling[2], Austin T. Humphries[3,4], Nicholas A. J. Graham[1], Christina C. Hicks[1], Nadine Marshall[5], Tim R. McClanahan[6]

1 Australian Research Council Centre of Excellence for Coral Reef Studies, James Cook University, Townsville, Queensland, Australia, 2 Earth to Ocean Research Group, Simon Fraser University, Burnaby, British Columbia, Canada, 3 Coastal Research Group, Rhodes University, Grahamstown, South Africa, 4 Coral Reef Conservation Project, Wildlife Conservation Society, Mombasa, Kenya, 5 Ecosystem Sciences, Commonwealth Scientific and Industrial Research Organisation, Townsville, Queensland, Australia, 6 Marine Programs, Wildlife Conservation Society, Bronx, New York, United States of America

Abstract

There is an increasing need to evaluate the links between the social and ecological dimensions of human vulnerability to climate change. We use an empirical case study of 12 coastal communities and associated coral reefs in Kenya to assess and compare five key ecological and social components of the vulnerability of coastal social-ecological systems to temperature induced coral mortality [specifically: 1) environmental exposure; 2) ecological sensitivity; 3) ecological recovery potential; 4) social sensitivity; and 5) social adaptive capacity]. We examined whether ecological components of vulnerability varied between government operated no-take marine reserves, community-based reserves, and openly fished areas. Overall, fished sites were marginally more vulnerable than community-based and government marine reserves. Social sensitivity was indicated by the occupational composition of each community, including the importance of fishing relative to other occupations, as well as the susceptibility of different fishing gears to the effects of coral bleaching on target fish species. Key components of social adaptive capacity varied considerably between the communities. Together, these results show that different communities have relative strengths and weaknesses in terms of social-ecological vulnerability to climate change.

Editor: Sam Dupont, University of Gothenburg, Sweden

Funding: The Australian Research Council and the United Nations Food and Agricultural Organization (FAO) funded this research. This paper is based on an unpublished report to FAO led by JEC titled "Social-ecological vulnerability of coral reef fisheries to climatic shocks." The funders had no role in study design, data collection and analysis, decision to publish, or preparation of the manuscript.

Competing Interests: The authors have declared that no competing interests exist.

* E-mail: joshua.cinner@jcu.edu.au

Introduction

Millions of the world's poorest people depend on the ecosystem goods and services provided by coral reefs [1]. Coral reefs are particularly important for fishing and tourism, but also contribute to coastal protection and in some places have significant cultural values. Coral reefs are one of the most productive and biologically diverse aquatic environments on Earth, yet they are also one of the most ecologically sensitive to climatic change [2,3], are currently undergoing large-scale changes [4,5]. Consequently, evaluating the links between the social and ecological and dimensions of vulnerability to climate change is a priority for reducing difficult-to-reverse impacts on coral reefs and increasing human food security [6,7].

Climate change is affecting coral reefs through alterations in the long-term mean environmental conditions, inter-annual cycles, and seasonality, and the frequency of extreme climate events [8]. The increasing frequency of extreme climatic events can affect fish habitat, productivity, and distribution, as well as impact directly on fishing operations and the physical infrastructure of coastal communities [9]. Extreme events such as high-intensity cyclones and increased sea surface temperatures can have profound impacts on coral reef ecosystems and the communities that depend on them [10,11]. For example, elevated sea temperature events can cause corals to bleach and die. This can alter the goods and services that coral reefs provide by changing the species compositions of fish and potentially reducing reef fisheries productivity, and consequently harming reef-dependent people [6,12,13,14,15]. The current era of rapid anthropogenic-driven climate change has the potential to undermine coral-reef associated livelihoods [7].

An increasingly critical aspect of sustaining coral reefs and the livelihoods of dependent people is understanding the vulnerability of particular reefs and their associated human communities to climate change impacts [16,17]. Vulnerability, in the context of social and environmental change, is defined as the state of susceptibility to harm from perturbations [18]. Understanding vulnerability in coral reef fisheries is complicated because there is considerable heterogeneity in: 1) places that experience climate change-related events such as coral bleaching; 2) the ways that coral reef ecosystems are affected by and can recover from these impacts; 3) the ways that societies and individuals are impacted by these changes; and 4) the capacity of people to cope with and adapt to these changes. Knowledge about how vulnerable a system is, and the specific conditions that make it vulnerable, can help to provide a foundation for developing key actions that minimize the impacts of environmental change on people.

The conceptual framework of vulnerability to climate change provides a basis for operationalizing and assessing the vulnerability of linked social and ecological systems [19,20]. A framework promoted by the Intergovernmental Panel on Climate Change (IPCC) [8] has been widely adopted for vulnerability assessments [21] (Figure S7 in File S1). The framework suggests that the extent to which people's livelihoods are vulnerable to the impacts of climate change is dependent on: 1) their exposure to climate impacts (i.e. if impacts are felt in their location); 2) their sensitivity (i.e. the extent to which their livelihood is affected by an impact); and 3) their capacity to adapt to the likely impacts [16,18,19,20,22,23,24].

Exposure is the degree to which a system is stressed by climate, such as the magnitude, frequency, and duration of a climatic event such as temperature anomalies or extreme weather events [18,25]. In a practical sense, exposure is the extent to which a region, resource, or community experiences change [8]. For fishing communities, exposure captures how much the resource they depend on will be affected by environmental change. In tropical reef fisheries, exposure can vary depending on factors such as oceanographic conditions, prevailing winds, and latitude, which can increase the likelihood of being impacted by events such as cyclones or coral bleaching [26]. Sensitivity, in the context of environmental change, is the susceptibility of a defined component of the system to harm, resulting from exposure to stresses [18]. The sensitivity of social systems depends on economic, political, cultural and institutional factors that allow buffering or attenuation of change. For example, social systems are more likely to be sensitive to climate change if they are highly dependent on a climate-vulnerable natural resource. Sensitivity can confound (or ameliorate) the social and economic effects of climate exposure. Adaptive capacity is a latent characteristic that reflects peoples' ability to anticipate and respond to changes, and to minimize, cope with, and recover from the consequences of change [22]. For example, people with low adaptive capacity may have difficulty adapting to change or taking advantage of the opportunities created by changes in the availability of ecosystem goods and services stimulated by climate change or changes in management.

The above examples illustrate the three dimensions of social vulnerability, but they also have ecological components. For example, the sensitivity of ecological systems to climate change can include physiological tolerances to change and/or variability in physical and chemical conditions (i.e. temperature, pH, etc.), such as certain corals that are highly sensitive to increases in sea temperatures. This creates the need to evaluate both systems and, therefore, a new multi-disciplinary literature on the vulnerability of linked social-ecological systems to climate change has emerged [16,23,27,28]. The central idea behind linked or coupled social-ecological systems is that human actions and social structures profoundly influence ecological dynamics, and vice-versa [18,29].

Modified Vulnerability Framework

Here, we use a case study from the Kenyan coral reef fishery to operationalize a modification of the IPCC vulnerability framework. Our aim is to improve on previously developed applications of vulnerability in fisheries [e.g. 30,31] by explicitly considering both social and ecological dimensions of vulnerability. Our specific modification to the IPCC framework entails linking two vulnerability sub-models: one represents the components of ecological vulnerability to exposure to climate change, while the other represents social vulnerability to changes in the ecological system (Figure 1). In our modified framework, the potential impact of climate change on ecological systems results from the physical exposure to climatic stressors combined with the sensitivity of those ecosystems. Whether these potential impacts are fully experienced in the long term depends on the potential of the

ecosystem to recover its basic structure and function in response to impacts. Thus, the combination of ecological exposure, ecological sensitivity, and ecological recovery potential (together what we refer to as ecological vulnerability) result in the degree to which climate change will impact on the continued supply of ecosystem goods and services. In turn, this ecological vulnerability represents the exposure of the socioeconomic domain to climate threats. The overall social-ecological vulnerability is then a result of the sensitivity of socioeconomic systems to ecological vulnerability, and the adaptive capacity of the society to adapt to such impacts (Figure 1).

The Social and Ecological Context of the Kenyan Case Study

Our Kenyan case study demonstrates how assessments of exposure, sensitivity, and adaptive capacity can be undertaken for both social and ecological subsystems and provide an overall assessment of system vulnerability. Kenya presents an interesting case study to evaluate social-ecological vulnerability for four key reasons. First, in a comparison of vulnerability across five Western Indian Ocean countries, Kenyan sites are the most vulnerable overall [30], but there is considerable spread in both sensitivity and adaptive capacity. Indeed, much of the variability encountered in the region is contained within Kenya. Second, Kenya is at the frontline of climate change- its reefs were severely impacted by the 1998 El Nino-related coral-bleaching event. Temperature records suggest that the scale of this temperature anomaly was unprecedented [32,33] and resulted in high levels of coral mortality in the northern Indian Ocean [34]. Consequently, extreme climate events are a current reality rather than a distant possibility. Third, people in coastal Kenya are heavily dependent on fisheries and other natural resources for their livelihoods [35]. Fishing in Kenya is typically conducted from the beach to the fringing reef within the sand, coral, and seagrass habitats of the fringing reef lagoon. Last, Kenya has a range of marine resource governance regimes, ranging from large national marine parks enforced by paramilitary organizations to largely open access areas where regular use of destructive beach seine nets damage marine habitats. In between

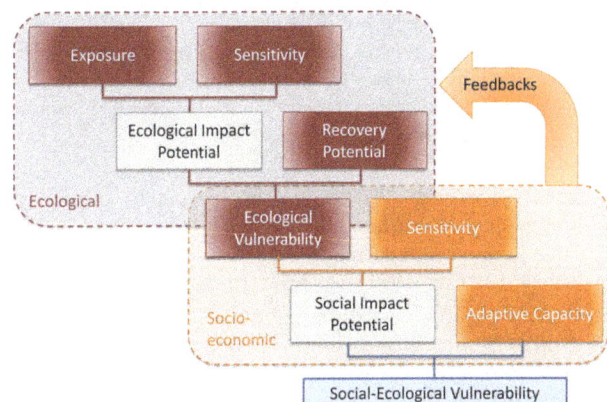

Figure 1. Heuristic framework for linked social-ecological vulnerability. In the ecological domain, ecological exposure and ecological sensitivity create impact potential. The impact potential and the ecological recovery potential together form the ecological vulnerability, or exposure in the social domain. This ecological vulnerability combined with the sensitivity of people form the impact potential for society. The social adaptive capacity and the impact potential together create social-ecological vulnerability. Adapted from Marshall et al. [40,53].

are community controlled co-managed areas called Beach Management Units (BMUs) [36,37]. In recent years, BMUs have started developing community-based fishery closures. Together this governance spectrum presents an opportunity to examine whether, and how different, governance regimes have the potential to influence vulnerability.

Materials and Methods

Ecological Sampling

In 2009, 2011, and 2012 we surveyed 15 ecological sites associated with 10 coastal communities along the Kenyan coast, including heavily fished reefs, reefs within small, recently established community co-managed fisheries closures ("tengefus" in Swahili), and reefs in larger, well established no-take National Marine Parks managed by the Kenya Wildlife Service (Figure 2). All reef surveys were conducted in shallow back-reef flat habitat or shallow reef slope (<4 m). This depth was chosen partly because the extensive shallow back reef habitats along the Kenyan coast make up the majority of the available reef habitat. More importantly though, this back-reef lagoonal system is where the majority of the reef fishery activities are concentrated, meaning our ecological and social data are tightly coupled. Surveys were conducted in 2011 and 2012, with the exception of the Kisite Marine National Park, which was surveyed in 2009 (marked as Shimoni Park in Figure 2). At each site, we used standard underwater survey methods to evaluate coral reef benthic habitat and associated reef fish communities (Methods S1 in File S1).

Ecological Indicators of Vulnerability

We developed metrics to explain key aspects of the ecological exposure, ecological sensitivity, and recovery potential of coral reef ecosystems to the impacts of climate change-associated coral bleaching (Table 1):

1) Ecological exposure to coral bleaching was described by a previously published multivariate model of how temperature, light, currents, tidal variation, chlorophyll, and water quality combine to create environmental conditions that make a site susceptible to coral bleaching impacts [26,38]. Higher exposure values indicated environmental conditions that were more likely to result in thermal stress and subsequent coral bleaching, while lower values indicated sites that were less likely to experience thermal stress and coral bleaching (Methods S1 in File S1).

2) We estimated the ecological sensitivity of a site to coral bleaching using two indicators: i) the susceptibility of the coral community to bleaching; and ii) the susceptibility of the fish community to population declines associated with coral habitat loss from bleaching (Table 1, Methods S1 in File S1).

3) At each site, we collected information on ten ecological indicators of recovery potential (Table 1, Methods S1 in File S1). These were: 1) hard coral cover; 2) the ratio of coral to macroalgae cover; 3) coral size distribution; 4) coral richness; 5) fish biomass; 6) fish species richness; 7) substrate complexity; 8) fish size distribution; 9) herbivore (fishes and sea urchins) diversity; and 10) an index of herbivore grazing relative to algal production. These indicators were weighted based on the scientific evidence that they contribute to recovery [39] (Table 1, Methods S1 in File S1).

We normalized each indicator between 0 and 1 (Methods S1 in File S1). Normalized indicators were averaged into composite metrics of sensitivity and recovery using an evidence-weighted framework based on expert opinion that evaluated the strength of evidence in support of each indicator [39] (Table 1). Ecological vulnerability was then estimated as [ecological exposure+ecological sensitivity] − Recovery Potential.

Socioeconomic Data Collection

In 2010, we employed a combination of surveys targeted at resource users' (fishermen, fish sellers, etc.) households and semi-structured interviews with key informants (community leaders, resource users, and other stakeholders) to gather information about their sensitivity and adaptive capacity to changes in the coral reef system. We triangulated results in each study site. In total, we conducted 310 household surveys, 9 key informant interviews, 10 community leader interviews, and 10 organizational leader interviews. All interviews were conducted in Swahili by trained interviewers. Respondents for the household surveys were randomly selected from lists of resource users provided by local leaders. Lists were cross-referenced with other fishermen for accuracy. We sampled 38% of approximately 810 resource users from our study sites. Key informant interviews were conducted using three semi-structured interview forms to specifically target: 1) knowledgeable fishers; 2) community leaders; and 3) fishery landing site leaders. Key informants were selected using non-probability sampling techniques.

Social Indicators of Vulnerability

Based on all of these survey types, we generated 13 socioeconomic indicators, which we separated into social sensitivity and adaptive capacity measures (Table 2, Methods S1 in File S1). We developed a metric of sensitivity based on two key aspects: 1) the level of dependence on marine resources [31,40]; and 2) data on how susceptible the catch composition of different gears were to climate change impacts [41,42]. Information on these two aspects of sensitivity was combined into a single metric of social sensitivity (see Methods S1 in File S1 for detailed description). We modified the adaptive capacity index developed in McClanahan et al. [43] and Cinner et al. [30]. Based on both the household surveys and key informant interviews described above, we examined 11 social indicators of local-scale adaptive capacity (Table 2 and S5 in File S1). These were combined into a single, un-weighted metric of social adaptive capacity (Methods S1 in File S1).

Analysis

We compared the three aspects of ecological vulnerability (ecological exposure, ecological sensitivity, and ecological recovery potential) across the three management groups (fished reefs, tengefus, and no-take marine reserves) using a one-way analysis of variance. We described the multivariate relationships among the ecological exposure, ecological sensitivity, and ecological recovery potential indicators of ecological vulnerability using a correlation-based Principal Components Analysis (PCA) on Euclidean distances among indicators. We visualized the differences among the three components of ecological vulnerability using a bubble plot, where ecological sensitivity was plotted against ecological recovery potential and ecological exposure was indicated by the size of the points. These three components were combined into a metric of ecological vulnerability, which was then used as the exposure component of the social-ecological vulnerability (Figure 1, equation 1) as follows:

$$V_{Soc.Ecol} = V_{Ecol} + S_{Soc} - -AC_{Soc}$$

Where V is vulnerability, E is exposure, S is sensitivity, AC is

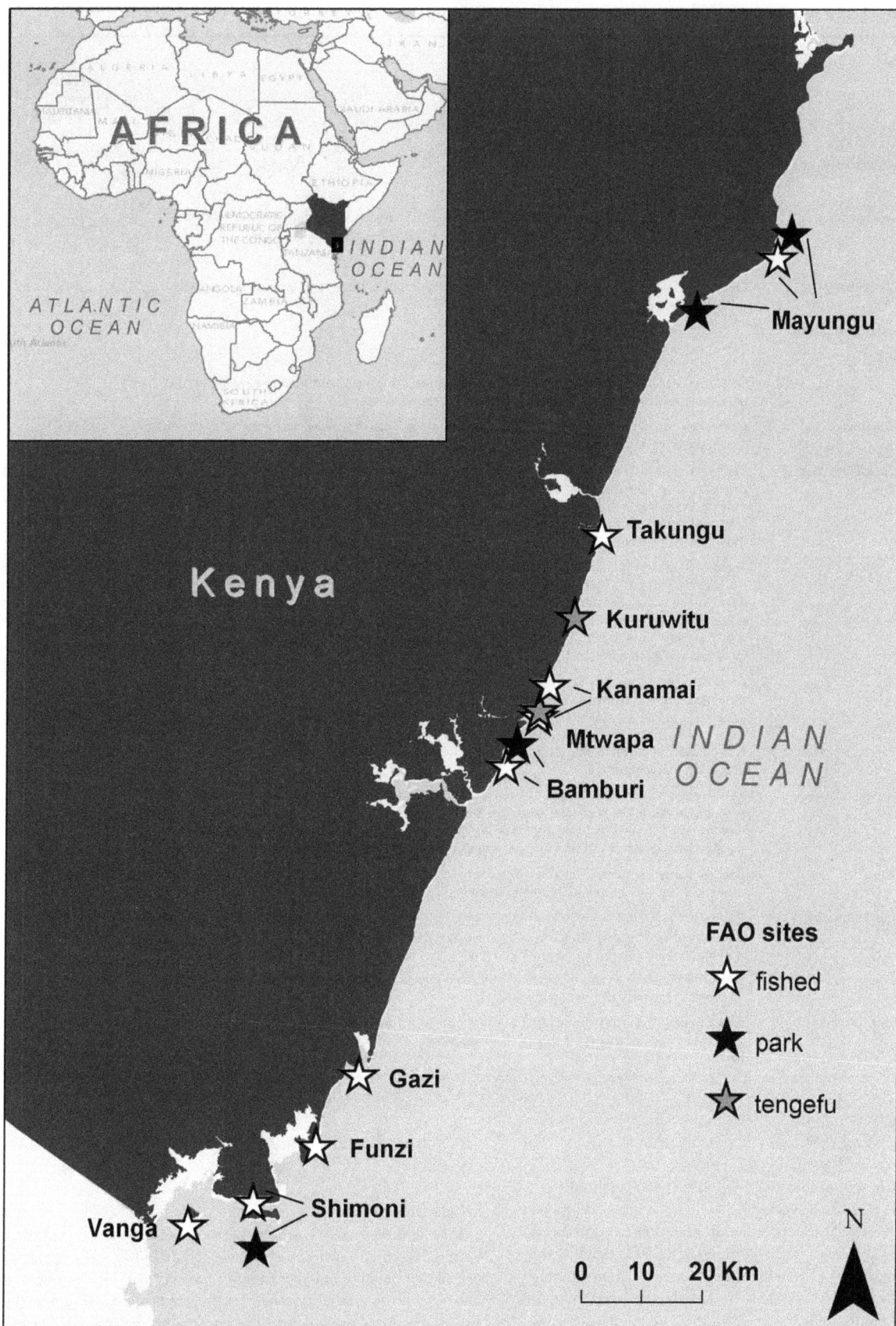

Figure 2. Map of study sites.

Table 1. Indicators of ecological sensitivity and ecological recovery potential.

	Statement of evidence	Weight of scientific evidence (−5 to 5)
Ecological sensitivity indicators		
Coral		
Coral bleaching susceptibility	Some species (e.g. branching or plating corals) are often severely impacted by disturbance and a high abundance of these species confers higher sensitivity	4.07
Fish		
Fish bleaching susceptibility	Certain fish species are more heavily impacted by disturbance and a high abundance of these species confers higher sensitivity	3.2
Recovery potential indicators		
Autotrophs/Corals		
Coral cover	Coral cover is linked to increased resilience and recovery but most field studies showing no correlation between coral cover pre- or post-disturbance with recovery rates.	2.27
Coral to macroalgae cover	Macroalgae is a significant factor limiting the recovery of corals following disturbance by increasing competition for benthic substrate, allelopathy and by trapping sediment that smothers coral recruits.	3.37
Calcifying to non-calcifying cover	Calcifying organisms are important for reef framework (e.g., processes of settlement, recruitment and cementation of reef structure) and more calcifying organisms relative to non-calcifying organisms are expected to increase or accelerate recovery following disturbances. However, the interactive effects of settlement induction, competition and increased predation make the influence unclear.	1
Coral size distribution	There is scientific evidence that evenness across size classes increases recovery. An even distribution across size classes indicates a recovering community of coral recruits, juveniles and adult colonies, whereas the under-representation of juvenile colonies suggests recruitment failure and a suppressed recovery rate. Moreover, the lack of large adult coral colonies may limit spawning stock and indicate environmental stress that has caused partial colony mortality and fragmentation.	2.5
Coral richness	Coral richness is expected to promote recovery, however there is limited evidence that coral diversity promotes recovery following disturbance.	2.5
Heterotroph/Fish		
Fish biomass	Fish biomass in indicates the status of the fish stock, its potential growth, and is a proxy for ecological metabolism	4.5
Herbivore grazing rate relative to algal production	Most studies have linked increased herbivory to reduced macroalgal cover and an increase in coral recruitment despite higher corallivory. One study has gone further and shown that increased herbivore biomass led to a reversal in the reef trajectory from one of coral decline to coral recovery. Relative importance of fish and urchins varies geographically and with fishing intensity.	3.32
Fish species richness	Species richness is often used as a proxy for functional redundancy and is expected to promote ecological recovery by avoiding undesired ecological states.	3.5
Substrate complexity (rugosity)	Evidence that habitat complexity promotes recovery for corals occurs at small-scale sediment tiles but has not been scaled up. There is good evidence that habitat complexity promotes refuge and recovery for fish	1.52
Fish size distribution	Large individuals in an assemblage indicate more even size-spectra and can increase fecundity to promote recovery of fish communities.	4
Herbivore functional diversity	Experimental evidence indicates that the presence of a diverse guild and functional groups of herbivores (reef fishes, sea urchins) can enhance coral recovery.	2.46

Weight of scientific evidence examines the consistency and type of evidence for each component, following the method of McClanahan et al. [39].

adaptive capacity, $_{Soc}$ is social, $_{Ecol}$ is ecological, and $V_{Ecol} = E_{Ecol} + S_{Ecol}$ − Recovery Potential$_{Ecol}$.

Following Cinner et al. [30], we used two techniques to examine social-ecological vulnerability. First, we developed a quantitative vulnerability score using an equation to combine the three contributing indices (each normalized to 0–1 scale) (Social-ecological vulnerability = [ecological exposure+social sensitivity] − social adaptive capacity). Secondly, to visualize differences in key components of vulnerability, we plotted the three dimensions on a bubble plot, where social sensitivity was plotted against social adaptive capacity and ecological vulnerability (i.e. social exposure)

was indicated as the size of the points (larger point = higher exposure).

Ethics Statement

JEC obtained ethics approval from James Cook University's human research ethics committee (ID#H4331). We obtained verbal consent from participants before conducting surveys. During verbal consent, participants were informed about the survey, its purpose, and how the data would be utilized. Written consent from participants was not obtained because of low literacy rates in many of our field sites, which meant that participants may not have fully understood what they signed. Verbal consent was

Table 2. Indicators of social adaptive capacity.

Indicator	Description	Bounding
Human agency *"HumanAgency"*	Recognition of causal agents impacting marine resources (measured by content organizing responses to open-ended questions about what could impact the number of fish in the sea)	Binomial: 0; 1
Access to credit* *"AccessCredit"*	Measured as whether the respondent felt he or she could access credit through formal institutions or informal means (e.g., family, friends, middlemen/dealers)	Binomial: 0; 1
Occupational mobility *"OccupMob"*	Indicated as whether the respondent changed jobs in the past five years and preferred their current occupation	Binomial: 0; 1
Occupational multiplicity *"OccupMult"*	The total number of person-jobs in the household	Continuous: 1st quartile = 1; 3rd quartile = 3
Social capital *"SocialCapital"*	Measured as the total number of community groups the respondent belonged to	Continuous: min = 0; max = 3
Material style of life *"MSL"*	A material style of life indicator measured by factor analyzing whether respondents had 15 material possessions such as vehicle, electricity and the type of walls, roof, and floor	Continuous: 1st quartile; 3rd quartile
Gear diversity *"GearDiv"*	Technology (measured as the diversity of fishing gears used); 8) infrastructure	Binomial: 0 = 1 gear; 1 = more than 1 gear
Community infrastructure *"CommInfrastr"*	Infrastructure	Continuous: min = 0; max = 26
Trust* *"Trust"*	Trust- measured as an average of Likert scale responses to questions about how much respondents trusted community members, local leaders, police, and local government	Continuous: min = 0.8; max = 5
Capacity to change[2012] *"CapacityChange"*	Capacity to anticipate change and to develop strategies to respond (measured by content organizing responses to open ended questions relating to a hypothetical 50% decline in fish catch)	Binomial: 0; 1
Debt*[2012] *"NoDebt"*	Measured as whether or not the respondent was presently in debt of more than 1 week's salary (this indicator negatively contributed to adaptive capacity so we took the inverse).	Binomial: 0 = in debts; 1 = not in debts

2012 = only used for 2012 analysis.
* = New indicators added to the adaptive capacity compare to previous.

authorized by the James Cook University human ethics panel. Permission for social and ecological research in Kenya was provided by the National Secretary for Science and Technology (research clearance permit # NCST/RRI/12/1/BS/209).

Results

Ecological Aspects of Vulnerability

The ecological indicators were highly variable across the 15 study sites (Table S6 in File S1). Sites included degraded reefs with low coral abundance (<1% absolute live coral cover, Takaungu), limited coral diversity (13 genera, Kuruwitu), low reef fish biomass (<100 kg/ha, Kanamai, Takaungu, RasIwatine), limited herbivore grazing diversity (<0.01 Simpson index of Acanthurids, Siganids, and Scarids, Kanamai, RasIwatine), and herbivore grazing rates that were substantially less than estimated rates of algal production (>100 kg/day deficit, Mayungu, Takaungu). More intact reefs had higher coral cover (>50%, Mradi), more diverse coral assemblages (25 genera, Changai, Kisite), and more productive fish communities (~1600 kg/ha reef fish biomass, Kisite) with greater herbivore diversity (~0.7 Simpson index, Mombasa) and higher herbivore grazing relative to algal production (>50 kg/day surplus, Changai, Kisite).

The wide range of ecological condition across the 15 coral reef sites in Kenya led to considerable spread in the composite ecological vulnerability index (Table S6 in File S1, Figure 3). Ecological vulnerability ranged from 0.42 to 0.79 (mean 0.64±0.11 SD, vulnerability index scaled between 0 and 1). The three facets of ecological vulnerability (ecological exposure, ecological sensitivity and ecological recovery potential; Table S7

in File S1) were not strongly correlated, suggesting these different components of ecological resilience are not related (Pearson correlation coefficients: ecological exposure to ecological sensitivity, r = −0.46, ecological exposure to ecological recovery potential, r = −0.15, ecological sensitivity to ecological recovery potential, r = 0.11). Overall, fished sites had marginally higher ecological vulnerability than sites within tengefus and no-take marine reserves (one-way Analysis of Variance, F = 3.2, df = 2,12, P = 0.08; Table S7 in File S1; Figure 3).

The two principal components axes explained 61.7% of the variation among indicators across the sites (Figure 4). Management did not distinguish exposure because some fished reefs, community-managed tengefus, and government no-take marine reserves were associated with high levels of exposure (upper-right quadrant of Figure 4). Fished reefs were associated with higher climate sensitivities of coral and fish assemblages (bottom quadrants). Recovery potential indicators separated into two groups. Herbivore diversity, rugosity, fish biomass, and coral size were associated with the no-take marine reserves (upper-left quadrant of Figure 4), while coral richness, hard coral cover, and higher rates of herbivore grazing relative to algal production were associated with one tengefu, Mradi, and some fished reefs (lower-left quadrant).

There was a wide spread of the three facets of ecological vulnerability across different types of fisheries management. Variable exposure, high sensitivity, and low recovery potential to coral bleaching events resulted in higher ecological vulnerability scores for some fished sites, one tengefu (Kuruwitu) and some no-take marine reserves (upper right of Figure 5). Other no-take reserves and one tengefu (Mradi) were associated with lower

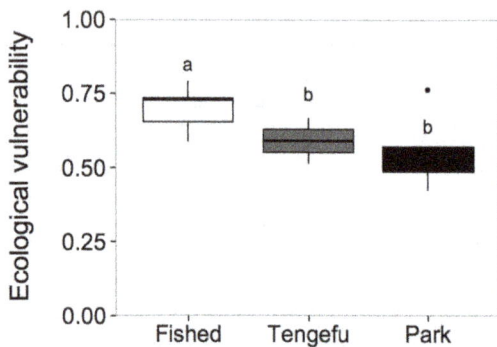

Figure 3. Ecological vulnerability on 17 Kenyan reefs across three types of fisheries management: open access fished reefs, community-managed 'tengefu', and National Marine Parks. One-way Analysis of Variance suggests fished reefs are marginally more vulnerable to climate change than *tengefus* and no-take parks (one-way ANOVA, P = 0.08). Letters indicate where significant differences exist across management groups). The different colours of bars represent different management types corresponding to those in Figure 2.

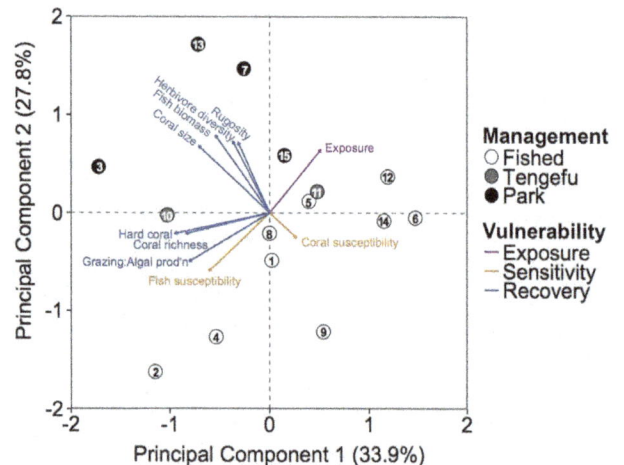

Figure 4. Principal components analysis of ecological vulnerability. Eigenvectors describe normalized indicators of exposure, sensitivity and recovery potential. Points indicate reefs within different management groups (white – fished; grey – community co-managed areas; black – no-take marine reserves). Numbers indicate study sites (see Table 1).

ecological vulnerability due to low ecological sensitivity and high ecological recovery potential, despite medium to high exposure (Table S6 and S7 in File S1; Figure 5).

Social Aspects of Vulnerability

Social sensitivity. We focused our sampling on direct resource users, meaning that the resource dependence aspects of sensitivity had relatively little variation between communities (ranging from 0.23–.35; Table S1 in File S1). Our analysis of the gear use side of sensitivity produced some counter-intuitive results and highlighted some key research gaps (Table S2 and S3 in File S1). In particular, we found that the sensitivity of certain gears types to the impacts of coral bleaching events varied considerably (Fig. S5 in File S1). In particular, the species targeted by traps and beach seine nets in the Kenyan fishery were expected to decline as a result of bleaching-induced mortality. However, available information to date suggested that the species targeted by other gears may actually demonstrate short-term increases in abundance as a result of bleaching mortality (Fig. S5 in File S1). However, we only had species-specific responses for 48–69% of the catch abundance (Fig. S3–S4, Table S4 in File S1) and many of the species-specific responses were supported by only one study (Fig. S6 in File S1).

Social adaptive capacity. The ten communities displayed considerable variation in many of the indicators of adaptive capacity that we measured (Table S8 in File S1), particularly access to credit, debt, human agency, capacity to change, social capital, community infrastructure, and material style of life. For example, the proportion of respondents not in debt (recorded as more than one week's typical earnings) ranged from 55–90%. Alternatively, several of the indicators displayed little variation between the highest and lowest values, particularly, occupational mobility, gear diversity, and trust.

We ran a PCA based on the co-variance matrix (because the units were all on the same scale) (Figure 6). Visual inspection of scree plots revealed that the first three Principal Components (PCs), which explained 82% of the variance (Table S9 in File S1), could be used. Social capital, capacity to change, access to credit, community infrastructure, gear diversity, and Material Style of Life (MSL) all had substantial factor loadings on PC 1 (Table S10 in File S1). MSL, occupational multiplicity, and community

infrastructure dominated PC2, but gear diversity and access to credit also had substantial loadings on that PC. Interestingly, MSL and community infrastructure loaded negatively on PC2, while gear diversity and occupational multiplicity loaded positively. Human agency loaded highly on the PC3 (Table S10 in File S1). Trust did not load highly on any of the components, primarily because there was little variation in trust between communities. Although there was substantial variation in trust at the individual level, community-level means and standard errors were relatively similar (Table S8 in File S1).

Social-Ecological Vulnerability

Our measure of social-ecological vulnerability comprised three components: 1) social exposure (which is ecological vulnerability; Fig. 1); 2) social sensitivity; and 3) social adaptive capacity. We used a bubble plot to visualize social-ecological vulnerability at our study sites (Figure 7). This visualization helped show how the vulnerability of our communities compared to each other and helped demonstrate which component(s) of vulnerability contributed the most to their vulnerability, so that specific actions could be taken for each of them. For example, Takaungu had a high vulnerability mainly due to its high exposure and low adaptive capacity, even though its sensitivity was low. Actions to improve the vulnerability of this community might focus on increasing adaptive capacity (it is harder to have actions that can reduce exposure). Vanga had a high vulnerability primarily because of its high sensitivity. Actions to improve the vulnerability of this community might focus on decreasing sensitivity.

Discussion

Our modification to the IPCC vulnerability assessment framework provides a conceptual model for considering both socioeconomic and ecological dimensions in an integrated assessment of system vulnerability. Integration between socioeconomic and ecological systems highlights and exposes the codependency between the systems components; where vulnerability is visibly and quantitatively influenced by each component. Facing the growing threat of climate change and because of the inter-

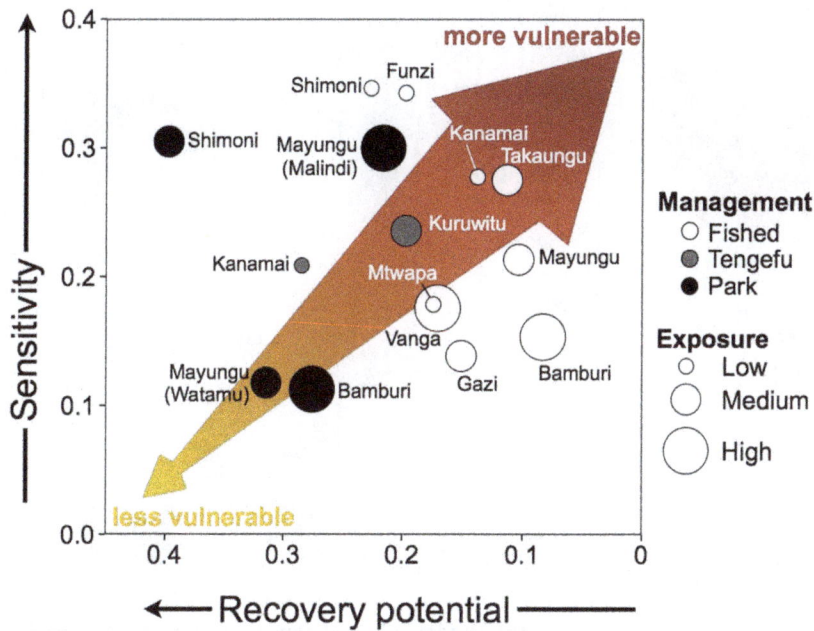

Figure 5. Ecological vulnerability of Kenyan coastal communities to the impacts of coral bleaching on reef fisheries. Ecological sensitivity is plotted against ecological recovery potential (note: axis is reversed) and ecological exposure is indicated by bubble size. The arrow highlights less vulnerable to more vulnerable communities.

dependencies between people and ecosystems, understanding the linkages is increasingly important for effective management. Nevertheless, there are few examples quantifying this integrated understanding of vulnerability in the literature.

Integrated vulnerability analyses can be used to identify status, trends, and possible opportunities for adaptation in the face of climate change. In particular, our study exposes the role of local level management in influencing the sensitivity and recovery

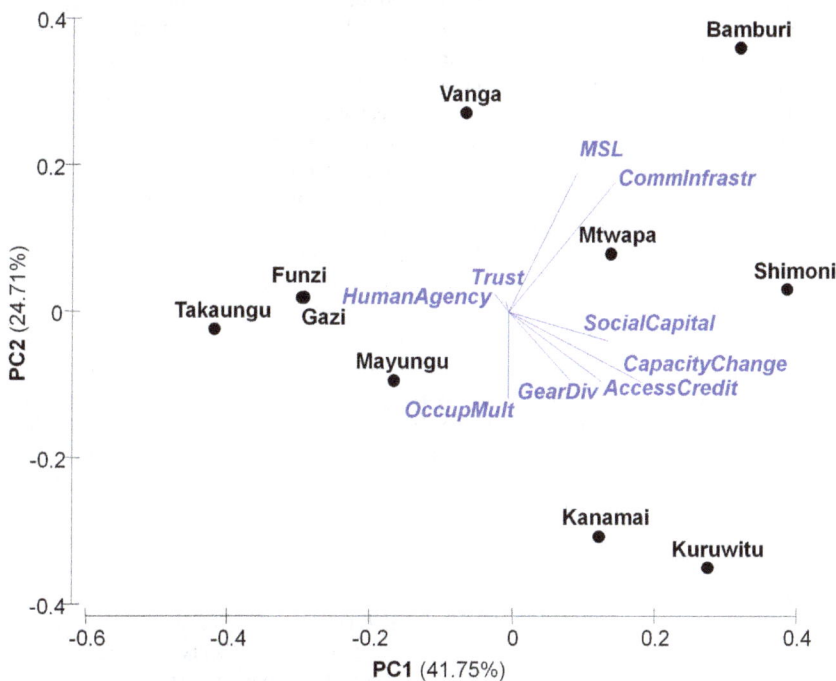

Figure 6. PCA of the 9 social adaptive capacity indicators analysed at an aggregate community level. The blue vectors represent the 9 social adaptive capacity indicators: Material style of life (MSL), Community Infrastructure (CommInfrastr), Trust, Social capital, Human Agency, Capacity to change (CapacityChange), Gear diversity (GearDiv), Access to Credit (AccessCredit) and Occupational Multiplicity (OccupMult) (No Debt and Occupational mobility not included). The black dots represent the 10 communities.

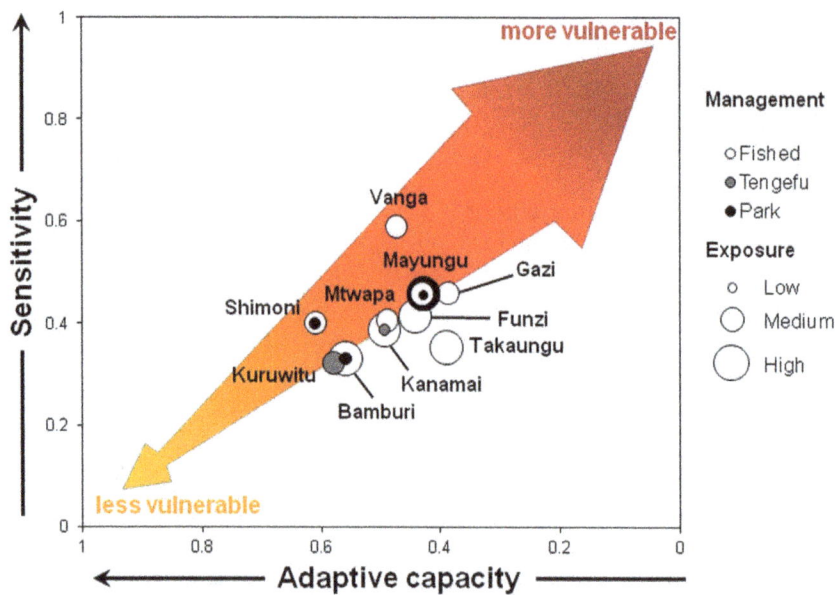

Figure 7. Social-ecological vulnerability of Kenyan coastal communities to the impacts of coral bleaching on reef fisheries. Social sensitivity is plotted against social adaptive capacity (note: axis is reversed) and ecological vulnerability is indicated by bubble size. The arrow highlights less vulnerable to more vulnerable communities. Note that some sites (such as Shimoni) may have more than 1 ytpe of management present, indicated by overlapping dots.

potential of corals and associated fish assemblages, which ultimately reduces exposure in the social domain. This is in contrast to ecological exposure, which can only be reduced by international action to reduce carbon emissions. Likewise, social adaptive capacity and social sensitivity are also amenable to policy actions at local and national scales [30].

This case study uses a diagnostic approach and supports the argument that one-size-fits-all or panaceas to adaptation planning are unlikely to succeed [44,45]. Our methods and results highlight how specific aspects of adaptive capacity and sensitivity can determine the strengths and weaknesses that contribute to high or low vulnerability. This should be useful for adaptation planning that takes advantage of existing capacities and can strengthen the identified weaknesses. By examining the types of vulnerability (exposure, sensitivity and adaptive capacity) that different communities have (e.g. Figure 7), the most appropriate policy priorities become apparent (Table 3). For example, social systems might be buffered from ecological degradation through local management strategies that increase ecological recovery potential (e.g. through marine protected areas or gear-based management). Likewise, social sensitivity could be decreased by promoting the use of gears less likely to be negatively impacted by coral bleaching (e.g. handlines), or by creating supplemental livelihood activities. Adaptive capacity is perhaps the component of vulnerability most amenable to influence, and may be a useful focus for adaptation planning. Some aspects of adaptive capacity, such as infrastructure development, can be directly and predictably enhanced by physical development projects, while other livelihood or cognitive dimensions are not so amenable to enhancement by central government. Alternatively, some aspects of building adaptive capacity, such as skills development, support for micro-credit schemes, and poverty alleviation may be best delivered by NGOs and development organizations working in conjunction with local and national governments.

An important finding of our research highlights that there may be tradeoffs inherent in specific aspects of adaptive capacity,

particularly those associated with people's flexibility and assets. This finding is supported by studies of livelihood diversification, which have found occupational specialization with increasing socioeconomic development [35,46]. Specialization within industries such as the fishing industry occurs as the result of capital being secured in vessels and equipment [47,48]. This increases the efficiency of the operation, decreases the price of the product, and maintains social status [46]. Yet, resource-users that target only a few species, or are reliant on a single resource are severely restrained in their ability to be flexible and adapt to changes in the resource relationship. Specialist behaviour is typical of regions in which resources are abundant and predictable and the system is regarded as 'stable'. However, the 'stable' system is not necessarily resilient in the face of change. Thus, in areas where resources are less predictable, a 'generalist' or risk-spreading strategy may be more resilient. Generalists or resource-users that target more than one species can exhibit a more flexible nature since they can interchange between resource types as the need arise.

A surprising result from this study is that, based on available information to date, it appears as though temperature-induced coral mortality has the potential to result in modest short-term increases in catches for some gear users. For example, the algae that often grow over dead corals could promote the abundance of certain types of low-trophic herbivorous fishes that are targeted by certain gears. Thus, sensitivity is not always negative; climate change could impact some fish species, some gear, and some people positively. However, a degraded and algal covered reef is unlikely to sustain fish populations after the structural complexity of the reef has declined. Likewise, targeting the species that eat algae may hinder prospects of reef recovery after a bleaching event. Thus, we do not view these potential selective short-term increases in catch as a sustainable fisheries benefit from climate change. Additionally, our initial investigation of the impacts of temperature-induced mortality on reef fishers examined likely changes to *in situ* abundance of commonly targeted reef fishes, but does not consider status or trends of key fisheries parameters such

Table 3. Possible policy responses to influence different types of social-ecological vulnerability.

Vulnerability component	Potential to influence	Possible policy actions for enhancement
Social Exposure (i.e. Ecological Vulnerability)	Medium	Develop local level management to increase ecological recovery potential and ecological sensitivity (e.g. marine protected areas, gear based management).
Social Sensitivity		
Gear sensitivity	High	Promote the use of gears less likely to be negatively impacted by coral bleaching (e.g. hand lines)
Occupational sensitivity	Medium	Develop supplemental livelihood activities
Social adaptive capacity		
Capacity to Change livelihood	Low	Skills and capacity building
Access Credit	High	Microcredit schemes, support for community savings
Community Infrastructure	High	Infrastructure development projects in rural areas
Gear Diversity	Low	Training, gear provision
Trust	Low	Eradication of corruption
Occupational Multiplicity	Low	Support for economic growth
Wealth (MSL)	Low	Poverty alleviation plans and pro-poor growth policies
Recognition of Human Agency	Medium	Education and participation in research
Social Capital	Medium	Support for community initiatives/organizations

as catch per unit effort, biomass, trophic structure, or catchability that are often used to estimate yields. Future studies may incorporate these types of fisheries parameters in estimates of gear sensitivity. Lastly, our results here should not be generalized to how other reef fisheries may respond to further bleaching events. Our analysis could produce extremely different results somewhere like Papua New Guinea, where many of the species captured by artisanal fishers are more reef associated and the starting condition of the fishery is much better [41,43]. A limitation of the approach we employed is that we were unable to examine changes in catch sensitivity over time. A key concept in fisheries is that catch compositions can change over time.

Our study is the most comprehensive of its kind, particularly for reef fisheries. However, there are several caveats about our approach and methodology that are important to acknowledge. We are aware that our index of vulnerability is limited to the effect of a single climate change impact (coral bleaching), through a single impact pathway (impact on fisheries). In reality, climate change is a multifaceted threat that will comprise multiple interacting impacts that will also be mediated or extenuated by other social and economic trends. The impacts of climate change on fisheries through coral bleaching are hard to discern and may be overwhelmed by: i) existing trends such as overexploitation; ii) climate impacts affecting other aspects of the ecosystem (e.g. seagrasses); iii) or socioeconomic characteristics, such as demographics, migration and the provision of food and employment from agriculture. In addition, the novel indexes we use here incorporate multiple sources of uncertainty about the nature of exposure, sensitivity and adaptive capacity and this high level of uncertainty needs to be recognized by adaptation prioritization and planning efforts.

Our methodology also assumes future sensitivity and adaptive capacity based on a snapshot of current conditions. Clearly, climate impacts or other external forces such as development projects (e.g. the proposed port development project in Lamu in northern Kenya) could result in substantial economic and social restructuring of surrounding coastal communities in ways that would profoundly alter social sensitivity and social adaptive capacity. Likewise, this study focuses on impacts on currently targeted species, which could be altered by climate change. For example, climate anomalies in Peru that severely impacted the dominant anchovy fishery also created opportunities for exploitation of different species in different areas, which were taken up fishers who had spatial and technological flexibility to exploit them [49]. Additionally, our ecological research is focused on coral reef fish species, although certain non-coral associated (e.g. *Leptoscarus vaigiensis* and *Siganus sutor*), pelagic and semi-pelagic (e.g., *Sphyraena barracuda*), and non-fish resources (e.g. lobsters and octopus) are also significant fishery resources supporting livelihoods and food security. Despite these caveats, we present a first step to understanding vulnerability, and highlight the importance of maximizing use of all available data when assessing the vulnerability of a place.

This study advances the application of climate change impact and adaptation theory to empirical data, and identifies several key gaps requiring further research. For example, our socioeconomic study focused on direct resource users with only limited information about the broader socioeconomic context, which can be a significant driver of social adaptive capacity. An understanding of the broader socioeconomic context within which resource users are embedded may further progress our understanding of how resource dependent people can be assisted so as to minimise their vulnerability to future climate changes. Similarly, the relative importance of different components of adaptive capacity for adapting to different types and magnitudes of impacts over time is not well understood. For example how can we understand the tradeoff between infrastructure and wealth resources with development and the loss of occupational flexibility? We also recognize that future research will need to consider the susceptibility of fish to climate impacts other than coral bleaching (e.g. ocean acidification), and there is a need to ascertain the species-specific responses to bleaching of five key fishes that makes up a large proportion of the catch (Methods S1 in File S1).

Conclusions

The modified IPCC vulnerability assessment framework provides a useful model for assessing adaptive capacity and sensitivity of social systems that are exposed to changes in the condition of the ecological system upon which they depend. In applying this modified model to resource-dependent communities in Kenya, we are able to derive useful insights into the relative magnitude and key sources of vulnerability to potential climate changes and to consider possible strategies that can minimise vulnerability. The framework allows us to simplify assessments and consider heterogeneity within: 1) places that experience climate change-related events such as coral bleaching; 2) the ways that coral reef ecosystems are affected by and can recover from these impacts; 3) the ways that societies and individuals are impacted by these changes; and 4) the capacity of people to cope with and adapt to these changes. Overall, indicators of ecological exposure, ecological sensitivity, and ecological recovery potential are different facets of ecological vulnerability, which provides justification to our effort to identify indicators describing these different aspects of the vulnerability. Although focusing on small-scale fisheries that operate in coral reef systems, the vulnerability assessment, framework, and survey we develop are adaptable to other kinds of fishery or natural resource dependent systems. Likewise, the framework could be adapted to explore vulnerability to other kinds of environmental, economic, or social stresses and could be complemented by qualitative social science research methodologies [50,51,52].

Supporting Information

File S1 Contains: Methods S1. Figure S1. Ecological indicators compared across sites in the western Indian Ocean sites (n = 482), Kenya (n = 214), and the 15 Kenyan sites included in this study (Labelled Kenya BMU in this figure). Box plots show 25% and 75% quartiles (box) with median (line) and outliers. **Figure S2.** Comparison between indicator values normalized to Kenya 2% and 98% percentiles, vs. Western Indian Ocean regional site 2% and 98% percentiles. The red line indicates the 1:1 line. **Figure S3.** Relative contribution in fish abundance from catch data of species, genus, family level data and species with no data. **Figure S4.** Relative abundance of species targeted by gear type. Species are coloured as to whether we have species level data (black), genus level averages (dark grey), family level averages (light grey), or no data (white) on their response to coral mortality. **Figure S5.** Average fish response to coral decline of each gear using only species data, or species and genus data, or species, genus and family data, ±SE. **Figure S6.** Relative abundance *response to decline of fish species targeted by gear type. This figure illustrates the influence of each species on the results and helps to identify critical research directions. The colour indicates the number of study in the global database of species response to coral loss that were used for each species: green for more than 1

study, red for only 1 study, and black where genus data were used. **Figure S7.** Intergovernmental Panel on Climate Change (IPCC) conceptual framework of vulnerability to climate change. **Table S1.** Occupational sensitivity scores by community. A score of 1 would mean all respondents depended on marine resources and had no livelihood alternatives, while a score of 0 would mean that none of the respondents had marine resource based livelihoods. **Table S2.** Average percent change in abundance of fish per percent decline in coral cover by gear type, using species and genus data (and also without Lethrinus nebulosus). **Table S3.** Gear sensitivity scores by community. **Table S4.** Missing information on five species creates a significant gap in our understanding on how species respond to coral mortality. Column 1 shows the relative abundance of the five critical species without species-specific data on responses to coral mortality by gear type. Column 2 shows existing species level data by gear type. Column 3 shows the proportion of catch data that we would have species-specific understandings of if just five species were studied. **Table S5.** Spearman correlations between the 11 adaptive capacity indicators (correlations conducted at the community scale). **significant at 0.01, *significant at 0.05. **Table S6.** Ecological vulnerability indicators of exposure, sensitivity and recovery potential for 15 ecological sites. Detailed description of the rational for indicators and how indicators were calculated can be found in Table 1 and the Methods. **Table S7.** Dimensions of ecological vulnerability for 17 coral reef sites in Kenya. Ecological vulnerability was calculated from normalized and weighted indicators as (Exposure+Sensitivity) – Recovery Potential. Sites are ranked from most vulnerable to least vulnerable. **Table S8.** The 11 adaptive capacity indicators aggregate values at community level shown as % or mean ± standard deviations. **Table S9.** Eigenvalues and percentage of variation explained by the different PCs. **Table S10.** Factor loadings of adaptive capacity indicators. Factor loadings above 0.4 (in bold) on any given Principal Component are generally considered to contribute substantially to that Component. **Table S11.** Absolute factor loadings, weights and normalised weights of each adaptive capacity indicator.

Acknowledgments

We would like to thank Orjan Bodin, Tim Daw, Eddie Allison, Cassandra DeYoung, and Tessa Hempson for assistance and constructive comments. We are also sincerely grateful to the coastal communities that participated in this study.

Author Contributions

Conceived and designed the experiments: JEC CH ESD ATH NAJG CCH NM TRM. Performed the experiments: JEC CH ESD ATH NAJG CCH NM TRM. Analyzed the data: JEC CH ESD ATH NAJG CCH NM TRM. Contributed reagents/materials/analysis tools: JEC CH ESD ATH NAJG CCH NM TRM. Wrote the paper: JEC CH ESD ATH NAJG CCH NM TRM.

References

1. Donner SD, Potere D (2007) The inequity of the global threat to coral reefs. Bioscience 57: 214–215.
2. Paulay G, editor (1997) Diversity and distribution of reef organisms. New York: Chapman & Hall. 298–353 p.
3. Reaka-Kudla M (1997) The global biodiversity of coral reefs: a comparison with rainforests. In: Reaka-Kudla M, Wilson DE, Wilson EO, editors. Biodiversity II: Understanding and Protecting Natural Resources. Washington, D.C.: Joseph Henry/National Academy Press. 83–108.
4. Bruno JF, Selig ER (2007) Regional decline of coral cover in the Indo-Pacific: Timing, extent, and subregional comparisons. PLoS ONE 2: e711.
5. Paddack MJ, Reynolds JD, Aguilar C, Appeldoorn RS, Beets J, et al. (2009) Recent region-wide declines in Caribbean reef fish abundance. Current Biology 19: 590–595.
6. Hughes TP, Baird AH, Bellwood DR, Card M, Connolly SR, et al. (2003) Climate change, human impacts, and the resilience of coral reefs. Science 301: 929–933.
7. Hughes S, Yaun A, Max L, Petrovic N, Davenport F, et al. (2012) A framework to assess national level vulnerability from the perspective of food security: The case of coal reef fisheries. Environmental Science and Policy 23: 95–108.
8. IPCC (2007) Climate change 2001: impacts, adaptation, and vulnerability. GenevaSwitzerland: Intergovernmental Panel on Climate Change. 21 p.

9. Sumaila UR, Cheung WWL, Lam VWY, Pauly D, Herrick S (2011) Climate change impacts on the biophysics and economics of world fisheries. Nature Climate Change.

10. Elsner JB, Kossin JP, Jagger TH (2008) The increasing intensity of the strongest tropical cyclones. Nature 455: 92–95.

11. Madin JS, Hughes TP, Connolly SR (2012) Calcification, storm damage, and population resilience of tubular corals under climate change. PLoS ONE 7: e46637.

12. Hoegh-Guldberg O (1999) Climate change, coral bleaching and the future of the world's coral reefs. Marine and Freshwater Research 50: 839–866.

13. Westmacott S, Cesar H, Pet Soede L, Linden O (2000) Coral bleaching in the Indian Ocean: socio-economic assessment effects. In: Cesar H, editor. Collected Essays on Economics of Coral Reefs. Kalmar, Sweden: CORDIO. 94–106.

14. Graham NAJ, Wilson S, Jennings S, Polunin N, Robinson J, et al. (2007) Lag effects in the impacts of mass coral bleaching on coral reef fish, fisheries, and ecosystems. Conservation Biology 21: 1291–1300.

15. MacNeil A, Graham N (2010) Enabling regional management in a changing climate through Bayesian meta-analysis of a large-scale disturbance. Global Ecology and Biogeography 19: 412–421.

16. Folke C (2006) Resilience: the emergence of a perspective for social-ecological systems analysis. Global Environmental Change 16: 253–267.

17. McClanahan TR, Castilla JC, White A, Defeo O (2009) Healing small-scale fisheries and enhancing ecological benefits by facilitating complex social-ecological systems. Reviews in Fish Biology and Fisheries 19: 33–47.

18. Adger NW (2006) Vulnerability. Global Environmental Change 16: 268–281.

19. Adger WN (2000) Social and ecological resilience: are they related? Progress in Human Geography 24: 347–364.

20. Kelly PM, Adger WN (2000) Theory and practice in assessing vulnerability to climate change and facilitating adaptation. Climatic Change 47: 325–352.

21. Bell JD, Johnson JE, Hobday AJ (2011) Vulnerability of tropical pacific fisheries and aquaculture to climate change. Auckland, NZ: Secretariat of the Pacific Community.

22. Adger WN, Vincent K (2005) Uncertainty in adaptive capacity. CR Geoscience 337: 399–410.

23. Gallopìn G (2006) Linkages between vulnerability, resilience, and adaptive capacity. Global Environmental Change 16: 293–303.

24. Smit B, Wandel J (2006) Adaptation, adaptive capacity and vulnerability. Global Environmental Change 16: 282–292.

25. Cutter SL (1996) Vulnerability to environmental hazards. Progress in Human Geography 20: 529–539.

26. Maina J, Venus V, McClanahan TR, Ateweberhan M (2008) Modelling susceptibility of coral reefs to environmental stress using remote sensing data and GIS models in the western Indian Ocean. Ecological Modelling 212: 180–199.

27. Nelson DR, Adger WN, Brown K (2007) Adaptation to environmental change: Contributions of a resilience framework. Annual Review of Environment and Resources 32: 395–419.

28. Adger WN, Hughes TP, Folke C, Carpenter SR, Rockstrøm J (2005) Social-ecological resilience to coastal disasters. Science 309: 1036–1039.

29. Hughes T, Bellwood D, Folke C, Steneck RS, Wilson J (2005) New paradigms for supporting the resilience of marine ecosystems. Trends in Ecology & Evolution 20: 380–386.

30. Cinner JE, McClanahan TR, Graham NAJ, Daw TM, Maina J, et al. (2012) Vulnerability of coastal communities to key impacts of climate change on coral reef fisheries. Global Environmental Change 22: 12–20.

31. Allison EH, Perry AL, Badjeck MC, Adger WN, Brown K, et al. (2009) Vulnerability of national economies to the impacts of climate change on fisheries. Fish and Fisheries 10: 173–196.

32. Saji NH, Goswami BN, Vinayachan ndran PN, Yamagata T (1999) A dipole mode in the tropical Indian Ocean. Nature 4001: 360–363.

33. Nakamura N, Kayanne H, Iijima H, McClanahan TR, Behera SK, et al. (2011) Footprints of IOD and ENSO in the Kenyan coral record. Geophys Res Lett 38: L24708.

34. Ateweberhan M, McClanahan TR, Graham NAJ, Sheppard C (2011) Episodic heterogeneous decline and recovery of coral cover in the Western Indian Ocean. Coral Reefs 30: 739–752.

35. Cinner J, Bodin O (2010) Livelihood diversification in tropical coastal communities: a network-based approach to analyzing 'livelihood landscapes'. PLoS ONE 5: e11999.

36. Cinner JE, McClanahan TR, MacNeil MA, Graham NAJ, Daw TM, et al. (2012) Comanagement of coral reef social-ecological systems. Proceedings of the National Academy of Sciences 109: 5219–5222.

37. Cinner JE, Daw TM, McClanahan TR, Muthiga N, Abunge C, et al. (2012) Transitions toward co-management: The process of marine resource management devolution in three east African countries. Global Environmental Change 22: 651–658.

38. Maina J, McClanahan TR, Venus V, Ateweberhan M, Madin J (2011) Global gradients of coral exposure to environmental stresses and implications for local management. PLoS ONE 6: e23064.

39. McClanahan TR, Donner SD, Maynard JA, MacNeil MA, Graham NAJ, et al. (2012) Prioritizing key resilience indicators to support coral reef management in a changing climate. PLoS ONE 7: e42884.

40. Marshall NA, Marshall PA, Tamelander J, Obura DO, Mallaret-King D, et al. (2010) A framework for social adaptation to climate change: sustaining tropical coastal communities and industries. Gland Switzerland: IUCN. 36 p.

41. Cinner JE, McClanahan TR, Graham NAJ, Pratchett MS, Wilson SK, et al. (2009) Gear-based fisheries management as a potential adaptive response to climate change and coral mortality. Journal of Applied Ecology 46: 724–732.

42. Pratchett MS, Hoey AS, Wilson SK, Messmer V, Graham NAJ (2011) Changes in biodiversity and functioning of reef fish assemblages following coral bleaching and coral loss. Diversity 3: 424–452.

43. McClanahan TR, Cinner JE, Maina J, Graham NAJ, Daw TM, et al. (2008) Conservation action in a changing climate. Conservation Letters 1: 53–59.

44. Ostrom E (2007) A diagnostic approach for going beyond panaceas. Proceedings of the National Academy of Sciences 104: 15181–15187.

45. Ostrom E, Cox M (2010) Moving beyond Panaceas: A multi-tiered diagnostic approach for social-ecological analysis. Environmental Conservation 37: 1–13.

46. Daw TM, Cinner JE, McClanahan TR, Brown K, Stead SM, et al. (2012) To fish or not to fish: factors at multiple scales affecting artisanal fishers' readiness to exit a declining fishery. PLoS ONE 7: e31460.

47. Daw TM, Adger N, Brown K, Badjeck MC (2009) Climate change and capture fisheries. Climate change implications for fisheries and aquaculture. Overview of current scientific knowledge. Rome: Food and Agricultural Organization. 95–135 p.

48. Grandcourt EM, Cesar H (2003) The bio-economic impact of mass coral mortality on the coastal reef fisheries of the Seychelles. Fisheries Research 60: 539–550.

49. Badjeck M-C, Allison EH, Halls AS, Dulvy NK (2010) Impacts of climate variability and change on fishery-based livelihoods. Marine Policy 34: 375–383.

50. Béné C, Evans L, Mills D, Ovie S, Raji A, et al. (2011) Testing resilience thinking in a poverty context: Experience from the Niger River basin. Global Environmental Change 21: 1173–1184.

51. Schwarz A-M, Béné C, Bennett G, Boso D, Hilly Z, et al. (2011) Vulnerability and resilience of remote rural communities to shocks and global changes: Empirical analysis from Solomon Islands. Global Environmental Change 21: 1128–1140.

52. Mills D, Béné C, Ovie S, Tafida A, Sinaba F, et al. (2011) Vulnerability in African small-scale fishing communities. Journal of International Development 23: 308–313.

53. Marshall N, Tobin R, Marshall P, Gooch M, Hobday A (2013) Social vulnerability of marine resource users to extreme weather events. Ecosystems DOI: 10.1007/s10021-013-9651-6.

The 50 Most Important Questions Relating to the Maintenance and Restoration of an Ecological Continuum in the European Alps

Chris Walzer[1]*, Christine Kowalczyk[1], Jake M. Alexander[2], Bruno Baur[3], Giuseppe Bogliani[4], Jean-Jacques Brun[5], Leopold Füreder[6], Marie-Odile Guth[7], Ruedi Haller[8], Rolf Holderegger[9], Yann Kohler[10], Christoph Kueffer[2], Antonio Righetti[11], Reto Spaar[12], William J. Sutherland[13], Aurelia Ullrich-Schneider[14], Sylvie N. Vanpeene-Bruhier[15], Thomas Scheurer[16]

1 Research Institute of Wildlife Ecology, Department of Integrative Biology and Evolution, University of Veterinary Medicine, Vienna, Austria, 2 Institute of Integrative Biology, Swiss Federal Institute of Technology Zurich, Switzerland, 3 Department of Environmental Sciences, University of Basel, Basel, Switzerland, 4 Department of Earth and Environmental Sciences, University of Pavia, Pavia, Italy, 5 Unité Ecosystèmes Montagnards, National Research Institute of Science and Technology for Environment and Agriculture, Saint Martin d'Hères, France, 6 Institute of Ecology, University of Innsbruck, Innsbruck, Austria, 7 Ministry of Ecology, Sustainable Development and Energy, Paris, France, 8 Research and Geoinformation, Swiss National Park, Zernez, Switzerland, 9 Research Unit Biodiversity and Conservation Biology, Swiss Federal Institute for Forest, Snow and Landscape Research, Birmensdorf, Switzerland, 10 Task Force Protected Areas - Permanent Secretariat of the Alpine Convention, Chambery, France, 11 Partner-innen in Umweltfragen, Liebefeld, Switzerland, 12 Swiss Ornithological Institute, Sempach, Switzerland, 13 Conservation Science Group, Department of Zoology, University of Cambridge, Cambridge, United Kingdom, 14 International Commission for the Protection of the Alps-CIPRA International, Schaan, Liechtenstein, 15 Écosystèmes Méditerranéens et Risques, National Research Institute of Science and Technology for Environment and Agriculture, Aix-en-Provence, France, 16 Swiss Academy of Sciences, Berne, Switzerland

Abstract

The European Alps harbour a unique and species-rich biodiversity, which is increasingly impacted by habitat fragmentation through land-use changes, urbanization and expanding transport infrastructure. In this study, we identified the 50 most important questions relating to the maintenance and restoration of an ecological continuum – the connectedness of ecological processes across many scales including trophic relationship and disturbance processes and hydro-ecological flows in the European Alps. We initiated and implemented a trans-national priority setting exercise, inviting 48 institutions including researchers, conservation practitioners, NGOs, policymakers and administrators from the Alpine region. The exercise was composed of an initial call for pertinent questions, a first online evaluation of the received questions and a final discussion and selection process during a joint workshop. The participating institutions generated 484 initial questions, which were condensed to the 50 most important questions by 16 workshop participants. We suggest new approaches in tackling the issue of an ecological continuum in the Alps by analysing and classifying the characteristics of the resulting questions in a non-prioritized form as well as in a visual conceptualisation of the inter-dependencies among these questions. This priority setting exercise will support research and funding institutions in channelling their capacities and resources towards questions that need to be urgently addressed in order to facilitate significant progress in biodiversity conservation in the European Alps.

Editor: Raphaël Arlettaz, University of Bern, Switzerland

Funding: Funding for this project was provided in part by the "Platform Ecological Network" of the Alpine Convention (http://www.alpconv.org/en/organization/groups/WGEcologicalNetwork/default.html) and the ECONNECT project funded by the EU within the framework of the European Territorial Cooperation Alpine Space Programme and co-funded by the European Regional Development Fund (www.econnectproject.eu). RH was supported by the CCES-GENEREACH project of the ETH domain. The funders had no role in study design, data collection and analysis, decision to publish, or preparation of the manuscript.

Competing Interests: The authors acknowledge that one of the co-authors (AR) is employed by a private company (PiU GmbH).

* E-mail: chris.walzer@fiwi.at

Introduction

The European Alps span eight countries, from the Mediterranean shores of Southern France to Slovenia and link with adjacent mountain ranges such as the Carpathians, Balkans and Apennines. The Alps harbour an extraordinary diversity of habitats, plants and animals including many endemics. They are considered as one of the most important regions for the preservation of biodiversity in Europe [1]. However, the Alps are also the home

and workplace of up to 14 million people and a holiday destination for more than 100 million tourists each year. Anthropogenic land-use changes, urbanization and the development of transport infrastructure have fragmented the ecological continuum - the connectedness of ecological processes across many scales including trophic relationship and disturbance processes and hydro-ecological flows of the Alps [2,3]. Plant and animal populations become increasingly isolated on ever-smaller remnant habitat patches in a human-dominated landscape. Isolation leads to increased vulnerability in the face of stochastic events and decreased genetic diversity [4]. Enhancing ecological connectivity is an integral part of the European Commission's ambitious 2020 Biodiversity Strategy (http://ec.europa.eu/environment/nature/ecosystems/index_en.htm). During the past few years, major efforts have been undertaken to maintain and foster pan-alpine biodiversity. The "Platform Ecological Network" of the Alpine Convention (http://www.alpconv.org/en/organization/groups/WGEcological Network/default.html), the "Ecological Continuum Initiative" (http://www.alpine-ecological-network.org/about-us/ecological-continuum-initiative) and the Alpine Space project "ECON-NECT" (www.econnectproject.eu) are three examples of such initiatives that also aim at enhancing ecological connectivity across the Alpine range.

ECONNECT is a project funded by the EU within the framework of the European Territorial Cooperation Alpine Space Programme and co-funded by the European Regional Development Fund. The project envisions an enduringly restored and maintained ecological continuum, consisting of inter-connected landscapes and protected areas, across the Alpine Arc region, where biodiversity will be conserved for future generations and the resilience of ecological processes will be enhanced. Employing an integrated and multidisciplinary approach, the project has provided an Alpine-wide overview on areas important for ecological connectivity by integrating quantitative and qualitative information on selected sites and the level of interconnectivity among them [5]. Additionally, social, legal and economic barriers to connectivity have been identified and proposals made on how to overcome them [6]. However, in the course of the project it became evident that relevant knowledge and information is missing despite being considered essential for on-going and future projects. This knowledge gap includes consistent scientific definitions, evidence-based assessment methods, evaluation of impacts on the ecological continuum in the Alps as well as best practice measures. Similarly, other authors have recently addressed an analogous lack of evidence in respect to the design and implementation of ecological corridors, an important conservation intervention in an ecological continuum [7]. To address and surmount this knowledge deficit, the three initiatives mentioned above formed a small core group comprising both researchers and practitioners in order to start a process identifying these knowledge gaps.

Sutherland et al. 2011 [8] introduced the method of a "priority setting exercise" to identify research needs of practitioners and policy makers. Identifying research needs is essential in furthering evidence-based conservation [9,10]. Priority setting exercises have been successfully applied in the UK, USA, Switzerland and even on a global scale tackling such diverse issues as biodiversity, global agriculture and national environmental policy [11–14]. A priority setting exercise identifies existing knowledge, supports research and helps funding institutions channel their capacities and resources towards questions that need to be urgently answered in order to facilitate significant progress in a respective field: in our case biodiversity conservation in the European Alps.

In this study, we identified the 50 most important questions relating to the maintenance and restoration of an ecological continuum in the European Alps. We initiated and implemented a trans-national priority setting exercise, inviting researchers, practitioners, NGOs, policy makers and other stakeholders from the Alpine region to participate. The exercise was composed of an initial call for pertinent questions, a first online evaluation of the received questions and a final discussion and selection process during a joint workshop. Furthermore, we suggest new approaches in tackling the issue of an ecological continuum in the Alps by analysing and classifying the characteristics of the resulting questions in a non-prioritized form as well as a visual conceptualisation of the inter-dependencies among these questions.

Results

The non-prioritized list of the 50 most important questions concerning an ecological continuum in the Alps is shown in tables 1 and 2. The resulting questions were individually classified broadly in nature-, people- and management contexts (NC, PC, MC) based on a concept previously introduced by Worboys et al. [15]. Each context area has three sub-topic areas to which the questions are finally attributed. For reasons of clarity, we assign each question to only one context area, but provide the second interrelated context and dependencies in brackets where appropriate.

The largest proportion of questions (46%) was attributed to the nature context. The nature context questions are rather evenly distributed between the three subtopics. This is followed by the management context (44%) where by far the largest proportion of questions relates to the legislation, policy and planning needs subtopic (63%). Finally the people context makes up a mere 10% of the total questions. From the 50 questions, by far the largest fraction (60%) was formulated as "how" questions, followed by "what" (26%) and "which" (14%) questions. Consequently, most attention was given to transformation processes aiming at practices to improve the current situation in Alpine connectivity.

Visualization of Interactions, Inter-relations and Dependencies of the 50 Final Questions

We created a graph in the form of a web to illustrate the "location" of each question within the framework of the nine sub-topics in the nature-, people- and management-context (Figure 1). Each context and its assigned questions are shown in the same colour, whereas the sub-topics in each context area are given in varying shades of the same colour (NC: green; PC: red; MC: blue). The sub-topics designate the edges of the web, while the questions from one context area are arranged in a connected cycle in the middle of the web to illustrate their interactions. Questions, which were also assigned to a second sub-topic of another context area, are interlinked with a grey line. Bold arrows indicate that this particular question has to be answered first before a subsequent one (to which the arrow points) can be addressed. In some cases, answers from two previous questions are required for the solution of a third question (e.g. question 8 depends on the answers from questions 1 and 5) whereas in other cases, two or three questions depend on the answer of a single question in advance (e.g. questions 2, 19 and 46 depend on the answer of question 15).

Overall, 27 questions were assigned to a second context area due to their transdisciplinary nature. The sub-topics 6 "economic, social and political needs" (PC) and 9 "tools, incentives, knowledge" (MC) showed the most inter-relations to other context areas (8 and 7 links, respectively).

Discussion

The aim of this study was to identify and analyse gaps of knowledge with respect to achieving, restoring and maintaining an ecological continuum in the European Alps. In order to identify information gaps, we implemented an alpine wide "priority setting exercise" as previously described by Sutherland et al. [12]. This exercise was a useful and efficient tool to compile inputs from various researchers, practitioners, administrators, stakeholders and policy makers from different countries with a relatively low initial

Table 1. Non-prioritized list of the 50 most important questions.

01	Which landscape elements and land use types enhance or moderate gaps in connectivity?
02	How are corridors best implemented; with clearly spatially defined borders or as functional units integrated in wide ecological continuums?
03	How do major land use changes affect ecological connectivity across the Alps?
04	What is the relative importance of climate/land-use change to changes in the ecological continuum of Alpine regions?
05	Which indicators reflect the changes in connectivity that result from climate or human induced changes in Alpine landscapes?
06	How important is connectivity in maintaining key ecosystem services?
07	How can ecological connectivity maintain the adaptive capacity of ecosystems in the face of environmental change?
08	Which of the habitat types important for landscape connectivity are most affected by climate change
09	How does alternative energy production impact on connectivity and natural habitats?
10	What is the best method to design corridors for multiple species?
11	How severe is the current lack of connectivity between populations of alpine species?
12	What are indicators for a multi-species continuum?
13	What impacts do various seasonal leisure activities (including low-impact practices) have on ecological connectivity across the Alps?
14	How can wilderness areas (wildlife, recreation, tourism) contribute to ecological connectivity?
15	What is an effective set of indicators (i.e., for species and habitats) that can be used to evaluate and monitor ecological connectivity at different scales?
16	How does the return of large carnivores affect ecosystems in the Alpine ecological network?
17	What is the impact of gene flow through an ecological continuum on genetic adaptation to climate change?
18	How does the ecological continuum allow shifts in species distribution to keep pace with climate change?
19	Are artificially engineered ecological networks a threat or a benefit to endemic species?
20	What are the consequences for both genetic and species diversity if the system of natural barriers changes?
21	How will future changes in species distribution affect connectivity and fitness among interacting species?
22	How much gene flow fostered by connectivity is beneficial to populations and species without disrupting local adaptations?
23	How can the spread of invasive species and diseases be minimized, while ensuring connectivity for native species?
24	How do elements of the ecological network affect human welfare and perception?
25	How can agricultural and silvicultural land use be optimised in order to promote and conserve ecological connectivity?
26	How can connectivity for biodiversity and ecosystem conservation become and be managed as a public good?
27	How do demographic changes in the Alps affect the future ecological continuum?
28	How do the aims of ecological connectivity and tourism conflict?
29	What is the most effective way to employ the different categories of protected areas to ensure connectivity and the provision of ecosystem services in the Alps?
30	How can we use and integrate existing instruments and programmes to enhance trans-sectoral funding for ecological connectivity?
31	How can ecological connectivity be integrated into spatial and infrastructural planning and legislation at various administrative levels?
32	How can legal and conceptual tools stimulate the development of trans-border connectivity?
33	How is it possible to harmonise contradictory, competing spatial sectoral policies in order to enhance connectivity?
34	Which policy-measures are necessary to safeguard the ecological network beyond protected areas?
35	Which of the existing sectoral funding systems have a positive and which have a negative effect on connectivity?
36	What incentives for agriculture and forestry are needed to maintain and restore ecological connectivity in different Alpine areas?
37	Which strategy, integration or segregation, is more appropriate for promoting ecological connectivity in different alpine areas?
38	How can we effectively manage areas heavily affected by tourism in order to maintain their function within an ecological continuum?
39	How can we enhance sharing of theoretical and empirical good practice knowledge amongst and between sectors?
40	How can the management of protected areas better incorporate functional relationships with surrounding areas?
41	Which specific restoration measures can increase connectivity?
42	What kind of monitoring is needed to evaluate the long-term efficiency of connectivity measures in the face of dynamic anthropogenic change?
43	How can an alpine-wide, accessible and effective connectivity data platform be created?
44	How can databases for existing or emerging bio- and geo-data be improved for the promotion of connectivity projects in the Alps?
45	What is the effectiveness of different methods (e.g. sensor data) to monitor the consequences of structural connectivity or its elements across different spatial and temporal scales?
46	What is the effectiveness of different methods to record the effectiveness of functional connectivity or its elements across different spatial and temporal scales?
47	How can we use evidence-based education to increase public awareness of ecological networks?
48	How can methods of conflict resolution be adapted and/or used to mitigate concerns and obstruction to ecological networks?
49	How should we integrate spatial and temporal dynamics into the realization of the Alpine ecological continuum?
50	How can the species and habitat approaches to designing ecological connectivity be integrated into the process of landscape planning?

Table 2. Non-prioritized list of the 50 most important questions concerning an ecological continuum in the Alps classified into nature-, people- and management contexts.

No.	NATURE CONTEXT			PEOPLE CONTEXT			MANAGEMENT CONTEXT		
	Structural and functional connectivity	Habitat connectivity	Evolutionary process and connectivity	Natural land: social, cultural and spiritual values	Life support needs	Economic, social and political needs	Legislation, policy and planning needs	Management support needs	Tools, incentives, knowledge
1	NC								
2	NC								MC (depends on question 15)
3	NC					PC			
4	NC								
5	NC								MC
6	NC				PC				
7	NC								
8		NC (depends on questions 1 and 5)							
9		NC			PC				
10		NC							MC (depends on question 12)
11		NC							
12		NC							X
13		NC				PC			
14		NC					MC		
15		NC							X
16		NC							
17			NC						
18			NC						
19			NC						MC (depends on questions 11 and 15)
20			NC						
21			NC						
22			NC						
23			NC						
24				**PC**					
25	NC				**PC**				
26	NC				**PC**				MC
27	NC					**PC**			
28		NC				**PC**			
29	NC						**MC**		
30							**MC** (depends on question 35)		
31								X	

Table 2. Cont.

No.	NATURE CONTEXT			PEOPLE CONTEXT			MANAGEMENT CONTEXT		
	Structural and functional connectivity	Habitat connectivity	Evolutionary process and connectivity	Natural land: social, cultural and spiritual values	Life support needs	Economic, social and political needs	Legislation, policy and planning needs	Management support needs	Tools, incentives, knowledge
32						PC	X		
33						PC	X		
34							X		
35							X		
36							X		
37							X		
38						PC		MC	
39								MC	
40						PC		MC	
41	NC								MC
42						PC			MC
43									MC
44									MC
45	X								MC
46	NC (depends on question 15)			PC					MC
47									MC
48						PC			MC
49									MC
50		NC							MC

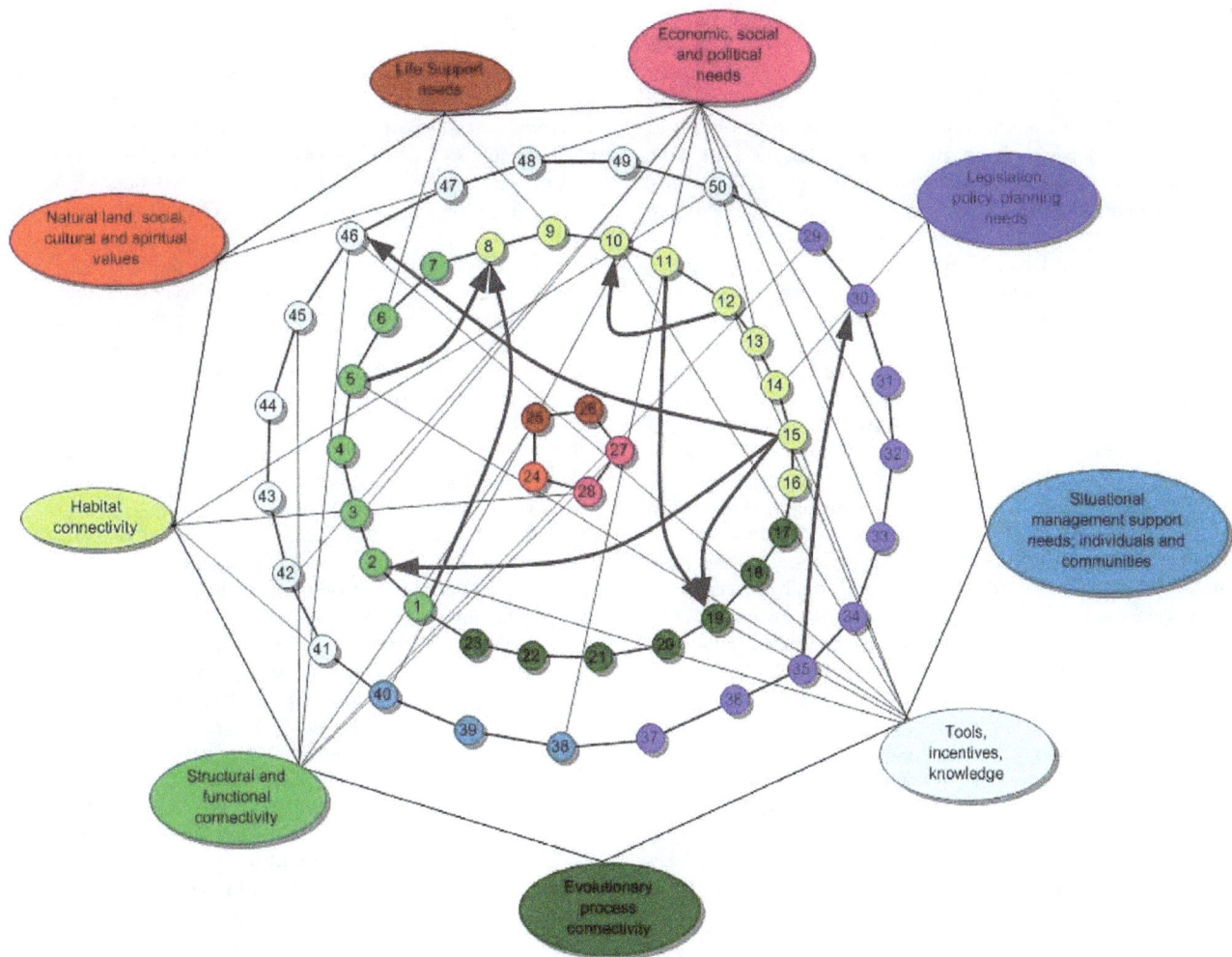

Figure 1. Web of the 50 most urgent questions concerning an ecological continuum in the European Alps. The nine sub-topics of the three context areas (nature: green; people: red; management: blue) mark the edges of the web. The questions of each context area are interlinked and marked in the same colour as the sub-topics they were assigned to. Questions (numbers; see text) which also address another context area are further linked with a second sub-topic, highlighting the interactions. Bold arrows indicate that one or more questions need to be solved before a subsequent question can be answered.

effort. The major part of the process was performed via email communication and was administered part-time by one person. We feel that this resource-saving method is a strong argument in favour of this approach, especially given the generally limited resources for connectivity conservation [16,17].

The output of priority setting exercises depends on a number of factors, such as the selection of participants, the number of individuals effectively involved in formulating the questions and on the voting system [11–13]. One selection bias that could arise as a consequence of listing some 400+ questions is that an individual performing the selection process is almost certainly going to become tired as he proceeds down the list. In order to address this possible bias in future exercises, it would be prudent to provide the numbered questions in a individualized random sequence to the participants. Decisions claim to be objective, however, since people's choices are subjective, the only option is to make them "to the best of one's knowledge". Only a constant review-process from the organisational core team, consisting preferably of several people from different fields, and continuous discussion with and feedback from other colleagues can minimize the inherent bias. Recently Braunisch et al. [14] have reported on a two-step

evaluation of a similar priority setting exercise that elegantly addresses bias through a process of rating and speciation of the questions. The composition of the workshop participants exemplifies this problem. Although we invited six policy makers to join the initiative, only two generated initial questions and only one participated in the workshop. The problem of a non-representative group of experts missing valuable issues has been pointed out by previous authors [14,18]). In future efforts of this type, it could be beneficial to provide more information to the potential participants in order to alleviate any concerns as to information from discussions being individually attributed and distributed outside the framework of the actual exercise. Clearly stating, and explaining if necessary, that the Chatham House Rule: "When a meeting, or part thereof, is held under the Chatham House Rule, participants are free to use the information received, but neither the identity nor the affiliation of the speaker(s), nor that of any other participant, may be revealed". (http://www.chathamhouse. org/about-us/chathamhouserule) will be adhered to is in our view beneficial to encourage openness and the sharing of information. Furthermore, as previously stated by Wuelser et al. [19] effective transdisciplinary problem framing and research orientation exercises require a careful specification of context and purpose. Possibly, this exercise was unsatisfactorily framed with a generic "identifying knowledge gaps". Without a clearly defined output context it can be difficult for an individual to select the most important questions from the bulk of initially proposed questions.

A further issue that hindered and complicated the exercise is the fact that the Alps are distributed across eight countries, 28 regions, 98 provinces and 5 spoken languages. Participants originated from five alpine countries (in the collection phase complemented by one participant from the UK) and divergent cultural backgrounds. In contrast to most previous studies that were English-language based, this study involved four different languages (English, German, French and Italian), which made the succinct formulation of questions particularly difficult. We tried to alleviate the fact that English was a foreign language to all of the participants by consulting with a native speaker during the entire process. Similarly, Braunisch et al. [14] had to deal with multiple languages in their study and in contrast to this study opted to translate questions into the two main local languages (French and German). Moreover, we tried to address uncertainties in the formulation of questions by making direct queries to the main author of a question if needed and encouraged all participants to do the same during the evaluation process and the workshop. It is interesting to note that numerous questions are formulated in very generic terms. While this is possibly a result of language constraints, it can, in these authors' view also be seen as a consequence of the nonspecific use of the term "connectivity" and the wide-ranging interpretation and understanding of the term. In contrast to a previous study in Switzerland (Braunisch et al.) it is interesting to note the absence of species- specific questions and the significant integration of questions related to ecological theory (14%) in this study. While the significance of this theoretical basis has been recognized in this exercise it is clear that these questions are inherently difficult to address and implement into action [20] We analysed the interactions, inter-relations and dependencies of the final questions that resulted from the priority setting exercise in order to suggest approaches to answer these questions. Therefore, we classified the questions into a nature-, people- and management context, applying a concept previously introduced by Worboys et al. [15]. The majority of the questions (54%) could be assigned to more than one context area, indicating the transdisciplinary character of the results. It is important to note that the resulting list is neither prioritized in respect to the individual questions nor in respect to the assigned contexts. The nature context though listed first does not necessarily take precedence over the people and management contexts in framing this problem. It is interesting to note that while only 4 questions were attributed to the people context, these were highly interlinked to other questions clearly demonstrating in this case the pivotal role of economic, social and political needs. Braunisch et al. have previously showed a similar importance of questions reconciling biodiversity conservation with societal and economic goals for Switzerland [14] though, the goal of this exercise was a non-prioritized list, eleven questions where directly dependent on each other, suggesting that a chronological order must be considered when addressing these problems. This fact inherently creates some prioritization. In order to answer question 8 "Which of the habitat types important for landscape connectivity are most affected by climate change?", we first need to understand question 1 "Which landscape elements and land use types enhance or moderate the gaps in connectivity?". Furthermore, the answer to question 5 "Which indicators reflect the changes in connectivity that result from climate or human induced changes in Alpine landscapes?" is also essential to adequately address question 8

above. Hence, question 8 depends on the results of two other questions.

Similarly question 15 "What is an effective set of indicators (for species and habitats) that can be used to evaluate and monitor connectivity at different scales?" is a prerequisite in addressing 3 other questions: namely question 2 "How are corridors best implemented; with clearly spatially defined borders or as functional units integrated in wide ecological continuums?", question 19 "Are artificially engineered ecological networks a threat or benefit to endemic species?" and question 46 "What methods can be used to record the effectiveness of functional connectivity, or its elements, across different spatial and temporal scales?". Essential questions like question 15, dealing with indicators and methodological approaches, should be given clear priority, as their answer is a precondition for scientifically sound answers to other, subsequent questions.

Solving certain questions in a consecutive order indicates a potential time lag between putting forward a question and actually answering it. This time lag may lead to a change of the initial situation, which would imply that the results of the preceding question are no longer suitable for solving the initial issue. This time lag in connectivity conservation makes it impossible to "come back" to the initial situation after the extended process of research, decision-making and implementation. This clearly indicates that there is an asymmetric drift of the initial problem and its proposed solution due to the general inertia in decision-making processes dealing with environmental issues. As holds true for numerous environmental problems, most notably global climate change, we suggest that the conservation and restoration of an ecological continuum is a "wicked problem" [21]. The example of question 19 clearly demonstrates several characteristics of a "wicked" and possibly even a "super-wicked problem" [22].

The gaps of knowledge, in conserving and restoring connectivity emphasized in this exercise make evident that these involve a highly dynamic and interconnected process rather than a simplistic and straightforward approach. It appears essential to reconcile the dynamic and complex nature of the problem with the problem solving approaches. Inadequate simplification of the interdependencies will possibly lead to results that are not relevant in forming policy [19]. Furthermore, our results indicate that maintaining and restoring ecological connectivity in the Alps is most likely a "super-wicked problem", and this implies the need for novel approaches in addressing this issue. As has been previously suggested by other authors, we also feel strongly that the usual backward looking method of investigating the past and generating selective and singular predictions, is only sufficient for "tame problems" but inadequate for a highly dynamic and interconnected process such as ecological connectivity [21,22].

In order to address the complex issue of an Alpine ecological continuum, it appears necessary to apply a forward reasoning approach which identifies possible future scenarios and integrates uncertainties [23]. It is somewhat surprising that questions concerning how ecological connectivity is affected and can be managed make up the largest percentage (60%) of the generated questions. Authors from the field of transdisciplinary research have termed knowledge related to this type of question "transformation knowledge" [24,25]. These questions deal with the genesis and future development of a problem and subsequently with the interpretation and perception of the problem in the "real world". "What" questions address determining factors of connectivity, and answers to such questions provide "system knowledge". Finally, "which" questions address desired goals and better practices. This has been termed "target knowledge". Each of these knowledge forms has specific challenges and notably "system knowledge" must confront uncertainties. It is essential to understand that solutions are only possible when the other postulated forms of knowledge, "target-" and "transformation knowledge", are integrated into the solution-mix. Another method introduced by Sutherland et al. [26,27], called "horizon scanning", an exercise that focuses on potential threats and opportunities in the future rather than identifying present knowledge gaps, may supplement and enhance this approach.

The visual "chaos" and multi-structural character of our results, shown in the "web of questions" (Figure 1), reflects the sectoral structure of society, governance and administration with respect to environmental problems in general. To overcome this, an integrative transdisciplinary approach is necessary. What appears to be missing in order to find a starting point to address the problem of the Alpine ecological continuum, is a common strategy or vision [5,15]. In the authors' view, this is also supported by the fact that the largest percentage of the formulated questions questioned the manner, condition or quality of ecological connectivity. This exemplifies the necessity of generating "system knowledge" and confronting uncertainties. Total conformity among all actors in the search for a common denominator is unrealistic and cannot be a realistic goal, as Worboys et al. [15] have pointed out, but a clear vision that "expresses the joint

aspirations of leaders, managers and participants in the initiative, without closing off avenues for constructive debate and disputation" to support and sustain connectivity conservation may be a starting point. Possibly, ecological connectivity can constitute a common "anchor" for trans-sectoral deliberations on biodiversity conservation. However, in order to not become overburdened by the complexity of the issue, it appears essential to address the inherent complexity within a well-reflected investigational framework [19].

For this type of study to provide guidance and contribute towards conservation-action implementation the results must be disseminated accordingly. As has been pointed out previously, bridging the gap of knowledge between research and conservation practice cannot be achieved with unidirectional platforms [28,29]. While other authors have suggested that new platforms of bidirectional knowledge dissemination must be developed these authors believe that it is more efficient to employ and if necessary adapt existing information platforms. The results from this study will be disseminated over existing platforms, most importantly the "Platform of Ecological Networks" of the Alpine Convention. This platform encompasses representatives of the Alpine countries, protected areas, Alpine institutions and experts. It strives to provide a bidirectional link between policy makers, the scientific

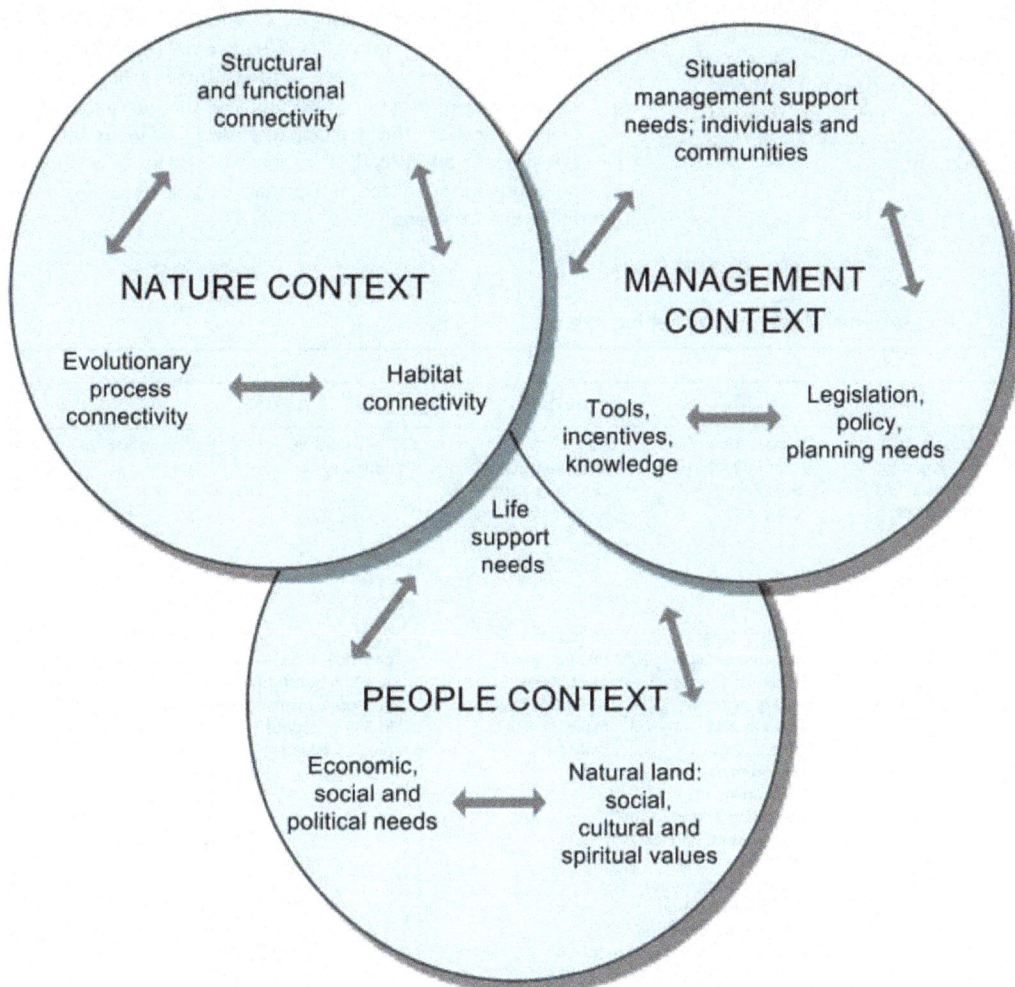

Figure 2. The three inter-related context areas of connectivity conservation adapted from Worboys et al. (2010). Every context area consists of three different sub-topics which interact with each other.

community and practitioners and encourages more efficient cooperation with other sectors [30]. As has been pointed out by Braunisch et al. [14], integrating scientifically trained and transdisciplinary-cognizant personnel that can translate science and ease the exchange of information could significantly enhance the effectiveness of such platforms. In these authors view, an initial task of the platform could be to organize and facilitate research and conservation-action activities centred on the inter-dependent questions identified in this study. It is the authors' opinion that this priority setting exercise and the subsequent dissemination of results will support research and funding institutions in channelling their capacities and resources towards questions that need to be urgently addressed in order to facilitate significant progress in biodiversity conservation in Europe and specifically in the Alps. Furthermore, the definition of 50 most important questions is an important first step towards a common and harmonized approach in maintaining and enhancing ecological connectivity across the heterogeneous Alpine arch.

Methods

Priority Setting Exercise to Identify Gaps of Relevant Knowledge and Generating the most Urgent Questions

A core team of five researchers and practitioners initiated and facilitated the entire exercise. The priority setting process was divided into four steps: initiation, assembling and organizing initial questions, pre-workshop voting and workshop voting. The method has been previously described in detail in various publications [11–13]. Therefore, we only briefly outline the main steps of the process. Further information on characteristics of the participants and the voting process are given in Table 3.

(1) Initiation. Based on the individual networks of the core team and striving to depict a representative cross-section of the main players in Alpine biodiversity conservation, we invited 48 institutions including researchers, conservation practitioners, NGOs, policymakers and administrators from the Alpine area to participate in this initiative.

(2) Collating of initial questions. 25 of the invited 48 institutions (52%) generated 484 initial questions. The core team collated and assigned these questions to twelve main topics for clarity (Table 3).

(3) Pre-workshop voting. In order to reduce the set of questions, we distributed the initial-questions-list to all participants for a first evaluation. Rephrasing and adding of questions was actively encouraged at this stage. In preparation for the workshop phase, participants were then asked to select a maximum of 55 questions from the total of 484 questions submitted. Participants were also asked to only select questions from areas in which they had sufficient knowledge base. A total of 385 questions received at least one vote and an additional 15 questions were proposed at this stage.

(4) Workshop. Final discussion and voting took place during a two-day workshop in December 2010 in Switzerland. Unfortunately, only 16 participants were able to attend. The remaining 400 questions assigned to the twelve topic areas (median: 33; range: 24–46/topic area) were discussed in four sessions. In each session, three groups worked simultaneously to reduce and rephrase the questions. In the plenary session the participants decided to reduce the questions to 50 instead of 55 due to overlaps. The final 50 questions were selected in a plenary session by majority vote following discussion.

Table 3. Characteristics and voting systems of the priority setting exercise.

Pre-workshop				Workshop		
Participating countries	Characteristics of participating organisations	Number of participating organisations[1]/ number of individuals involved in generating initial questions	12 main topics to which initial 484 questions were assigned to	Number and characteristics of workshop participants (organisations/individuals)[1]	Pre-workshop voting system	Workshop voting system[1]
Austria, Germany, France, Italy, Liechtenstein, Switzerland	Researcher (n = 12), practitioners (n = 6), NGOs (n = 4), policy makers (n = 2), network (n = 1)	26 organisations/ 109 individuals generating 484 initial questions	Climate change, protected areas, indicators, species biology, natural networks and barriers, spatial development and legal constraints, monitoring and data management, habitat management and land use, participation and communication, tourism and recreation, economics and ecosystem services, multidisciplinary approaches and common strategy	14 organisations research (n = 8), practitioners (n = 4), NGOs (n = 1), policy maker (n = 1)/16 individuals	Votes per participant: 55/484 × 100% questions per topic, resulting in 385 questions plus 15 added questions	Voting system for sessions: 50/400×100% questions per topic as top priority, 2 questions per topic as second priority. Voting system for plenary session: 50 top priority minus the questions (dismissed in discussion) plus top ranked second priority questions, resulting in 50 final questions

[1]including core team.

II) Classification and Analysis of the Questions

The core team ascribed individual questions from the final non-prioritized list of 50 questions, to three interacting context areas according to a slightly modified, previously published, management framework for connectivity conservation: nature-, people- and management contexts. Each context area has 3 interacting sub-topic areas, We subsequently assigned each question to one or more sub-topic area within the individual context areas (Figure 2) [15]. We selected the context area most likely concerned when addressing the question and the context area that would be affected by the outcome. This approach was given preference over linking questions to a context area with respect to the cause of the issue or question. For instance, question 5 "Which indicators reflect the changes in connectivity that result from climate or human induced changes in Alpine landscapes?" was assigned to "Structural and functional connectivity" and linked to "Tools, incentives, knowledge", since the solution ("indicators" = tool, knowledge) provides answers to a particular problem ("changes in connectivity"). It is not a question concerning the causes of this problem ("climate or human induced changes").

All questions were formulated either as (i) what - asking for information about something, (ii) which - asking about choice and (iii) how – questioning manner, condition or quality in order to further breakdown and classify the outcomes.

Finally, we established a simple non-prioritized list that attributes all 50 questions to the three context areas and respective sub-topics (Tables 1 and 2). The majority (54%) of the final 50 questions could be assigned to two context areas due to their transdisciplinary nature. Furthermore, we identified direct dependencies of individual questions on other questions, meaning that a specific question can only be solved if another question is answered beforehand. Finally, we created a graphic conceptualisation showing the interactions, inter-relations and dependencies of the questions.

Acknowledgments

The authors thank the following persons for the submission of questions and comments: Michaela Kuenzl and Anne-Katrin Heinrichs (Nationalpark Berchtesgaden, Germany); Kerstin Lehmann (Bundesamt für Naturschutz, Federal Ministry for the Environment, Nature Conservation and Nuclear Safety, Germany); Kathrin Renner (European Academy of Bozen - EURAC); Daniela Pauli and Marc Gessner (Swiss Biodiversity Forum); Janine Bolliger (Swiss Federal Institute of Technology/Swiss Federal Institute for Forest, Snow and Landscape Research WSL); Christina Kubalek and Natalia Razumovsky (WWF Austria); Gabor von Bethlenfalvy (Federation of Associations for Hunting and Conservation of the EU (FACE)); Anke Hahn (Transeconect); Anne-Sophie Croyal (Conseil general de L'Isere, France); Michael Proschek Hauptmann (Umweltdachverband, Austria); Riccardo Santolini (Univ. of Urbino, Italy); Karl Buchgraber (LFZ Raumberg-Gumpenstein, Austria); Rebecca Drury (FIWI, Austria), Marianne Badura (BLUE!). Renate Hengsberger for editing support.

Author Contributions

Implemented the study: CW C. Kowalczyk YK TS LF. Developed the study design: CW C. Kowalczyk WJS TS. Provided significant editorial inputs: JMA BB C. Kueffer R. Holderegger AR RS. Participated in question generation, question evaluation and workshop discussions: CW C. Kowalczyk JMA BB GB JJB LF MOG R. Haller R. Holderegger YK AR RS AUS SNVB TS. Analyzed the data: C. Kowalczyk CW. Wrote the paper: CW C. Kowalczyk TS.

References

1. Kohler Y, Scheurer T, Ullrich A (2009) Ecological networks in the Alpine Arc. Innovative approaches for safeguarding biodiversity. J Alp Res 97. doi: 10.4000/rga.808.

2. Lindenmayer D, Fischer J (2006) Habitat fragmentation and landscape change: an ecological and conservation synthesis. Melbourne: CSIRO Publishing.

3. Soulé ME, Mackey BG, Recher HF, Williams JE, Woinarski JC, et al. (2006) The role of connectivity in Autralian conservation. In: Crooks KR, Sanjayan M, editors. Conservation biology 14: connectivity conservation: Cambridge University Press pp. 649–675.

4. Keller LF, Waller DM (2002) Inbreeding effects in wild populations. Trends Ecol Evol 17: 230–241. doi: 10.1016/S0169-5347(02)02489-8.

5. Walzer C, Angelini P, Füreder L, Plassmann G, Renner K, et al. (2011) Webs of life - Alpine biodiversity needs ecological connectivity - Results from the ECONNECT project. Milan Grafica Metelliana.

6. Füreder L, Waldner T, Ullrich-Schneider A, Renner K, Streifeneder T, et al. (2011) Econnect - Policy recommendations. Innsbruck: STUDIA Universitätsbuchhandlung und -verlag. 16 p.

7. Beier P, Gregory AJ (2012) Desperately Seeking Stable 50-Year-Old Landscapes with Patches and Long, Wide Corridors. PLoS Biol 10: e1001253. doi: 10.1371/journal.pbio.1001253.

8. Sutherland WJ, Fleishman E, Mascia MB, Pretty J, Rudd MA (2011) Methods for collaboratively identifying research priorities and emerging issues in science and policy. Methods Ecol Evol 2: 238–247. doi: 10.1111/j.2041-210X.2010.00083.x.

9. Milner-Gulland EJ, Fisher M, Browne S, Redford KH, Spencer M, et al. (2010) Do we need to develop a more relevant conservation literature? Oryx 44: 1–2. doi: 10.1017/S0030605309991001.

10. Sutherland BJ, Pullin AS, Dolman PM, Knight TM (2004) The need for evidenced-based conservation. Trends Ecol Evol 19: 305–308. doi: 10.1016/j.tree.2004.03.018.

11. Sutherland WJ, Armstron-Brown S, Armsworth PR, Brereton T, Brickland J, et al. (2006) The identification of 100 ecological questions of high policy relevance in the UK. J Appl Ecol 43: 617–627. doi: 10.1111/j.1365-2664.2006.01188.x.

12. Sutherland WJ, Adams WM, Aronson RB, Aveling RB, Blackburn TM, et al. (2009) One hundred questions of importance to the conservation of global biological diversity. Conserv Biol 23: 557–567. doi: 10.1111/j.1523-1739.2009.01212.x.

13. Sutherland WJ, Albon SD, Allison H, Armstrong-Brown S, Bailey MJ, et al. (2010) The identification of priority policy options for UK nature conservation. J Appl Ecol 47: 955–965. doi: 10.1111/j.1365-2664.2010.01863.x.

14. Braunisch V, Home R, Pellet J, Arlettaz R (2012) Conservation science relevant to action: A research agenda identified and prioritized by practitioners. Biol Conserv 153: 201–210. doi: 10.1016/j.biocon.2012.05.007.

15. Worboys GL, Francis WL, Lockwood M (2010) Connectivity Conservation Management - A global Guide; Worboys GL, Francis WL, Lockwood M, editors. London, Washington DC: Earthscan. 382 p.

16. Bennett G (2004) Linkages in practice, a review of their conservation value. Gland: IUCN Publication Services.

17. Morrison S, Reynolds M (2006) Where to draw the line: integrating feasibility into connectivity planning. In: Crooks KR, Sanjayan M, editors. Conservation biology 14: connectivity conservation: Cambridge University Press pp. 536–554.

18. Cooke SJ, Danylchuk AJ, Kaiser MJ, Rudd MA (2010) Is there a need for a '100 questions exercise' to enhance fisheries and aquatic conservation, policy, management and research? Lessons from a global 100 questions exercise on conservation of biodiversity. J Fish Biol 76: 2261–2286. doi: 10.1111/j.1095-8649.2010.02666.x.

19. Wuelser G, Pohl C, Hirsch Hadorn G (2012) Structuring complexity for tailoring research questions to sustainable development: a framework. Sustain Sci 7: 81–93. doi: 10.1007/s11625-011-0143-3.

20. Fazey I, Fischer J, Lindenmayer DB (2005) What do conservation biologists publish? Biol Conserv 124: 63–73. doi: 10.1016/j.biocon.2005.01.013.

21. Rittel H, Webber M (1973) Dilemmas in a general theory of planning. Policy Sci 4: 155–169.

22. Lazarus RJ (2009) Super Wicked Problems and Climate Change: Restraining the Present to Liberate the Future. Cornell L Rev 94: 77. Available: http://ssrn.com/abstract = 1302623.

23. Bernstein S, Lebow RN, Stein JG, Weber S (2000) God Gave Physics the Easy Problems. Eur J Int Rel 6: 43–76. doi: 10.1177/1354066100006001003.

24. Hirsch Hadorn G, Hoffmann-Riem H, Biber-Klemm S, Grossenbacher W, Joye D, et al. (2008) The emergence of transdisciplinarity as a form of research. In: Hirsch Hadorn G, Hoffmann-Riem H, Biber-Klemm S, Grossenbacher W, Joye D et al., editors. Handbook of transdisciplinary research. Heidelberg: Springer. pp. 19–39.

25. ProClim/CASS (1997) Visions by Swiss researchers. Research on sustainability and global change - visions in science policy by Swiss Researchers. Berne: CASS/SANW. 33 p.

26. Sutherland WJ, Clout M, Cote IM, Daszak P, Depledge MH, et al. (2010) A horizon scan of global conservation issues for 2010. Trends Ecol Evol 25: 1–7. doi: 10.1016/j.tree.2009.10.003.

27. Sutherland WJ, Woodroof HJ (2009) The need for environmental horizon scanning. Trends Ecol Evol 24: 523–527. doi: 10.1016/j.tree.2009.04.008.

28. Arlettaz R, Mathevet R (2010) Dossier « Le réveil du dodo III » - Biodiversity conservation: from research to action. Natures Sciences Sociétés 18: 452–458. doi: 10.1051/nss/2011009.

29. Roux DJ, Rogers KH, Biggs HC, Ashton PJ, Sergeant A (2006) Bridging the science-management divide: Moving from unidirectional knowledge transfer to knowledge interfacing and sharing. Ecol Soc 11: 4 online. Available: http://www.ecologyandsociety.org/vol11/iss1/art4/.

30. Ecological Networks in the European Alps. Available: http://www.alpine-ecological-network.org/.

7

Correlation between Thermodynamic Efficiency and Ecological Cyclicity for Thermodynamic Power Cycles

Astrid Layton[1]*, John Reap[2], Bert Bras[1,4], Marc Weissburg[3,4]

1 George W. Woodruff School of Mechanical Engineering, Sustainable Design and Manufacturing, Georgia Institute of Technology, Atlanta, Georgia, United States of America, 2 School of Business and Engineering, Quinnipiac University, Hamden, Connecticut, United States of America, 3 School of Biology, Georgia Institute of Technology, Atlanta, Georgia, United States of America, 4 Center for Biologically Inspired Design, Georgia Institute of Technology, Atlanta, Georgia, United States of America

Abstract

A sustainable global community requires the successful integration of environment and engineering. In the public and private sectors, designing cyclical ("closed loop") resource networks increasingly appears as a strategy employed to improve resource efficiency and reduce environmental impacts. Patterning industrial networks on ecological ones has been shown to provide significant improvements at multiple levels. Here, we apply the biological metric cyclicity to 28 familiar thermodynamic power cycles of increasing complexity. These cycles, composed of turbines and the like, are scientifically very different from natural ecosystems. Despite this difference, the application results in a positive correlation between the maximum thermal efficiency and the cyclic structure of the cycles. The immediate impact of these findings results in a simple method for comparing cycles to one another, higher cyclicity values pointing to those cycles which have the potential for a higher maximum thermal efficiency. Such a strong correlation has the promise of impacting both natural ecology and engineering thermodynamics and provides a clear motivation to look for more fundamental scientific connections between natural and engineered systems.

Editor: Vishal Shah, Dowling College, United States of America

Funding: This material is based upon work supported by the National Science Foundation under Grant Nos. CMMI-0600243 and 0628190, and CBET-0967536. Any opinions, findings, and conclusions or recommendations expressed in this material are those of the authors and do not necessarily reflect the views of the National Science Foundation. The funders had no role in study design, data collection and analysis, decision to publish, or preparation of the manuscript.

Competing Interests: The authors have declared that no competing interests exist.

* E-mail: alayton6@gatech.edu

Introduction

1.1 Motivation: Ecology and Industrial Networks

A sustainable global community, one that meets the needs of the current generation without sacrificing those of future generations [1] requires the successful integration of environment and engineering. In the public and private sectors, designing cyclical ("closed loop") resource networks increasingly appears as a strategy employed to improve resource efficiency and reduce environmental impacts [2,3]. Multiple structural and material flow metrics that one might use to aid in network design exist [4]. These metrics quantify commonsense imperatives to reduce and reuse, but they contain limited, if any, information about sustainable thresholds. Some metrics even hold the potential to mislead [5]. One approach that can improve the efficient use of resources at multiple levels and simultaneously meet sustainable thresholds involves patterning industrial networks on ecological ones [4,6,7]. Decades ago, the potential for transferring ecological principles to human systems was recognized as a way to increase the efficient use of energy and resources and reduce waste [8]. In 1989 Frosch and Gallopoulos proposed to convert the traditional manufacturing model, one composed of linear industrial chains of activities, to an integrated model they deemed an 'Industrial Ecosystem' [9]. Such a system would use lessons learned from biology to optimize the use of raw materials and energy while minimizing waste through the redefining of effluents as raw material for neighboring processes. Since then, ecological systems have provided analogies for sustainable engineering and industrial systems [4,7], but there have been few attempts to translate core ecological principles into industrial practice (but cf. [10]). Attempts to organize human systems into more ecologically-realistic patterns continue to be based on the "waste equals food" concept (but cf. [11]) where the output of a given system component (e.g. industry) provides the input for another. While better than previous models, this type of organization does not accurately reproduce the connections patterns of ecosystems where full benefits from the analogy could be realized [6]. In this paper we explore if there are similar advantages for thermodynamic networks.

"To be ultimately sustainable, biological ecosystems have evolved over the long term to be almost completely cyclical in nature, with 'resources' and 'waste' being undefined, since waste to one component of the system represents resources to another." – Jelinski, et al. [12]

In 1969, Odum recognized that ecological systems, particularly mature ones, are associated with a high degree of internal recycling of energy and materials, such that the amount of new inputs into the system is small compared to what is transformed among the system components [8]. Human systems in contrast (e.g. agricultural ones) are geared for production rather than efficiency, resembling young rather than mature natural systems. Odum has suggested mimicking mature systems would help shift the focus of human systems from production to efficiency. One desirable property of mature systems is a complex food-web structure; a proliferation of connections between species that exchange material and energy by consuming one another [13]. The extent to which principles derived from ecological systems may be applied in other contexts is unclear. If we can connect the structural properties of ecological networks to well understood physical principles, such as the Laws of Thermodynamics, we might gain sufficient insight to apply ecological lessons to the engineering and development of resource networks [9].

1.2 Cyclicity and Thermodynamic Cycles

In this paper we use 28 familiar thermodynamic power cycles of increasing complexity to explore trends in network structure defined by the ecological metric cyclicity [13,14]. Cyclicity is an older metric reintroduced by Fath and Halnes that measures the presence of cyclic (closed loops as opposed to linear) pathways in a system [13]. Unlike the cycling index (CI), a similar metric which also quantifies the amount of cycling in the system, cyclicity needs no knowledge of flow magnitude, only flow path [8,15]. Flow magnitude information can be quite complex, if not impossible, to acquire thus cyclicity greatly increases the usefulness and simplicity of the metric as. Cyclicity, which represents what is also known as "strongly connected components" in ecology and graph theory, "refers to the subset of species for which energy can flow from one another and back" [16]. The connections in a system between species, or 'actors,' are organized in a matrix form, from which the systems 'cyclicity' is calculated. The higher the cyclicity of the system the more interconnected its components. High cyclicity values relate strongly to the overall proportion of the energy retained vs. that which is lost by the system, which may lead to more robust and efficient engineered systems. Fath and Halnes calculated the cyclicity of a number of ecosystems and saw values ranging from 1.62 for the Coachella Valley ecosystem (made up of 30 actors) in Southern California to 14.17 for a mangrove ecosystem with 94 actors [14]. Our results point to the maximum thermal efficiency increasing with cyclicity. So it appears that thermal efficiency, a result of the First Law of Thermodynamics, correlates to a very high degree with an ecological metric based solely on the construction of the system.

Ideal Rankine and Brayton cycles composed the 28 power cycles used. The ideal Brayton cycle is used to model the gas turbine engine and the ideal Rankine cycle is the simplest representation of the vapor power cycles utilized by the electric power generating industry. The inclusions of feedwater heaters, regeneration, reheating and intercooling are all standard ways of increasing the thermal efficiency of the Rankine and Brayton cycles [17]. All of these changes increase the number of times the initial energy in the system is cycled, so it may be reused to reduce the potential heat or work lost and required, thereby decreasing the dependence on outside power. This seems to align with the circuitous structure of food webs favored by nature. As cyclicity is a measure of the existence and *strength* of this internal structural cycling of energy [13,14,16] we test if cyclicity can also be used as a measurement tool in thermodynamic power systems, while we explore potential associations with both traditional measures of efficiency and the structure of engineered systems.

Methods

2.1 Conversion to Energy Flow Networks

To uncover the internal cycling present in the system we must first use the network approach in thermodynamics to construct a graphical model revealing system topology, referred to here as an energy flow network [18]. In this approach mechanical components are considered 'nodes' in the network representing the power cycle (a node is a system component that receives and-or transmits energy). Connections between nodes occur when energy embodied in the working fluid as well as internal exchanges of work and heat flow from one node to another. Work and heat entering the cycle from outside are not considered. We analyzed 20 standard variations on the ideal Rankine cycle and 8 standard variations on the ideal Brayton cycle. Only one of the ideal cycles is covered here in detail as the procedure was the same for all cycles used. Figure 1b recasts the familiar equipment diagram of an ideal Rankine cycle with one open feedwater heater, seen in Figure 1a, as a set of nodes joined by energy exchanges. Starting in the lower left corner of Figure 1a, one sees that energy, in the form of shaft work, at Pump 1 enters the system raising the energetic state of the working fluid above that found at State 1 (the reference state for this energy flow network), this translates into the link between node 1 and node 2 in Figure 1b. Energy carried by the working fluid flows to the open feedwater heater where it combines with another energy flow in the form of steam bled from the turbine. The network continues the transferring, adding and subtracting of energy as the working fluid moves between ideal components. With the power cycles recast as energy flow networks, we need only to write the structural adjacency matrix and compute its maximum real eigenvalue to determine cyclicity for each cycle.

2.2 Structural Matrix

A structural adjacency matrix ($\mathbf{A}$), analogous to a connectivity matrix [14], is concerned only with the structural information (links and nodes) of a network and defines the pathways that exist by which material and energy flows from one compartment to another. It is blind to information such as flow rate, quality, and the type of working fluid. A link exists as long as some physical quantity directly joins two nodes (mapped by Figure 1b). The adjacency matrix captures flow *direction*. Row space contains information about flow *to* a node, the 'predator' in nature, while column space contains information about flow *from* a node, the 'prey' in nature.

The adjacency matrix in Figure 2a is a structural depiction of the network in Figure 1. The matrix is a binary representation of the connections in the system such that $a_{ij} = 1$ if there is a connection from j to i, and is zero otherwise [14]. For example, energy flowing from Node 1 to Node 2 in Figure 1b is documented by placing a value of 1 in the second row of the first column in the matrix $\mathbf{A}$ of Figure 2a. Flow from Node 2 to Node 3 is indicated by a 1 at [3,2] and so on.

2.3 Maximum Eigenvalue

With the power cycles now in matrix form, cyclicity is found by calculating the maximum real eigenvalue (λ_{max}) for each corresponding adjacency matrix. The eigenvalues of a matrix are mathematically defined as the solutions to equation 1: the determinant of the quantity of the matrix in question minus the eigenvalues times the identity matrix of the equivalent size, all equal to zero. The result of equation 1 is a set of eigenvalues (which may be both real and imaginary); MATLAB's "*eigs*" function was used to execute this task (MATLAB R2011b, Atlanta,

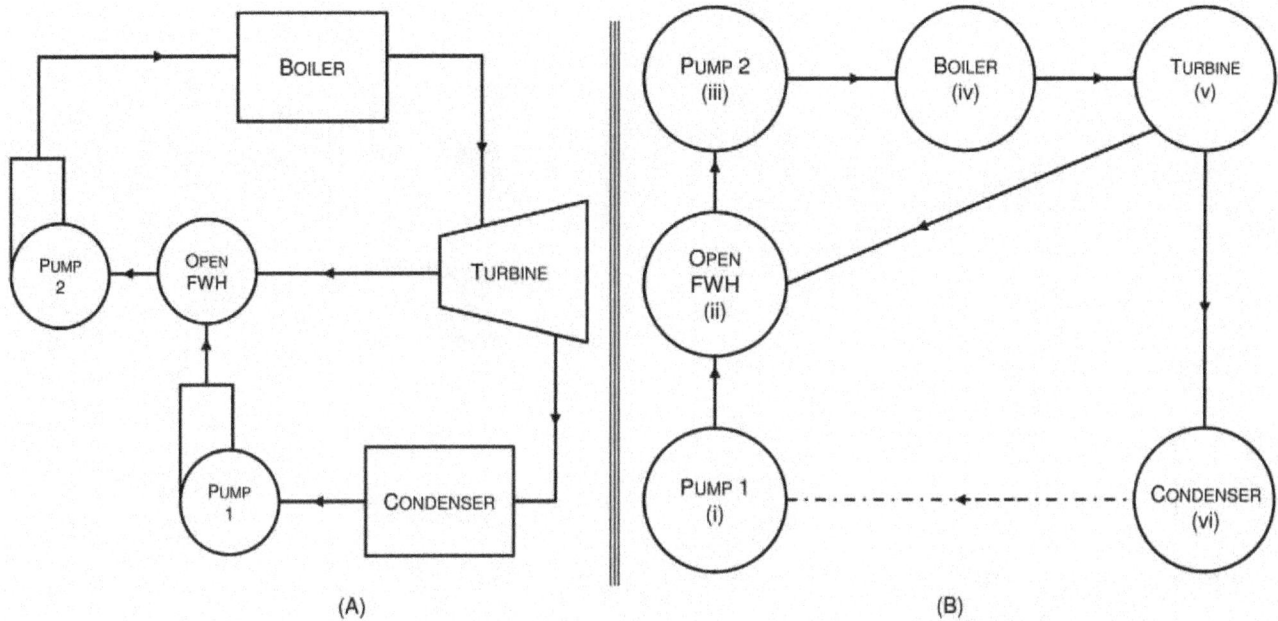

Figure 1. Ideal Rankine power cycle with one open feed water heater redrawn as energy flow networks following thermodynamic network theory [26]. Note that the link between the condenser (Node vi) and Pump 1 (Node i) is not a physical flow of energy. Since State 1 acts as an energetic reference state for the network, working fluid returning to that reference state only closes the *material* loop; energy embodied in the working fluid leaving the condenser is rejected to the surroundings.) *(a) Energy, in the form of heat and work and carried by the working fluid, flows to and from the mechanical components of the idealized equipment diagram for a power cycle. (b) The system is simplified with the mechanical components modeled as 'nodes' connected by flows of energy in the energy flow diagram.*

Georgia). The maximum real eigenvalue in this set is the cyclicity of matrix **A**, as shown by Borrett et al. [19]. λ_{max} is a measure of the proliferation of pathways that connect two nodes in a network. There is a greater potential for flows to remain within the system as pathways proliferate, λ_{max} is indicative of the resulting internal cycling [17]. The Rankine cycle seen in Figure 1 and represented by the matrix in Figure 2a results in a cyclicity value of 1 ($\lambda_{max} = 1$) as seen in Figure 2d.

$$\det(A - \lambda I) = 0 \qquad (1)$$

Cyclicity can be either 0, 1 or greater than 1. This is illustrated in Figure 3, which is based on the similar figure by Fath and

Halnes [14,20]. Zero cyclicity indicates that no internal cycles are present, Figure 3a. Therefore energy traveling through the system never passes through a component twice. A value of one indicates 'weak cycling,' meaning only simple closed loop pathways exist, Figure 3b. Values of greater than one indicate that the system is made up primarily of complex looped pathways, Figure 3c, the larger the cyclicity the more complex and numerous the paths are between components.

The proof presented by Borrett et al. (2007) for the use of eigenvalues to determine the cyclicity (what Borrett et al. call "pathway proliferation rate") of a system combines results from graph theory and linear algebra [19]. The proof uses the Perron-Frobenius theorem, which guarantees that there is only one real eigenvalue that is greater than or equal to all other eigenvalues

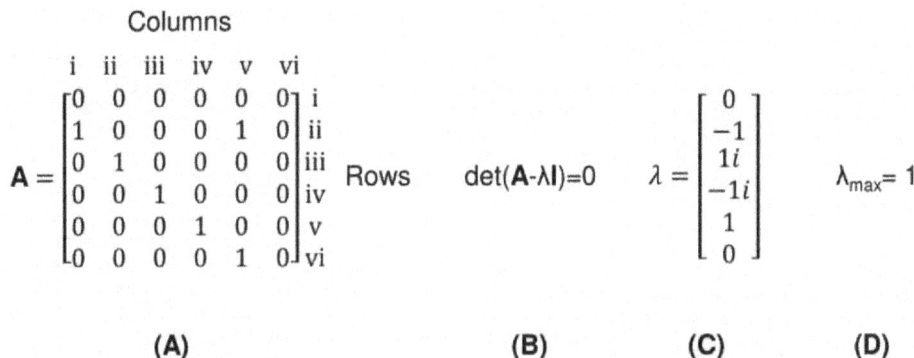

Figure 2. The process for calculating the cyclicity of the 6 component Rankine cycle from Figure 1. (a) Labeled adjacency matrix for the ideal Rankine cycle with one open feed water heater – rows represent flow *to* a node, columns *from* a node. (b) Equation for the calculation of the eigenvalues for the adjacency matrix. (c) Eigenvalues. (d) Maximum real eigenvalue, or the cyclicity, of the cycle.

(A)

$$A = \begin{bmatrix} 0 & 0 & 0 \\ 1 & 0 & 0 \\ 0 & 1 & 0 \end{bmatrix} \qquad \begin{aligned} |A - \lambda I| &= 0 \\ \lambda_{max} &= 0 \end{aligned}$$

(B)

$$A = \begin{bmatrix} 0 & 0 & 1 \\ 1 & 0 & 0 \\ 0 & 1 & 0 \end{bmatrix} \qquad \begin{aligned} |A - \lambda I| &= 0 \\ \lambda_{max} &= 1 \end{aligned}$$

(C)

$$A = \begin{bmatrix} 0 & 0 & 1 \\ 1 & 0 & 0 \\ 1 & 1 & 0 \end{bmatrix} \qquad \begin{aligned} |A - \lambda I| &= 0 \\ \lambda_{max} &= 1.32 \end{aligned}$$

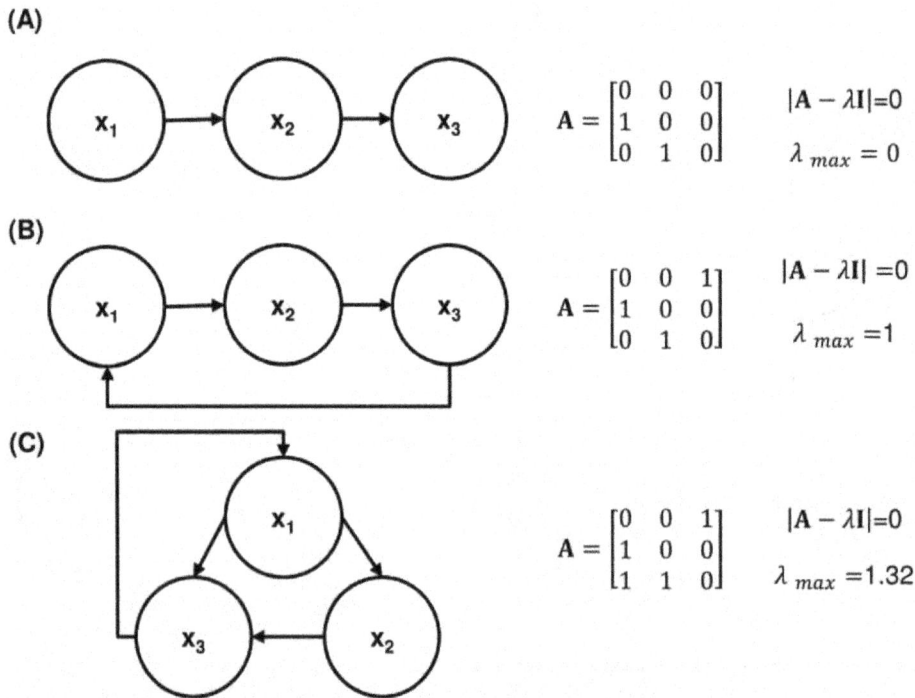

Figure 3. Examples of the three types of internal structural cycling based on cyclicity (eigenvalues). (a) No cycling $\lambda_{max} = 0$, (b) weak cycling $\lambda_{max} = 1$, (c) and strong cycling $\lambda_{max} > 1$ [10].

($\lambda_1 \geq \lambda_i$ for $i = 2...n$) in adjacency matrices associated with a network where it is possible to reach every node from every other node (i.e. a strongly connected network) [19]. In strongly connected networks only the maximum (dominant) eigenvalue is left to represent the pathway proliferation rate of the system as the limit of the number of indirect links (pathways between two nodes which consist of more than one link) goes to infinity. Disconnected networks, those networks which have no internal cycling, will have a cyclicity value of zero. Weakly connected networks, those which have cycles made up of one link (self-loops) or have cycling only if link-direction is ignored, may have a maximum eigenvalue of either 1 or 0. [19] Most food webs are composed of networks where large subsets of "nodes" are strongly connected such that the dominant eigenvalue is greater than one, indicating the existence of multiple cyclic pathways.

2.4 Thermal Efficiency

All thermal efficiencies (η_I in equation 2) and pertinent state point data were calculated using Engineering Equation Solver (EES) version V8.881-3D. The maximum and minimum cycle temperatures and pressures or pressure ratios were kept constant throughout the modified cycles for consistency, as described in Table 1. Extraction pressures for the feedwater heaters were chosen on a per cycle basis to maximize the thermal efficiency of each cycle. The work and heat externally supplied to the power cycle, W_{in} and Q_{in} respectively, and the work produced by the power cycle, W_{out}, were calculated based upon enthalpies (h) at pertinent inlet and exit points (outlined by equations 3–5). For more information on calculating work, heat, and the thermal efficiency for thermodynamic power cycles please see a thermodynamic reference book such as Sonntag, Borgnakke, and van Wylen's *Fundamentals of Thermodynamics* [17].

$$\eta_I = \frac{\sum_i (W_{out,\,i} + W_{in,\,i})}{\sum_i (Q_{in,\,i})} \qquad (2)$$

$$W_{in,\,i} = (h_{exit} - h_{inlet})_{compressor,\,pump} \qquad (3)$$

$$W_{out,\,i} = (h_{exit} - h_{inlet})_{turbine} \qquad (4)$$

$$Q_{in,\,i} = (h_{exit} - h_{inlet})_{boiler,\,combustor} \qquad (5)$$

Results

Analysis of 28 variations on the ideal Brayton and Rankine cycles shows a positive correlation between cyclicity and the maximum thermal efficiency. The compiled values for cyclicity

Table 1. Specified state point data for all ideal Rankine and Brayton cycle analyses.

Rankine Cycles - water	Brayton Cycles - air
$T_{min} = 318.9$ K	$T_{min} = 288.2$ K
$T_{max} = 873.2$ K	$T_{max} = 1273$ K
$P_{pump1,\,input} = 10$ kPa	$P_{compresser,\,input} = 100$ kPa
$P_{boiler,\,input} = 15000$ kPa	$r_p = 10$ (pressure ratio)

Table 2. Thermal efficiency and cyclicity values for 20 (R1–R20) Ideal Rankine power cycles evaluated under the same conditions.

Cycle	Thermal Efficiency (η_I)	Cyclicity (λ_{max})
(R1) Basic Rankine	0.430	0
(R2) Rankine with reheat	0.451	1
(R3) Rankine with 1 closed FWH trapped condensate	0.453	1
(R5) Rankine with 1 open FWH	0.463	1
(R6) Rankine with 2 open FWHs	0.472	1.15
(R8) Rankine with 1 closed FWH pumped condensate	0.453	1.17
(R7) Rankine with 3 open FWHs	0.476	1.21
(R9) Rankine with 1 open and 1closed FWH	0.476	1.30
(R10) Rankine with 4 open FWHs	0.479	1.24
(R11) Rankine with 5 open FWHs	0.480	1.25
(R12) Rankine with 6 open FWHs	0.482	1.26
(R13) Rankine with 7 open FWHs	0.482	1.27
(R14) Rankine with 8 open FWHs	0.483	1.27
(R15) Rankine with reheat and 1 open FWH	0.470	1.27
(R16) Rankine with reheat and 2 open FWH	0.483	1.33
(R17) Rankine with reheat and 3 open FWH	0.488	1.43
(R18) Rankine with reheat and 4 open FWH	0.491	1.44
(R19) Rankine with reheat and 5 open FWH	0.492	1.45
(R20) Rankine with reheat and 6 open FWH	0.493	1.45

*FWH, feed water heater.

and thermal efficiency, as well as the specific modifications made to the Brayton and Rankine cycles can be found in Tables 2 and 3. Supporting Figures S1–S6 offer additional insights into the modifications made. The results of these two tables are displayed in Figure 3. The Brayton cycle, by design, gives higher thermal efficiencies than the Rankine cycle, and modifications to the Brayton cycle produce a much larger increase in thermal efficiency than for the Rankine cycle; the addition of one extra component in each (reheat in the Rankine cycle, **R2** in Table 2, and regeneration in the Brayton cycle, **B2** in Table 3) results in a 16.8% increase in thermal efficiency for the Brayton cycle but only

a 4.7% increase for the Rankine cycle. Both are desirable, even a small increase in efficiency in practice is highly sought after.

The vapor power cycles utilized for the generation of 90% of all electric power used throughout the world are modeled by the Rankine cycle [21,22]. The Brayton cycle is used to model the gas turbine engine. The theoretical upper bound for the efficiency of these and any other real or ideal heat engines is the Carnot efficiency, equation 6. The Carnot efficiency represents the maximum possible work that may be done between any two temperatures and is independent of the working substance used or any particular design feature of the engine. One could continue to increase the number of links added thereby increasing the cyclicity; however, the Carnot efficiency (η_C) will not be reached. The Carnot efficiency, although physically unattainable, is useful in that it gives us an upper limit to strive for. If the efficiency of a real engine is significantly lower, then additional improvements may be possible. More information on efficiencies and power cycles can be found in any thermodynamic reference book, for example *Fundamentals of Thermodynamics* by Sonntag, Borgnakke, and van Wylen [17]. The Carnot efficiency for the Rankine and Brayton cycles analyzed are 0.635 and 0.774 respectively. We will specify all thermal efficiencies as either maximum Rankine or Brayton cycle efficiencies or Carnot efficiency. The Carnot efficiency creates a ceiling which will lead to a logarithmic-type relationship relating cyclicity to the maximum thermal efficiency if infinite data points were used. Modifications made to real world systems, which must deal with irreversabilities (also known as losses, such as friction), will eventually become cost ineffective in that the addition of feedwater heaters, regeneration, reheating and intercooling will no longer increase cycle efficiency, for example once 8 feedwater heaters are in place in a Rankine cycle [23].

$$\eta_C = 1 - \frac{T_{min}}{T_{max}} \qquad (6)$$

There is a clear lack of data points between the values of zero and one for cyclicity in the Rankine cycles due to the nature of cyclicity being zero, 1, or greater than 1. This constraint makes it impossible to drastically increase the R^2 value, or coefficient of determination, by obtaining data between the cyclicity values of zero and 1. Including all cycle points (Figure 4) R^2 values for the linear trend lines are 0.988 and 0.768 for Brayton and Rankine cycles respectively. The R^2 value, for the Rankine cycle increases to 0.818 if we focus on those cycles which are greater than or equal to one (the Brayton

Table 3. Thermal Efficiency And Cyclicity Values 8 (B1–B8) Ideal Brayton Power Cycles Evaluated Under The Same Conditions [27].

Cycle	Thermal Efficiency (η_I)	Cyclicity (λ_{max})
(B1) Basic Brayton	0.482	1.00
(B2) Brayton with Regeneration	0.563	1.22
(B3) Brayton with regeneration, intercooling, and reheat (2 turbines)	0.685	1.39
(B4) Brayton with regeneration, intercooling, and reheat (3 turbines)	0.718	1.46
(B5) Brayton with regeneration, intercooling, and reheat (4 turbines)	0.733	1.50
(B6) Brayton with regeneration, intercooling, and reheat (5 turbines)	0.742	1.52
(B7) Brayton with regeneration, intercooling, and reheat (6 turbines)	0.748	1.53
(B8) Brayton with regeneration, intercooling, and reheat (7 turbines)	0.751	1.54

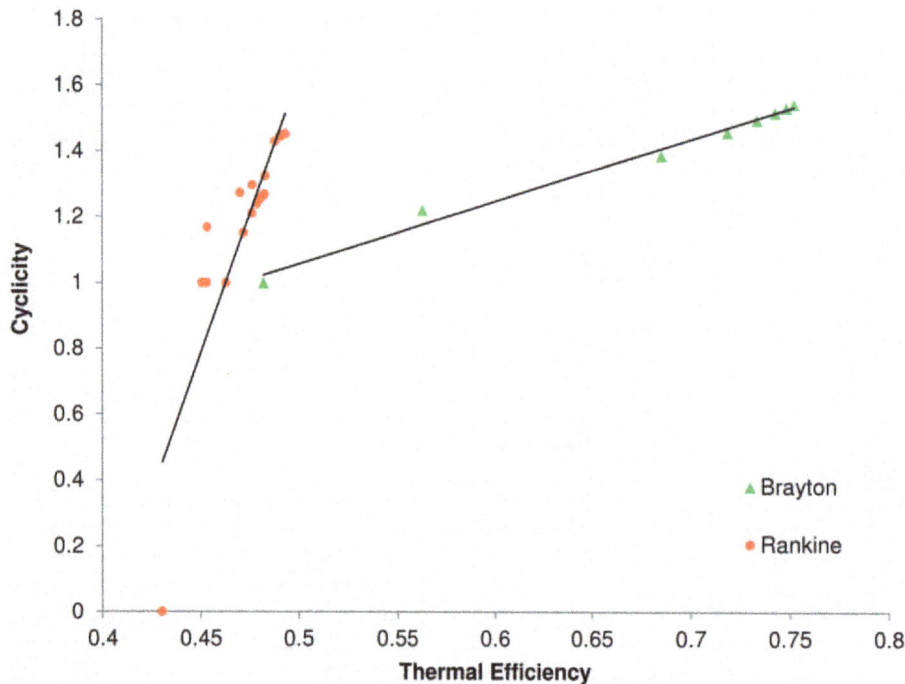

Figure 4. Maximum Thermal Efficiency vs. Cyclicity for all 28 Power Cycles with linear trend lines. Note: All cycles described here are ideal and optimized for maximum thermal efficiency; changes in kinetic and potential energy from one point to another have been neglected as well as losses in connections between components, such as friction losses in pipes, turbulence, and flow separation.

cycles all contain some amount of internal structural cycling and therefore are unaffected by this refocusing).

Discussion

We conclude from this analysis that the structural method for computing energy cyclicity accurately predicts maximum thermal efficiency for both Rankine and Brayton power cycles. The correlation between cyclicity and maximal thermal efficiency ranges from 0.88 to 0.99, suggesting an extremely strong relationship between these two measures of efficiency. This suggests that increasing the cyclicity (a biological metric) in energetic networks is associated with, or perhaps partially driven by, the maximization of thermodynamic work (an engineering 'metric'). Alternate power cycle models should be analyzed to further validate the positive relationship between cyclicity and maximum thermal efficiency. From an immediately practical perspective, the benefit of verifying this connection is in determining the relative potential efficiencies of the power cycles. When comparing two modifications to the same cycle it is a great deal easier to calculate cyclicity than to carry out a complete thermodynamic analysis. If cycle A has a higher cyclicity than cycle B, the correlation found here would lead the investigator to believe that cycle A has the potential for a higher maximum thermal efficiency. Establishing this correlation, we can now take advantage of the ecological strategies that we know increase cyclicity, use analogous solutions in human problems, and investigate the extent to which current solutions employing such principles function more effectively.

Our analysis also suggests the two power cycles differ in the extent to which each may be improved by changing the connectivity of its components. The efficiency of the Brayton cycle is extremely sensitive to how interconnected its components are with respect to the transfer of energy. The linear

trend lines and coefficients of determination in Figure 4 reveal that less than 2% of the thermal efficiency of a Brayton cycles depends on things other than the internal structural cycling of energy. The thermal efficiency for a Rankine cycle is somewhat less affected by its structural cyclicity, leaving about 23% of the efficiency to depend on other factors. This too may be an area for further study to help clarify the connection for use in engineering design.

Nature's networks and mankind's power cycles must both obey the Laws of Thermodynamics, but connecting the two often proves less than straightforward. Although it is well appreciated that thermodynamic constraints affect energy flow in ecological systems [16], ecological systems have been challenging to explain using equilibrium thermodynamic methods. To alleviate this problem, a non-equilibrium perspective is currently in use. This perspective emphasizes the capacity of such complex systems to dissipate energy internally such that they are able to maintain their organization in a physical gradient [24,25]. Systems with greater structural complexity (such as more mature ecosystems) cycle more energy internally and are associated with stronger physical gradients [24,25]. Examining power cycles allows us to test the correlation between non-equilibrium and equilibrium thermodynamic measures by computing both cyclicity and thermodynamic efficiency in the same system. The compatibility of both equilibrium and non-equilibrium approaches is shown by the observation that greater cyclicity produced via structural complexity is associated with increases in thermodynamic efficiency.

Finally, our results also suggest additional structural parallels between efficient human vs. natural systems, aside from relationships between structural complexity (number of links) and efficiency. Odum, in his paper *The strategy of ecosystem development* in 1969, observed that the cycling of energy in food webs increases with system maturity, with the bulk of the biological energy flow following detritus pathways [8]. He cites for example a mature forest, where less than 10% of the annual net production is consumed (by grazing) in a living state, most is used as dead matter (detritus) through delayed and complex pathways. Detrital pathways, particularly in mature forests, are composed of low quality energy inputs since the dominant plant biota contain large amounts of relatively refractory structural material. The additional linkages in the modified Brayton and Rankine cycles (**R2–R20** and **B2–B8**) are put in place to increase thermodynamic efficiency. The added linkages cycle low quality energy (energy entering the system at node 1 is of the highest quality and energy leaving the system is of the lowest quality) through the system, energy which would otherwise be discarded (**R1** and **B1**).

New possibilities and questions appear in the field of industrial ecology and power systems design if the link between cyclicity and thermodynamic efficiency withstands further analysis. Maximization of system work becomes an important goal when aiming to base closed loop industrial systems on ecological ones. One may ask, what is system work in a natural ecosystem? What is the analogy between the average heat input temperature of a thermodynamic power cycle and measurable quantities in an ecosystem? Although answering these answers may or may not yield better system designs, it is doubtful that one would ask the questions were it not for an apparent maximum thermal efficiency-cyclicity correlation. Other analyses will most likely continue to show the importance of cyclical connections to the efficient use and production of energy and matter. Additional cycles, including and beyond thermodynamic ones, should be investigated to broaden the positive relationship seen here to one between any network structure and its efficiency. As the resources that current systems are based on continue to diminish, engineering can only benefit from a greater theoretical structure establishing biology and nature as a source of principles, inspiration and guidance.

Supporting Information

Figure S1 Basic Rankine cycle idealized equipment diagram for a power cycle (a), energy flow diagram (b).

Figure S2 Rankine cycle with one open feed water heater idealized equipment diagram for a power cycle (a), energy flow diagram (b).

Figure S3 Rankine cycle with two open feed water heaters idealized equipment diagram for a power cycle (a), energy flow diagram (b).

Figure S4 Basic Brayton cycle idealized equipment diagram for a power cycle (a), energy flow diagram (b).

Figure S5 Brayton cycle with regeneration (i.e. counterflow heat exchanger) idealized equipment diagram for a power cycle (a), energy flow diagram (b).

Figure S6 Brayton cycle with regeneration (i.e. counterflow heat exchanger), intercooling, and reheat (2 turbines) idealized equipment diagram for a power cycle (a), energy flow diagram (b).

Acknowledgments

We thank S.M. Ghiaasiaan, S.M. Jeter, and S.R. Borrett for discussions.

Author Contributions

Conceived and designed the experiments: AL JR. Performed the experiments: AL. Analyzed the data: AL. Contributed reagents/materials/analysis tools: AL. Wrote the paper: AL JR BB MW. Manuscript reviews, revisions, and discussion: BB MW.

References

1. Brundtland GH (1987) Our Common Future, Report Of The United Nations World Commission On Environment And Development. Oxford: Oxford University Press.
2. Ehrenfeld J, Gertler N (1997) Industrial Ecology in Practice: The Evolution of Interdependence at Kalundborg. Journal of Industrial Ecology 1: 67–79.
3. EU (2003) Directive 2002/96/EC of the European Parliament and of the Council of 27 January 2003 on Waste Electrical and Electronic Equipment. 13.2.2003 ed: Official Journal of the European Communities. pp. L 37/24–38.
4. Graedel TE, Allenby BR (1995) Industrial Ecology. Englewood Cliffs: Prentice Hall.
5. Naish J (2008) Lies…Damned Lies…and Green Lies. The Ecologist 38: 36–39.
6. Reap JJ (2009) Holistic Biomimicry: A Biologically Inspired Approach to Environmentally Benign Engineering. Atlanta: Georgia Institute of Technology.
7. Frosch RA (1992) Industrial Ecology: A Philosophical Introduction. Proceedings of the National Academy of Sciences of the United States of America 89: 800–803.
8. Odum EP (1969) The Strategy of Ecosystem Development. Science 164: 262–270.
9. Frosch RA, Gallopoulos NE (1989) Strategies for Manufacturing. Scientific American. 144–152.
10. Garmestani AS, Allen CR, Mittelstaedt JD, Stow CA, Ward WA (2006) Firm size diversity, functional richness, and resilience. Environmental and Development Economics 11: 533–551.
11. Hardy C, Graedel TE (2002) Industrial Ecosystems as Food Webs. Journal of Industrial Ecology 6: 29–38.
12. Jelinski LW, Graedel TE, Laudise RA, McCall DW, Patel CKN (1992) Industrial Ecology: Concepts and Approaches. Proceedings of the National Academy of Sciences of the United States of America 89: 793–797.
13. Fath BD (2007) Structural food web regimes. Ecological Modelling 208: 391–394.
14. Fath BD, Halnes G (2007) Cyclic energy pathways in ecological food webs. Ecological Modelling 208: 17–24.
15. Finn JT (1976) Measures of ecosystem structure and function derived from analysis of flows. Journal of Theoretical Biology 56: 363–380.
16. Allesina S, Bodini A, Bondavalli C (2005) Ecological subsystems via graph theory: the role of strongly connected components. Oikos 110: 164–176.
17. Sonntag RE, Borgnakke C, van Wylen GJ (2003) Fundamentals of Thermodynamics. Wiley. 816 p.
18. Oster G, Perelson A, Katchalsky A (1971) Network Thermodynamics. Nature 234: 393–399.
19. Borrett SR, Fath BD, Patten BC (2007) Functional integration of ecological networks through pathway proliferation. Journal of Theoretical Biology 245: 98–111.
20. Fath BD (1998) Network analysis: foundations, extensions, and applications of a systems theory of the environment. Athens: University of Georgia. 176 p.
21. Wiser WH (2000) Energy Resources: Occurrence, Production, Conversion, Use. New York: Springer-Verlag.
22. Muller F, Leupelt M (1998) Eco targets, goal function and orientors. Berlin: Springer-Verlag. 623 p.
23. Kadem L (2007) Vapor and combined power cycles. MECH 351: Thermodynamics II. Algiers: Universite Des Sciences et de la Technologie Houari Boumedienne. 6.
24. Schneider ED, Kay JJ (1994) Life as a manifestation of the Second Law of Thermodynamics. Mathermatical and Computer Modelling 19: 25–48.
25. Ho MW (1998) The Rainbow and the Worm. Singapore: World Scientific.
26. Lewis ER (1995) Network thermodynamics revisited. BioSystems 34: 47–63.
27. Brokowiski ME (1998) Improving an air-standard power cycle. In: University N, editor.

8

The Dynamics of Nestedness Predicts the Evolution of Industrial Ecosystems

Sebastián Bustos[1,2], Charles Gomez[3], Ricardo Hausmann[1,2,4], César A. Hidalgo[1,5,6]*

1 Center for International Development, Harvard University, Cambridge, Massachusetts, United States of America, **2** Harvard Kennedy School, Harvard University, Cambridge, Massachusetts, United States of America, **3** Graduate School of Education, Stanford University, Stanford, California, United States of America, **4** Santa Fe Institute, Santa Fe, New Mexico, United States of America, **5** The MIT Media Lab, Massachusetts Institute of Technology, Cambridge, Massachusetts, United States of America, **6** Instituto de Sistemas Complejos de Valparaíso, Valparaíso, Chile

Abstract

In economic systems, the mix of products that countries make or export has been shown to be a strong leading indicator of economic growth. Hence, methods to characterize and predict the structure of the network connecting countries to the products that they export are relevant for understanding the dynamics of economic development. Here we study the presence and absence of industries in international and domestic economies and show that these networks are significantly nested. This means that the less filled rows and columns of these networks' adjacency matrices tend to be subsets of the fuller rows and columns. Moreover, we show that their nestedness remains constant over time and that it is sustained by both, a bias for industries that deviate from the networks' nestedness to disappear, and a bias for the industries that are missing according to nestedness to appear. This makes the appearance and disappearance of individual industries in each location predictable. We interpret the high level of nestedness observed in these networks in the context of the neutral model of development introduced by Hidalgo and Hausmann (2009). We show that the model can reproduce the high level of nestedness observed in these networks only when we assume a high level of heterogeneity in the distribution of capabilities available in countries and required by products. In the context of the neutral model, this implies that the high level of nestedness observed in these economic networks emerges as a combination of both, the complementarity of inputs and heterogeneity in the number of capabilities available in countries and required by products. The stability of nestedness in industrial ecosystems, and the predictability implied by it, demonstrates the importance of the study of network properties in the evolution of economic networks.

Editor: Luís A. Nunes Amaral, Northwestern University, United States of America

Funding: The authors acknowledge the generous support of Alejandro Santo Domingo, Asahi Broadcast Corporation Chair, Media Lab consortium, MPower Foundation and Standard Bank. The funders had no role in study design, data collection and analysis, decision to publish, or preparation of the manuscript.

Competing Interests: The authors received commercial funding through the Asahi Broadcast Corporation Career Development Chair, Standard Bank and Santo Domingo Foundation.

* E-mail: hidalgo@mit.edu

Introduction

One of the best-documented findings of biogeography is that rare species inhabit predominantly diverse patches, while ubiquitous species tend to inhabit both, diverse and non-diverse locations [1–4]. In ecology, the term *nestedness* is used to refer to this feature, which has been observed numerous times in geographic patterns [1–4] and mutualistic networks [5–8]. In the case of mutualistic networks, nestedness implies that ecosystems are composed of a core set of interactions to which the rest of the community is attached [5]. The nestedness of interaction networks also implies that specialist species interact mostly with generalist species, and because generalist are less fluctuating [9], nestedness can help enhance the survival of rare species [10]. Nestedness has also been shown to enhance biodiversity [11] and overall ecosystem stability [12], and therefore, it is considered an important structural property of interaction networks in ecology.

Nestedness, however, is a general network measure that can be used to characterize non-biological ecosystems, such as global and local economies. In fact, in the past, the nestedness of economic systems has been described for interaction networks, connecting industries to other industries, such as the input-output matrices introduced half a century ago by Leontief [13], or the supply relationships in the New York Garment industry [14,15].

Here, we study the dynamics of economic geographic, instead. We look at the presence and absences of industries across a wide range of locations and show that (i) nestedness tends to remain stable; (ii) it can be used to predict the location of industrial appearances and disappearances; and (iii) can be accounted for by a simple model.

In recent years, the structure of industry-location networks has received a wide range of attention. A country's level of income is tightly connected to the mix of products that they export [16–18], as measured by their Economic Complexity Index or ECI [16,17]. The ECI is a structural measure of the network connecting countries to the products that they export that estimates the amount of productive knowledge embedded in a country [16] from information on who exports what. Countries that have an income that is lower than what would be expected from their ECI, such as China, India and Thailand, tend to grow faster than those that have an income that exceeds what would be expected from their current level of economic complexity, such as Greece and

Portugal [16,17]. Hence, what countries export, as proxied by the ECI, is a strong leading indicator of economic growth.

In the past, the network connecting countries to the products that they export has been used to identify related varieties [19–21]. Here, products that tend to collocated, or co-exported, are connected with a strength that grows with the probability of co-export. Colocation networks, like the product space [20], have been used to show that the productive structure of countries, and regions, evolve as these move from the products that they do to others that are close by in this network. The use of colocation data provides an alternative to more data intensive methods, such as networks connecting industries based on labor flows, labor similarities [22] or plant level data [23]. This is because labor and plant level data lacks standardized international coverage and therefore cannot be used for international comparisons.

The evolution of a country's product mix, however, is highly path dependent [16,20]. Here, we look at the nestedness of the industry location network and show that deviations from nestedness can help predict these path dependencies for both, industrial appearances and disappearances. These predictions add to our ability to explain the evolution of a country's product mix, and therefore, variations in cross-country levels of income. Moreover, we show that the high level of nestedness observed in the data can be reproduced using a simple model when we assume that the heterogeneity of capabilities available in a country, or required by a product, is large.

The paper is structured as follows. First, we study the nestedness of the industry locations matrix and find it to be highly stable over time. We do this by using Almeida-Neto et al's NODF [24,25] (and Atmar and Patterson's Temperature metric [26,27] in the SM). We asses the stability of nestedness by comparing it with both, static and dynamic null models, showing that the observed level, and stability of the network's nestedness, is larger than what would be implied by these null models.

Next, we show that deviations from nestedness are associated, respectively, with increases and decreases in the probability that an industry will appear or disappear at a given location. Finally, to provide an explanation of the observed phenomena we generalize the model recently introduced by Hidalgo and Hausmann [16,28] to show that this model can account for both, the high level of nestedness values, and their stability.

Together, these results illustrate the relevance of nestedness for the evolution of industrial ecosystems and shows that a simple model can account for the high level of nestedness observed in economic networks.

Data and Methods

The ideal data to study the patterns of economic geography would consist of plant level information, collected for all countries, with high spatiotemporal resolution, and following a disaggregate standardized classification covering all economic sectors. Unfortunately, such data is not available. Instead, we use yearly trade data connecting 114 countries to 772 different products. Here, products are classified according to the SITC-4 rev2 classification. We use data from 1985 to 2009 to approximate the evolution of the global patterns of production. Going forward, we refer to this as the country-product network. We consider a country to be connected to a product if that country's exports per capita are larger than 25% of the world's exports per capita in that product for at least five consecutive years. These thresholds reduce the noise in the country product data coming from re-exports and helps make sure that a country is connected to the products that they export substantially and consistently. In Materials S1 we

check for the robustness of our results by using a different definition of presences and absences based on Balassa's [29] Revealed Comparative Advantage (RCA), and find the results to be robust to this alternative definition of presences.

We note two important limitations of international trade data. First, it does not include products that are produced and consumed domestically. This is because it only considers a product once it has crossed an international border. Second, trade data is limited to goods, and therefore does not include any data on services. Despite these limitations, trade data is good for international comparisons because it is collected in a standardized classification that makes data for different countries comparable.

At the domestic level we use information on the tax residence of Chilean firms collected by Chile's *Servicio de Impuestos Internos* (SII), which is the equivalent of the United States Internal Revenue Service (IRS). Going forward, we refer to this dataset as the municipality-industry network. The municipality-industry network contains information on 100% of the firms that filed value-added and/or income taxes in Chile between 2005 and 2008. This data comprises firms from all economic sectors, whether they export or not, and whether they produce goods or services. The municipality-industry network consists of the universe of Chilean firms (nearly 900,000), which are classified into 700 different industries and assigned to each of Chile's 347 municipalities. Here we consider an industry to be present in a municipality if one or more firms, filing taxes under that industrial classification, declare that municipality as their tax residency.

Finally, we note that the Chilean tax data has the limitation that the tax residency of a firm can differ from the location of all of its operations. Going forward, we take the fact that our results hold in both, international trade and domestic tax data, as an indication that they are not driven by the limitations of these datasets and that they represent a natural characteristic of the economic networks underlying them. For more details on both datasets see the SM.

Results

Figures 1 **a** and **b** show the matrices of the country-product and the municipality-industry networks (Respectively NODF = 70.81 and NODF = 83.35. We note that NODF = 100 indicates perfect nestedness and NODF = 0 indicates no nestedness, [30]). Here, the red lines indicate the diversity of each country and the ubiquity of each product -the number of locations where it is present- (see SM). These lines are used as a guide to indicate where presences would be expected to end if the nestedness of these networks were to be perfect. They can be thought of a simplified extinction line [27]. Figure 1 **c** and **d** show their corresponding Bascompte et al. null models [5]. In the Bascompte et al. null model, the probability to find a presence in that same cell of the matrix is equal to the average of the probability of finding it in that row and column in the original matrix. The figures show that nestedness of the original networks is clearly larger than that of their respective null models, showing that industrial ecosystems are more nested than what would be expected for comparable networks (respective null model NODF of 35.0±0.6 and 46.5±0.3, errors are 99% confidence intervals calculated from 100 implementations of the null model).

Next, we study the temporal evolution of nestedness. In the case of the country-product network, where a larger time series is available (1985–2009), the percentage of presences almost doubled during the observation period (Figure 1 **e**), going from less than 15% to nearly 25%. In the case of the municipality-industry network, presences went up from 22.9% to 25.7% between 2005

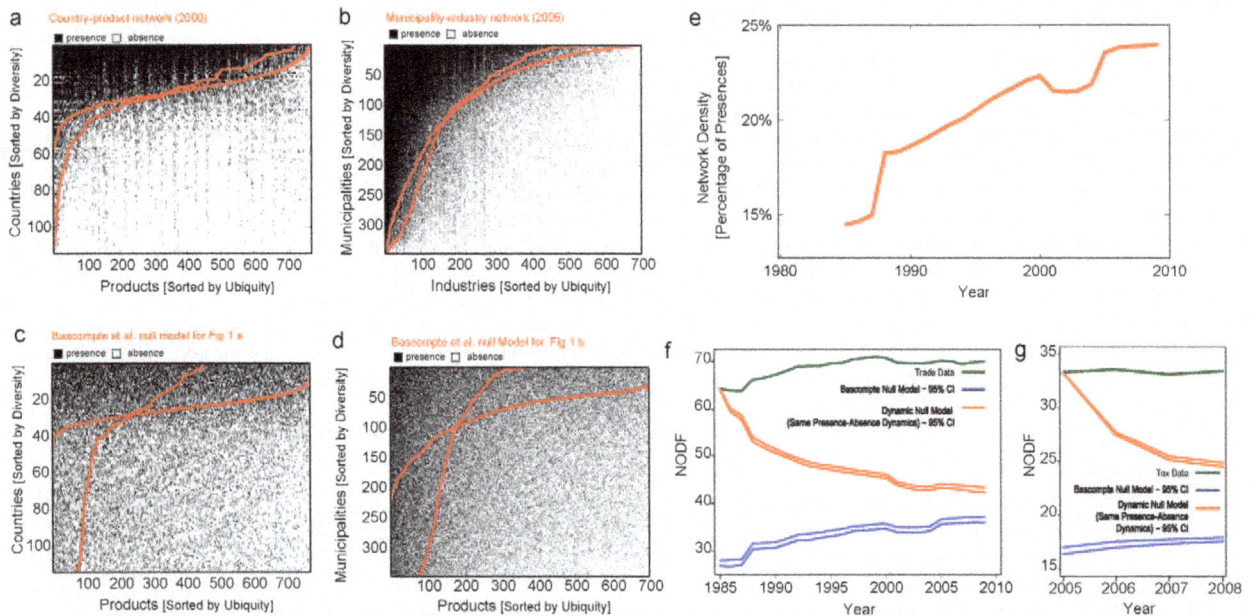

Figure 1. The nestedness of international and domestic economies. a Country-product network for the year 2000. **b** Municipality-industry network for the year 2005. **c** Bascompte et al. null model for the matrix shown in **a**. **d** Bascompte et al. null model for the matrix presented in **b**. In **a–d** red lines indicate the diversity of a location and the ubiquity of an industry (see full text for details). **e** Evolution of the density, or fill, of the country-product network between 1985 and 2009. **f** Evolution of the NODF of the country-product network between 1985 and 2009 (green), its corresponding Bascompte et al. null model (blue, upper and lower lines indicate 95% conf. intervals), and that of a matrix that started identical to that for 1985, but that was evolved by considering an equal number of appearances and disappearances than in the original data (red, upper and lower lines indicate 95% conf. interval). **g** Same as **f** but for the municipality-industry network (see SM for results with Atmar and Patterson's temperature metric).

and 2008. The nestedness of both, the country-product and the municipality-industry networks, however, remained relatively stable during this period as measured by NODF (green lines in Figure 1 **f–g** and SM).

We test the constancy of these networks' nestedness by comparing them with two null models. The first one is an ensemble of null models [5] calculated for each respective year (blue lines in Figure 1 **f–g**). This shows that the nestedness of the empirical networks is always significantly higher than their randomized counterpart. Then, we show that a network subject to the same exact turnover dynamics would have lost its nestedness during the observation period. We do this by starting with the empirically observed network and simulate its evolution by sequentially adding and subtracting a number of links equal to the one gained or lost by the original network. We do this following the probability distributions defined by the Bascompte et al null model [5] to make sure that these additions and subtractions keep the degree sequence of the network close to the original one. Otherwise, the lost of nestedness could be a consequence of changes in the underlying distributions. This dynamic null model represent a strong control, since it preserves the exact density of the network and also its turnover dynamics, as the number of links that appeared and disappeared each year, in each country, and for each product is exactly that observed in the original data. The dynamic model, however, does not preserve nestedness, showing that its stability comes from the specific way in which links appear and disappear from the network, and not due to a more trivial dynamics. In fact, when the appearance and disappearance of the links are chosen differently, the nestedness of the network quickly evaporates (red line in Figure 1 **f–g**). This allows us to conclude that the stability of nestedness observed in

these networks is higher than what would be expected from a null model with the same general turnover dynamics.

Could the stability of nestedness be used to predict appearances and disappearances? In the past, nestedness has been used to make prediction of the biota available in ecological patches, albeit not in economic networks [2,31]. For the country-product network we consider as an appearance an increase in exports per capita from less than 5% of the world average to more than 25%. To make sure that we are capturing structural changes and not mere fluctuations, we ask the increase in exports per capita of a country to be from less than 5%, for five consecutive years, to more than 25% sustained for at least 5 years. Hence, our final year of observation is 2005. Conversely, we count disappearances as a decrease in exports per capita of a country from 25% or more of the world's average to 5% or less (also sustained for at least 5 years). For the municipality-industry network we count appearances as changes from zero industries to one or more, and disappearances as changes from one or more industries to zero.

Figure 2 **a–d** visualizes the position in these networks' adjacency matrices of the industries that were observed to appear (green) and disappear (orange) in the intervening period. We predict these appearances and disappearances by fitting each observation in the industry-location network using a probit model that considers information on the diversity of the location and the ubiquity of the industry for the initial year (see SM). This represents a parameterization of nestedness and is similar to previous approaches that have used nestedness to make predictions [2,31]:

$$M_{c,p,t} = \alpha k_{c,t} + \beta k_{p,t} + \gamma\left(k_{c,t} \times k_{p,t}\right) + \varepsilon_{c,p,t} \qquad (1)$$

Here $M_{c,p,t}$ is the industry-location network's adjacency matrix, $k_{c,t}$ is the diversity of location c at time t (defined as its degree centrality or $k_{c,t} = \sum_p M_{c,p,t}$), $k_{p,t}$ is the ubiquity of product p at time t (defined as its degree centrality or $k_{p,t} = \sum_c M_{c,p,t}$), and where we have also added an interaction term taking the product between diversity $(k_{c,t})$ and ubiquity $(k_{p,t})$. The error term is represented by $\varepsilon_{c,p,t}$. We find that all coefficients are highly significant, meaning that a model that would only consider diversity or ubiquity, or both of them without an interaction term, would not be as accurate.

In general, we find that the probit regression accurately explains presences and absences (average Efron's pseudo-$R^2 = 0.53 \pm 0.02$ for the country-product network and 0.54 ± 0.01 for the municipality-industry network). Here, however, we use the deviance residuals of this regression to predict future appearances and disappearances. Negative residuals, represent unexpected absences [2] and are used to rank candidates for new appearances. Positive deviance residuals, on the other hand, represent unexpected presences [2] and are used to rank the likelihood that an industry will disappear in the future. (Figures 2 **e–h**).

But how accurate are these predictions? We quantify the accuracy of predictions by using the area under the Response Operator Characteristic curve or ROC curve [32,33]. An ROC curve plots the true positive rate of a prediction as a function of its false positive rate. The Area Under the Curve, or AUC, is commonly used to measure the accuracy of the prediction criterion [32,33]. A random prediction will find true positives and false positives at the same rate, and therefore will give an AUC of 0.5. A perfect prediction, on the other hand, will find all true positives before hitting any false positive and will be characterized by an AUC = 1. Figures 2 **i–l** show the ROC curves obtained when the appearances and disappearances shown in Figures 2 **a–d** are predicted using the deviance residuals obtained from (1) for data on the initial year. In all cases, the ROC curves of these predictions (in blue), have an area that is significantly larger than the one expected for a random prediction (in red), showing that nestedness can help predict which links in these industry-location networks are more likely to appear or disappear.

Finally, we extend this analysis to all pairs of years. Figures 3 **a** and **b** show the number of events (appearances or disappearances) for each pair of years for the international trade data. As expected, there are fewer events for pairs of years that are close by in time. Also, we note that the number of appearances is larger than that of disappearances, a fact that is consistent with the observed increase in the density of the network. Figure 3 **c** shows the AUC value obtained for each pair of years, showing that for the country product network, disappearances (Fig. 3 **b**) are predicted much more accurately than appearances.

The time series data available for Chile's municipality-industry network is much more limited. Hence, we show the average number of events (Figure 3 **d**), and the average AUC for networks separated by a given number of years (Figure 3 **e**). Here, we find that predictions of appearances and disappearance are both remarkably strong, and that there is no statistically significant difference in the predictability of both kinds of events.

To conclude this section, we look at the position in the network's adjacency matrix of appearances and disappearances. If the stability of nestedness is related to the location in this matrix of industrial appearances and disappearances, then appearances should be closer to the diversity-ubiquity line than random appearances. By the same token, disappearances should be farther away. For each event, we estimate its distance to the diversity and the ubiquity lines illustrated in figures 1 **a–d** and figures 2 **a–d** using,

$$D = Sign((c',p')_i) \min \left(\frac{\vec{I}_p - (c',p')_i}{N_c}, \frac{\vec{I}_c - (c',p')_i}{N_p} \right) \quad (2)$$

Here $\vec{I}_c$ and $\vec{I}_P$ are respectively the lines of diversity and ubiquity (i.e. the red lines in Figure 1 a–d), $(c',p')_i$ is the position in the adjacency matrix of the ith event, and N_c and N_p are respectively the number of locations and industries in the network. We use N_c and N_p to normalize the maximum possible vertical and horizontal distances to 1 and thus make sure that the measure is less sensitive to the rectangularity of the different matrices. The $\|$ operator represents the Euclidean distance and $Sign((c',p')_i) = 1$ if the position of the event is outside of the nested area defined by both $\vec{I}_c$ and $\vec{I}_P$ and -1 otherwise (see SM). As a benchmark comparison we consider an equal number of appearances and disappearances, but draw these from a random set of eligible positions in the adjacency matrix.

Figures 3 **f–i** compare the distributions of distances (D) with those associated with an equal number of random appearances or disappearances. We find that appearances tend to lie significantly closer to the diversity/ubiquity lines than what would be expected for an equal number of random events (ANOVA F = 59,935, p-value = 0 for the country-product network and ANOVA F = 10895 p-value = 0 for the municipality-industry network). In the case of disappearances, the opposite holds true. The observed appearances tend to be mostly located outside of the nested area defined by the diversity/ubiquity lines. Our random expectation, however, would be for disappearances to come mostly from the highly populated area inside the diversity/ubiquity lines. Once again, differences between observations and null model expectations are highly significant for both networks (ANOVA F = 6246 p-value = 0 for the country-product network and ANOVA F = 6463 p-value = 0 for the municipality-industry network).

Finally, we show that a modified version of the neutral development model introduced in [17], and solved analytically in [28], can be used to explain both, the observed level of nestedness and its stability. This neutral development model consists of three simple assumptions;

(i) Products require a set of non-tradable inputs, or capabilities, to be produced.

(ii) Locations are characterized by a set of capabilities.

(iii) Locations can only produce the products for which they have all the required capabilities.

The model is formalized by introducing three mathematical objects: two matrices and one operator. P_{pa} is a matrix that is 1 if product p requires capability a, and 0 otherwise. C_{ca} is a matrix that is 1 if location c has capability a, and zero otherwise. Finally (iii) provides a way of mapping C_{ca} and P_{pa} into M_{cp}, since it implies that $M_{cp} = 1$ if the set of capabilities required by a product is a subset of the capabilities available in a location. Mathematically (iii) can be expressed as the following operator:

$$M_{c,p} = 1 \text{ if } \sum_a P_{p,a} = \sum_a C_{c,a} P_{p,a} \text{ and } M_{c,p} = 0 \text{ otherwise. } (3)$$

More details about the model can be found in [28].

To compare the model to the data we need to assume the form of $C_{c,a}$ and $P_{p,a}$. In [28] the model was solved analytically by assuming that both, $C_{c,a}$ and $P_{p,a}$ were random matrices. This means that each location has a capability with probability r and

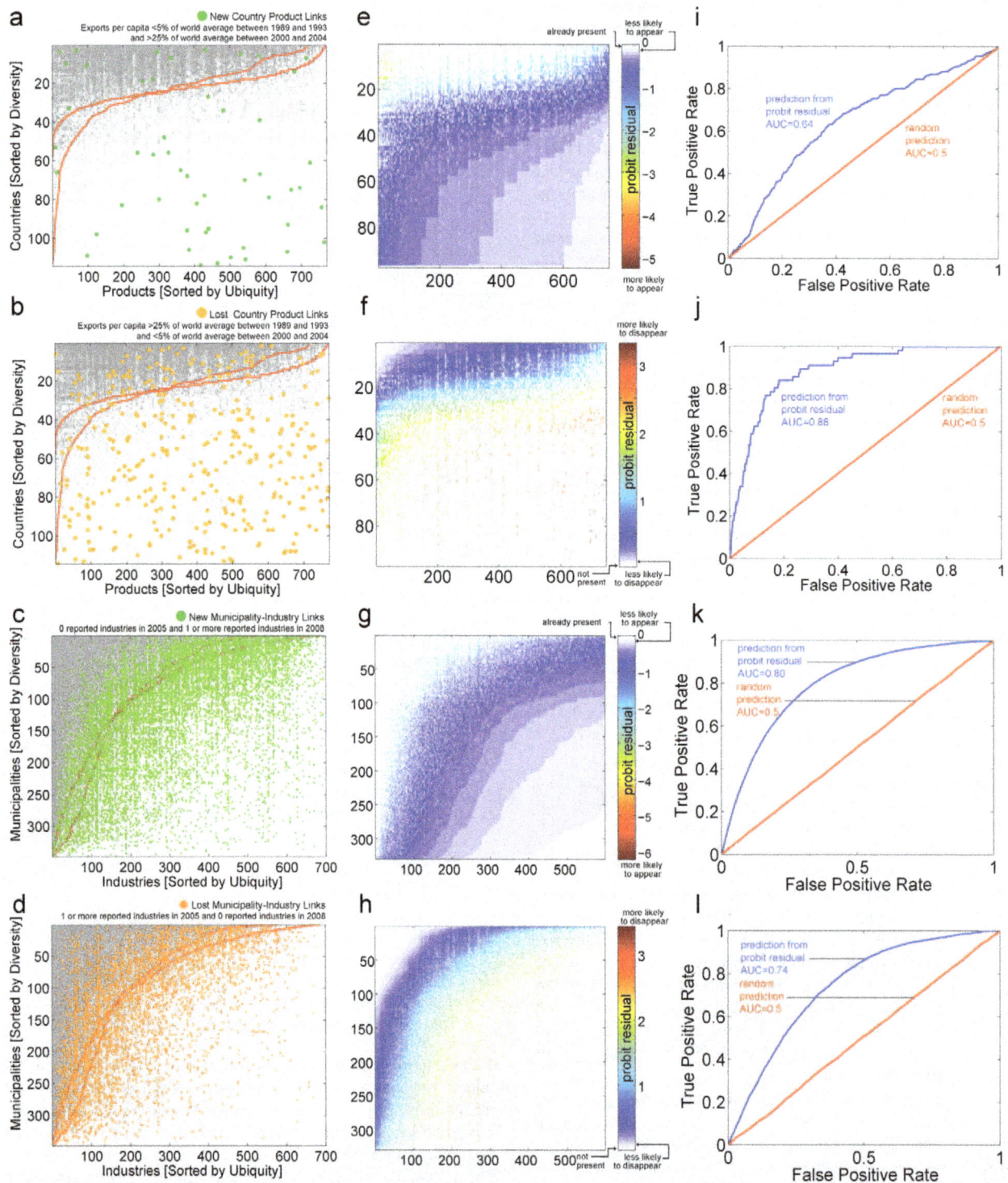

Figure 2. Nestedness predicts appearing and disappearing industries. a The country-product network for the year 1993 is shown in grey. Green dots show the location of industries that were observed to appear between 1993 and 2000. **b** Same as **a**, but with the industries that disappeared in that period shown in Orange. **c** The municipality-industry network is shown in grey and green dots show the location of industries that were observed to appear between 2005 and 2008. **d** Same as **c**, but with the industries that disappeared in that period shown in Orange. **e–h** Deviance residuals of the regression presented in (1) applied to the presences-absences shown in **a–d**. **i–l** ROC curves summarizing the ability of the deviance residuals shown in **e–h**, to predict the appearances and disappearances highlighted in **a–d**.

that products require a capability with a probability q. From this we can trivially deduce that the number of capabilities available in a random country, or required by a random product, follows a

binomial distribution. Because of this, we call this implementation of the neutral model: the binomial model. The third and final parameter that needs to be specified is the number of capabilities

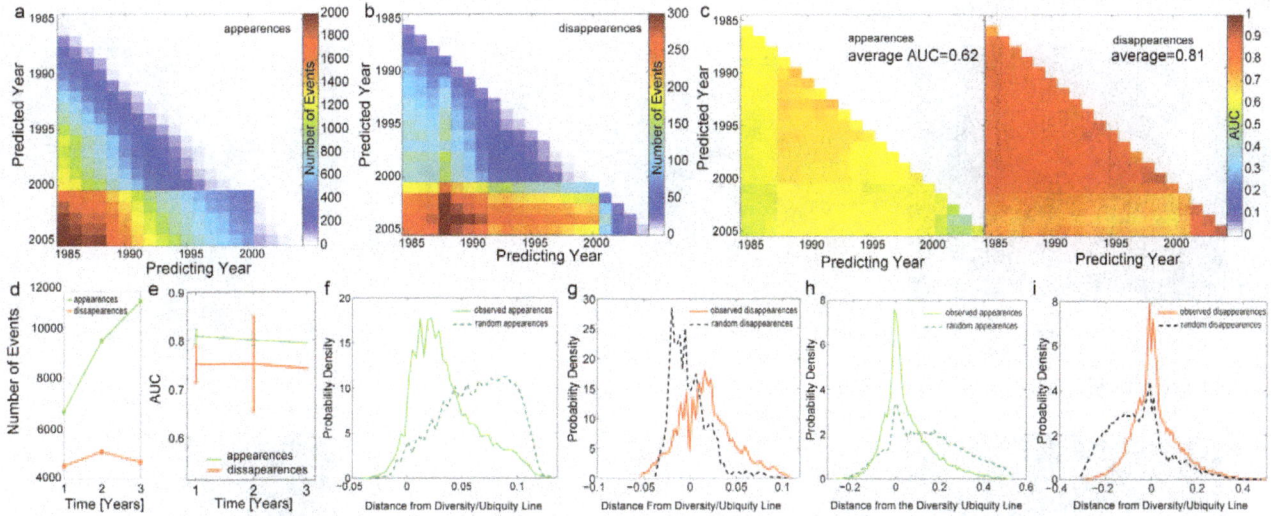

Figure 3. Predicting appearances and disappearances using nestedness. a Number of appearances for every pair of years in the country-product network. **b** Number of disappearances for every pair of years for the country-product network. **c** Accuracy of the predictions for each pair of years measured using the Area Under the ROC Curve (AUC). **d** Average number of appearances and disappearances for the Chilean data (error bar smaller than symbol). **e** Average accuracy of the predictions for the municipality-industry network. Error bars indicate 99% confidence intervals. **f** Distribution for the distance to the diversity-ubiquity line obtained for the observed appearances and for an equal number of random appearances. **g** Same as **f** but for disappearances. **h** Same as **f**, but for the municipality-industry network. **i** Same as **h** but for disappearances.

required by a product (N_a). This is because the number of locations N_c, and the number of products N_p, is fixed to match the number of locations and products observed in the data.

Effectively, the binomial model has two free parameters. This is because it is always possible to determine r, q or N_a once the fill of the $M_{c,p}$ matrix is known. The binomial model has been shown to reproduce the distribution of diversities, ubiquities, co-exports, and the relationship between diversity and ubiquity of the country-product network using $N_a = 80$, $r = 0.87$ and $q = 0.18$. In addition to the binomial model we consider an alternative form that has the same number of parameters. We call this the uniform model, since in this case the number of capabilities that a country has is distributed uniformly between 0 and R and the number of capabilities that a product requires is distributed uniformly between 0 and Q. Hence, in this model country c has a capability a with probability equal to $r_c = \min(1, R \times c/N_c)$. We take the minimum to ensure r_c is upper bounded by 1. In the uniform model, allowing values of R larger than one allows having a small number of fully diversified countries.

Figures 4a and 4b illustrate the binomial model and the uniform model, respectively. For both models, we show their respective $C_{c,a}$ and $P_{p,a}$ matrices together with their resulting country-product network $M_{c,p}$. We find that in both cases the resulting $M_{c,p}$ matrices are significantly more nested than the null model, yet the nestedness emerging from the uniform model is considerably larger, resembling closely the values observed for the country-product network. This comes from the fact that countries with a diverse capability endowment are likely to make a wide range of products, whereas countries with few capabilities will only be able to make those products that require few capabilities. This last observation is implied by assumption (iii), and is therefore true for both, the binomial and the uniform model. Yet, the large degree of heterogeneity among countries and products present in the uniform model enhances the nestedness implied by the complementarity assumption.

Figure 4 c compares the nestedness of the country-product network with the one found for the neutral models and null model.

Here we plot nestedness as a function of the fill of the network since this is a good proxy for time and the neutral models and null model do not have an explicit time dimension. We implement this comparison by generating an initial $P_{p,a}$ matrix that is kept constant during the procedure. In the binomial model we choose $q = 0.18$, and for the uniform model we take $Q = 0.21$. We interpret this as an assumption that productive technologies change slowly during the time frames considered, and therefore, the increases in diversification observed in the empirical network comes from locations catching up to produce the products that more diversified locations were already making. To create $M_{c,p}$, we generate 100 $C_{c,a}$ matrices for 200 different values of r and R. For the binomial model we consider values of r between 0.9 and 0.95, while for the uniform model we consider values of R between 0.9 and 1.07. In both cases we set the total number of capabilities in the system to $N_a = 80$. These values are chosen to ensure that the fills of the modeled $M_{c,p}$ matrices are close to the ones observed in the original data. The analysis shows that the nestedness of the $M_{c,p}$ matrices implied by the neutral model matches the ones observed in the economic networks only for the uniform model. In the context of assumptions (i)–(iii), we interpret this result as evidence that heterogeneity in the distribution of capabilities available in a country, or required by a product, are needed to generate the high levels of nestedness observed in these economic networks.

Discussion

In this paper we showed that industry-location networks are nested, just like industry-industry networks [13–15], or their biological counterparts [1–4,26,27]. Using time series data for both, international and domestic economies, we showed that the nestedness of these networks tends to remain constant over time and that this empirical regularity can be used to predict the pattern of industrial appearances and disappearances over time. Moreover, we showed that the high level of nestedness observed in the world can be accounted for by a simple model, but only if we

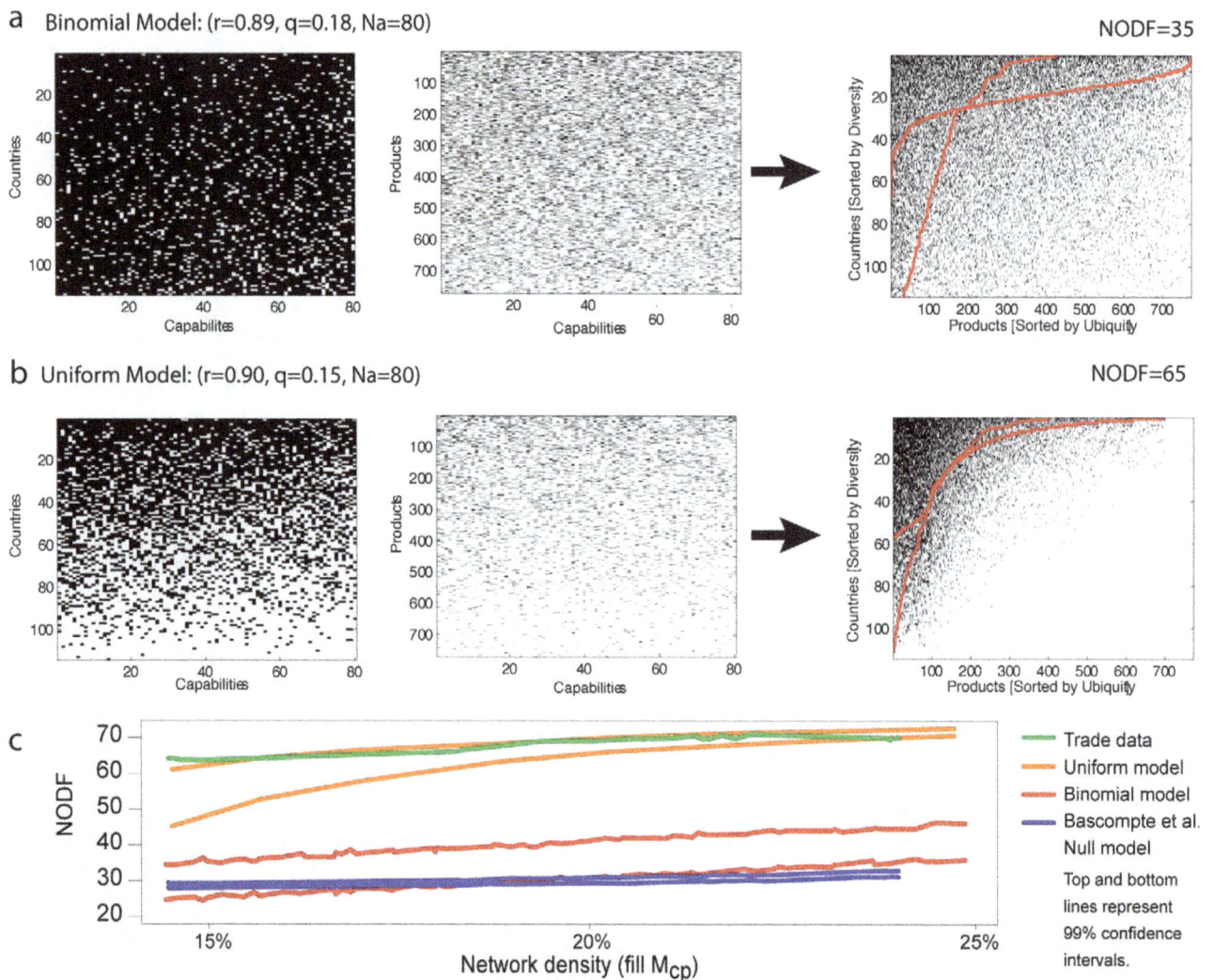

Figure 4. Modeling nestedness. a Illustration of the binomial model. From left to right; C_{ca}, P_{pa} and the resulting M_{cp}. b Illustration of the uniform model. From left to right C_{ca}, P_{pa} and the resulting M_{cp}. c NODF as a function of matrix fill for the country-product network (green), the uniform model (orange), the binomial model (red), and the Bascompte et al null model (blue).

assume a relatively large degree of heterogeneity in the number of capabilities present in a country or required by a product.

The strong link between biological and industrial ecosystems opens a variety of questions. First, is the geographical nestedness described in this paper a consequence of industry-industry nestedness, or are these independent phenomena? Second, are the mechanisms generating nestedness at the global level the same that generate nestedness at the national level?

In this paper we showed that the geographical nestedness of industries holds at both, the global and at the national scale. This is certainly not the case for biological ecosystems, since the biota of the artic is not a subset of that of the rain forest. The fact that the nestedness of industrial ecosystems holds at scales as large as that of the world economy suggests that the coupling between international economies is strong. This highlights the importance of understanding the global economy as a unified ecosystem, since after all, its nestedness suggests that it appears to be working as one.

The predictability implied by nestedness, on the other hand, has important implications in a world where income is connected to the mix of products that a country makes [17,18]. Ultimately,

the dynamics implied by nestedness could represent a fundamental constraint to the speed at which international incomes could either converge or diverge.

More research will certainly need to be done on both, the causes of the structures and the time patterns that were uncovered in this paper. This will require strengthening the bridge between the natural and social sciences because, if there is something that the nestedness of economies show, is that humans tend to generate patterns in social systems that strongly mimic those found in nature [34,35].

Supporting Information

Materials S1 Additional information on the data and methods used throughout the manuscript.

Acknowledgments

We thank Jordi Bascompte, Frank Neffke, Javier Galeano, Pablo Marquet, Daniel Stock & Muhammed Yildirim for their comments.

Author Contributions

Conceived and designed the experiments: CAH SB. Performed the experiments: CAH SB. Analyzed the data: CAH SB. Wrote the paper: CAH SB RH CG.

References

1. Hulten E (1937) Outline of the history of Artic and Boreal biota during the Quaternary Period, Lund University.
2. Ulrich W, Almeida M, Gotelli NJ (2009) A consumer's guide to nestedness analysis. Oikos 118, 3–17. doi:10.1111/j.1600-0706.2008.17053.x
3. Darlington PJ, (1957) Zoogeography: the geographical distribution of animals. Wiley.
4. Daubenmire R (1975) Floristic plant geography of eastern Washington and northern Idaho. Journal of Biogeography 2.
5. Bascompte J, Jordano P, Melian CJ, Olesen JM (2003) The nested assembly of plant-animal mutualistic networks. Proceedings of the National Academy of Sciences of the United States of America 100, 9383–9387, doi:10.1073/pnas.1633576100
6. Dupont YL, Hansen DM, Olesen JM (2003) Structure of a plant-flower-visitor network in the high-altitude sub-alpine desert of Tenerife, Canary Islands. Ecography 26, 301–310. doi:10.1034/j.1600-0587.2003.03443.x
7. Ollerton J, Johnson SD, Cranmer L, Kellie S (2003) The pollination ecology of an assemblage of grassland asclepiads in South Africa. Annals of Botany 92, 807–834, doi:10.1093/aob/mcg206
8. Gilarranz LJ, Pastor JM, Galeano J (2011) The architecture of weighted mutualistic networks. Oikos, 001–009. doi:10.1111/j.1600-0706.2011.19592.x
9. Turchin P, Hanski I (1997) An empirically based model for latitudinal gradient in vole population dynamics. American Naturalist 149, 842–874, doi:10.1086/286027
10. Jordano P (1987) Patterns of mutualistic interactions in pollination and seed dispersal – Connectance, dependence asymmetries and coevolution. American Naturalist 129, 657–677, doi:10.1086/284665
11. Bastolla U, Fortuna MA, Pascual-García A, Ferrera A, Luque B, et al. (2009) The architecture of mutualistic networks minimizes competition and increases biodiversity. Nature 458, 1018–U1091, doi:10.1038/nature07950
12. Bascompte J, Jordano P, Olesen JM (2006) Asymmetric coevolutionary networks facilitate biodiversity maintenance. Science 312, 431–433, doi:10.1126/science.1123412
13. Leontief WW (1965) The Structure of the U.S. Economy. Scientific American 212, 25–35.
14. Saavedra S, Reed-Tsochas F, Uzzi B (2009) A simple model of bipartite cooperation for ecological and organizational networks. Nature 457, 463–466, doi:10.1038/nature07532
15. Saavedra S, Stouffer DB, Uzzi B, Bascompte J (2011) Strong contributors to network persistence are the most vulnerable to extinction. Nature 478, 233–U116, doi:10.1038/nature10433
16. Hausmann R, Hidalgo CA, Bustos S, Coscia M, Chung S, et al. (2011) The Atlas of Economic Complexity. Puritan Press, Cambridge MA.
17. Hidalgo CA, Hausmann R (2009) The building blocks of economic complexity. Proceedings of the National Academy of Sciences of the United States of America 106, 10570–10575, doi:10.1073/pnas.0900943106
18. Hausmann R, Hwang J, Rodrik D (2007) What you export matters. Journal of Economic Growth 12, 1–25, doi:10.1007/s10887-006-9009-4
19. Teece DJ, Rumelt R, Dosi G, Winter S (1994) Understanding corporate coherence – Theory and Evidence. Journal of Economic Behavior & Organization 23, 1–30.
20. Hidalgo C, Klinger B, Barabasi A, Hausmann R (2007) The product space conditions the development of nations. Science 317, 482–487.
21. Bryce DJ, Winter SG, (2009) A General Interindustry Relatedness Index. Management Science 55, 1570–1585, doi:10.1287/mnsc.1090.1040
22. Neffke F, Svensson-Henning M (2008) Revealed relatedness: Mapping industry space. Working Paper Series 08.19, Papers in Evolutionary Economic Geography, Utrecht University, Utrecht, the Netherlands.
23. Neffke F, Svensson-Henning M, Boschma R (2011) How Do Regions Diversify over Time? Industry Relatedness and the Development of New Growth Paths in Regions. Economic Geography 87, 237–265, doi:10.1111/j.1944-8287.2011.01121.x
24. Almeida M, Guimaraes PR, Lewinsohn TM (2007) On nestedness analyses: rethinking matrix temperature and anti-nestedness. Oikos 116, 716–722, doi:10.1111/j.2007.0030-1299.15803.x
25. Almeida-Neto M, Guimaraes P, Guimaraes PR, Loyola RD, Ulrich W (2008) A consistent metric for nestedness analysis in ecological systems: reconciling concept and measurement. Oikos 117, 1227–1239, doi:10.1111/j.2008.0030-1299.16644.x
26. Atmar W, Patterson BD (1993) The measure of order and disorder in the distribution of species in fragmented habitat Oecologia 96, 373–382, doi:10.1007/bf00317508
27. Patterson BD, Atmar W (1986) Nested subsets and the structure of insular mammalian faunas and archipielagoes Biological Journal of the Linnean Society 28, 65–82, doi:10.1111/j.1095-8312.1986.tb01749.x
28. Hausmann R, Hidalgo CA (2011) The network structure of economic output. Journal of Economic Growth 16, 309–342, doi:10.1007/s10887-011-9071-4
29. Balassa B (1986) Comparative advantage in manufactured goods – A reappraisal. Review of Economics and Statistics 68, 315–319
30. Guimaraes PR, Guimaraes P (2006) Improving the analyses of nestedness for large sets of matrices. Environmental Modelling & Software 21, 1512–1513, doi:10.1016/j.envsoft.2006.04.002
31. Maron M, Mac Nally R, Watson DM, Lill A (2004) Can the biotic nestedness matrix be used predictively? Oikos 106, 433–444, doi:10.1111/j.0030-1299.2004.13199.x
32. Bradley AP (1997) The use of the area under the roc curve in the evaluation of machine learning algorithms. Pattern Recognition 30, 1145–1159, doi:10.1016/s0031-3203(96)00142-2
33. Zweig MH, Campbell G (1993) Receiver-operating characteristic (ROC) plots – A fundamental evaluation tool in clinical medicine. Clinical Chemistry 39, 561–577.
34. Bettencourt L, West G (2010) A unified theory of urban living. Nature 467, 912–913.
35. West GB, Brown JH, Enquist BJ (1997) A general model for the origin of allometric scaling laws in biology. Science 276, 122–126

9

Macro-Invertebrate Decline in Surface Water Polluted with Imidacloprid: A Rebuttal and Some New Analyses

Martina G. Vijver[1]*, Paul J. van den Brink[2,3]

1 Institute of Environmental Sciences (CML), Leiden University, Leiden, The Netherlands, 2 Alterra, Wageningen University and Research centre, Wageningen, The Netherlands, 3 Wageningen University, Wageningen University and Research centre, Wageningen, The Netherlands

Abstract

Imidacloprid, the largest selling insecticide in the world, has received particular attention from scientists, policymakers and industries due to its potential toxicity to bees and aquatic organisms. The decline of aquatic macro-invertebrates due to imidacloprid concentrations in the Dutch surface waters was hypothesised in a recent paper by Van Dijk, Van Staalduinen and Van der Sluijs (PLOS ONE, May 2013). Although we do not disagree with imidacloprid's inherent toxicity to aquatic organisms, we have fundamental concerns regarding the way the data were analysed and interpreted. Here, we demonstrate that the underlying toxicity of imidacloprid in the field situation cannot be understood except in the context of other co-occurring pesticides. Although we agree with Van Dijk and co-workers that effects of imidacloprid can emerge between 13 and 67 ng/L we use a different line of evidence. We present an alternative approach to link imidacloprid concentrations and biological data. We analysed the national set of chemical monitoring data of the year 2009 to estimate the relative contribution of imidacloprid compared to other pesticides in relation to environmental quality target and chronic ecotoxicity threshold exceedances. Moreover, we assessed the relative impact of imidacloprid on the pesticide-induced potential affected fractions of the aquatic communities. We conclude that by choosing to test a starting hypothesis using insufficient data on chemistry and biology that are difficult to link, and by ignoring potential collinear effects of other pesticides present in Dutch surface waters Van Dijk and co-workers do not provide direct evidence that reduced taxon richness and abundance of macroinvertebrates can be attributed to the presence of imidacloprid only. Using a different line of evidence we expect ecological effects of imidacloprid at some of the exposure profiles measured in 2009 in the surface waters of the Netherlands.

Editor: Christopher Joseph Salice, Texas Tech University, United States of America

Funding: These authors have no support or funding to report.

Competing Interests: For transparency reasons, we mentioning the following: PvdB's chair was cofunded between 2008 and 2011 by the following pesticide producers, Bayer, which produces imidacloprid and Syngenta. We feel that this cofunding provides no compete of interest since we don't claim that imidacloprid poses less risks or toxicity than stated in the Van Dijk et al. (2013) as in the current paper we only criticized their methodology. This current work has not been funded. Sponsors thus had no role in study design, data collection and analysis, decision to publish, or preparation of the manuscript.

* E-mail: vijver@cml.leidenuniv.nl

Introduction

The Netherlands is one of the world's foremost agricultural producers, with 2/3 of the total land mass devoted to agriculture or horticulture. Land use is highly intensive in terms of output per hectare or head of livestock [2]. To achieve such high outputs a vast range of agricultural chemicals are used, including fertilizers, veterinary drugs, pesticides and biocides. Different pesticides are used depending on the crop that is grown on the land. There are several routes that pesticides may enter surface waters. Pesticides may be washed into ditches and rivers by rainfall; surface waters can be contaminated by direct overspray or via runoff and leaching from agricultural fields [3]. Emission to surface waters (and thus pesticide residue concentrations) is dictated by many factors such as distance of the crop from the ditch and the mode of application, weather conditions and so on.

Neonicotinoids are the first new class of insecticides to be introduced in the last 50 years. The neonicotinoid imidacloprid is currently one of the most widely used insecticides in the world [4]. Recently, imidacloprid has received much negative attention: The use of certain neonicotoids has been restricted in some countries due to evidence of an unacceptably high risk of toxicity to bees, but this restriction was not in effect in the Netherlands at the time of writing this paper. On April 29, 2013, the European Union passed a two-year ban on the use of three neonicotinoids: European law restricts the use of imidacloprid, clothianidin, and thiamethoxam on flowering plants for two years unless compelling evidence comes out that proves that the use of the chemicals is environmentally safe [5]. This ban is partially, restricted to some applications in specific crops and likely covers 15% of the total use of the three neonicotinoids in the Netherlands [6]. Temporary suspensions had previously been enacted in countries such as France, Germany, Switzerland and Italy. In March 2013, a review of 200 studies on neonicotinoids was published by Mineau and Palmer [7], calling for a ban on neonicotinoid use as seed treatments because of their toxicity to birds, aquatic invertebrates, and other wildlife. The EPA – USA is now re-evaluating the safety of neonicotinoids.

Van Dijk and co-workers [1] aimed to assess the specific relationship between imidacloprid residues in Dutch surface waters, and the abundance of non-target macro-invertebrate taxa.

As also stated by the authors, finding a statistical relationship between those two datasets does not necessarily reflect causality, because there could be other factors (e.g. other pesticide residues, other local habitat factors) which drive observed patterns of abundance. We have some fundamental criticisms on the way the data were analysed and the results were interpreted, and we feel that this can be challenged by existing data. Therefore as a response to the paper of Van Dijk et al [1], and by using additional data, we explore their two key assumptions: 1) residues of pesticides other than imidacloprid, that are collinear with imidacloprid exposure either do not exist or have negligible effects on macroinvertebrate abundance and 2) that imidacloprid concentrations can be extrapolated successfully over 160 days and at a 1 km^2 spatial scale.

Materials and methods

Data collection and treatment

Data on pesticides concentrations in surface water in the Netherlands were obtained from the Dutch Pesticides Atlas. [8]. This is an online tool from which Dutch monitoring data can be collected and processed into a graphic format. Here, data of all pesticide active ingredients and metabolites (n = 634) collected in 2009 were used, since this data set is contiguous with the data used by Van Dijk et al. [1]. Only one year was selected since it can be expected that the correlations between pesticide occurrences will be year-specific, so this correlation should also be assessed for each year specifically. The 2009 dataset covered 302111 individual measurement records of which 19693 measurements exceeded the reporting limit (LOR). The measurements were performed on 4816 samples obtained from 723 different locations. The sample by pesticide dataset is characterised by missing values (90% of entries) and below LOR values (9% of all entries). This is a result of the fact that every water manager has his own suite of pesticides that is sampled, measured and evaluated. The selection of this suite of pesticides is based on the crops and land-use in their region. This selection of pesticides to be monitored improves the efficiency of the monitoring efforts of the individual water managers but yields a data set that has missing values and with many < LOR values when the data of multiple water managers are combined into one. To obtain frequency distributions of the imidacloprid concentrations, data from 2010 and 2011 have also been used.

Environmental quality standards (EQS) of all pesticides were as follows: for imidacloprid the annual average-EQS value (AA-EQS) is 0.067 µg/L (database value set 2-6-2010), and the maximum allowable concentration (MAC-EQS) is 0.2 µg/L (database value set 2-6-2010) as specified by the European Water Framework Directive. In addition, in the Netherlands, the maximum permissible concentration (MPC) of 0.013 µg/L is an important additional criterion (database value set 8-10-2008).

For all samples in which a pesticide could not be detected or quantified, the database substitutes a value of lower than the LOR. The values of reporting limits vary across samples (unique location x time). In our calculations these measurements below LOR are set as zero. We chose to do so, as choosing any other value below LOR would be arbitrary. Moreover, if not taking zero as a value, any other chosen value will result in relatively high toxicity at intensively measured surface waters even if the pesticides are not applied in that area since all measurements results in a lowest value possible of being below the LOR. These types of assumptions are inherent when working with data sets based on monitoring efforts.

Collinearity of imidacloprid concentrations with concentrations of other pesticides

Collinearity refers to a linear relationship between two explanatory variables, meaning that one can be linearly predicted from the others with a non-trivial degree of accuracy. Collinearity was determined on the data set of 2009 measurements restricted to all samples with at least one measurement above the LOR. The reduced data contained measured values for 18% of the samples, of which 8% of the total were measurements above the LOR. In order to assess the correlation between the concentrations of different pesticides we needed a sample by pesticide matrix with as little missing values as possible. From this gappy database, the largest closed data sets were extracted using Principal Component Analysis [9]. For this, measured values in the database were coded as one and missing data by zero. After running the PCA, the species-by-substance matrix was sorted, based on the scores of the substances and samples on the first principal component. Using this approach, it was possible to extract closed data sets by extracting groups of samples with the same score on the first principal component. Four data sets could be extracted that contained more than 100 samples in which the same pesticides were measured. One data set did not include imidacloprid and was not taken into account. The remaining three matrices contained 114, 108 and 191 samples, 27, 51 and 54 pesticides, with 11, 11 and 13% of the measurements above the LOR for data set 1, 2 and 3, respectively. All sampling points of data set 1 were within the provinces of Utrecht and Gelderland while all sampling points of data set 2 and 3 were located in the province of South Holland.

The log((1000 * conc) +1) transformed pesticide concentration values were analysed with Principal Component Analysis (PCA) using the Canoco5 computer programme [10], (see Zafar et al. [11] for the rationale of the transformation]. The pesticide data were centred and standardised for each pesticide. The graphical pictures based on orthogonal coordinate systems describe optimal variance in a dataset. Points that are clustered near each other have a strong correlation. PCA [9] transforms data to a new coordinate system such that the greatest variance by any projection of the data comes to lie on the first coordinate (called the first principal component), the second greatest variance on the second coordinate [12].

Calculating multi substance PAF

The potential affected fraction (PAF) is a common way to express ecotoxicological risks [13]. Following this approach, measured pesticides concentrations were translated into PAF using the species sensitivity distribution (SSD) approach. Toxicity data for each pesticide was obtained from De Zwart [14], and based on acute median effect concentrations (EC50) as derived in the laboratory (database eTox, RIVM as described in [14]). The eTox database consists mainly of data entries from the ECOTOX EPA database. The SSD for imidacloprid is given in Figure 1, and includes 41 different species from 7 different taxonomic groups. Underlying data including references are given in Table S1 of the Supplementary Information. The full database used for the multi substance PAF (mSPAF) calculations contained data of 496 different pesticides with 75 different modes of action. To quantify the ecological impacts due to imidacloprid concentrations amongst all other pesticide concentrations as measured in the surface waters, the mSPAF was calculated. Firstly, all concentrations of individual pesticides measured over one month per location were aggregated using the maximum measured value. Secondly individual pesticide concentrations were compared to the toxicity data resulting in the PAF. Thirdly, pesticides were grouped based on their mode of action. The PAF's of the pesticides with a similar mode-of-action were added using a concentration addition equation. In this equation, each substance concentration is divided by its effect concentration, ECxa, i.e., the concentration of a that represents a standard effect expressed as EC50 for endpoint x. This gives: Emix (Cmix) = (Ca/ECxa) + (Cb/ECxb) + …... In which Emix(Cmix) is the summed ratio of the mixture components at the exposure concentration of each chemical (Cx). Fourthly, the different pesticides groups with dissimilar mode-of-action were added using a response addition equation. In response addition, the toxicity of the substances in the mixture can be predicted from the product of the fractional effects of the mixture components. This gives Emix (Cmix) = 1 − ((1 − E(Ca)) * (1 − E(Cb)) * …... In which Emix(Cmix) is the calculated effect of the mixture, Ca the exposure concentration of substance a, and E(ca) the effect of substance a at concentration Ca.

Both models for mixture toxicity are described in Hewlett and Plackett [15]. Chemicals with an unknown mode-of-action were treated according to a unique mode-of-action. As a result an msPAF value per month per monitoring location was derived. In this study we reported the maximum msPAF of the year 2009. The quantification of the relative contribution of imidacloprid on the total chemical pressure as expressed by msPAF was based on acute toxicity data as insufficient chronic toxicity data were available in the literature.

Pairwise combinations of samples taken within 1 km and 160 days

Datasets on imidacloprid concentrations and abundances of macroinvertebrates were linked to each other by Van Dijk and co-workers [1] by using the criteria ≤1 km distance and ≤ 160 days

of time difference. We performed pairwise comparisons of imidacloprid measurements to determine whether imidacloprid concentrations at sites that meet these criteria, matched successfully. Therefore, all imidacloprid measurements were extracted from the 2009 data set. All sampling sites were first ranked on their x coordinate and the difference in distance with the next sample was assessed (using Pythagoras theorem). All site combinations which yielded a difference less than 1 km were extracted. The same procedure was performed using a ranking based on the y-coordinate. The site combinations from both queries were combined. This procedure is not exhaustive since two sites that are not ranked next to each other can also be closer to 1 km from each other, but is likely to find most combinations. The imidacloprid concentrations of all samples taken at the paired sites were compared to each other when the samples were taken within 160 days. The result of the comparison were categorised into: 1) two measurements below the LOR, 2) one measurement below and one above the LOR (0% matching), 3) two measurements above the LOR, of which the number of sample pairs that matched 100% (based on one decimal) was also noted. The analysis resulted in 37 pairs of sites containing a total of 260 observations and 584 concentration measurement pairs being evaluated.

Time series of imidacloprid exposure

For each sampling site it was determined how often imidacloprid samples were analysed. For 34 sampling sites 10 or more samples were analysed, of which imidacloprid was not detected in any of the samples at 14 sites (41%), and in less than half of the samples at 28 sites (82%). The concentration dynamics of the

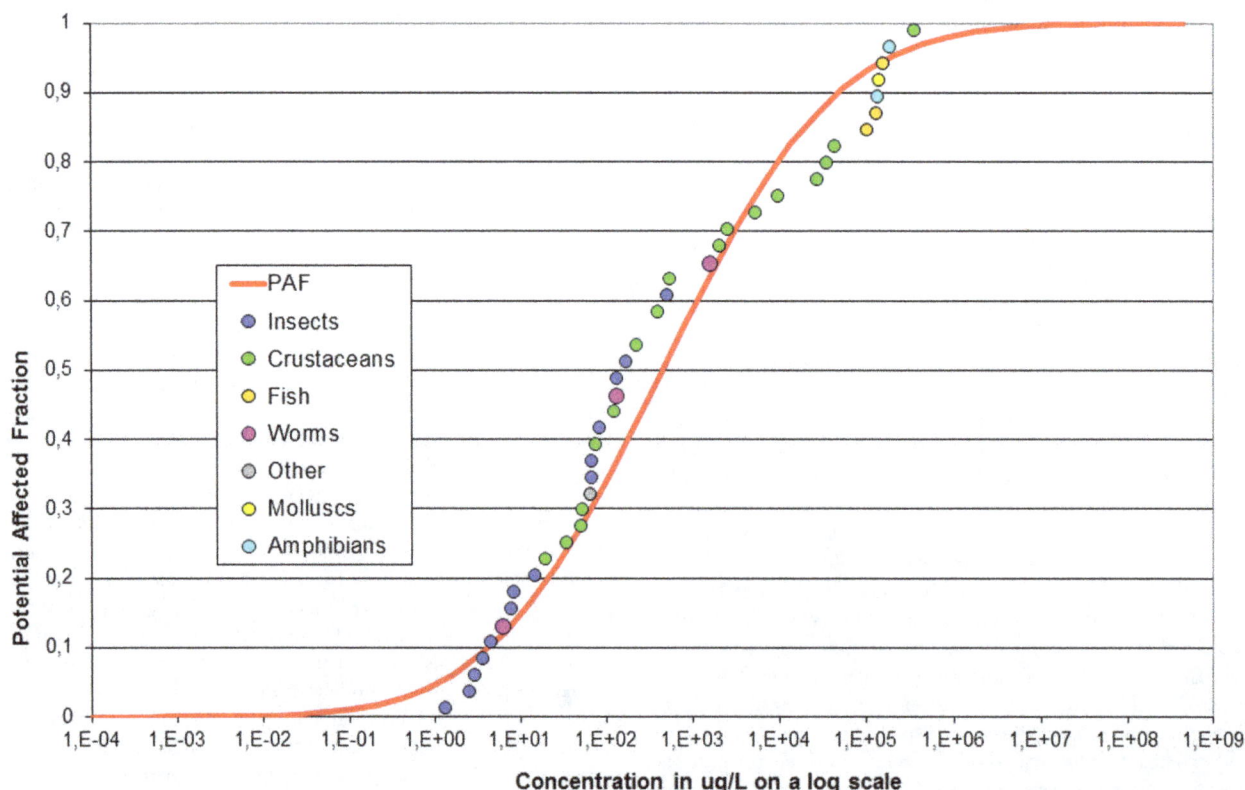

Figure 1. The Species Sensitivity Distribution of imidacloprid based on acute toxicity data. The data consist of 7 different taxomonic groups and 41 species. EPA database downloaded at Oct 23th 2013.

remaining 18% of the sites were plotted to evaluate whether chronic concentrations of imidacloprid may be expected.

Cumulative frequency of maximum imidacloprid concentrations

The measured maximum concentration of each site was compared with threshold concentrations based on the findings of Roessink et al. [16], i.e. the chronic EC10 of the mayfly species *Caenis horaria* and *Cloeon dipterum* ($\approx$0.03 μg/L) and the different environmental quality standards. In order to remove the within-site sample dependency, for each sampling site the maximum imidacloprid concentration was extracted. The analysis resulted in 225 negative measurements (below the LOR) and 226 positive measurements (above the LOR).

MPC exceedances of imidacloprid compared to other pesticides

Since only for a restricted number of pesticide AA-EQS and MAC-EQS values have been set in the WFD, we used the (Dutch) MPC standard to compare exceedance frequencies between pesticides. For this comparison, both the magnitude of exceedance as well as the frequency of exceedance was incorporated. Firstly, the exceedance of the MPC of an individual pesticide concentration was derived per measuring location. Secondly, the degree of standard exceedance was weighted according to the following classes: 0 ($\leq$MPC); 1 ($>$ MPC and $\leq$ 2 x MPC), 2 ($>$ 2 x MPC and $\leq$ 5 x MPC) and 5 ($>$ 5 x MPC exceedance). Thirdly, the exceedance classes were summed over all measuring locations per year. Fourthly, pesticides were ranked on the basis of the weighted number of monitoring sites at which the MPC for the compound was exceeded, i.e. corrected for the number of monitoring sites by taking the percentage of sites that show an exceedance of the MPC. Compounds monitored at fewer than ten sites were ignored.

Results and Discussion

For many locations pesticide concentrations have been found to exceed the MPC in 2009 (see Fig. 2). Figure 2 shows that throughout the entire country more than one pesticide exceeds their respective quality standard, so this exceedance is not a common regionally problem. The maximum amount of pesticides exceeding their MPC in one sample is 35. From this it can be concluded that a single pesticide is not likely to drive solely the macro-invertebrate quality, rather all pesticides exceeding the quality standards should be considered.

Collinearity of imidacloprid concentrations with other pesticides

Figure 3A clearly shows that imidacloprid exposure is highly correlated with all chemicals placed on the right, lower side of the diagram, like carbendazim and DEET and to a lesser extend with the large group of chemicals which have a high loading with the horizontal axis, which explains almost double the amount of variance compared to the vertical axis. The results of the second data set (Fig. 3B) show that imidacloprid is placed in the centre of a large group of pesticides placed in the middle of the diagram, since it was measured only in a few samples (7% of the total). The results of the third data set shows a high occurrence of imidacloprid above the LOR (78% of all samples), with concentrations strongly collinear with those that have a high loading on the horizontal axis which explains almost triple the amount of variance of the vertical one (Fig. 3C). The results of the first and third data set show that the contribution of imidacloprid toxicity in surface waters cannot

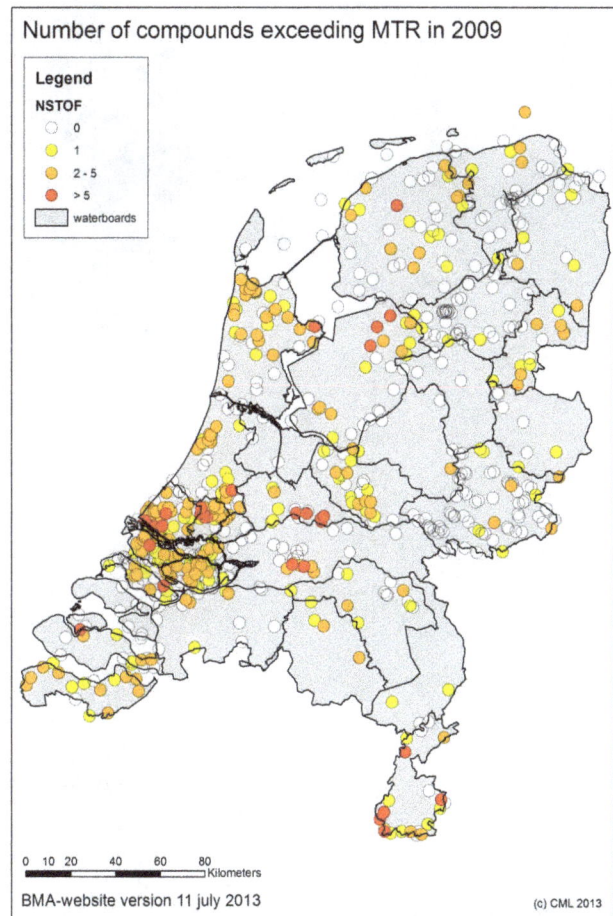

Figure 2. Number of pesticides exceeding the MPC in 2009. All monitoring locations in the Dutch surface waters with one (yellow); two till five pesticides concentrations (orange); and > five different pesticides (red) exceeding their MPC-values are depicted. Locations were measurements were performed but no exceedances were found are depicted in white.

easily be separated from the toxicity arising from other co-occurring pesticides, or indeed any other co-occurring chemical or physical stressing agent.

The correlations derived from the PCA-plots (Fig. 3) can also be explained from the fact that the active ingredient imidacloprid currently has several authorizations in 38 different products (database ctgb.nl [17], accessed 21-5-2013). The professional use ranges from the use in crops grown in glasshouses such as all different vegetables and in open systems for different bulbs of flowers, potatoes and sugarbeets. Imidacloprid is also registered for use in fruit trees including apple and pear trees. Generally, more than one pesticide is used to protect a specific crop from pest attack. Thus, depending on the land use type, imidacloprid is invariably emitted to surface waters in combination with other pesticides that are authorized to be used on those crops.

Imidacloprid contribution in the msPAF

The potentially affected fraction of the aquatic species by the measured pesticides is higher than 5% in 11 locations (reflecting 1.2 % of all monitoring sites) in the Netherlands in the year 2009. The maximum level that we determined based on the msPAF was 23% in the province of South-Holland. Imidacloprid contributed

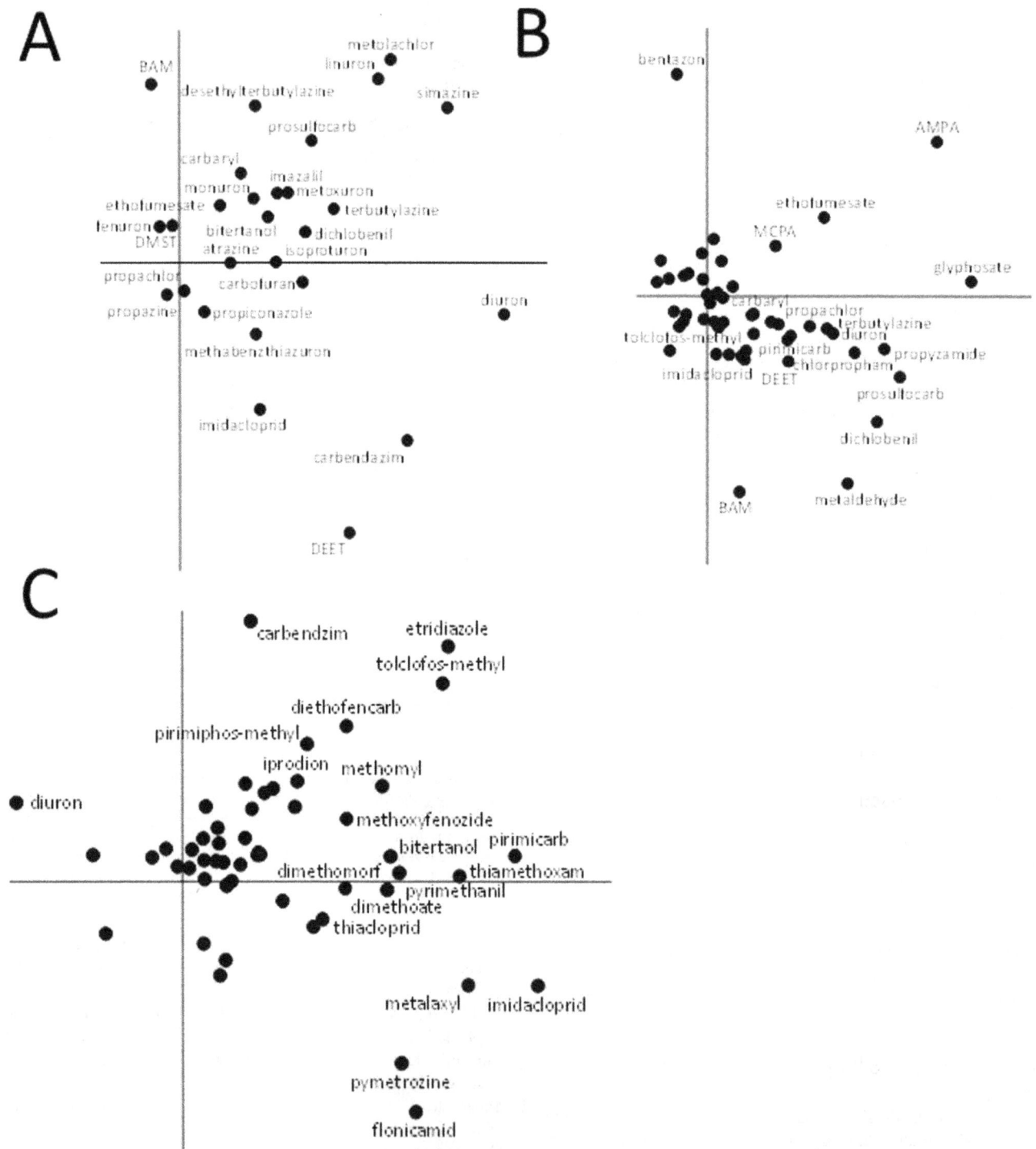

Figure 3. Results of the PCA analysis on data set 1 (A), 2 (B) and 3 (C). The PCA diagram of data set 1 displays 51% (33% on horizontal axis and 18% on vertical one) of the variation in chemical concentrations between the sites while 34% is displayed for data set 2 (21% on horizontal axis and 13% on vertical one) and 38% for data set 3 (28% on horizontal axis and 10% on vertical one).

in 8 out of 11 cases to this potential risk (Table 1). The relative contribution compared to other pesticides as measured at the same location at the same sampling time is rather modest and varied with a maximum of 21% at one location. Note that this calculation was based on acute toxicity data only, so likely is an underestimation of the potential risks that include both acute as chronic effects. From Table 1, it can be deduced that depending on

location, the contribution of specific individual active ingredients differs.

Pairwise combinations of samples

Imidacloprid measurements performed within a time window of 160 days which were taken at sites closer than 1 km from each other were compared. By this pairwise analysis we investigate if

Table 1. Contribution of imidacloprid to the msPAF at locations where msPAF > 5%.

x-coordinate	y-coordinate	Province	Total msPAF of measured pesticides (%)	Relative contribution of imidacloprid to the total msPAF of measured pesticides (%)
N 51 46 39.9	E 4 16 36.7	South Holland	22.53	0
N 52 1 29.6	E 4 30 24.7	South Holland	13.85	7.59
N 51 43 11.8	E 4 16 1.5	Zealand	12.48	0.002
N 51 52 33.5	E 4 10 26.2	South Holland	10.11	0
N 51 46 38.6	E 4 33 19.3	South Holland	9.91	0.009
N 51 45 0.4	E 4 25 46.2	South Holland	9.44	0
N 52 31 7.8	E 4 40 36.5	North Holland	9.25	0.014
N 51 57 10.2	E 4 15 8.8	South Holland	7.09	21.04
N 51 50 20	E 4 35 16.7	South Holland	6.61	0.001
N 51 21 52	E 4 2 10.1	Zealand	6.36	11.49
N 52 41 42.6	E 6 53 54.9	Drenthe	5.64	0.011

selected pairs of imidacloprid concentrations match with each other, and subsequently can be used to accurately link biological effect data and imidacloprid concentrations. Table 2 shows that in 39% of the comparisons there was no match in the presence of imidacloprid above the LOR, while only in 23% of the cases imidacloprid was present above the LOR in both samples. The remaining 38% of comparisons showed two measurements below the LOR. So when imidacloprid is found in at least one of the samples there is a large probability (62%) of not finding imidacloprid in the other site, which hampers the extrapolation of imidacloprid over a time window of 160 day and over a distance of 1 km (Table 2). We, therefore, conclude that the criteria used by Van Dijk et al. [1] to link chemical with biological observations result in a large probability (46%) of linking a site where imidacloprid was detected with a site, where the biological sample was taken, where actually no imidacloprid could be detected. The alternative, i.e. the first measurement being below the LOR and the second one above also has a relatively high probability (34%) (Table 2). Especially in a water-rich country such as the Netherlands, that has more than 350.000 km of ditch systems [18], it should be noted that sampling locations taken within 1 km, not necessarily have a hydrological connection with each other.

Imidacloprid dynamics

The concentration dynamics of imidacloprid (reflecting the concentrations of imidacloprid at the sampling locations with 10 or more samples taken in 2009 and with detection above the LOR in

at least 50% of those samples) are shown in Figure 4. In all but two (Fig. 4B and 4C) of these sampling sites the 28d, EC10 values for *C. horaria* and *C. dipterum* are exceeded for a period longer than 28 days, so at these sites chronic effects of imidacloprid exposure on mayflies can be expected. Also all standards are exceeded for some time in most of the sampling sites, with Fig. 4G showing the largest exeedence for a site near Boskoop in the province of South Holland. It should be noted that these 7 sites only constitute a small percentage (18%) of the total number of sites with 10 or more observations, so likely these exposure patterns represent the worst-cases of the exposure patterns at sites with 10 or more observations. Since we don't know whether there is a bias to measure imidacloprid more intensively at sites where exposure is expected we cannot extrapolate this to the whole population of sites.

Maximum concentrations of imidacloprid

Figure 5 shows the cumulative frequency of the all concentration measurements on the maximum level of imidacloprid for the years 2009, 2010 and 2011. The below LOR measurements are indicated at the 0.001 µg/L level and constituted 50, 53 and 55% of the maximum concentrations in 2009, 2010 and 2011, respectively. The results in Figure 4 show that peak concentrations of imidacloprid in the Dutch surface waters often exceeds the chronic effect concentrations of mayfly as determined in the chronic single species studies by Roessink et al. [16], as well as the three standards. In 2011 the MPC, 28d, EC10, AA-EQS and

Table 2. Result of the comparison of imidacloprid concentrations in samples taken in 2009 at sampling sites closer than 1 km and within 160 days.

Category	# sample pairs	% of total comparisons	% when 1st observation is above LOR	% when 1st observation is below LOR
Two below LOR	217	38		66
One below and above LOR	223	39	46	34
Two above LOR	134	23	54	
100% matching measurements	10	1.7		

LOR = analytical reporting limit.

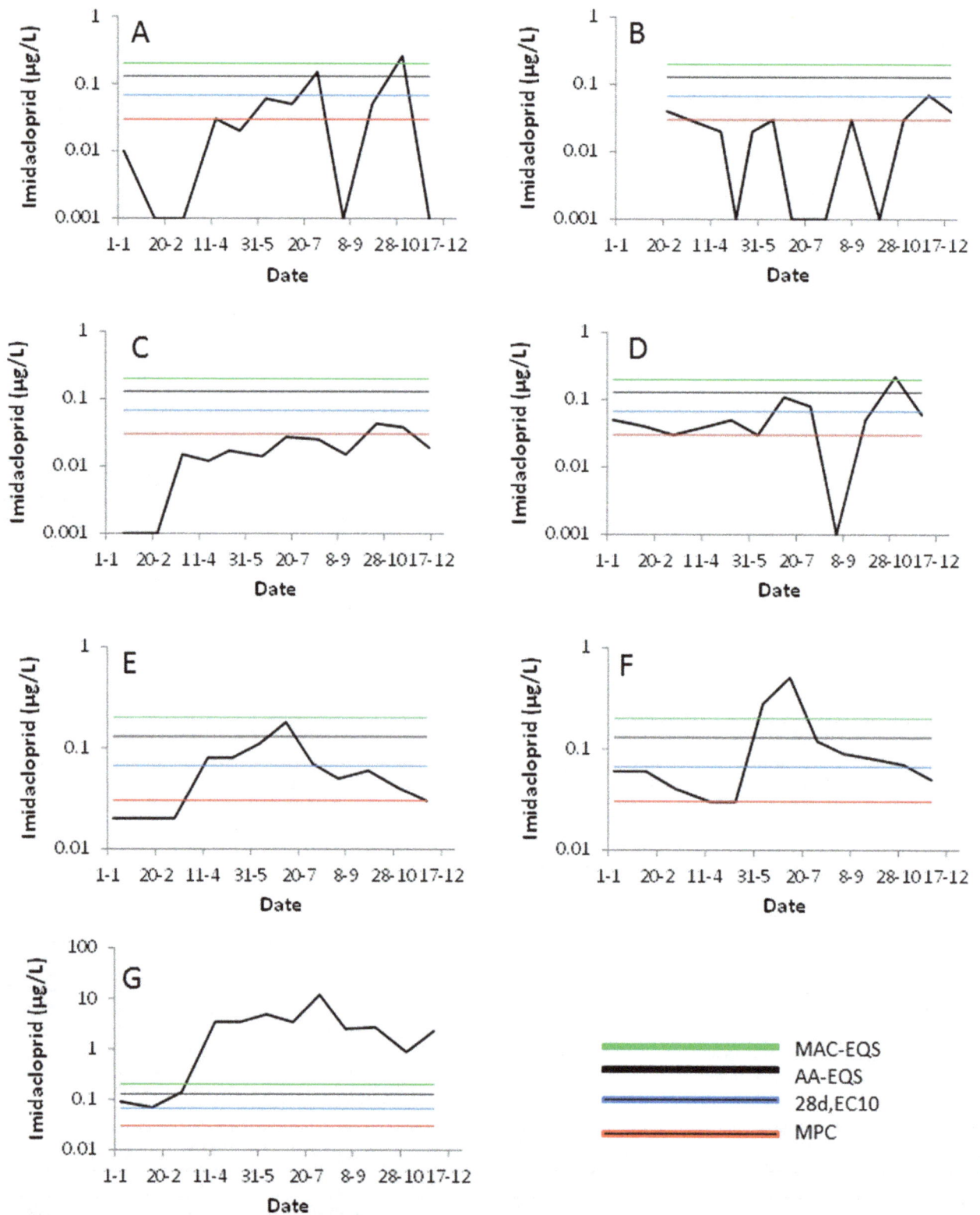

Figure 4. Concentration dynamics at the selected sampling sites (see text for procedure). The sampling sites 4A through 4G have X,Y coordinates of 108313,456412, 105888,455853, 103707,455196, 105927,453177, 170370,518957, 106781,503700 and 105079,453602, respectively. The horizontal lines denotes the MAC-EQS, the AA-EQS, the 28d, EC10 value for the mayflies C. horaria and C. dipterum (Roessink et al., 2013) and the MPC (top to bottom).

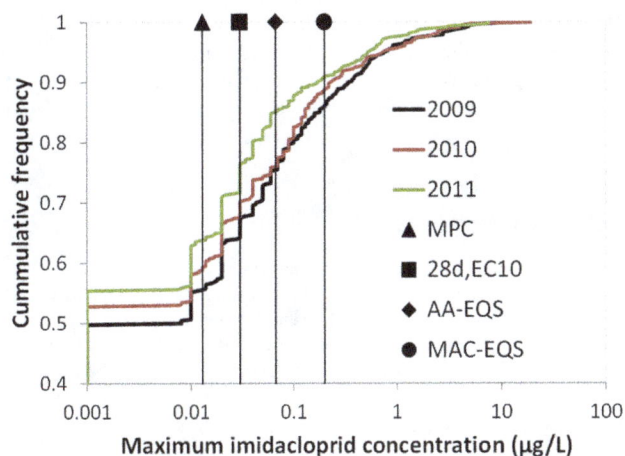

Figure 5. The cumulative frequency of the maximum imidacloprid concentrations of the sampling sites in 2009, 2010 and 2011, together with three standards and the 28d, EC10 of Cloeon dipterum and Caenis horaria.

MAC-EQS threshold values are exceeded by 36, 28, 15 and 9% of the maximum concentrations at the sampling sites, respectively. Since the Hazardous Concentration 5% based on 96h,EC10 values of 0.083 µg/L [16] corresponds more or less with the AA-EQS, acute effects of imidacloprid exposure cannot be excluded at a relatively large proportion of the sites ($\approx$15%). The maximum concentration is of course not a good predictor for the time weighted average concentration of 28d which should ideally be compared with the chronic threshold value of 0.03 µg/L. Still, when combining the results of the time-series (Fig. 3) and the exceedance of this threshold value by the maximum concentrations (Fig. 4) chronic effects of imidacloprid on insects like mayflies may be expected at a vast proportion of sites, with 28% being the most conservative estimate and 5% being the best guess. This 5% is calculated by multiplying the 28% chance of exceeding the threshold value by the maximum concentration and 15% chance of having above LOR measurements at more than 50% of the samples taken at a particular site where imidacloprid is measured at least 10 times. The comparison of the standards with the ecotoxicological threshold value for mayflies also suggests that the

MAC-EQS and AA-EQS are not fully protective for acute and chronic effects on insect taxa, respectively.

Exceedances of environmental quality standards

As stated in the Van Dijk et al [1] paper, in 2009 imidacloprid frequently exceeds quality standards for surface waters: 111 and 62 times for the AA-EQS and the MAC-EQS respectively [8,18]. In addition to the probability of exceeding a standard, also the magnitude of exceedance is important since it is likely that at higher magnitudes the ecological effects are more severe and maybe even last longer. Table 3 shows the compounds that exceeded the MPC most frequently in 2009, ranked according to degree of exceedance.

Imidacloprid was predicted to have a relatively large impact on the ecosystems compared to other pesticides, and gained third place in the Top 10 pesticides violating the environmental quality standards in respect to frequency and magnitude of exceedance. The number of measurements is high, as is also the number of locations from which the samples are taken. This means that monitoring is quite intensive for this compound, and surely covers many different surface waters belonging to different water managers and covering the geographical distribution of the different water types in the Netherlands. Although less intensively measured – a factor 5 to 10 – Table 3 also shows that other pesticides exceed the MPC more often. Thus although imidacloprid poses a significant ecological risk to surface waters in the Netherlands, it is not the only potential cause of degradation in macroinvertebrate abundance, as many other pesticides mentioned in Table 3 also exceed the MPC frequently (and in cases by orders of magnitude) and thus undoubtedly contribute to overall stress regime. It is a common flaw in ecological studies to selectively interpret individual causal agents within stressor regimes as the sole cause of observed phenomena, leading to erroneous conclusions.

Conclusion

Imidacloprid is one of several pesticides that can be detected in surface waters draining agricultural areas at levels frequently exceeding environmental quality standards. Despite this, we show here that key assumptions made by Van Dijk et al. [1] specifically relating to imidacloprid toxicity are not supported by observational data and, therefore, their assessment is unsuitable to determine threshold levels of effects. Specifically, the validity of

Table 3. Top10 pesticides exceeding the MPC in the Netherlands in the year 2009.

Pesticides name	No. of monitoring sites	% Exceedance	No. of measurements	% Exceedance
Captan	38	47	194	13
desethyl-terbuthylazin	63	37	299	10
Imidacloprid	451	44	2133	28
Triflumuron	24	21	142	4
Dicofol	24	17	142	3
Omethoaat	31	16	169	3
Foraat	51	14	313	2
Captafol	15	27	29	14
Fipronil	69	12	230	7
Pyraclostrobin	66	17	341	7

No. = number. The ranking of pesticides is based on frequency and magnitude of exceedances.

two assumptions: 1) that imidacloprid levels are not correlated with toxic levels of other pesticides residues and 2) that chemical exposure data can be extrapolated over a 1 km distance and 160 day time window are here shown to be highly questionable. The ecological status of field sites can be attributed to a complex suite of stressors resulting from a range of anthropogenic practices in the highly managed landscape of the Netherlands, of which pesticides are just one factor, and imidacloprid only one of many pesticides being applied, albeit an important one in terms of ecological risks. We therefore propose that any risk assessment should base the ecological threshold values not solely on field observations but also largely rely on the results of controlled experiments, since these types of experiments allow a full control of separating the imidacloprid stress from other stressors.

Supporting Information

Table S1 Acute toxicity values of imidacloprid (source eTox database, EPA database downloaded Oct 23th

2013). Legend: Species selected for the toxicity test were given with their scientific name and with their species group. Toxicity data were given as log10 effect concentrations at which 50% of the organisms showed adverse effects. The scientific papers from which those data are collected are given.

Acknowledgments

The authors thank Donald Baird for his critical comments and language suggestions. We thank Dick de Zwart for providing the eTox database. All pesticides measurements compared to the different EU and MPC quality standards can be found and freely downloaded at www. bestrijdingsmiddelenatlas.nl [8].

Author Contributions

Conceived and designed the experiments: MGV PJB. Performed the experiments: MGV PJB. Analyzed the data: MGV PJB. Contributed reagents/materials/analysis tools: MGV PJB. Wrote the paper: MGV PJB.

References

1. Van Dijk TC, Van Staalduinen MA, Van der Sluijs JP (2013) Macro-invertebrate decline in surface water polluted with imidacloprid. PLOS ONE 8 (5) e62374.
2. Vijver MG, De Snoo GR (2012) Overview of the state-of-art of Dutch surface waters in the Netherlands considering pesticides. (chapter (9) In: The impact of pesticides, M. Jokanovic (ed.) AcedemyPublish.org, WY, USA. ISBN: 978-0-9835850-9-1.
3. Vijver MG, Van 't Zelfde M, Tamis WLM, Musters CJM, De Snoo GR (2008) Spatial and Temporal Analysis of Pesticides Concentrations in Surface Water: Pesticides Atlas. J Environ Sci Health Part B 43: 665–674.
4. Yamamoto I (1999) "Nicotine to Nicotinoids: 1962 to 1997". In Yamamoto, Izuru; Casida John. Nicotinoid Insecticides and the Nicotinic Acetylcholine Receptor. Tokyo: Springer-Verlag. pp. 3–27 ISBN: 443170213X.
5. McDonald-Gibson C (29 April 2013). *The Independent.* Retrieved 1 May 2013.
6. Van Vliet J, Vlaar LNC, Leendertse PC (2013) Toepassingen, gebruik en verbod van drie neonicotinoïden in de Nederlandse land en tuinbouw. CLM 825- 2013. Available: www.clm.nl. Accessed 2013 May 5.
7. Mineau P, Palmer C (2013) The impact of the nation's most widely used insecticides on birds. Neonicotinoid Insecticides and Birds. American Bird Conservancy. Available: http://www.abcbirds.org/abcprograms/policy/toxins/neonic_final.pdf.
8. Dutch pesticides atlas website. Available: http://www.bestrijdingsmiddelenatlas. nl, version 2.0. Institute of Environmental Sciences (CML) at Leiden University and Waterdienst of the Dutch Ministry of Infrastructure and Environment. Accessed 2013 Oct 23.
9. Jolliffe IT (2002) Principal Component Analysis, Series: Springer Series in Statistics, 2nd ed., Springer, NY. ISBN 978-0-387-95442-4.
10. Ter Braak CJF, Šmilauer P (2012) Canoco reference manual and user's guide: software for ordination, version 5.0. Microcomputer Power, Ithaca, USA, 496 pp.
11. Zafar MI, Belgers JDM, Van Wijngaarden RPA, Matser A, Van den Brink PJ (2012) Ecological impacts of time-variable exposure regimes to the fungicide azoxystrobin on freshwater communities in outdoor microcosms. Ecotoxicol 21:1024–1038.
12. Van Wijngaarden RPA, Van den Brink PJ, Oude Voshaar JH, Leeuwangh P (1995) Ordination techniques for analyzing response of biological communities to toxic stress in experimental ecosystems. Ecotoxicol 4: 61–77.
13. Posthuma L, Suter GW II, Traas TP (eds) (2002) Species Sensitivity Distributions in Ecotoxicology. Lewis Publishers, Boca Raton, FL, USA.
14. De Zwart D (2005) Ecological effects of pesticide use in the Netherlands: Modeled and observed effects in the field ditch. Integrated Environmental Assessment and Management 1:123–134.
15. Hewlett PS, Plackett RL (1959) A unified theory for quantal responses to mixtures of drugs: non-interactive action. Biometrics 15:591–610.
16. Roessink I, Merga LB, Zweers HJ, Van den Brink PJ (2013) The neonicotenoid imidacloprid shows high chronic toxicity to mayfly nymphs. Environ Toxicol Chem 32: 1096 – 1100.
17. Statistics Netherlands. Available: http://www.statline.cbs.nl. Accessed 2013 May 21.
18. De Snoo GR, Vijver MG (eds) (2012) Bestrijdingsmiddelen en waterkwaliteit. Universiteit Leiden, 180 pp., ISBN: 978-90-5191-170-1.

Representation of Ecological Systems within the Protected Areas Network of the Continental United States

Jocelyn L. Aycrigg[1]*, Anne Davidson[1], Leona K. Svancara[2], Kevin J. Gergely[3], Alexa McKerrow[4], J. Michael Scott[5]

1 National Gap Analysis Program, Department of Fish and Wildlife Sciences, University of Idaho, Moscow, Idaho, United States of America, 2 Idaho Department of Fish and Game, Moscow, Idaho, United States of America, 3 United States Geological Survey Gap Analysis Program, Boise, Idaho, United States of America, 4 United States Geological Survey Gap Analysis Program, Raleigh, North Carolina, United States of America, 5 Department of Fish and Wildlife Sciences, University of Idaho, Moscow, Idaho, United States of America

Abstract

If conservation of biodiversity is the goal, then the protected areas network of the continental US may be one of our best conservation tools for safeguarding ecological systems (i.e., vegetation communities). We evaluated representation of ecological systems in the current protected areas network and found insufficient representation at three vegetation community levels within lower elevations and moderate to high productivity soils. We used national-level data for ecological systems and a protected areas database to explore alternative ways we might be able to increase representation of ecological systems within the continental US. By following one or more of these alternatives it may be possible to increase the representation of ecological systems in the protected areas network both quantitatively (from 10% up to 39%) and geographically and come closer to meeting the suggested Convention on Biological Diversity target of 17% for terrestrial areas. We used the Landscape Conservation Cooperative framework for regional analysis and found that increased conservation on some private and public lands may be important to the conservation of ecological systems in Western US, while increased public-private partnerships may be important in the conservation of ecological systems in Eastern US. We have not assessed the pros and cons of following the national or regional alternatives, but rather present them as possibilities that may be considered and evaluated as decisions are made to increase the representation of ecological systems in the protected areas network across their range of ecological, geographical, and geophysical occurrence in the continental US into the future.

Editor: Kimberly Patraw Van Niel, University of Western Australia, Australia

Funding: The National Gap Analysis Program at the University of Idaho is supported by the United States Geological Society Gap Analysis Program under grant #G08A00047. The url: gapanalysis.usgs.gov. The agreement mentioned above supported JA and AD to do the study design, data collection and analysis as well as the decision to publish and preparation of the manuscript. LS was supported by Idaho Department of Fish and Game to help with the study design, data analysis, and preparation of the manuscript. AM and KG were funded by United States Geological Survey GAP to help with study design, data collection and analysis. KG is the program officer for this agreement and he has been involved with the study design, data collection, and data analysis for this manuscript. JMS is retired and he helped with the study design, decision to publish and preparation of the manuscript. The funders had a role in the study design, data collection, and data analysis, but not in the decision to publish or preparation of the manuscript.

Competing Interests: The authors have declared that no competing interests exist.

* E-mail: aycrigg@uidaho.edu

Introduction

Traditionally, a mix of opportunity, available resources, and agency-specific conservation priorities are the foundation upon which networks of protected areas are developed over time [1–4]. This has led to a protected areas network in the continental US cultivated for multiple purposes including protecting biological resources, such as vegetation communities [5–8]. Often, to respond to conservation issues, such as habitat loss, the protected areas network is expanded by establishing new protected areas or enlarging existing ones [9–13]. However, with increasing land-use intensification the opportunities for expanding such networks are dwindling [4,14]. Furthermore, with the imminence of climate change along with increased loss and fragmentation of vegetation communities, the exigency of protecting areas that represent the full suite of vegetation communities and therefore the species found therein, has increased [15–17].

The conservation community has increasingly focused on landscape levels for national decision making, but the lack of relevant and consistent data at a national scale has been an impediment [18–20]. Most public land management agencies, even those with the broadest authorities to protect natural resources have yet to implement ecosystem-scale approaches, perhaps due to lack of relevant data [21,22]. However, the impediment that once prevented a national-scale approach to protected areas management in the continental US has recently been overcome with the availability of national-level data for vegetation communities, classified to ecological systems [23], and a protected areas database for the US [24]. Ecological systems are groups of vegetation communities that occur together within similar physical environments and are influenced by similar ecological processes (e.g., fire or flooding), substrates (e.g., peatlands), and environmental gradients (e.g., montane, alpine or subalpine zones) [23,25]. Ecological systems represent vegetation communities with spatial scales of tens to thousands of hectares and temporal scales of 50–100 years. They represent the habitat upon which vertebrate species rely for survival. The Protected Areas Database of the US (PAD-US) represents public land ownership and conservation lands (e.g., federal and state lands), including privately protected areas that are voluntarily provided (e.g. The Nature Conservancy) [24]. Each land parcel within PAD-US is assigned a protection status that denotes both the intended level of biodiversity protection and indicates other natural, recreational and cultural uses (Table 1) [24]. Together, these databases provide the foundation for assessing the representation of vegetation communities in the continental US within the protected areas network and thereby informing decision making at the national level.

The protected areas network within the continental US is often viewed as one of our best conservation tools for securing vegetation communities and the species they support into the future [26–29]. An inherent assumption behind a network of protected areas is that protection of vegetation communities will also protect the species that rely on them, including invertebrate and vertebrate species, many of which little is known of their life history or habitat requirements [11,30,31]. For our analysis, we narrowly defined a protected area as an area of land having permanent protection from conversion of natural land cover and a mandated management plan in operation to maintain a natural state within which disturbance events may or may not be allowed to proceed without interference and/or be mimicked through management (Table 1) [24]. Furthermore, we defined a protected areas network as a system of protected areas that increase the effectiveness of *in situ* biodiversity conservation [32]. Lastly, we defined biodiversity as a hierarchy from genes to communities encompassing the interdependent structural, functional, and compositional aspects of nature [33].

The questions of how much of a vegetation community to protect and what approach is best for systematically protecting vegetation communities have been discussed at length [34,35]. No single solution or specific amount of area has been established to

meet both policy targets and biological conservation needs [35]. Most recently the Convention on Biological Diversity set a target of 17% for terrestrial areas in the Aichi Biodiversity Targets described within the Strategic Plan 2011–2020 [36]. The Aichi Biodiversity Targets also attempt to address biological needs by stating that areas protected should be ecologically representative [36]. Representation of vegetation communities is often put forth as a goal of conservation planning because the aim is to protect something of everything in order to conserve the evolutionary potential of the entire protected areas network [34,37,38]. The US has not explicitly addressed the representation of vegetation communities within the protected areas network; however, Canada has used representation targets to structure their protected areas network [39–41]. Even though climate change will likely alter what is represented within Canada's protected areas network, starting from a representative group of protected vegetation communities provides a foundation for climate change adaptation [40,41].

Numerous assessments of the US protected areas network and its effectiveness at conserving vegetation communities have all concluded the network is falling short [15,20,42–48]. Each assessment used the best data available at the time, but in all cases, extent, resolution, and consistency of the data were limited. Shelford [42] conducted the first assessment of protected areas in the US in 1926. His aim was to study the native biota of North America, which started with inventorying the existing protected areas and how their vegetation communities had been modified from pre-settlement conditions. Later, Scott et al. [15] found that 302 of 499 (~60%) mapped vegetation communities within the US had <10% representation within protected areas. Dietz and Czech [20] found the median percentage of area protected within the continental US was 4% for the ecological analysis units they defined.

We recently have had the opportunity to evaluate the representation (i.e., saving some of everything) and redundancy (i.e., saving more than one of everything) of ecological systems within the existing protected areas network for the continental US. This opportunity was possible because of the availability of a complete ecological systems database for the continental US and a comprehensive database of the current protected areas network. Hence, we can now assess how well the protected areas network

Table 1. Description of protection status categories in the Protected Areas Database for US [24].

Protection status	Description	Example
Lands managed to maintain biodiversity (i.e., protected areas network)	An area of land having permanent protection from conversion of natural land cover and a mandated management plan in operation to maintain a natural state within which disturbance events may or may not be allowed to proceed without interference and/or be mimicked through management.	Yellowstone National Park, Wyoming
Lands managed for multiple-use, including conservation	An area having permanent protection from conversion of natural land cover for the majority of the area, but subject to extractive uses of either a broad low-intensity type (e.g., logging) or localized intense type (e.g., mining). Protection of federally listed endangered and threatened species throughout the area may be conferred.	Kaibab National Forest, Arizona
Lands with no permanent protection from conversion, but may be managed for conservation	An area with no known public or private institution mandates or legally recognized easements or deed restrictions held by the managing entity to prevent conversion of natural habitats to anthropogenic habitat types. Conversion to unnatural land cover throughout is generally allowed and management intent is unknown.	Fort Irwin, California

Protection status denotes the intended level of biodiversity protection and indicates other natural, recreational, and cultural uses. These designations emphasize the managing entity rather than the land owner because the focus is on long-term management intent. Therefore an area gets a designation of permanently protected because that is the long-term management intent.

encompasses the ecological and evolutionary patterns and processes that maintain ecological systems and thereby the species that depend on them [37]. Additionally, based on the Aichi Biodiversity Targets within the Strategic Plan 2011–2020 of the Convention on Biological Diversity, we can evaluate the current protected areas network in the continental US in context of meeting the suggested 17% target for terrestrial areas [36].

If the current protected areas network is falling short of conserving vegetation communities then what potential alternatives might be available to address those shortfalls? One such alternative is to replace protected areas that contribute minimally to conservation of vegetation communities with those with greater conservation value [49]. The goal would be to increase the overall biodiversity protection of the entire protected areas network. This approach proposed by Fuller et al. [49] could be attractive because the sale of protected areas with less conservation value could go towards acquiring new ones. Fuller et al. [49] proposed this approach in Australia where a protected areas network has been systematically designed with broad representation of Australia's vegetation types [49]. The protected areas network in the continental US has not been systematically designed [2,4]. Would this approach be feasible if the criteria for determining the contribution to conservation (i.e., cost-effectiveness analysis) could be agreed upon consistently across the continental US?

Another alternative to address the current protected areas network's shortfall could be to expand the network in area and number of protected areas [9,11,13]. A national assessment would be needed to identify vegetation communities not represented or under-represented within the existing protected areas network and a national conservation plan would be developed to prioritize acquisition of these vegetation communities to increase their representation on protected lands [50,51]. There are approximately 300 million hectares of public and private lands with no permanent protection on which native vegetation communities occur [23,24]. Could the representation of vegetation communities within the protected areas network be increased by prioritizing acquisition within these lands with no permanent protection?

A third alternative for addressing the protected areas network's shortcomings might be to increase the emphasis of maintaining biodiversity on some public and private lands currently managed for multiple-use (Table 1). Swaty et al. [52] found that in addition to the 29% of the continental US land area that has been converted by human use; there were an additional 23% of non-converted lands with altered vegetation structure and composition, which likely are lands managed for multiple-use. The protected areas network is comprised of approximately 50 million hectares in the continental US, while there are about 140 million hectares of public and private lands managed for multiple-use [24]. Vegetation communities that are currently not represented or underrepresented within the current protected areas network may have representation on the approximately 140 million hectares of land managed for multiple-use [20,24]. Could, therefore, an emphasis on maintaining biodiversity on a strategically targeted subset of lands managed for multiple-use be used to effectively expand the representation of vegetation communities within the entire protected areas network?

From a conservation management perspective for the US, the Department of Interior (DOI) has established a framework of Landscape Conservation Cooperatives (LCC) with the mission of landscape-level planning and management [53]. This national framework further supports the need for nationally consistent databases and analyses. We focused our analysis on alternative ways to potentially increase the representation of ecological systems in the protected areas network of the continental US.

Specifically we asked (1) how well are ecological systems represented in the protected areas network relative to their occurrence in the continental US, including with regards to soil productivity and elevation, (2) how alternative approaches may potentially increase the representation of ecological systems in the protected areas network, and (3) how Landscape Conservation Cooperatives (LCC), the new landscape unit for conservation initiatives, can be used to regionally assess conservation status of ecological systems.

Materials and Methods

Data Description

We used the National Gap Analysis Program (GAP) Land Cover [23] and US Geological Survey GAP's (USGS-GAP) Protected Areas Database of the US (PAD-US 1.0) [24] as the national datasets for our analyses. The land cover data contains 3 nested hierarchical levels of vegetation communities. Level I contains 8 groupings, based on generalized vegetative physiognomy (e.g., grassland, shrubland, forest), while Level II has 43 groupings representing general groups of ecological systems based on physiognomy and abiotic factors (e.g., lowland grassland and prairie, alpine sparse and barren). The third hierarchical level contains 551 map classes, including 518 ecological systems. We focused on the non-modified, non-aquatic classes at each level (Level I: 5 classes, Level II: 37 classes, and Level III: 518 ecological systems).

The National GAP Land Cover was compiled from the Southwest, Southeast, Northwest, and California GAP land cover data completed during 2004–2009 [23]. We incorporated data from LANDFIRE (www.landfire.gov) for the Midwest and Northeast. These national land cover data were based on consistent satellite imagery (Landsat Thematic Mapper (TM) and Enhanced Thematic Mapper (ETM)) acquired between 1999 and 2001 in conjunction with digital elevation model (DEM) derived datasets (e.g., elevation, landform) and a common classification system (i.e., ecological systems) to model natural and semi-natural vegetation [54–56]. The resolution is 30-m and typically the minimum mapping unit is 1 ha. Regional accuracy assessments and validations have been conducted and, based on those, in general, forest and some shrub ecological systems typically had higher accuracies than rare and small patch ecological systems, such as wetlands [57,58].

PAD-US (Version 1.0) consists of federal, state, and voluntarily provided privately protected area boundaries and information including ownership, management, and protection status [24]. Protection status is assigned to denote the intended level of biodiversity protection and indicate other natural, recreational, and cultural uses (Table 1) [24]. In assigning protection status, the emphasis is on the managing entity rather than the owner and focuses on long-term management intent instead of short-term processes [11]. The criteria for assigning protection status includes perceived permanence of biodiversity protection, amount of area protected with a 5% allowance of total area for intensive human use, protection of single vs. multiple features, and the type of management and degree to which it is mandated [59]. The protection status ranges from lands managed to maintain biodiversity to lands with little or no biodiversity protection (Table 1). Lands managed for multiple-use, including conservation, are permanently protected, but allow for extractive uses, such as mining and logging. In the continental US, lands with no permanent protection are considered any land parcel not designated either of the other protection status categories. We included only lands permanently protected and managed to maintain biodiversity in our definition of the protected areas network.

We also used elevation data obtained from the National Elevation Dataset (NED) [60] and soil productivity. The National Elevation Dataset, a seamless dataset with a resolution of approximately 30 m, was the best available raster elevation data for the continental US [60]. We divided the National Elevation Dataset into 8 classes ranging from 0 to 4500 meters at 500-meter intervals. Soil productivity classes for the continental US were based on STATSGO data (http://soils.usda.gov/survey/geography/statsgo/). These data were reclassified into 8 soil productivity classes based on land capability classes (http://soils.usda.gov/technical/handbook) and ranged from very high to very low productivity.

To apply our analysis and results to current conservation management in the continental US, we used the LCC framework [53]. LCCs represent large area conservation-science partnerships between DOI and other federal agencies, states, tribes, non-governmental organizations (NGOs), universities, and other public and private stakeholders. Their intent is to inform resource management decisions to address landscape-level stressors, such as land use change, invasive species, and climate change [53].

Data Analysis

The PAD-US 1.0 [24] and LCC data [53] were converted to grids (i.e., 30×30 m cells) and combined with the National GAP Land Cover [23] using ArcGIS 9.3.1 (ESRI, Redlands, CA). To assess the protection of ecological systems relative to their occurrence, we calculated a frequency distribution of protected area sizes within the existing protected areas network. To evaluate how the size range of protected areas would change with the inclusion of land managed for multiple-use, we calculated a frequency distribution of the protected areas network with lands managed for multiple-use added in (Table 1). We also calculated the amount of area of land managed for multiple-use needed to meet the 17% Aichi Biodiversity Target. To assess least protected or most endangered ecosystems, we summarized within each hierarchical level of the National GAP Land Cover (i.e., Levels I, II, and ecological systems) the number, size, protection status, and ownership of land parcels within PAD-US, as well as their distribution among LCCs. At the broadest level (Level I), we calculated percent availability versus percent protected to gain insight into the representation of each system in the protected areas network. We used a comparison index line (i.e., 1:1 line) to indicate the relationship between percent availability and percent protected [61]. Similarly, we calculated the percent area of ecological systems protected (i.e., managed to maintain biodiversity), managed for multiple-use, and not permanently protected for soil productivity and elevation ranges by combining these data with PAD-US [24] using ERDAS Imagine 9.3 (Table 1).

The diversity of ecological systems across and redundancy within LCCs was calculated by counting the number of ecological systems occurring within each LCC. Diversity was defined as the number of ecological systems within each LCC, while redundancy was defined as the number of LCCs in which a single ecological system occurred [37]. For example, if an ecological system occurred in 2 LCCs, its redundancy value was 2. Unique ecological systems were those that occurred in a single LCC. Furthermore, we calculated the number and percent area protected of ecological systems by each protection status within each LCC. To assess whether lands were being protected at the same rate as those converted to human dominated classes, such as developed areas, cultivated croplands, orchards, vineyards, quarries, mines, gravel pits, oil wells, and pastures, we calculated the conservation risk index (CRI) for each LCC by dividing percent area converted by percent area managed to maintain biodiversity or percent area managed to maintain biodiversity and for multiple-use [23,62]. Finally, we summarized CRI values by protection status.

Results

The current protected areas network in the continental US covers approximately 10% of the total area in which ecological systems occur. Across about 30,000 protected areas, the mean size of an individual protected area was 1942 ha with a size range of approximately 25–2,500,000 hectares over all protected areas. The analysis of representation of the network shows that the distribution of ecological systems managed to maintain biodiversity (i.e., the distribution of the protected areas network) is skewed towards high elevation and low productivity soils (Figure 1A). Overall 68% of all 518 ecological systems have <17% of their area protected, which is a target suggested by the Aichi Biodiversity Target of the Convention of Biological Diversity [36] and most of the ecological systems with <17% protected occur at low elevation and in areas with moderate to high productivity soils (Figures 1B and 1C, Table S1).

In examining the percent available versus percent protected for lands managed to maintain biodiversity, only two of the five Level I land cover groups (sparse and barren; riparian and wetland) occurred above the 1:1 line indicating a greater percentage of these groups are protected in relation to their availability (Figure 2). Representation of Level II land cover groups was lowest for lowland grassland and prairie (xeric-mesic), but most groups had <17% protected (Figure 3). Out of 37 Level II groups, 11 fell at or above the 17% Aichi Biodiversity Target [36].

Ecological systems on lands managed for multiple-use and on lands with no permanent protection comprised 29% and 61%, respectively, of the total area of the continental US in which ecological systems occur. When lands managed for multiple-use were included as part of the protected areas network, the overall number of protected areas increased to about 88,000 with a size range of approximately 25–117,757,000 hectares.

When both lands managed to maintain biodiversity and for multiple-use were included all five Level I land cover groups occurred above the 1:1 line and all five occurred at or above the suggested 17% Aichi Biodiversity Target (Figure 2) [36]. The largest increases were within the shrubland, steppe, and savanna group, forest and woodland group, and sparse and barren group. The percent area of Level II land cover groups increased for all 37 groups when lands managed for multiple-use were added to lands managed to maintain biodiversity (Figure 3). The largest increases in percent area occurred within the lowland grassland and prairie (xeric-mesic) and sagebrush dominated shrubland. Out of 37 Level II groups, 33 fell at or above the 17% Aichi Biodiversity Target [36] when both lands managed to maintain biodiversity and multiple-use were included (Figure 3).

To meet the suggested 17% Aichi Biodiversity Target [36], approximately 9 million hectares (6.4%) of the 140 million hectares of public and private lands managed for multiple-use or 34 million hectares (11.3%) of the 300 million hectares of lands with no permanent protection would need to emphasize maintaining biodiversity or be acquired as part of the protected areas network (Table S1). Including lands managed for multiple-use with lands managed to maintain biodiversity, 98% of all ecological systems increased their percent area protected (Table S1). Using the suggested 17% Aichi Biodiversity Target [36], we found 32%

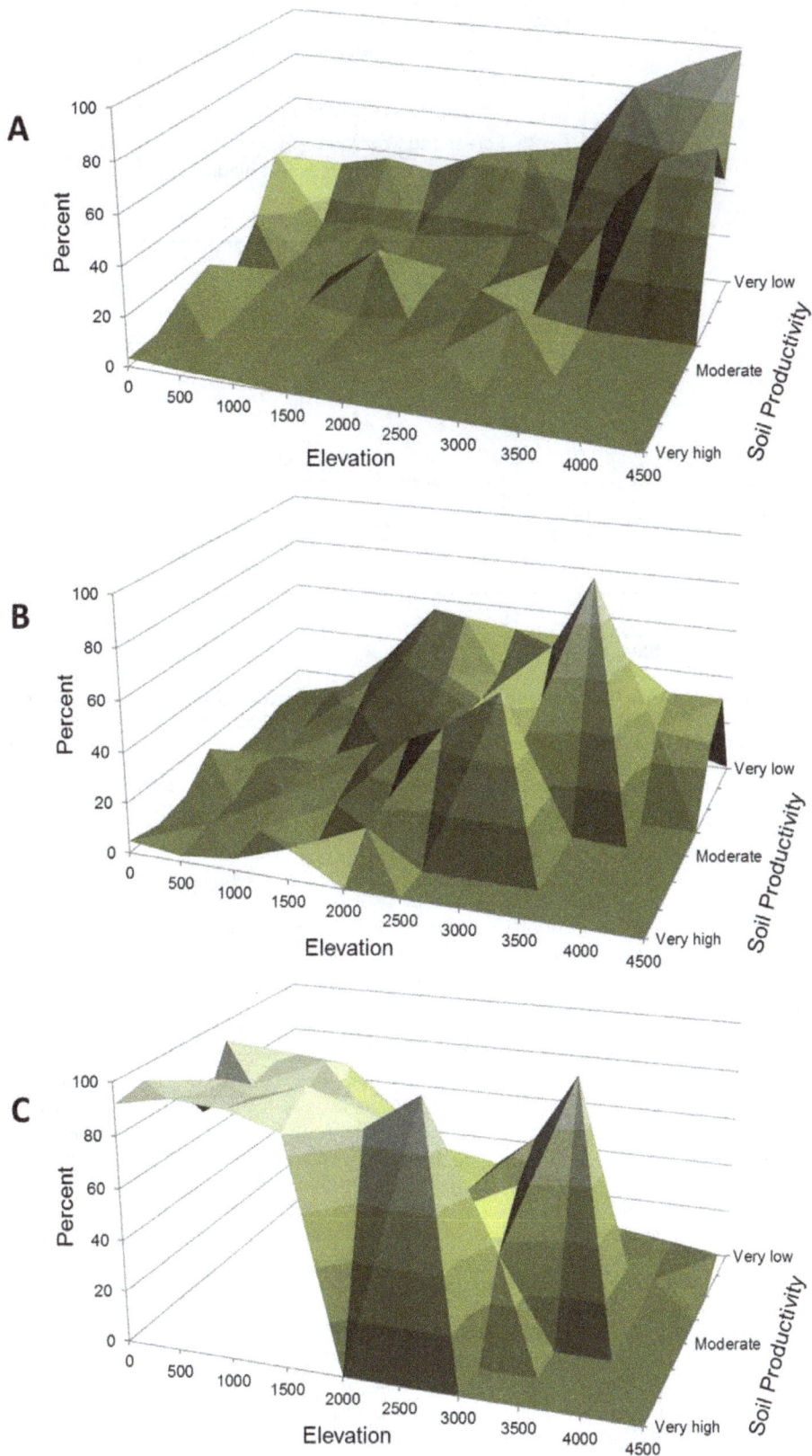

Figure 1. Percent area of ecological systems in relation to elevation, soil productivity, and protection status. Protection status designations include lands managed to maintain biodiversity (A), lands managed for multiple-use (B), and lands that have no permanent protection (C). See Table 1 for protection status descriptions. Percent area of ecological systems determined by combining data for elevation (meters) and soil productivity (http://soils.usda.gov/technical/handbook) with ecological systems grouped by protection status [23,24,60].

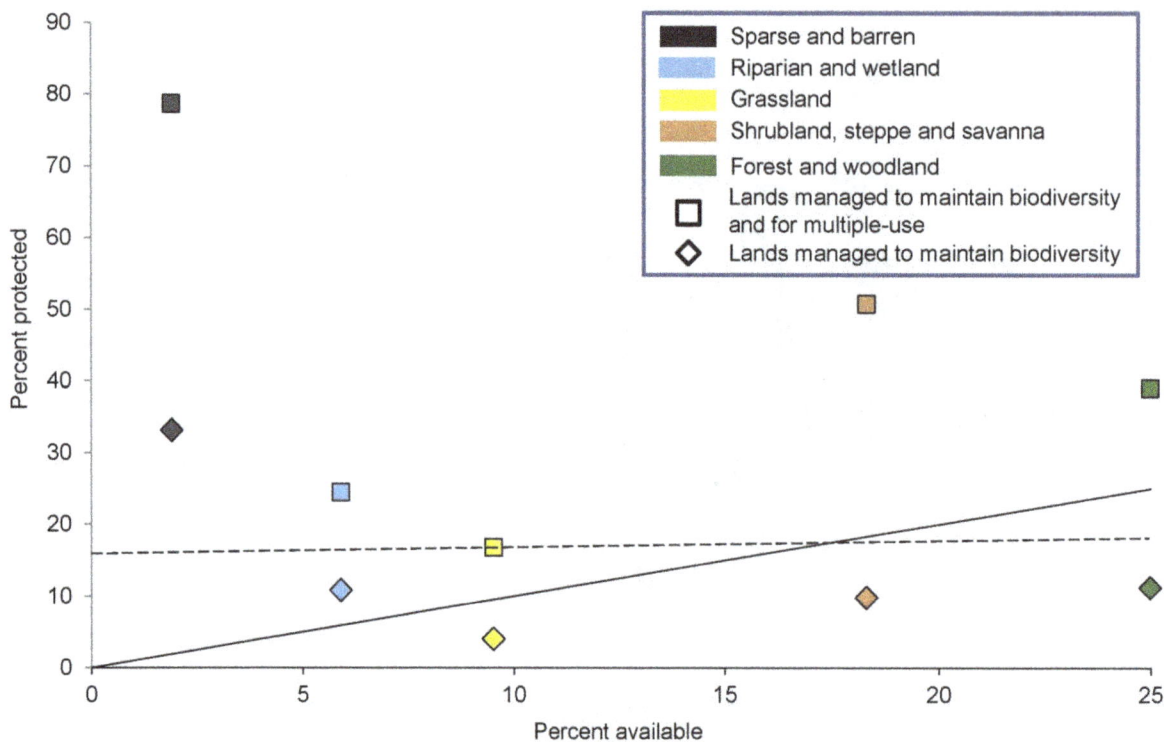

Figure 2. Percent protected and available for each Level I land cover group by protection status. Lands managed to maintain biodiversity (diamonds) are shown relative to lands managed to maintain biodiversity and for multiple-use (squares). See Table 1 for protection status descriptions. A comparison index line is shown, which indicates a 1:1 relation between percent availability and percent protected [61]. A value below the 1:1 line represents a Level I land cover group under-represented in the protected areas network, a value above represents a Level I land cover group well represented in the protected areas network, while a value on the line indicates a Level I land cover group available and protected equally [61]. For example, grassland, a Level I land cover group, has about 4% of its area managed to maintain biodiversity, but that increased to about 17% when lands managed for multiple-use were included [23,24]. A dashed line representing the 17% Aichi Biodiversity Target of the Convention on Biological Diversity is shown [36].

of all ecological systems met that target, but that increased to 68% when lands managed for multiple-use were included (Table S1).

Including lands managed for multiple-use in the protected areas network would result in dramatic geographic changes in the western US, but noticeable changes were also evident in northeastern US, Florida, the Appalachian mountains, and around the Great Lakes (Figure 4). Federal, state, and local governments as well as private entities manage lands to maintain biodiversity and for multiple-use (Figure 5). There are approximately 50 million hectares of lands managed to maintain biodiversity with Bureau of Land Management (BLM) and US Forest Service (USFS) managing about 29 million hectares, which is more than US Fish and Wildlife Service (USFWS), National Park Service (NPS), and all other federal land combined (Figure 5). Approximately 140 million hectares is managed for multiple-use in the continental US with BLM and USFS managing about 100 million hectares (Figure 5, Table S1).

Redundancy values for ecological systems occurring in LCCs ranged from 1–8, with redundancy values higher in LCCs in the west (Figure 6A). Ecological systems were highly diverse in 4 LCCs (Great Northern, Great Basin, Desert, and Gulf Coast Plain and Ozarks); however, only 1 had numerous unique ecological systems (Gulf Coast Plains and Ozarks; Figure 6B and Table 2). When including lands managed for multiple-use in the protected areas network, 7 out of the 16 LCCs in the continental US more than doubled the percent area protected (Table 2). Lands managed to maintain biodiversity represented between 0.6–17.0% of the area

of LCCs, adding lands managed for multiple-use increased that to 1.2–62.9% (Table 2). Eight out of 16 LCCs contained ecological systems that occurred only on lands managed for multiple-use or had no permanent protection (e.g., Great Plains, North Atlantic; Figure 7). The CRI values varied across LCCs with the Eastern Tallgrass Prairie and Big Rivers having the highest value (126.4) because almost 80% of its area was converted to human use (i.e., cultivated cropland) and the Desert and Southern Rockies having the lowest (0.2) because >10% of their area contained lands managed to maintain biodiversity (Figure 8). Including lands managed for multiple-use lowered the CRI for all LCCs and increased the number of LCCs meeting the suggested Aichi Biodiversity Target of 17% target from 1 to 7 (Figure 8) [36].

Discussion

Protection of Ecological Systems Relative to their Occurrence in the Continental US

The existing protected areas network in the continental US would need to capture a more representative complement of ecological systems if the US aims to meet the suggested Aichi Biodiversity Target of 17% for ecologically representative terrestrial areas [36]. The 518 ecological systems mapped in the continental US are disproportionately distributed by number, size, and protection status relative to elevation and soil productivity, which translates to an uneven representation of ecological systems within the protected areas network (Figure 1A) [15,63]. Soils with

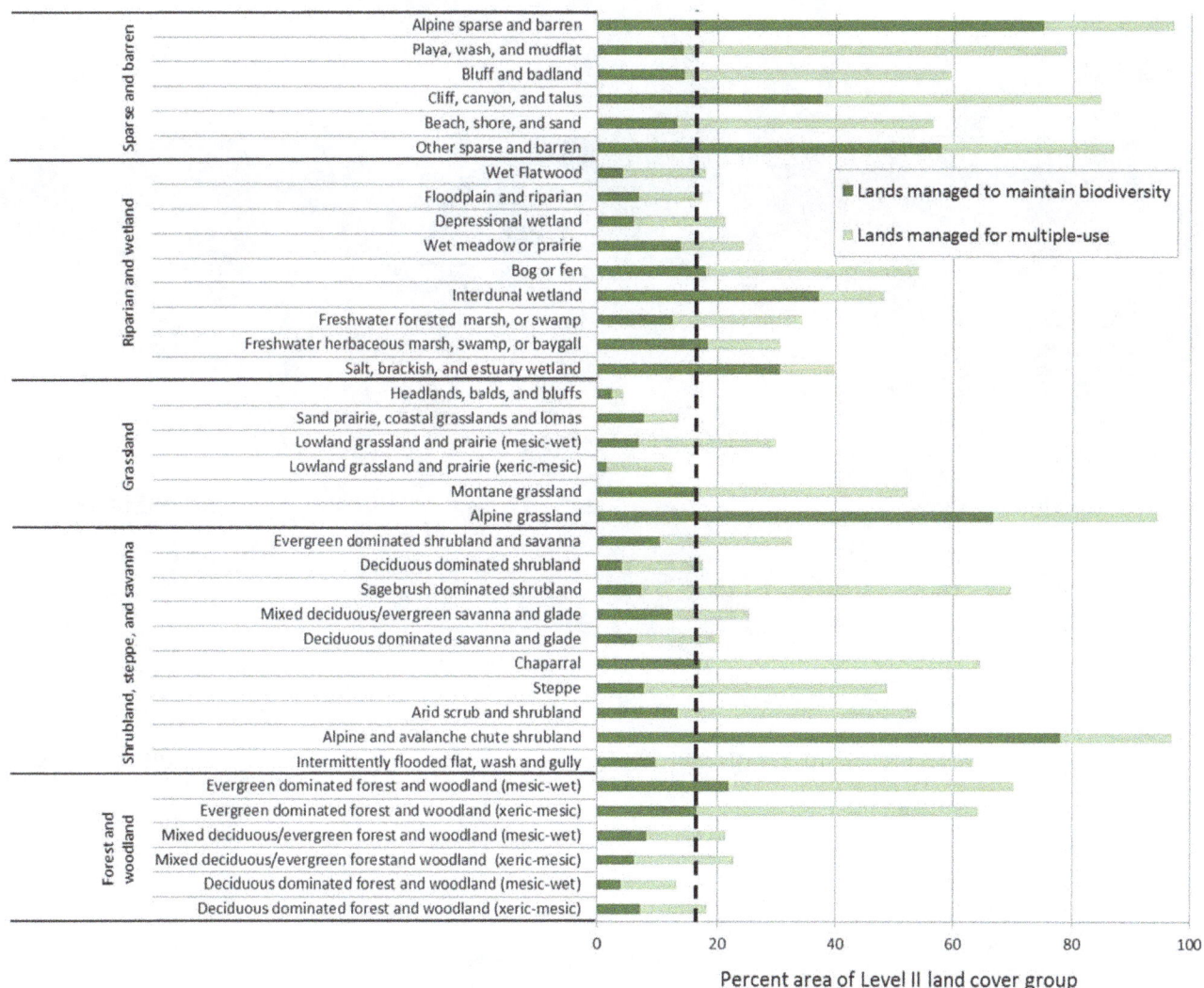

Figure 3. Percent area of Level II land cover groups by protection status. The Level II land cover groups are arranged by Level I land cover groups (see Table S1) [23]. Percent area for both lands managed to maintain biodiversity and lands managed for multiple-use are shown [24]. See Table 1 for protection status descriptions. A dashed line representing the 17% Aichi Biodiversity Target of the Convention on Biological Diversity is shown [36].

low productivity at high elevation are more likely to be found within the protected areas network; therefore ecological systems that occur in those areas are disproportionally represented in the network. Typically, low soil productivity at high elevations occurs in sparse and barren areas and these areas are well represented within the protected areas network (Figure 2) [15]. Capturing a broader range of elevation could be important to spatial patterns of biodiversity because ecological systems might shift with climate change, but the patterns of biodiversity will likely endure with geophysical features, such as elevation range [64]. How can the representation of ecological systems increase within the protected areas network of the continental US?

Alternatives for Increasing Representation and Conservation of Ecological Systems

Many alternatives exist for conserving ecological systems and successful conservation will likely come from employing one or more of them. One approach, presented earlier in the paper, would be to replace protected areas that are minimally contrib-

uting to conservation and have a high cost associated with protecting ecological systems within a specific protected area (i.e., least cost effective) with those having greater conservation value (i.e., more cost effective) to increase the overall biodiversity protection of the entire network [49]. Applying this approach could be challenging because public support for existing protected areas may make it difficult to convince those supporters to relinquish a protected area for the benefit of the entire network [8,65]. This approach, even though controversial because of the concept of giving up protected areas, could play a prominent role in addressing the impacts of climate change because of the potential opportunity to shift the distribution of ecological systems on current protected areas in response to shifts in temperature and precipitation [66,67].

Protected areas have long been downgraded, downsized, delisted, and degazetted and these practices are currently widespread [68,69]. Approximately 60 National Parks have been delisted and downgraded since the establishment of the National Park System in 1916 [68,70,71]. One of the major drivers of protected area degazettement, which is loss of legal protection for

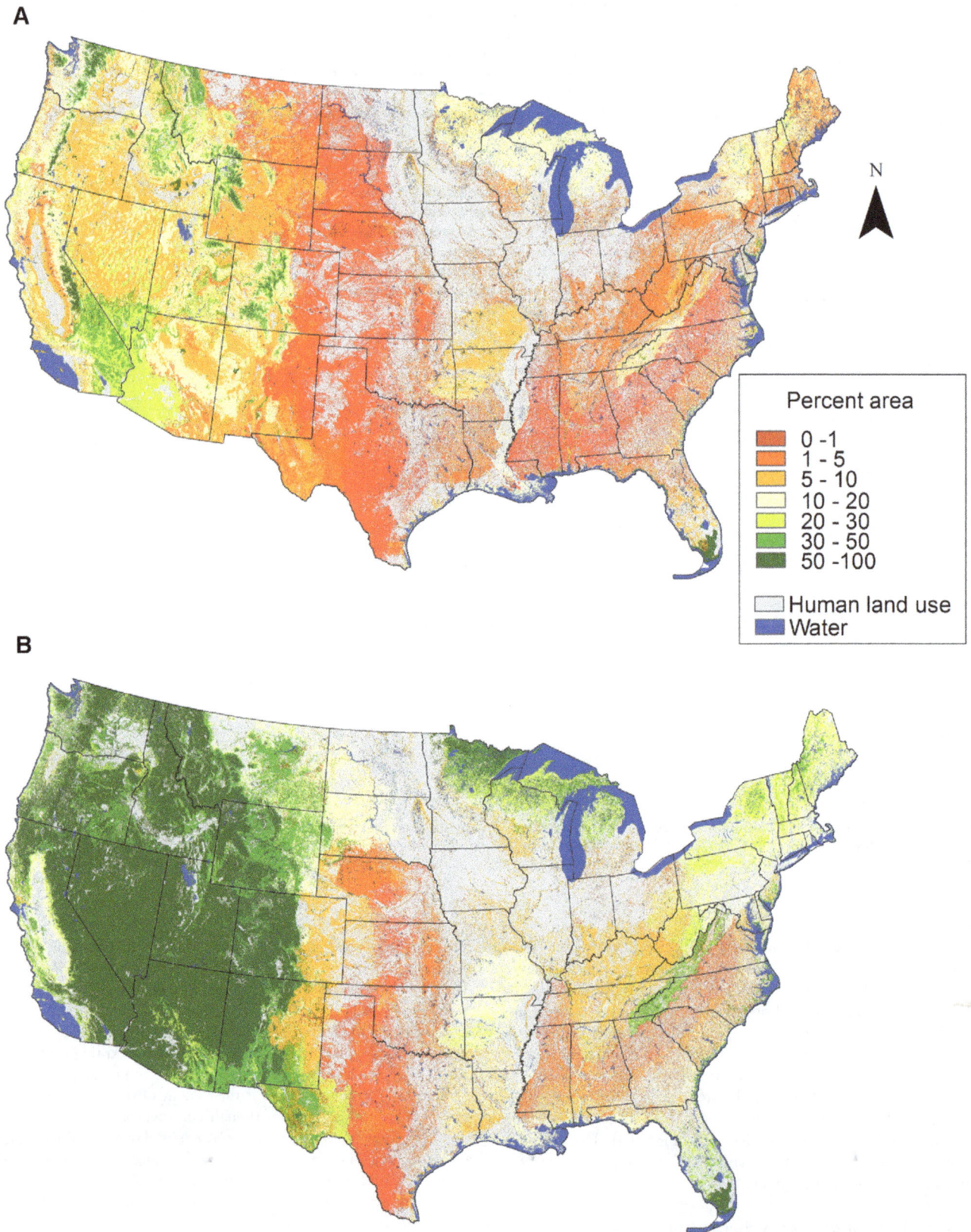

Figure 4. Percent area of ecological systems by protection status. Protection status designations are lands managed to maintain biodiversity (A) and lands managed to maintain biodiversity and multiple-use (B) for the continental US. Percent area is based on the area of each ecological system within each protection status divided by the total area of each ecological system [23,24]. See Table 1 for protection status descriptions. Only non-modified, non-aquatic ecological systems were included (n = 518; Table S1).

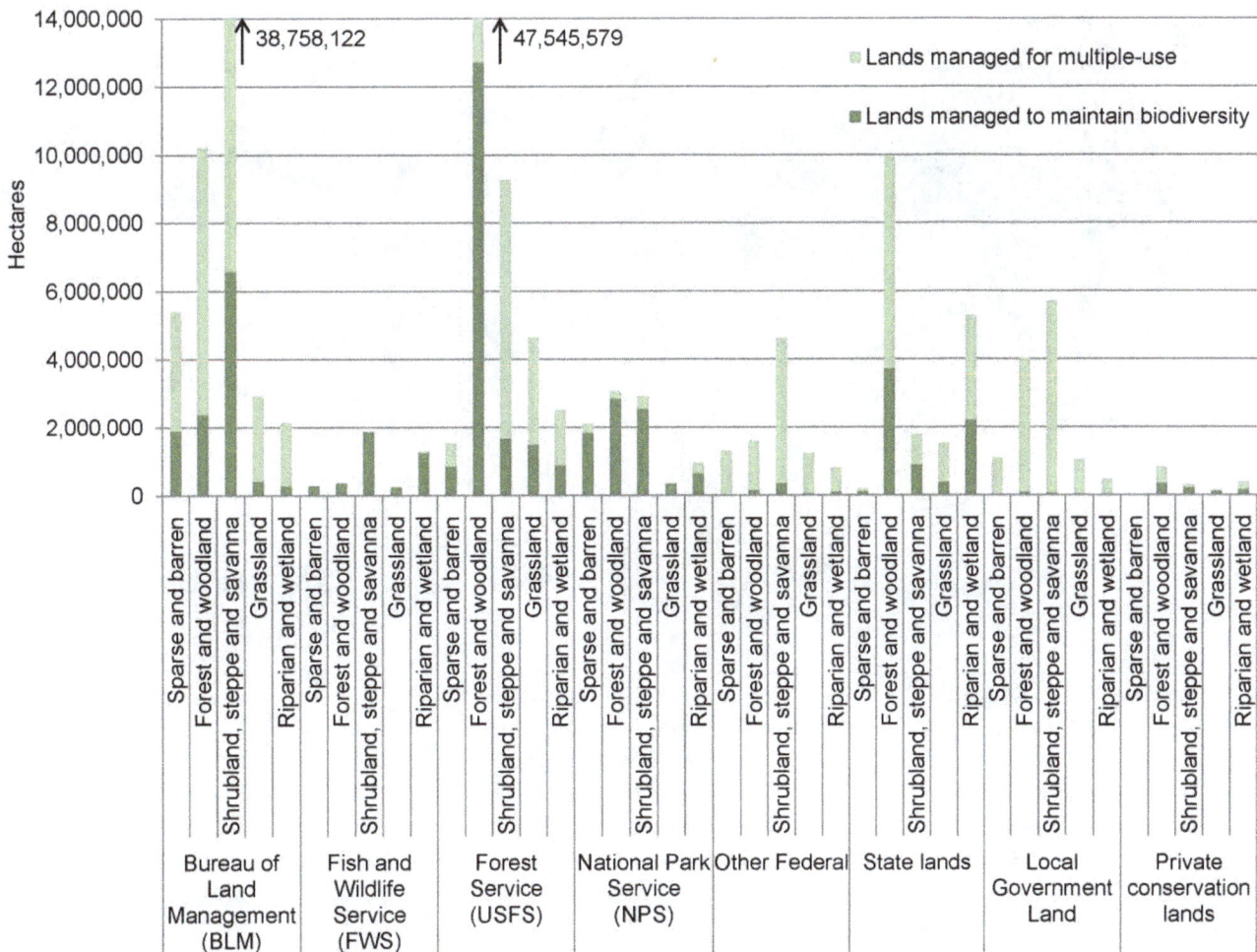

Figure 5. Area (ha) of Level I land cover groups by ownership and protection status. Ownership includes federal, state, and local governments as well as private conservation lands. See Table 1 for protection status descriptions. These values were for the continental US. Both BLM and USFS have areas of Level I land cover groups that fall outside the scale on this graph [23,24]. Values for those Level I land cover groups are shown.

an entire protected area, is access to and use of natural resources (e.g., commodity extraction) [69]. The impact on biodiversity protection because of access and use of natural resources is evident in Midwestern US where a low percent area of land is managed to maintain biodiversity and many areas are mapped as human land use (Figure 4). LCC's in the Midwest (i.e., Plains and Prairie Potholes, Great Plains, and Eastern Tallgrass Prairie and Big Rivers) have low diversity and few unique ecological systems (Figure 6B). A large percent of their area has been converted to human land use, which is reflected in high CRI values (Figure 8). To date, the ecological consequences of degazettement are unclear [69]. Both Fuller et al. [49] and Kareiva [8] believe degazettement would lead to a more dynamic and flexible approach to maintaining the current protected areas network, however it could depend on the level of systematic design used to establish the protected areas network.

Even though we did not specifically assess cost effectiveness of protected areas, our analysis could help inform the approach proposed by Fuller et al. [49]. A cost effectiveness analysis could be based on land ownership, protection status, and percent area converted to human modified systems. For example, the Great Basin LCC has potential for including some of the most cost effective protected areas because it has a low CRI value and

<10% of its area is converted. There is the potential to lower its CRI value and meet the suggested 17% Aichi Biodiversity Target [36] by increasing the percent of area managed to maintain biodiversity by 60% through emphasizing protection of biodiversity (Figure 8). The Great Basin LCC also contains ecological systems that occur only on lands managed for biodiversity (Figure 7) and has a high diversity of ecological systems even though only 1 is unique (Figure 6B). Other factors beyond land ownership, protection status, and percent area converted to human modified systems could be considered in efforts to assess the cost effectiveness of protected areas, such as representation of ecological systems and transaction costs. However, our analysis could help inform a conservation strategy for the continental US if the approach described by Fuller et al. [49] were implemented.

The second alternative for improving the conservation and representation of ecological systems described previously would be to increase the size (i.e., area or number) of our existing protected areas network through acquisition for the least protected, most endangered, or high priority ecological systems [50,51]. If a systematic approach for choosing new protected areas could increase the representation of elevation and soil productivity and thereby ecological systems then the network's ability to respond to varying conditions and future change could be strengthened

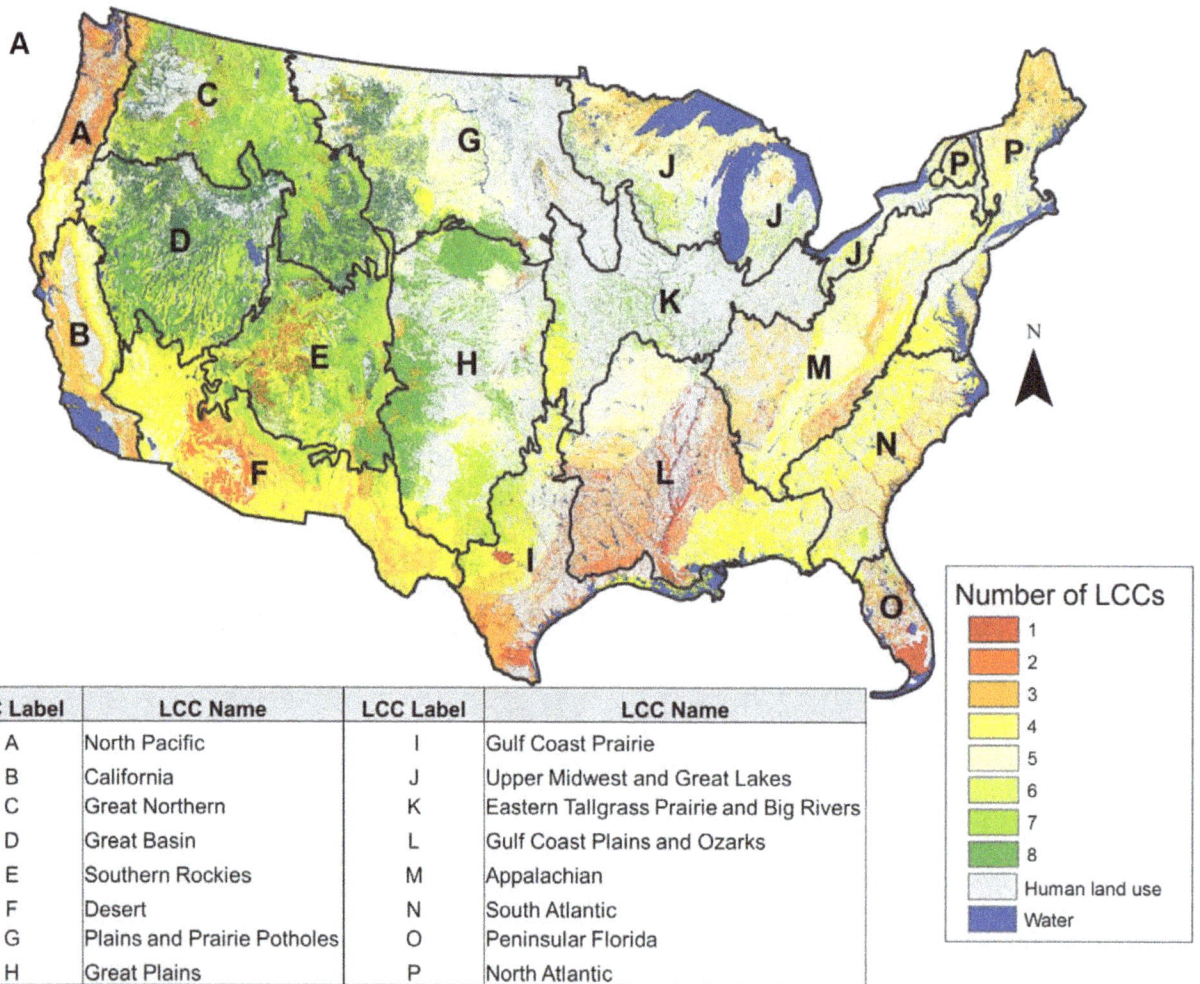

LCC Label	LCC Name	LCC Label	LCC Name
A	North Pacific	I	Gulf Coast Prairie
B	California	J	Upper Midwest and Great Lakes
C	Great Northern	K	Eastern Tallgrass Prairie and Big Rivers
D	Great Basin	L	Gulf Coast Plains and Ozarks
E	Southern Rockies	M	Appalachian
F	Desert	N	South Atlantic
G	Plains and Prairie Potholes	O	Peninsular Florida
H	Great Plains	P	North Atlantic

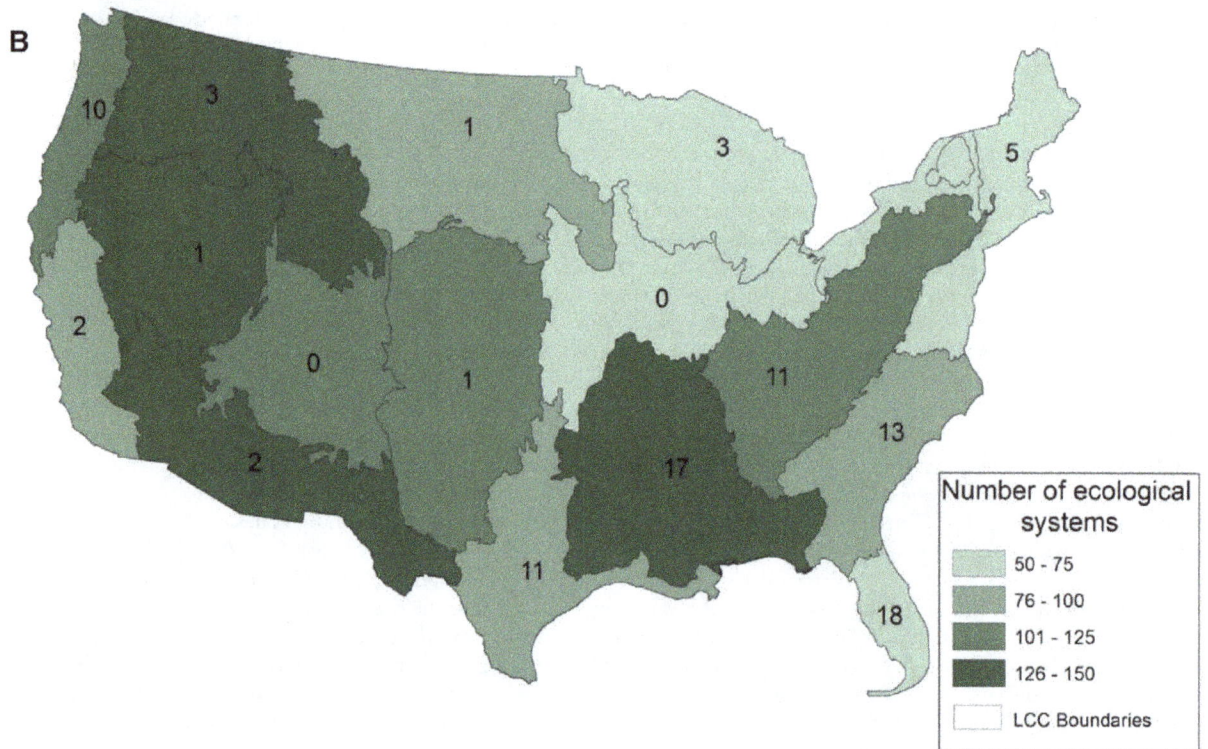

Figure 6. Redundancy, diversity, and uniqueness of ecological systems within Landscape Conservation Cooperatives (LCC). Redundancy measures the number of LCC's in which a single ecological system occurs (A) [23]. The higher the number of LCC's in which an ecological systems occurs the more redundancy displayed by that ecological system. For example, if an ecological system occurs in 2 LCCs, it has a redundancy value of 2. Diversity is the total number of ecological systems occurring with an LCC, which is shown by color shading of LCCs (B). Uniqueness is the number of ecological systems that occur in a single LCC, which is indicated by the number within each LCC (B). For example, the Great Northern LCC encompasses 126–150 ecological systems total, most of these occur in a total of 7 or 8 LCCs, but 3 are unique and only found in this LCC. Only non-modified, non-aquatic ecological systems were included (n = 518; Table S1). Each LCC is assigned a letter, which indicates the name of the LCC.

(Figure 1) [15,63]. Our results were similar to Scott et al. [15] because we found that ecological systems at lower elevations and higher soil productivity were under-represented within the current protected areas network (Figure 1). These areas could be prioritized if acquisition of new protected areas was employed for increasing protection of ecological systems. The least protected ecological systems and potentially most endangered (see Figure 8) are within all the Level I land cover groups except sparse and barren (Figures 2, 3, and 5, Table S1) and are located mostly in the Midwestern US (Figure 4). Prioritizing acquisition of the Level I land cover groups within the Midwestern US would increase the overall representation of ecological systems in the continental US. However, the feasibility of land acquisition for conservation is continually a challenge as resources for obtaining new protected areas are dwindling and competition for undeveloped private land is limiting expansion opportunities [4,14]. Furthermore, the support of policy makers for creating new protected areas could be perceived as ephemeral [72]. The idea of increasing the amount of protected land is attractive in part because of the perceived permanence associated with that protection. In other words, expanding the protected areas network reduces the risk of more land being converted to a state from which it might not recover (i.e., urban development), even though the immediate benefit to conservation is dependent upon management strategies employed.

A third alternative for improving the current protected areas network might be to take stock of our management within the current protected areas network and to evaluate the potential role of lands managed for multiple-use in conserving ecological systems. Our analysis found that increasing the emphasis on maintaining biodiversity on lands currently managed for multiple-use, which are permanently protected, but allow for extractive uses (e.g., mining and logging), offers an alternative for increasing the representation of ecological systems. However, much of the land managed for multiple-use has undergone ecosystem alteration and increased management or restoration may be needed to recover existing ecological systems [52]. If we increased the emphasis on maintaining biodiversity on some public and private lands managed for multiple-use, the total percent area of ecological systems protected could increase up to 39% in the continental US (lands managed to maintain biodiversity: 10%; lands managed for multiple-use: 29%). Geographically, the greatest potential for increased emphasis on maintaining biodiversity on lands managed for multiple-use is in the West, but also in the Northeast, South, and Midwest (Figure 4). To meet the suggested Aichi Biodiversity Target of 17% [36] increased emphasis on maintaining biodiversity would need to occur on 6.4% of the lands managed for multiple-use (Table S1). Even though lands managed for multiple-use occur on both public (i.e., federal, state, and local government) and private (i.e., non-governmental organization) lands, the potential for conservation efforts to increase the protection of

Table 2. Total number and unique number of ecological systems as well as percent area of ecological systems on lands managed to maintain biodiversity and for multiple-use within each Landscape Conservation Cooperative (LCC) in the continental US.

Landscape Conservation Cooperative (LCC)	Number of ecological systems	Number of unique ecological systems	Percent area of lands managed to maintain biodiversity	Percent area of lands managed for multiple-use
Appalachian	103	11	3.5	8.3
California	88	2	10.7	16.3
Desert	133	2	17.0	40.0
Eastern Tallgrass Prairie & Big Rivers	75	0	1.2	1.2
Great Basin	143	1	11.2	62.9
Great Northern	143	3	14.8	39.3
Great Plains	102	1	0.6	2.5
Gulf Coast Plains & Ozarks	148	17	3.5	4.9
Gulf Coast Prairie	95	11	1.3	1.4
North Atlantic	63	5	6.6	8.7
North Pacific	123	10	15.1	25.5
Plains & Prairie Potholes	95	1	2.4	10.6
Peninsular Florida	56	18	8.8	13.1
South Atlantic	97	13	2.8	4.0
Southern Rockies	116	0	14.1	50.6
Upper Midwest & Great Lakes	60	3	5.7	8.3

See Figure for location of LCC. See Table 1 for protection status descriptions. LCCs are listed alphabetically.

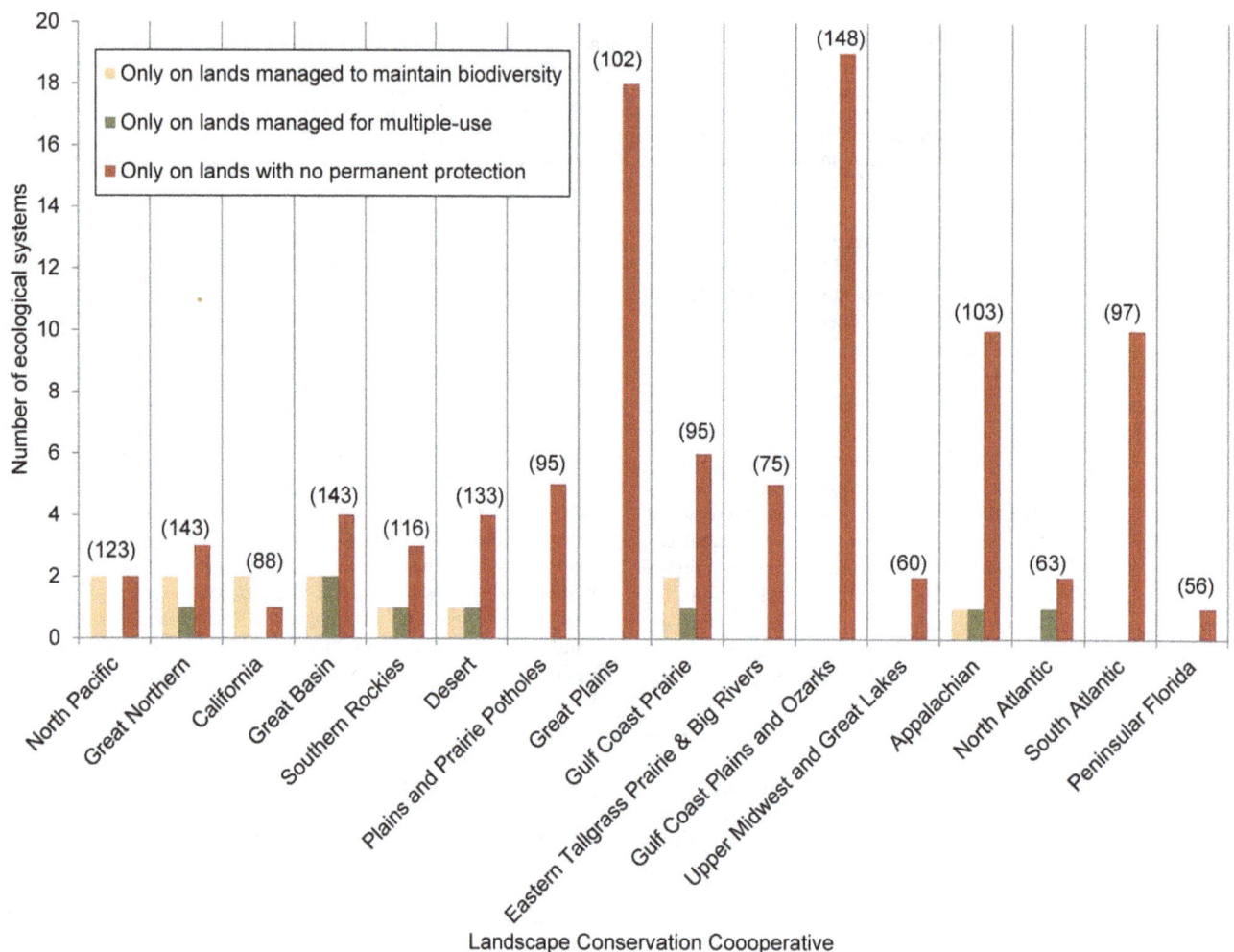

Figure 7. Number of ecological systems occurring only within each protection status by Landscape Conservation Cooperative (LCC). Ecological systems included occur only within the specified protection status [23,24]. The total number of ecological systems within each LCC is shown parenthetically. For example, the Great Plains LCC contains 102 ecological systems with 18 occurring only on lands with no permanent protection and none occurring on lands managed to maintain biodiversity or for multiple-use. See Table 1 for protection status descriptions. Only non-modified, non-aquatic ecological systems are included (n = 518; Table S1).

ecological systems on public lands is greater (i.e., quantitatively and geographically) (Figure 5).

To protect a broad representation of ecological systems within the continental US, opportunities within public land management agencies fall largely on lands managed by BLM and USFS (Figure 5). Both manage lands that maintain biodiversity, but the majority of the lands they manage are for multiple-use (Figure 5). However, if the US is to become less dependent on foreign energy sources and meet its own resource needs within its boundaries, then shifting management focus on even a small portion of lands currently managed for multiple-use could become a public lands dilemma. Lands managed for multiple-use provide multiple public benefits, including domestic energy production. [17,73,74].

In addition to the lands BLM manages for multiple-use, it has also designated 11 million hectares to the National Landscape Conservation System (NLCS), which is a network of conservation areas specifically aimed at conserving biodiversity [75]. The USFS manages over 17 million hectares of land managed to maintain biodiversity, which is more than USFWS, NPS, and other federal land management agencies combined (Figure 5). With BLM and USFS managing millions of hectares of land for maintaining

biodiversity, their role in protecting ecological systems is well established, and there may be potential to expand the protection and representation of ecological systems, for example, through the expansion of NLCS. In the past, administrative jurisdictional land transfers have occurred between land management agencies (e.g., BLM, USFWS, NPS, and USFS) [76–78]. Some of these land transfers have led to more emphasis on maintaining biodiversity.

Landscape Conservation Cooperatives Setting Priorities for Conservation of Ecological Systems

The framework and partnerships of the LCCs informs conservation at the landscape level, which will be needed to implement conservation across jurisdictional boundaries. Our analysis indicates that ecological systems in the East are less redundant and at more risk of conversion than those in the West (Figures 6 and 8). Because of this East-West dichotomy, increased conservation on some public and private lands may be important to the representation of ecological systems in the West, whereas increased public-private partnerships may play an important role in the East to increase the representation of ecological systems (Figures 4, 5, 6, 7, 8).

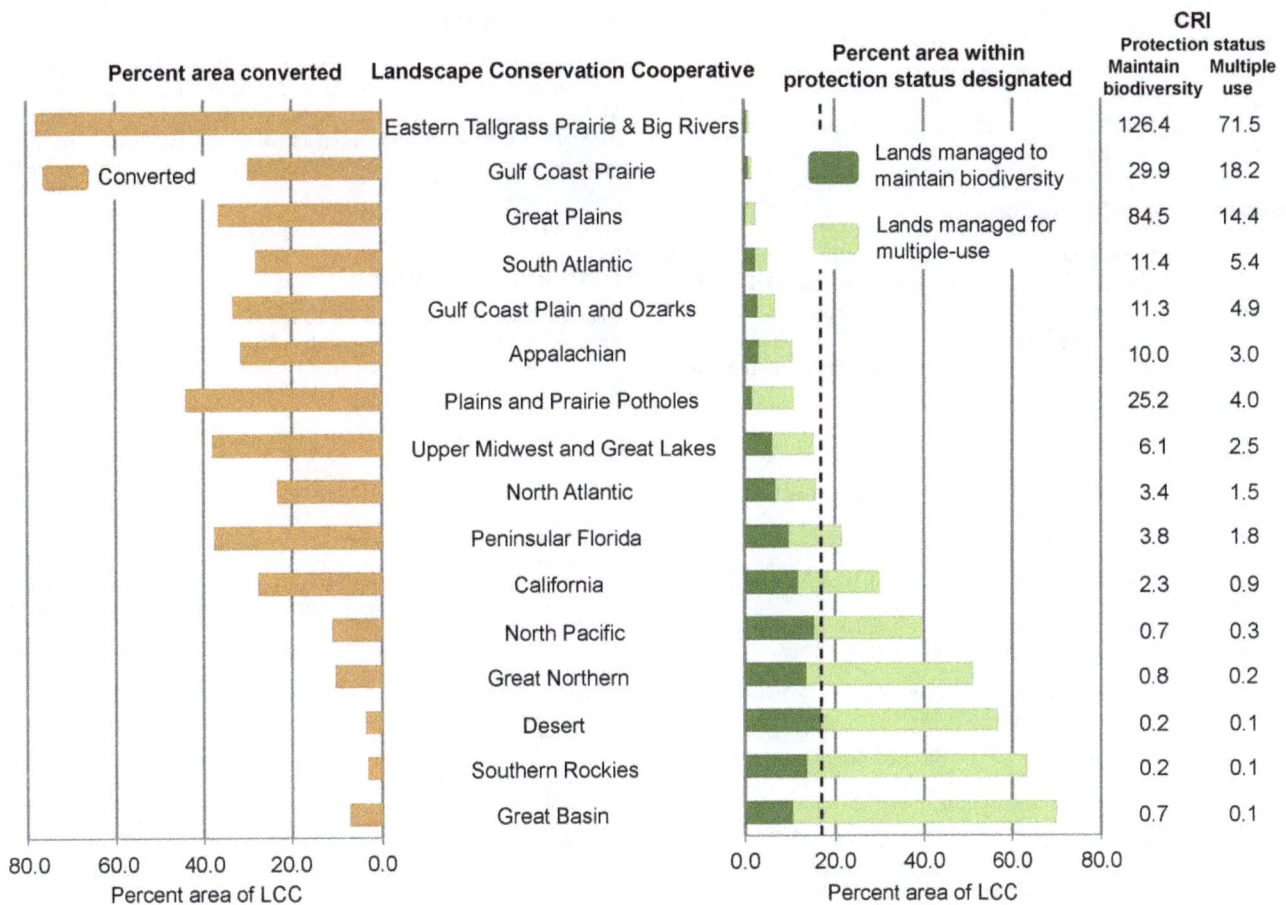

Figure 8. Percent area of Landscape Conservation Cooperative (LCC) protected or converted and its conversion risk index (CRI). CRI for each LCC is calculated by dividing percent area converted by percent area protected [62]. The CRI index is shown for lands managed to maintain biodiversity (i.e., labeled maintain biodiversity) as well as for lands managed to maintain biodiversity and multiple-use (i.e., labeled multiple-use) [23]. The LCCs are ordered by percent area within each protection status. See Table 1 for protection status descriptions. A dashed line representing the 17% Aichi Biodiversity Target of the Convention on Biological Diversity is shown [36].

Our research results highlighting low redundancy and unique ecological systems corroborate results from other studies [13,18]. In particular, the eastern US was identified as an ecoregion with high threats and irreplaceability value with regards to identifying conservation priorities [13,18]. For example, the Gulf Coast Plain and Ozarks LCC in southeastern US has high diversity and uniqueness, but low redundancy and a high conservation risk index (Figures 6 and 8). Within this LCC, there are few opportunities for increasing the representation of ecological systems on lands managed for multiple-use (Table 2, percent protected changes from 3.5% to 4.9%). An initial practical approach for conservation of ecological systems in this LCC, which contains many diverse and unique ecological systems, would be to engage both public and private conservation partners. In this case, our research results could serve as a catalyst for building public and private conservation partnerships. The larger scale perspective of LCCs provides a unique forum that previously did not exist for putting nationwide conservation planning at a scale that allows strategic emphasis on ecological systems that are in most need of added representation and protection.

There are numerous benefits to exploring alternatives for increasing the conservation and representation of ecological systems in the protected areas network. First, we can increase the number and area of ecological systems protected. Ecological systems represent a range of the habitats upon which many species rely; therefore we are increasing the protection of numerous species, including threatened, endangered, and species of concern. Second, we can increase the adaptability of ecological systems and the protected areas network to climate change impacts [79]. A wider range of environmental variables will enable ecological systems and the vertebrate species that rely on them to have room to shift their ranges in response to changes in climate. Third, we can increase the buffer area for all ecological systems and thereby reduce edge effects and increase the integrity of existing ecological systems. Lastly, we are more likely to capture the ecological processes that drive the pattern of ecological systems that we observe and allow for a more fully functional and robust protected areas network.

The current protected areas network for the continental US does not capture the full range of ecological systems or geophysical features (i.e., elevation and soil productivity). As a consequence, the species that rely on these ecological systems and geophysical features have fewer opportunities to adjust to changing environmental conditions. We have not assessed the pros and cons of using our alternatives for increasing the representation of ecological systems, but rather we have presented them as possibilities that may be considered and evaluated as decisions are made to conserve biodiversity. Each alternative may increase the representation of ecological systems, which can lead to protecting and securing habitat across a broader range of ecological, geographical, and geophysical occurrence of species. And may provide the greatest opportunity for evolutionary processes to persist regardless of imminent changes in the near, intermediate, and long term.

Supporting Information

Table S1 Area (ha) and percent area of ecological systems by protection status nested into Level I and II land cover groups [23,24]. All 5 Level I groups, 37 Level II groups, and 518 ecological systems are listed. See Table 1 for protection status descriptions. Only non-modified, non-aquatic ecological systems are included (n = 518).

References

1. Miller KR (1982) Parks and protected areas: considerations for the future. Ambio 11: 315–317.
2. Pressey RL (1994) Ad hoc reservations: forward and backward steps in developing representative reserve systems? Conserv Biol 8: 662–668.
3. Margules CR, Pressey RL (2000) Systematic conservation planning. Nature 405: 243–253.
4. Fairfax SK, Gwin L, King MA, Raymond L, Watt LA (2005) Buying nature: The limits of land acquisition as a conservation strategy, 1780–2004. Cambridge: The MIT Press. 357 p.
5. Pressey RL, Humphries CJ, Margules CR, Vane-Wright RI, Williams PH (1993) Beyond opportunism: key principles for systematic reserve selection. Trends Ecol Evol 8: 124–128.
6. Ando A, Camm J, Polasky S, Solow A (1998) Species distributions, land values, and efficient conservation. Science 279: 2126–2128.
7. van Jaarsveld AS, Freitag S, Chown SL, Muller C, Koch S, et al. (1998) Biodiversity assessment and conservation strategies. Science 279: 2106–2108.
8. Kareiva P (2010) Trade-in to trade-up. Nature 466: 322–323.
9. Harrison J, Miller K, McNeely J (1982) The world coverage of protected areas: development goals and environmental needs. Ambio 11: 238–245.
10. Ehrlich PR, Wilson EO (1991) Biodiversity studies: science and policy. Science 253: 758–762.
11. Scott JM, Davis F, Csuti B, Noss R, Butterfield B, et al. (1993) Gap analysis: a geographic approach to protection of biological diversity. Wildlife Monographs 123: 1–41.
12. Chape S, Blyth S, Fish L, Fox P, Spalding M (2003) 2003 United Nations list of protected areas. Available: http://www.unep.org/pdf/un-list-protected-areas.pdf. Access 29 February 2012.
13. Rodrigues ASL, Akçakaya HR, Andelman SJ, Bakarr MI, Boitani L, et al. (2004) Global gap analysis: priority regions for expanding the global protected-area network. Bioscience 54: 1092–1100.
14. McDonald RI (2009) The promise and pitfalls of systematic conservation planning. Proc Natl Acad Sci U S A 106: 15101–15102.
15. Scott JM, Davis FW, McGhee RG, Wright RG, Groves C, et al. (2001) Nature preserves: do they capture the full range of America's biological diversity? Ecol Appl 11: 999–1007.
16. Baron JS, Griffith B, Joyce LA, Kareiva P, Keller BD, et al. (2008) Preliminary review of adaptation options for climate-sensitive ecosystems and resources. Available: http://library.globalchange.gov/products/assessments/sap-4-4-preliminary-review-of-adaptation-options-for-climate-sensitive-ecosystems-and-resources. Accessed 29 February 2012.
17. Glicksman RL (2008) Ecosystem resilience to disruptions linked to global climate change: An adaptive approach to federal land management. Neb Law Rev 87: 833–892.
18. Brooks TM, Bakarr MI, Boucher T, Da Fonseca GAB, Hilton-Taylor C, et al. (2004) Coverage provided by the global protected-area system: is it enough? Bioscience 54: 1081–1091.
19. Estes JE, Mooneyhan DW (1994) Of maps and myths. Photogramm Eng Remote Sensing 60: 517–524.
20. Dietz RW, Czech B (2005) Conservation deficits for the continental United States: an ecosystem gap analysis. Conserv Biol 19: 1478–1487.
21. Noss RF, Cooperrider AY (1994) Saving Nature's legacy. Washington DC: Island Press. 416 p.
22. The President's Council of Advisors on Science and Technology (PCAST) (2011) Sustaining environmental capital: protecting society and the economy. Available: http://www.whitehouse.gov/administration/eop/ostp/pcast/docsreports. Accessed 2011 July 29.
23. US Geological Survey, Gap Analysis Program (USGS-GAP) (2010) National GAP Land Cover, Version 1. Available: http://gapanalysis.usgs.gov. Accessed 29 July 2011.
24. US Geological Survey, Gap Analysis Program (USGS-GAP) (2010) Protected Areas Database of the United States, Version 1.0. Available: http://gapanalysis.usgs.gov. Accessed 2011 July 29.
25. Comer P, Faber-Langendoen D, Evans R, Gawler S, Josse C, et al. (2003) Ecological systems of the United States: a working classification of US terrestrial systems. Available: http://www.natureserve.org/library/usEcologicalsystems.pdf. Accessed 29 July 2011.
26. Redford KH, Richter BD (1999) Conservation of biodiversity in a world of use. Conserv Biol 13: 1246–1256.
27. Sanderson EW, Jaiteh M, Levy MA, Redford KH, Wannebo AV, et al. (2002) The human footprint and the last of the wild. Bioscience 52: 891–904.
28. Hobbs RJ, Arico S, Aronson J, Baron JS, Bridgewater P, et al. (2006) Novel ecosystems: theoretical and management aspects of the new ecological world order. Global Ecol Biogeogr 15: 1–7.
29. Sodhi NS, Butler R, Laurance WF, Gibson L (2011) Conservation successes at micro-, meso-, and macroscales. Trends Ecol Evol 26: 585–594.
30. Noss RF, LaRoe III ET, Scott JM (1995) Endangered ecosystems of the United States: A preliminary assessment of loss and degradation. Available: http://biology.usgs.gov/pubs/ecosys.htm. Accessed 2010 July 22.
31. Bunce RGH, Bogers MMB, Evans D, Halada L, Jongman RHG, et al. (2012). The significance of habitats as indicators of biodiversity and their links to species. Ecol Indic http://dx.doi.org/10.1016/j.ecolind.2012.07.014. Accessed 2012 August 31.
32. Dudley N (2008) Guidelines for applying protected area management categories. Available: http://data.iucn.org/dbtw-wpd/edocs/paps-016.pdf. Accessed 2012 February 29.
33. Noss RF (1990) Indicators for monitoring biodiversity: a hierarchical approach. Conserv Biol 4: 355–364.
34. Tear TH, Kareiva P, Angermeier PL, Comer P, Czech B, et al. (2005) How much is enough? The recurrent problem of setting measurable objectives in conservation. Bioscience 55: 835–849.
35. Svancara LK, Brannon R, Scott JM, Groves CR, Noss RF, et al. (2005) Policy-driven versus evidence-based conservation: a review of political targets and biological needs. Bioscience 55: 989–995.
36. Convention on Biological Diversity Strategic Plan for Biodiversity 2011–><2020 including Aichi Biodiversity Targets. Available:.Accessed 2012 February 29.
37. Shaffer ML, Stein BA (2000) Safeguarding our precious heritage. In: Stein BA, Kutner LS, Adams JS, editors. Precious heritage: the status of biodiversity in the United States. New York: Oxford University Press. 301–321.
38. Groves CR (2003) Drafting a conservation blueprint: a practioner's guide to planning for biodiversity. Washington DC: Island Press. 458 p.
39. Scott D, Malcolm JR, Lemieux C (2002) Climate change and modelled biome representation in Canada's national park system: implications for system planning and park mandates. Global Ecol Biogeogr 11: 475–484.
40. Lemieux CJ, Scott DJ (2005) Climate change, biodiversity conservation and protected area planning in Canada. Can Geogr 49: 384–399.
41. Lemieux CJ, Beechey TJ, Gray PA (2011) Prospects for Canada's protected areas in an era of rapid climate change. Land use policy Available: doi:10.1016/j.landusepol.2011.03.008. Accessed 2011 April 29.
42. Shelford VE (1926) Naturalist's guide to the Americas. Baltimore: The Williams and Wilkins Company. 761 p.
43. Crumpacker DW, Hodge SW, Friedley D, Gregg WP (1988) A preliminary assessment of the status of major terrestrial and wetland ecosystems on Federal and Indian lands in the United States. Conserv Biol 2: 103–115.
44. Caicco SL, Scott JM, Butterfield B, Csuti B (1995) A gap analysis of the management status of the vegetation of Idaho (USA). Conserv Biol 9: 498–511.
45. Davis FW, Stine PA, Stoms DM, Borchert MI, Hollander AD (1995) Gap analysis of the actual vegetation of California 1. The Southwestern Region. Madroño 42: 40–78.

Acknowledgments

We thank C. Conway, D. Weinstein, R. White, and 2 anonymous reviewers for their comments that improved this manuscript. We also thank M. Croft, L. Duarte, J. Lonneker, K. Mallory, A. Radel, and G. Wilson for their help. Any use of trade, product, or firm names is for descriptive purposes only and does not imply endorsement by the US Government.

Author Contributions

Conceived and designed the experiments: JMS LS JA AD KG. Analyzed the data: AD JA. Wrote the paper: JLA LS AM AD JS KG.

46. Stoms DM, Davis FW, Driese KL, Cassidy KM, Murray MP (1998) Gap analysis of the vegetation of the intermountain semi-desert ecoregion. Great Basin Nat 58:199–216.

47. Scott JM, Murray M, Wright RG, Csuti B, Morgan P, et al. (2001) Representation of natural vegetation in protected areas: capturing the geographic range. Biodivers Conserv 10: 1297–1301.

48. Wright RG, Scott JM, Mann S, Murray M (2001) Identifying unprotected and potentially at risk plant communities in the western USA. Biol Conserv 98: 97–106.

49. Fuller RA, McDonald-Madden E, Wilson KA, Carwardine J, Grantham HS, et al. (2010) Replacing underperforming protected areas achieves better conservation outcomes. Nature 466: 365–367.

50. Langhammer PF, Bakarr MI, Bennun LA, Brooks TM, Clay RP, et al. (2007) Identification and gap analysis of key biodiversity areas: targets for comprehensive protected area systems. Available: data.iucn.org/dbtw-wpd/edocs/pag-015.pdf. Accessed 2012 February 29.

51. Kark S, Levin N, Grantham HS, Possingham HP (2009) Between-country collaboration and consideration of costs increase conservation planning efficiency in the Mediterranean Basin. Proc Natl Acad Sci U S A 106: 15368–15373.

52. Swaty R, Blankenship K, Hagen S, Fargione J, Smith J, et al. (2011) Accounting for ecosystem alteration doubles estimates of conservation risk in the conterminous United States. PLoS ONE 6: 1–10. DOI: 10.1371/journal.-pone.0023002.

53. Millard MJ, Czarnecki CA, Morton JM, Brandt LA, Shipley FS, et al. (2012) A national geographic framework for guiding conservation on a landscape scale. Journal of Fish and Wildlife Management 3: 175–183.

54. Lowry Jr JH, Ramsey RD, Boykin K, Bradford D, Comer P, et al. (2007) Land cover classification and mapping. In: Prior-Magee JS, et al., editors. Southwest Regional Gap Analysis Final Report. Available: http://fws-nmcfwru.nmsu.edu/swregap/report/swregap%20final%20report.pdf. Accessed 2011 November 16.

55. Zhu Z, Vogelmann J, Ohlen D, Kost J, Chen X, et al. (2006) Mapping existing vegetation composition and structure for the LANDFIRE Prototype Project. In: Rollins MG, Frame CK, editors. The LANDFIRE prototype project: nationally consistent and locally relevant geospatial data for wildland fire management Available: http://www.treesearch.fs.fed.us/pubs/24700. Accessed 2011 January 18.

56. Rollins MG (2009) LANDFIRE: a nationally consistent vegetation, wildland fire, and fuel assessment. International Journal of Wildland Fire 18: 235–249.

57. Sanborn (2006) GAP zone 1 vegetation mapping final report. Available: http://gap.uidaho.edu. Accessed 2011 July 29.

58. Lowry J, Ramsey RD, Thomas K, Schrupp D, Sajwaj T, et al. (2007) Mapping moderate-scale land-cover over very large geographic areas within a collaborative framework: a case study of the Southwest Regional Gap Analysis Project (SWReGAP). Remote Sens Environ 108: 59–73.

59. Crist PJ, Prior-Magee JS, Thompson BC (1996) Land management status categorization in gap analysis: a potential enhancement. In Brackney ES, Jennings MD, editors. Gap Analysis Bulletin 5. Available: http://www.gap.uidaho.edu/bulletins/5/LMSCiGA.html. Accessed 2010 March 29.

60. U. S. Geological Survey (2006) National elevation dataset. Available: http://ned.usgs.gov/. Accessed 2011 July 29.

61. Hazen HD, Anthamatten PJ (2004) Representation of ecological regions by protected areas at the global scale. Physical Geography 25: 499–512.

62. Hoekstra JM, Boucher TM, Ricketts TH, Roberts C (2005) Confronting a biome crisis: global disparities of habitat loss and protection. Ecol Lett 8: 23–29.

63. Groves CR, Kutner LS, Stoms DM, Murray MP, Scott JM, et al. (2000) Owning up to our responsibilities: who owns land important for biodiversity? In: Stein BA, Kutner LS, Adams JS, editors. Precious heritage: The status of biodiversity in the United States. New York: Oxford University Press. 399 p.

64. Andersen MG, Ferree CE (2010) Conserving the stage: climate change and the geophysical underpinnings of species diversity. PLoS ONE 5: 1–10. DOI:10.737/journal.pone.0011554.

65. Tversky A, Kahneman D (1974) Judgment under uncertainty: heuristics and biases. Science 185: 1124–1131.

66. Parmesan C (2006) Ecological and evolutionary responses to recent climate change. Annu Rev Ecol Evol Syst 37: 637–669.

67. Parmesan C, Yohe G (2003) A globally coherent fingerprint of climate change impacts across natural systems. Nature 421: 37–42.

68. Retti DR (1995) Our National Park System: caring for America's greatest natural and historic treasures. Urbana: University of Illinois Press. 293 p.

69. Mascia MB, Pailler S (2011) Protected area downgrading, downsizing, and degazettement (PADDD) and its conservation implications. Conservation Letters 4: 9–20.

70. Hogenauer AK (1991) Gone, but not forgotten: The delisted units of the US National Park System. The George Wright Forum 7: 2–19.

71. Hogenauer AK (1991) An update to "Gone, but not forgotten: The delisted units of the US National Park System. The George Wright Forum 8: 26–28.

72. US Department of Interior and US Department of Agriculture (2005) National Land Acquisition Plan. Available: http://www.fs.fed.us/land/staff/LWCF/Final%20DOI-USDA%20Land%20Acquisition%20Report%20to%20Congress.pdf. Accessed 2012 September 5.

73. Loomis JB (1993) Integrated public lands management. New York: Columbia University Press 474 p.

74. Thomas JW, Sienkiewicz A (2005) The relationship between science and democracy: public land policies, regulation, and management. Public Land and Resources Law Review 26: 39–69.

75. Darst CR, Huffman KA, Jarvis J (2009) Conservation significance of America's newest system of protected areas: National Landscape Conservation System. Natural Areas Journal 29: 224–254.

76. Towns E, Cook JE (1998) USDA, Forest Service, USDI, National Park Service: Notice of Transfer of Administrative Jurisdiction, Coconino National Forest and Walnut Canyon National Monument. Available: http://www.gpo.gov/fdsys/pkg/FR-1998-08-25/pdf/98-22723.pdf. Accessed 2012 August 31.

77. Stobaugh J (2003) Notice of Proposed Withdrawal and Opportunity for Public Meeting: Nevada. Available: http://www.gpo.gov/fdsys/pkg/FR-2003-07-09/pdf/03-17392.pdf. Accessed 2012 August 31.

78. Allred CS (2007) Public Land Order No. 7675: Transfer of Administrative Jurisdiction, Petrified Forest National Park Expansion, Arizona. Available: http://www.gpo.gov/fdsys/pkg/FR-2007-05-18/pdf/E7-9586.pdf. Accessed 2012 August 31.

79. Kujala H, Araújo MB, Thuiller W, Cabeza M (2011) Misleading results from conventional gap analysis – messages from the warming north. Biol Conserv 144: 2450–2458.

11

Effects of Sublethal Cadmium Exposure on Antipredator Behavioural and Antitoxic Responses in the Invasive Amphipod *Dikerogammarus villosus*

Pascal Sornom*, Eric Gismondi, Céline Vellinger, Simon Devin, Jean-François Férard, Jean-Nicolas Beisel

Laboratoire des Interactions, Ecotoxicologie, Biodiversité, Ecosystèmes, CNRS UMR 7146, Université de Lorraine, Metz, France

Abstract

Amphipods are recognised as an important component of freshwater ecosystems and are frequently used as an ecotoxicological test species. Despite this double interest, there is still a lack of information concerning toxic impacts on ecologically relevant behaviours. The present study investigated the influence of cadmium (Cd), a non-essential heavy metal, on both antipredator behaviours and antitoxic responses in the invasive amphipod *Dikerogammarus villosus* under laboratory conditions. Amphipod behaviour (i.e. refuge use, aggregation with conspecifics, exploration and mobility) was recorded following a 4-min test-exposure to 500 µg Cd/L with or without a 24-h Cd pre-exposure and in the presence or absence of a high perceived risk of predation (i.e. water scented by fish predators and injured conspecifics). Following behavioural tests, malondialdehyde (MDA) levels, a biomarker for toxic effect, and energy reserves (i.e. lipid and glycogen contents) were assessed. Cd exposures induced (1) cell damage reflected by high MDA levels, (2) erratic behaviour quantified by decreasing refuge use and exploration, and increasing mobility, and (3) a depletion in energy reserves. No significant differences were observed between 4-min test-exposed and 24-h pre-exposed individuals. Gammarids exposed to Cd had a disturbed perception of the alarm stimuli, reflected by increased time spent outside of refuges and higher mobility compared to gammarids exposed to unpolluted water. Our results suggest that Cd exposure rapidly disrupts the normal behavioural responses of gammarids to alarm substances and alters predator-avoidance strategies, which could have potential impacts on aquatic communities.

Editor: Kentaro Q. Sakamoto, Hokkaido University, Japan

Funding: The authors have no funding or support to report.

Competing Interests: The authors have declared that no competing interests exist.

* E-mail: pascal.sornom@umail.univ-metz.fr

Introduction

Aquatic ecosystems are constantly being exposed to chemical contaminants from industrial, agricultural and domestic sources. In recent decades, metals with no significant biological function, such as cadmium (Cd), have received particular attention due to their high ecotoxicity, even at very low concentrations, and their ability to bioaccumulate in many aquatic species [1]. Cd is a heavy metal toxicant that occurs naturally in the environment, in insignificant amounts; however, its impact is steadily increasing due to anthropogenic activities. Freshwater crustaceans are amongst the most sensitive of macroinvertebrate species to Cd [2]. This is especially so for gammarids, which are increasingly used as biological models in ecotoxicological studies. In addition to its ability to bioaccumulate and its adverse effects on survival, Cd has been shown to significantly affect an organism's behavioural patterns, including feeding, ventilation and locomotion [3–5]. Cd is also known to affect the transfer of chemical information between organisms [6]. Indeed, the phenomenon of Cd-induced info-disruption has been shown to impact on anti-predator behaviour in many aquatic species, including fish and crustaceans [7,8]. Although several studies have been devoted to the effects of heavy metals such as Cd on gammarids, the species most often used are either native or naturalised. Very little information is

available on the responses of invasive European amphipods to chemical stress, despite a number of species now reaching dominant levels in some European waters.

In recent decades, a number of exotic amphipod species have increased their ranges in Europe, spreading west from their native Ponto-Caspian region. One of these, *Dikerogammarus villosus*, has become well established in the River Moselle (north-eastern France) since its first appearance in 1999 [9]. Its invasive success has been helped by life-history traits, predatory behaviour and reproductive characteristics [10,11] that have resulted in *D. villosus* becoming one of the dominant freshwater amphipod species in many large European hydrosystems. Due to its recent wide distribution and high densities in European inland waters [12], *D. villosus* is rapidly becoming a classical model species used in ecotoxicological tests to develop biomarkers [13] or assess effects of pollutants [14,15].

Many tools have been developed in amphipod testing to estimate and predict the effects of contaminants on organisms, the most widely used ecotoxicological endpoints being survival, growth, food consumption and assimilation, moult frequency, reproduction, enzymatic biomarkers and osmoregulation. The assessment of sublethal ecotoxicity is of ecological relevance as mortality does not always occur in organisms exposed to pollutants. In such cases, behavioural changes are relevant tools for ecotoxicity testing and water quality monitoring [16]. Indeed, behavioural endpoints, previously described as "early warning responses" to toxicants and environmental stresses [17], are sensitive, fast and relatively easy to assess, and are cheap, non-invasive and useful indicators of sublethal exposure in both laboratory and field conditions. They are highly ecologically relevant and they have the potential to link physiological functions to ecological processes, e.g. locomotion is required not only to find food, to obtain mates and to migrate, but also to escape predation.

Amphipods constitute the prey of various upper trophic-level predators including other invertebrates, vertebrates and especially fish [18]. Hence, in the lower River Rhine, *D. villosus* rapidly became a regular food item for eel [19], while in Upper Lake Constance, five years after its first observation, *D. villosus* had readily been included into the diet of zoobenthivorous fish [20]. In aquatic environments, prey are able to assess the presence, activity and hunger of predators [21] through chemical signals [22], in addition to via visual, hydrodynamic or auditory cues [23,24]. Chemical cues can include the scent of the predator itself [25], that of its prey [26], or the scent emitted by injured conspecifics [27,28]. Earlier studies devoted to antipredator mechanisms in amphipods have highlighted their sensitivity to both fish scent [29] and injured conspecifics [30,31]. In order to avoid encounters with predators, amphipod prey can display chemical, structural or behavioural defences [32] such as increased refuge use and decreased locomotion activity, which have been shown to increase survival [30]. Further, gregarious behaviour, a form of cover-seeking, tends to decrease individual risk of predation in various organisms, including amphipods [33].

The aims of the present study were (1) to investigate antipredator responses of *D. villosus* to a predation alarm cue and (2) to test the hypothesis that Cd may interfere with the perception of predation risk chemical signals. In laboratory experiments, the perception of predation risk will be determined by measuring behavioural biomarkers such as refuge use, aggregation with conspecifics, environmental exploration and mobility. In parallel, we will measure malondialdehyde (MDA), a lipoperoxidation product used as a biomarker of toxic effect, and lipid and glycogen levels, to monitor energy expenditure in the organism. As gender has previously been shown to be a confounding factor in physiological biomarker responses assessment [34,35], all measurements were carried out separately on males and females.

Materials and Methods

Ethics Statement

All experiments therein have been conducted in accordance with current laws in France, including the European Convention for the Protection of Vertebrate Animals used for Experimental and Other Scientific Purposes, Strasbourg, 18.III.1986 (Annex B). All organisms unexposed to toxicants (i.e. unused and untested amphipods and all fish) were later released at their respective sampling sites within one week. The sampling site locations are not privately-owned or protected in any way and macroinvertebrate sampling at these locations did not require any specific permit. The field studies did not involve endangered or protected species. The bullhead *Cottus gobio* is listed in Annex 2 of Council Directive 92/43/EEC of 21 May 1992 on the Conservation of Natural Habitats and of Wild Fauna and Flora. In France, however, *C. gobio* is not a protected species and its use in scientific research does not require any specific permit. Fish sampling was performed by the Office National de l'Eau et des Milieux Aquatiques

(ONEMA), the French national agency for water and aquatic environments, a state-owned company licensed to conduct scientific fishing, allowing sampling and the transport of the animals to the laboratory. All experiments were supervised by the holder of a Level 1 authorisation to conduct experiments on animals, as approved by the French Ministry of Agriculture and Fisheries (Dr. Michael Danger).

Sampling and Housing

Specimens of *D. villosus* were collected using pond nets (5 mm mesh) during spring 2011, from an unpolluted station on the River Moselle in north-eastern France (Saulcy; 49°07′N and 6°10′E). We selected intermediate-sized adults of around15 mm in order to avoid potential bias induced by gammarid length/age. Similarly, ovigerous females were not sampled in order to standardise physiological condition. Specimens were brought to the laboratory in 3 L opaque plastic tanks filled with water taken on site within one hour of sampling. Leaves collected from the site were placed into each tank in order to reduce the stress induced by handling and transport. In the laboratory, specimens were maintained at a maximum density of 20 individuals/L in plastic aquaria (31 cm long, 28 cm wide, 14 cm high) filled with dechlorinated, UV-treated, and oxygenated tap water. Glass pebbles (15 mm diameter, convex side down) and artificial plants were added to provide refuges and minimise cannibalism. Gammarids were fed *ad libitum* with alder leaf disks and chironomids. The room's temperature was kept constant at 18±1°C and the photoperiod kept at 12 h light:12 h dark. Amphipods were housed under these conditions for 72 h before pre-exposures.

The bullhead *C. gobio* was chosen as the test predator as it is a native and widespread fish species in north-eastern France. This typical benthic-feeding fish is known to include amphipods in its diet and is frequently used to assess invertebrate antipredator behavioural responses [36,37]. Twenty fish of similar size (approx. 9 cm) were collected from the River Moselle in spring 2011 using a pond net (5 mm mesh). Bullheads were transported to the laboratory in a 30 L aerated opaque plastic tank filled with water collected on site within one hour of sampling. In the laboratory, specimens were maintained at a density that did not exceed one fish/L in aquaria (60×29×31 cm) filled with 6 L of dechlorinated, UV-treated and oxygenated tap water. To each housing unit, we added coarse gravel and stones (>1 cm diameter) and some artificial plants to minimise aggression between individuals. The aquaria were maintained under the same temperature and photoperiod conditions as those for amphipods and the fish were maintained under these conditions for one week prior to the start of experimentation. During this time, the fish were fed *ad libitum* with living *D. villosus* taken randomly from the sample site. Bullheads were deprived of food for 72 h prior to use for preparation of scented water.

Scented Water

Scented water was prepared using a protocol similar to that described in [38]. Six previously starved bullheads were placed in a plastic aquarium (31×28×14 cm) filled with 6 L of dechlorinated, UV-treated and oxygenated tap water for 24 h. Thirty amphipods were offered to the bullheads in order to obtain a mean of five predation events/fish/L over 24 h. The procedure was repeated for each test condition and was expected to provide a predation signal close to that observed under natural conditions, which includes both predator odour and chemical cues released by injured amphipods [31,39]. Behavioural tests were performed within 24 h following the preparation of scented water.

Cadmium Solutions

A toxic stock solution was prepared by dissolving cadmium chloride (CdCl$_2$) in dechlorinated and UV-treated tap water. The stock solution was stabilised with 0.1% of HNO$_3$ (65%, Suprapure, Merck). Cd-spiked test solutions were prepared through dilution of the stock solution in either unscented or previously prepared scented water in order to obtain a nominal concentration of 500 µg Cd/L.

Pre-exposure and Test Exposure Conditions

For all experiments, temperature and photoperiod were identical to those for amphipod housing. During pre-exposures, 120 specimens of *D. villosus* were placed in plastic aquaria (31×28×14 cm) filled with 6 L of oxygenated test solution, where they were kept unfed for 24-h prior to the behavioural tests. Amphipods were exposed to three conditions according to pre-exposure and test exposure: (1) Control - 24-h pre-exposure to Cd-free water (i.e. dechlorinated and UV-treated tap water) and 4-min test exposure to Cd-free water; (2) Cd-free pre-exposure - 24-h pre-exposure to Cd-free water and 4-min test exposure to 500 µg Cd/L; and (3) Cd pre-exposure - 24-h pre-exposure to 500 µg Cd/L and 4-min test exposure to 500 µg Cd/L. These conditions were replicated twice, both with and without the addition of chemical signals from the predator and injured conspecifics in test solution (scented water).

Description of the Experimental Units

Device #1. The first experimental unit comprised a plastic cylindrical tank (25 cm diameter, 31 cm high) that was open to the air and filled with 6 L of test solution (providing a water level of 11 cm; Figure 1A). A round Plexiglas plate (25 cm diameter, 13 mm high), drilled so as to create four equidistant groups of holes 2 cm from the periphery (hereafter called refuge areas), was placed at the bottom of the unit. Each refuge area comprised seven equidistant holes (8 mm diameter, 10 mm deep) inside which amphipods could take refuge by hiding its entire body.

Device #2. The second experimental unit used a similar plastic cylindrical tank as for device #1, also filled with 6 L of

test solution (Figure 1B). In order to test for aggregative behaviour of amphipods with conspecifics, four equidistant plastic cylindrical cages (4 cm diameter, 13 cm high, 1 mm mesh) were placed vertically on the bottom of the unit, 2 cm from the edge. Two diametrically opposed cages received 10 conspecifics (five individuals of each gender) and the two others were kept empty.

Behavioural Tests

Each experimental device was placed separately into an hermetic box (50×46×62 cm) fitted with three red-light emitting lamps and a digital video camera set to film the horizontal plan (see [34,38]). Red light was preferred to white light as amphipods display negative phototaxis. Before starting the test, a translucent plastic tube (3 cm diameter, 14 cm high), open to the air, was placed in the middle of the experimental unit and a single amphipod was introduced into the tube using a fine brush. After a 1-min acclimatisation period, the tube was removed and filming proceeded for 3-min at a rate of 20 frames per second. Each amphipod was used only once and experiments were conducted first with unscented and then with scented water. Each test was repeated 30 times per gender, representing 360 individuals for all experiments. The water was renewed after 5 amphipods had been tested and a 90° rotation of the experimental unit was performed in order to avoid a position effect. During aggregation tests with experimental device #2, the 10 conspecifics enclosed in the cages were not changed throughout the experiment to keep the level of attractiveness constant. At the end of each experiment, all tested amphipods were measured to the nearest mm (from the tip of the rostrum to the base of the telson) under a binocular magnifying glass (Leica MZ 125) fitted with an eyepiece micrometer (Zeiss, 10×). To test the potential effect of gender on the measured responses, amphipod gender was determined *a priori* based on the size of gnathopods (typically larger in males) and was confirmed *a posteriori* by examining the 7th ventral segment (genital papillae only present in males).

In the first experiment, device #1 allowed us to assess (1) hiding, by measuring the proportion of time spent prostrate inside holes; (2) environmental exploration, by counting the number of refuge

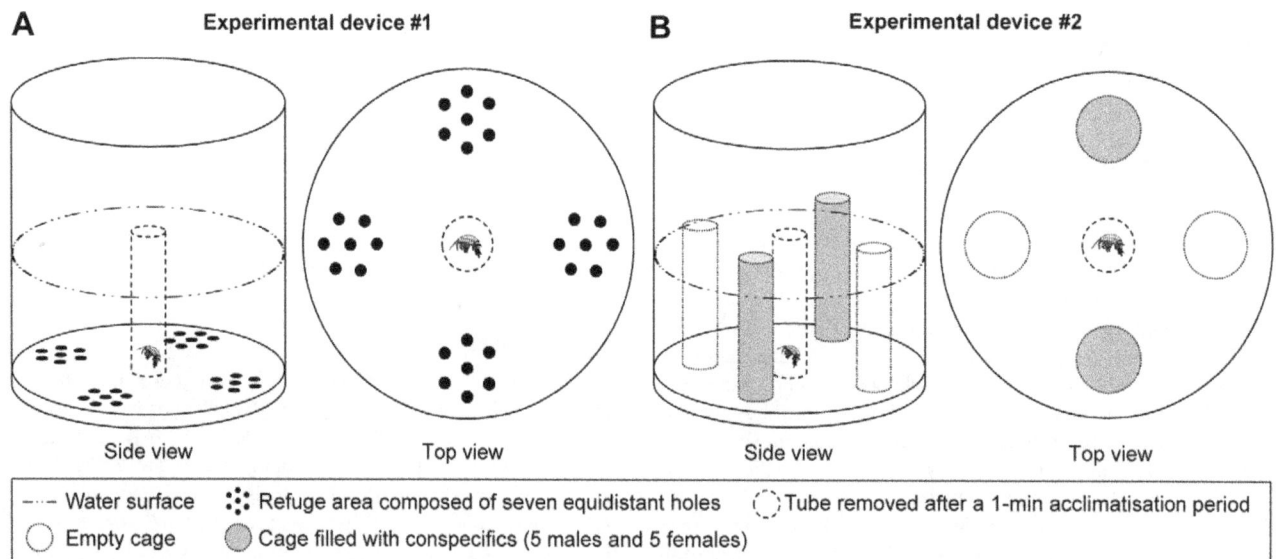

Figure 1. Schematic views of the experimental devices. A) device #1, used to assess hiding, exploration and mobility behaviour; B) device #2, used to assess hiding, aggregation and mobility behaviour.

areas used by a single amphipod; and (3) mobility, by measuring the proportion of time spent moving when outside of refuges. A refuge was considered as used by amphipods only when 80% of its body (excluding antennae) was inside a hole for a period greater than one second.

For the second experiment, we analysed the videos made using device #2 to assess (1) hiding, by first measuring the proportion of time spent close to all cages (i.e. both empty cages and cages filled with conspecifics); (2) aggregation, by focusing on periods during which amphipods were attached to cages and by measuring the proportion of time spent attached to cages containing conspecifics; and (3) mobility, by measuring the proportion of time spent moving when outside of refuges. Individuals were considered as hidden in a refuge area only when immobile and clinging to a cage for a period greater than one second.

Biomarker Measurement

After each behavioural test, ten pooled samples of 3 males and ten pooled samples of 3 females were prepared in order to analyse energy reserves and MDA levels [40]. Each pooled sample was frozen in liquid nitrogen and stored at −80°C awaiting biomarker measurement. Measurements were performed for females and males separately.

Sample preparation. Each pooled sample was homogenised with a manual Potter Elvejhem tissue grinder in 50 mM phosphate buffer KH_2PO_4/K_2HPO_4 (pH 7.6) supplemented with 1 mM phenylmethyl sulphonylfluoride (PMSF) and 1 mM L-serine-borate mixture as protease inhibitors and 5 mM phenylglyoxal as a γ-glutamyl transpeptidase inhibitor. The homogenisation buffer was adjusted to a volume of two times the wet weight of the pooled sample (i.e. 200 μL of homogenisation buffer for 100 mg of wet weight tissue). The total homogenate was divided into two parts in order to measure each of the different parameters. For each replicate, two independent measures were conducted for each biomarker.

Energy reserve assessment. Measurement of total lipids and glycogen content was adapted from [41]. Twenty μL of 2% sodium sulphate (w/v) and 540 μL of chloroform/methanol at 1:2 (v/v) were added to 40 μL of total homogenate. After 1 h on ice, samples were centrifuged at 3000×g for 5 min at 4°C. The resulting supernatant and pellet were used to determinate total lipid and glycogen content, respectively. One hundred μL of supernatant was transferred into a culture tube and placed into a dry bath at 95°C to evaporate the solvent, following which 200 μL of 95% sulphuric acid was added to each tube and left for 10 min. The culture tubes were then cooled on ice before addition of 4.8 mL of a vanillin-phosphoric acid reagent. After 10 min of reaction, absorbance was measured at 535 nm. Commercial cholesterol was used as a standard and lipid content was expressed in mg/mL. Total dissolution of the pellet was performed in 400 μL of deionized water for 10 min in an ultrasonic bath. One hundred μL of sample was placed into culture tubes and 4.9 mL of anthrone reagent added. The mixture was placed into a dry bath at 95°C for 17 min and then cooled on ice. Absorbance was measured at 625 nm, with glucose used as a standard and concentrations expressed in μg/mg wet weight.

Toxic effect biomarker. MDA levels were measured using a high-pressure liquid chromatography (HPLC) method adapted from [42], with UV detection set at 267 nm. Seventy μL of the total homogenate was deproteinised by diluting four-fold with 95% ethanol (HPLC grade) and cooling on ice for 1.5 h. The mixture was then centrifuged at 18,000×g for 30 min at 4°C. One hundred μL of the resultant supernatant was injected directly into a reversed-phase LiChrospher 100RP18-encapped HPLC col-

umn. Separation was conducted at 25°C and elution was carried out with sodium phosphate buffer (pH 6.5) containing 25% ethanol and 0.5 mM tetradecylmethylammoniun bromide as an ion-paring reagent. MDA levels were expressed in ng MDA/mg lipid.

Data Analysis and Statistics

As all data met assumptions of normality and homogeneity of variance, each response was analysed using ANOVA, followed by the post-hoc Tukey HSD test. All tests were performed with a 5% type I error risk, using R 2.9.0 Software. To verify if significant relationships existed between behavioural indices and individual lengths, correlations were performed and tested using Pearson correlation tests (not illustrated). As no significant differences were observed between males and females during behavioural tests, we decided to pool the genders for each condition in order to make the results more robust.

Results

Toxic Effect Biomarker

The results of three-factor ANOVA analysis are presented in Table 1. In device #1, Tukey HSD post-hoc tests indicated that, without predation risk, males and females from the Cd-free pre-exposure and Cd pre-exposure groups had more than twofold higher MDA levels than control gammarids (Figure 2A). No significant differences were observed between Cd-free pre-exposure and Cd pre-exposure groups. In the presence of predation risk, MDA levels in both genders of the Cd pre-exposure group showed a significant increase of more than threefold compared to the control. No significant differences were observed between the control group and the Cd-free pre-exposure group. In addition, no significant inter-sex differences were observed, whatever the pre-exposure/exposure conditions.

In device #2, Tukey HSD post-hoc tests indicated that, without predation risk, MDA levels were significantly increased by twofold in males of the Cd-free pre-exposure group and more than fivefold in both genders from the Cd pre-exposure group, when compared to the control (Figure 2B). Gammarids pre-exposed to Cd showed significantly higher MDA levels than those not pre-exposed. In scented water, MDA levels of both genders in the Cd-free pre-exposure and Cd pre-exposure groups increased significantly by more than twofold compared to the control. A significant difference between the Cd-free pre-exposure and Cd pre-exposure groups was only observed in males, however, with MDA levels being higher in the Cd pre-exposure group. In scented water, a significant inter-sex difference was only observed in the Cd pre-exposure group, with MDA levels being significantly higher in males.

Behavioural Responses

The results of two-factor ANOVA analysis are presented in Table 2. In device #1, Tukey HSD post-hoc tests indicated that, whether in unscented or scented water, hiding and exploration decreased significantly by more than threefold, while mobility increased significantly by more than 50%, in Cd-free pre-exposure groups compared to the control (Figures 3A, B, C). For the Cd pre-exposure groups, hiding and exploration decreased significantly by more than twofold and mobility increased significantly by more than 50% compared with the control. No significant difference was observed in hiding, exploration and mobility values between the Cd-free pre-exposure and Cd pre-exposure groups. In the control groups, values for exploration and mobility decreased significantly by more than twofold, while hiding increased

Table 1. Analysis of variance (three-factor ANOVA) in biomarkers for toxicity (MDA) and energetic reserves (lipid and glycogen) in *D. villosus*, according to gender, predation risk and cadmium (Cd) contamination.

Experimental unit	Parameter	Source of variation	DF	F	P
Device #1	MDA	Gender	1	4.57	**0.0349**
		Predation risk	1	13.88	**<0.001**
		Cd	2	223.16	**<0.001**
		Gender × Predation risk	1	0.002	0.9604
		Gender × Cd	2	0.40	0.6701
		Predation risk × Cd	2	161.15	**<0.001**
		Gender × Predation risk × Cd	2	0.06	0.9377
	Lipid	Gender	1	101.20	**<0.001**
		Predation risk	1	471.87	**<0.001**
		Cd	2	66.95	**<0.001**
		Gender × Predation risk	1	3.52	0.0635
		Gender × Cd	2	1.96	0.1462
		Predation risk × Cd	2	104.88	**<0.001**
		Gender × Predation risk × Cd	2	4.33	**0.0155**
	Glycogen	Gender	1	193.78	**<0.001**
		Predation risk	1	168.29	**<0.001**
		Cd	2	40.06	**<0.001**
		Gender × Predation risk	1	1.92	0.1688
		Gender × Cd	2	1.09	0.3402
		Predation risk × Cd	2	82.68	**<0.001**
		Gender × Predation risk × Cd	2	9.13	**<0.001**
Device #2	MDA	Gender	1	52.05	**<0.001**
		Predation risk	1	2.25	0.1364
		Cd	2	276.66	**<0.001**
		Gender × Predation risk	1	0.32	0.5715
		Gender × Cd	2	7.65	**<0.001**
		Predation risk × Cd	2	69.47	**<0.001**
		Gender × Predation risk × Cd	2	0.64	0.5273
	Lipid	Gender	1	106.75	**<0.001**
		Predation risk	1	338.02	**<0.001**
		Cd	2	82.21	**<0.001**
		Gender × Predation risk	1	8.11	**0.0052**
		Gender × Cd	2	1.76	0.1774
		Predation risk × Cd	2	64.05	**<0.001**
		Gender × Predation risk × Cd	2	8.03	**0.0006**
	Glycogen	Gender	1	205.04	**<0.001**
		Predation risk	1	93.58	**<0.001**
		Cd	2	95.04	**<0.001**
		Gender × Predation risk	1	7.59	**0.0069**
		Gender × Cd	2	14.37	**<0.001**
		Predation risk × Cd	2	72.01	**<0.001**
		Gender × Predation risk × Cd	2	3.54	**0.0323**

Significant values are shown in bold.

significantly by twofold, in scented water. No significant effect of predation risk was observed on antipredator behaviour in both the Cd-free pre-exposure and Cd pre-exposure groups, except that exploration decreased significantly by threefold in the Cd pre-exposure groups in scented water.

In device #2, Tukey HSD post-hoc tests indicated that, whether or not water was scented, hiding in the Cd-free pre-exposure groups decreased significantly by around 40%, while mobility values increased significantly by more than twofold, compared with the controls (Figures 3D, F). In the Cd pre-

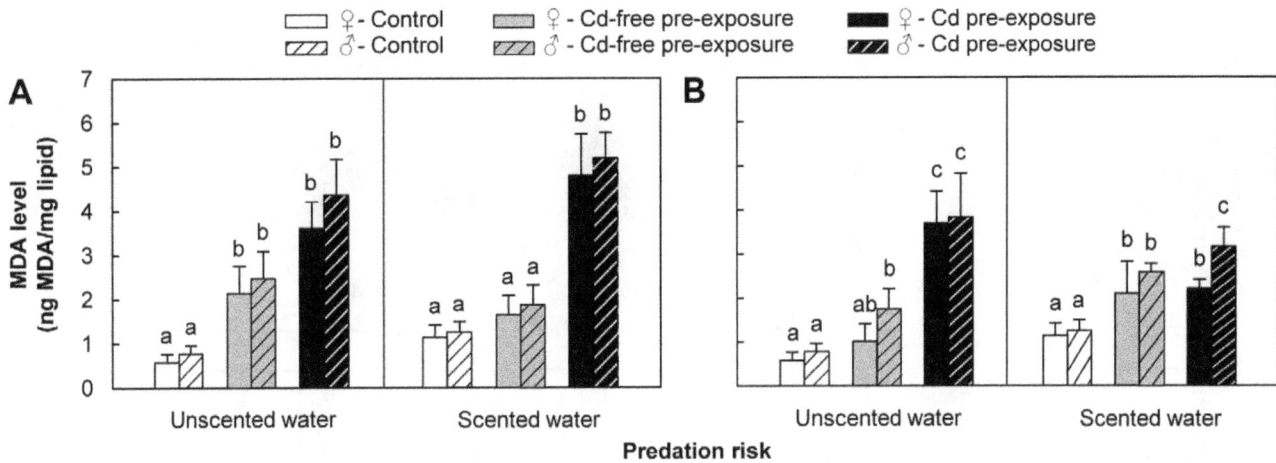

Figure 2. Malondialdehyde (MDA) levels measured in the presence/absence of predation risk. A) device #1; B) device #2. *D. villosus* females and males are represented by filled and hatched bars, respectively. MDA levels were measured following behavioural tests in which amphipods were exposed to the following conditions: Control = 24-h pre-exposure to Cd-free water +4-min test exposure to Cd-free water (white bars); Cd-free pre-exposure = 24-h pre-exposure to Cd-free water +4-min test exposure to 500 µg Cd/L (grey bars); Cd pre-exposure = 24-h pre-exposure to 500 µg Cd/L +4-min test exposure to 500 µg Cd/L (black bars). All data are means of N = 10± SD. Significant differences are indicated by different letters (Tukey HSD post-hoc test after ANOVA, P<0.05).

exposure groups, hiding values decreased significantly by more than two fold and mobility values increased significantly by more than 40% compared to the control. Significant differences between the Cd-free pre-exposure and Cd pre-exposure groups were only observed for hiding in scented water and for mobility

Table 2. Analysis of variance (two-factor ANOVA) in hiding, exploration and mobility behaviour in *D. villosus* measured with device #1, and hiding, aggregation and mobility behaviour measured with device #2, in response to predation risk and cadmium (Cd) contamination.

Experimental unit	Parameter	Source of variation	DF	F	P
Device #1	Hiding	Predation risk	1	8.22	**0.0095**
		Cd	2	63.57	**<0.001**
		Predation risk × Cd	2	1.74	0.1770
	Exploration	Predation risk	1	57.92	**<0.001**
		Cd	2	62.29	**<0.001**
		Predation risk × Cd	2	15.05	**<0.001**
	Mobility	Predation risk	1	65.48	**<0.001**
		Cd	2	70.74	**<0.001**
		Predation risk × Cd	2	8.13	**0.0156**
Device #2	Hiding	Predation risk	1	0.92	0.3382
		Cd	2	79.90	**<0.001**
		Predation risk × Cd	2	1.45	0.2367
	Aggregation	Predation risk	1	2.71	0.1007
		Cd	2	0.80	0.4519
		Predation risk × Cd	2	0.04	0.9603
	Mobility	Predation risk	1	6.48	**0.0113**
		Cd	2	22.98	**<0.001**
		Predation risk × Cd	2	2.31	0.1006

Significant values are shown in bold.

in unscented water. In both cases, values in the Cd pre-exposure groups were lower than in the Cd-free pre-exposure groups. No significant differences were observed in aggregation, whatever the test conditions (Figure 3E). No significant effect of predation risk was observed on antipredator behaviour in the three groups, except that, in the Cd-free pre-exposure groups, mobility was significantly lower in scented water.

In device #1, length ranges for *D. villosus*, respectively with and without predation risk, were 15.32–19.08 and 14.66–20.00 mm in the control group; 15.00–19.50 and 16.88–20.39 mm in the Cd-free pre-exposure group; and 16.38–19.63 and 14.75–19.00 mm in the Cd pre-exposure group. In device #2, length ranges of *D. villosus*, respectively with and without predation risk, were 15.32–19.08 and 15.97–19.21 mm in the control group; 15.96–20.76 and 14.63–20.63 mm in the Cd-free pre-exposure group; and 16.64–19.50 and 16.47–21.06 mm in the Cd pre-exposure group. No significant effect of gammarid length was observed for hiding, exploration or mobility in device #1, and for hiding, aggregation or mobility in device #2 (Pearson correlation tests, P>0.05; not illustrated).

Energy Reserves

Results for three-factor ANOVA analysis are presented in Table 1. In device #1, Tukey HSD post-hoc tests revealed that, without predation risk, lipid content in both the Cd-free pre-exposure and Cd pre-exposure groups decreased significantly by about 30% compared to the control (both genders). No significant differences were observed between the Cd-free pre-exposure and Cd pre-exposure groups (Figure 4A).

Values for both genders in the Cd pre-exposure group showed a significant threefold decrease in glycogen content compared to both the control and the Cd-free pre-exposure group. No significant differences were observed between the control and the Cd-free pre-exposure group (Figure 4B). In the presence of predation risk, lipid content was significantly higher only in females of the Cd-free pre-exposure group and decreased significantly by threefold for both genders in the Cd pre-exposure group, compared to the control. For both genders, glycogen content did not differ significantly from the control in

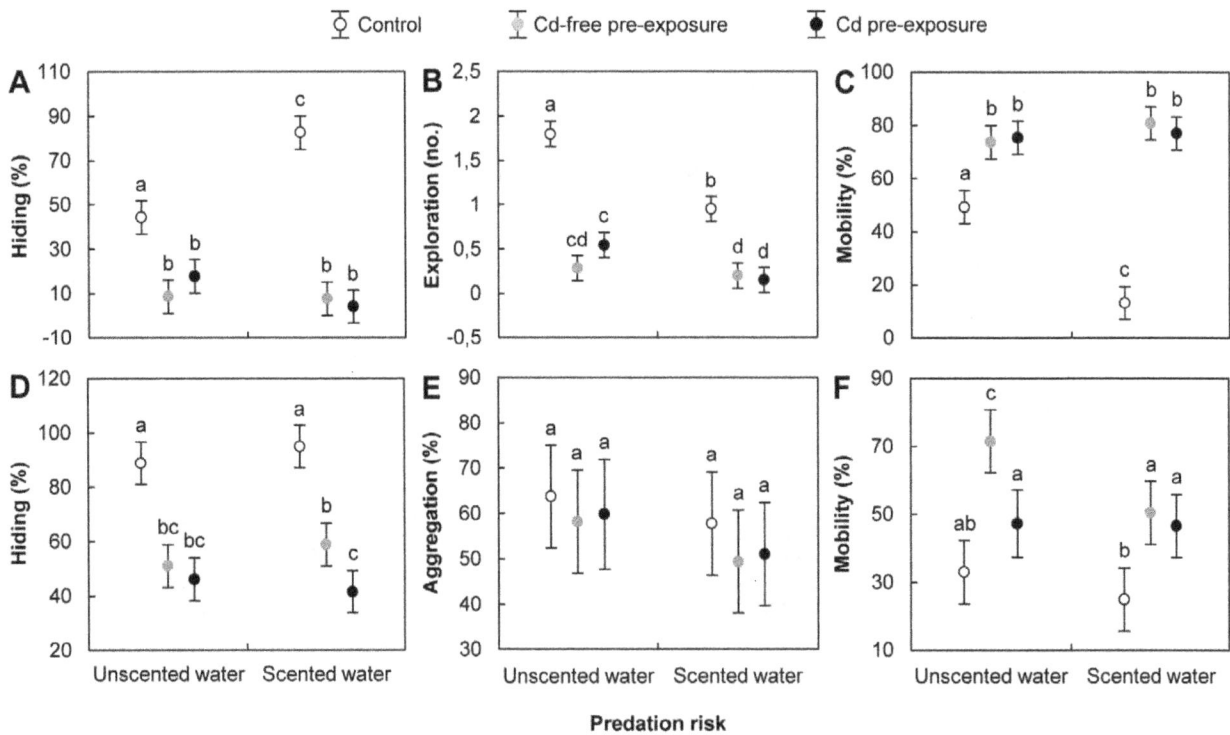

Figure 3. Behavioural responses measured in the presence/absence of predation risk. A) hiding in device #1 (i.e. proportion of time spent in holes); B) environmental exploration in device #1 (i.e. number of refuge areas used); C) mobility in device #1 (i.e. proportion of time spent moving when outside of refuges); D) hiding in device #2 (i.e. proportion of the time spent clinging to cages); E) aggregation in device #2 (i.e. proportion of the time spent clinging to cages filled with conspecifics); F) mobility in device #2 (i.e. proportion of the time spent moving when outside of refuges). *D. villosus* were exposed to the following conditions: Control = 24-h pre-exposure to Cd-free water +4-min test exposure to Cd-free water (white dots); Cd-free pre-exposure = 24-h pre-exposure to control water +4-min test exposure to 500 μg Cd/L (grey dots); Cd pre-exposure = 24-h pre-exposure to 500 μg Cd/L +4-min test exposure to 500 μg Cd/L (black dots). All data are means of $N = 60 \pm 95\%$ confidence intervals. Significant differences are indicated by different letters (Tukey HSD post-hoc test after ANOVA, $P < 0.05$).

both the Cd-free pre-exposure and Cd pre-exposure groups. Values for females in the Cd pre-exposure group, however, were significantly lower than those in the Cd-free pre-exposure group. In both scented and unscented water, significant inter-sex differences were observed for all responses (female energy reserves being higher than those of males), except for lipid content in the Cd pre-exposure group in scented water and for glycogen content in the Cd pre-exposure group for unscented water.

In device #2, Tukey HSD post-hoc tests revealed that, without predation risk, lipid content values for both genders decreased significantly by more than 25% in the Cd-free pre-exposure group, while in the Cd pre-exposure group, lipid content decreased by around 10% in males only compared to the control. A significant difference was observed in both genders between the Cd-free pre-exposure and Cd pre-exposure groups, with lipid content in the Cd pre-exposure group being higher (Figure 4C).

Glycogen content values decreased significantly by around 30% in males in the Cd-free pre-exposure group and decreased significantly by more than twofold for both genders in the Cd pre-exposure group, compared to the control. Individuals in the Cd pre-exposure group contained significantly less glycogen than those in the Cd-free pre-exposure group (Figure 4D). In scented water, lipid content decreased significantly by more than 20% in both genders in the Cd pre-exposure group, compared with both the control and the Cd-free pre-exposure group. No significant differences were observed, however, between the control group and the Cd-free pre-exposure group. Glycogen content in both the

Cd-free pre-exposure and Cd pre-exposure groups did not differ significantly from the control in both genders, though values for females from the Cd pre-exposure group were significantly lower than those of the Cd-free pre-exposure group. Whether scented or unscented water, significant inter-sex differences were found for all responses (female energy reserves being higher than those of males), except for glycogen content in the Cd pre-exposure group for both scented and unscented waters.

Discussion

Cadmium Responses

Toxic effect biomarker. Cd clearly had a significant effect on MDA values, with levels in *D. villosus* increasing upon exposure to the heavy metal in both experimental devices. A significant rise in MDA level has previously been reported for the amphipod *Gammarus pulex* after exposure to 300 μg Cd/L for 13 days [43]; and also in bivalves such as *Ruditapes decussatus* following exposure to 4, 40 and 100 g Cd/L for 28 days [44] and both *Crassostrea gigas* and *Mytilus edulis* exposed to 200 μg Cd/L for 21 days [45]. MDA content is a reliable indicator of cell damage as it is a final product of membrane lipid degradation (i.e. lipoperoxidation). Indeed, the higher the MDA level the higher the level of cell damage [46]. Cd is known to generate reactive oxygen species that may act as signalling molecules in the induction of gene expression and apoptosis [47] and deplete endogenous radical scavengers. Cd is also known to cause damage to a variety of transport proteins including Na^+/K^+-ATPase, as has been demonstrated for *G. pulex*

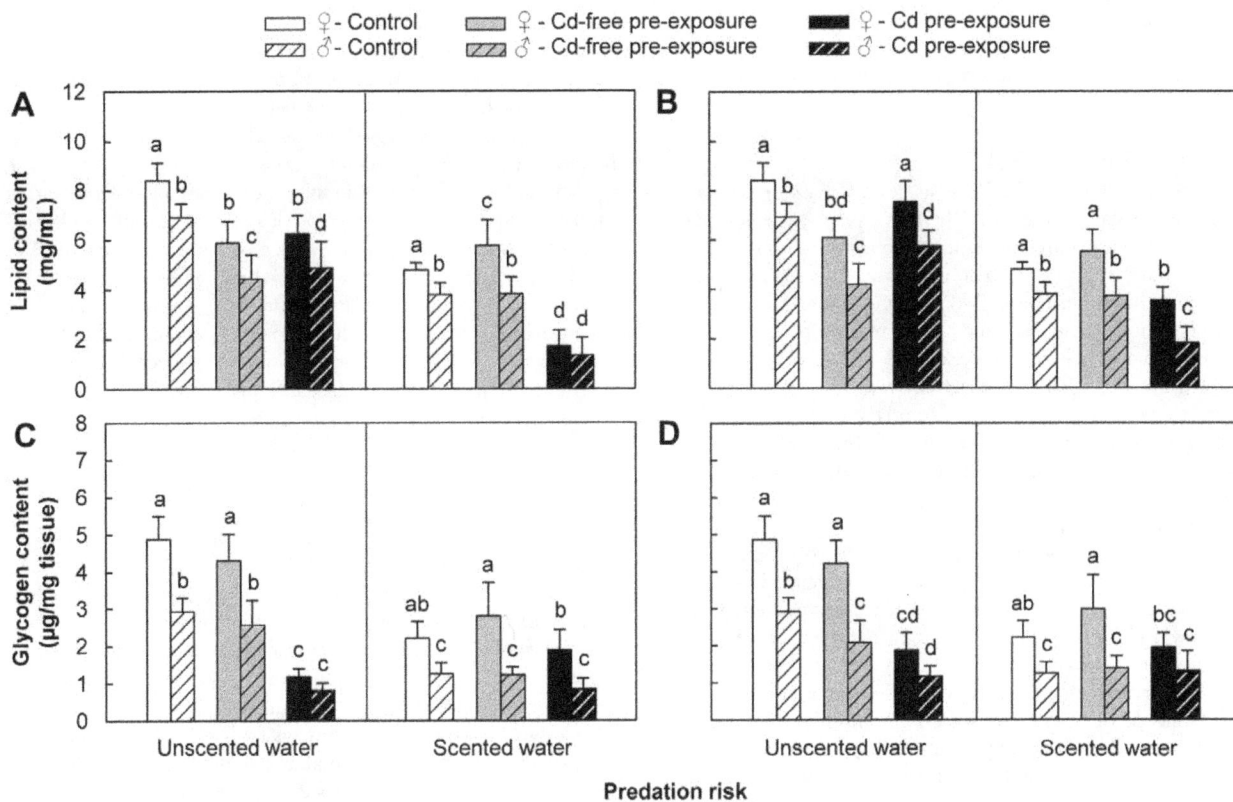

Figure 4. Energy reserves measured in the presence/absence of predation risk. A) lipid content in amphipods from device #1; B) lipid content in amphipods from device #2; C) glycogen content in amphipods from device #1; D) glycogen content in amphipods from device #2. *D. villosus* females and males are represented by filled and hatched bars, respectively. Energy reserves were measured following behavioural tests in which amphipods were exposed to the following conditions: Control = 24-h pre-exposure to Cd-free water +4-min test exposure to Cd-free water (white bars); Cd-free pre-exposure = 24-h pre-exposure to Cd-free water +4-min test exposure to 500 µg Cd/L (grey bars); Cd pre-exposure = 24-h pre-exposure to 500 µg Cd/L +4-min test exposure to 500 µg Cd/L (black bars). All data are means of $N = 10 \pm$ SD. Significant differences are indicated by different letters (Tukey HSD post-hoc test after ANOVA, $P < 0.05$).

[48]. Our results indicated no significant difference between males and females, suggesting that responses to stress may not be gender-dependant in *D. villosus*, contrary to what has been shown in other amphipod species exposed to various stressors [34,35]. In addition, our study highlighted the rapidity of adverse effects to Cd, with significant increases in MDA levels being measured directly after a 4-min test exposure to 500 µg Cd/L without Cd pre-exposure.

Behavioural responses. Our results revealed a significant increase in *D. villosus* locomotion in both experimental devices when exposed to Cd, with both refuge use and exploratory activity decreasing, and mobility increasing when outside of refuges. Such stress-induced responses can be interpreted as a rapid evasion tactic and, in the field, may represent the first line of defence that reduces the probability of contact with the toxicant (e.g. upstream movement). In contrast, a significant decrease in locomotor activity has been reported in *G. pulex* following 120-h exposure to 7.5 µg Cd/L [48], and in *G. lawrencianus* following 72-h exposure to 62 µg Cd/L [49], Cd concentrations that are ten- to fiftyfold lower than those used in this study. Differences to previous studies may be attributed to the assessment of a quasi-immediate/short-term locomotor response to Cd ecotoxicity in our experiments, which may reflect behaviour expressed under sudden discharges of highly concentrated toxic elements under natural conditions. Moreover, a 24-h pre-exposure to Cd did not result in any significant enhancement in behavioural response relative to that induced following a 4-min test exposure. Acclimation to Cd

may explain why 24-h pre-exposed individuals were as mobile as instantaneously-exposed individuals. Indeed, a resistance to acute Cd ecotoxicity has been demonstrated in *G. pulex* pre-exposed for 24 h to different concentrations [50]. The induction of metal-lothionein-like proteins, which can sequester Cd in a less toxic form, is one of the most plausible hypotheses to explain the physiological acclimation of gammarids to Cd [51].

Energy reserves. In our study, total lipid and glycogen contents were assessed to estimate energy reserves of individuals. Lipids are stored in fat bodies and used during starvation and/or reproduction periods [52] while glycogen generally provides energy resources for short-term activities [53]. Both can be easily catabolised to supply the energy necessary for antitoxic defences involved in defence mechanisms [54]. Our results revealed that Cd affected energy reserves, especially when *D. villosus* individuals were pre-exposed. A similar decrease in both lipid and glycogen content has also been highlighted in *G. roeseli* exposed to Cd [40], and a decrease in lipid content has been observed in in *Daphnia magna* following a 48-h exposure to Cd [55]. The decrease in energy content in organisms exposed to Cd could be explained by the energetic cost of tolerance offsetting the stress produced by the toxicant [54,56] as lipids and glycogen are both known to be first mobilised and rapidly used by organisms exposed to stressors [57]. A cumulative hypothesis might be the use of lipid and glycogen as an energy source for the increased locomotor activity observed in gammarids exposed to Cd. Our results also revealed higher energy reserves in females compared to males. These gender-based energy reserves differences can be explained by the reproductive process, oogenesis in females being more costly than spermatogenesis in males.

Responses to Predation Risk

Our results indicate that, in the absence of Cd, *D. villosus* were able to sense the alarm stimuli as, when exposed to chemical cues from a fish predator and/or injured conspecifics, individuals expressed a more cautious behavioural strategy by (1) increasing the time spent hiding in holes, (2) decreasing exploratory activity (i.e. visiting refuge areas less), and (3) decreasing mobility when outside of refuges. These responses are in accordance with a general decrease in locomotor activity and drift rate reported for several gammarid species exposed to predation risk [39,58,59]. Minimising displacement is believed to reduce the risk of being spotted by a potential predator, and especially by a fish predator. As hypothesised in [60], fish predators may be avoided by hiding in dense structures and/or by decreasing activity levels, as fish swim faster than amphipods. The absence of aggregative behaviour in *D. villosus*, an efficient means reducing individual risk of predation, is coherent with the species' aggressiveness and voracity towards potential prey, including conspecifics, reported by other authors [61,62].

In Cd spiked water, gammarids exposed to predation risk spent most time outside of refuges and were more mobile when outside of refuges. The absence of antipredator behavioural adjustments following Cd exposure could be explained by a disruption in the transfer of chemical information. The deregulation of internal information transfer signalling pathways by chemical pollutants is well known in various organisms across the taxonomic spectrum [63] and concerns not only chemical information within animals (i.e. endocrine disruption) but also chemical information transfer between organisms. Several studies have demonstrated disruption of chemically induced antipredator responses by pollutants such as pesticides [64], surfactants [63] and heavy metals, which have all been shown to impair predator avoidance behaviour in fish [65], aquatic snails [66], amphibians [67] and water fleas [68]. In fish, for example, Cd has been demonstrated to significantly affect (1) environmental exploration performance and behavioural reactions to conspecific chemical communication in *Brycon amazonicus* exposed for 96 h to 9.04 ± 0.07 µg Cd/L [6], and (2) the ability of *Oncorhynchus mykiss* to respond to natural pheromones after being exposed for 7 days to 2 µg Cd/L [65]. Even if the disruption mechanisms remain unresolved, three hypotheses can be proposed here: (1) a direct alteration to kairomone chemistry by Cd, (2) out-

competing of kairomone molecules by Cd at the receptor sites, or (3) interference by Cd with some process along the signal transduction pathway from kairomone reception to behavioural response. Such impairments to chemosensory functioning may have a strong effect on highly ecological functions. Organisms use chemical messengers not only to detect enemies but also to sense prey and to locate food, obtain mates, recognise close kin or mark a territory and impairment could lead to important ecological perturbations in populations inhabiting contaminated systems. We would suggest that assessing whether the absence of antipredator behavioural responses in Cd-exposed gammarids significantly compromises their survival when facing a predator should be the next step investigated. Additionally, more research is needed to determine the large-scale ecological implications of chemosensory dysfunction in metal-contaminated systems.

Conclusions

The results of this study demonstrate that 24-h exposure of *D. villosus* to Cd at non-lethal concentrations can (1) generate cell damage as revealed by higher MDA levels in exposed individuals, (2) induce short-term behavioural changes in locomotion, and (3) affect energy reserves. Our results also highlight that short-term Cd exposure can alter the olfactory-mediated behavioural responses of a freshwater gammarid to alarm substances from a predator or conspecifics. As a result, Cd could have important implications for predator avoidance strategies and, quite possibly, could affect other aspects of the animal–environment relationship, e.g. *D. villosus* prey population success, the leaf litter breakdown process, or toxicant biomagnification along the food chain. The mechanisms involved in Cd disturbance of olfactory functioning, depletion of energy reserves, and degradation of membrane lipid should be the subject of further studies.

Acknowledgments

We thank Céline Montazeau for her help in gammarid sampling and Romain Payot for his help in gammarid sampling and behavioural testing. The authors are also grateful to two anonymous referees for pertinent comments that helped to improve the manuscript and to Dr Kevin Roche for linguistic correction.

Author Contributions

Conceived and designed the experiments: PS EG JNB. Performed the experiments: PS EG CV. Analyzed the data: PS EG SD JNB. Contributed reagents/materials/analysis tools: PS EG CV. Wrote the paper: PS EG CV SD JFF JNB.

References

1. Wright DA, Welbourn PM (1994) Cadmium in the aquatic environment: a review of ecological, physiological, and toxicological effects on biota. Environ Rev 2: 187–214.

2. Williams KA, Green WJ, Pascoe D (1985) Studies on the acute toxicity of pollutants to freshwater macroinvertebrates. 1. Cadmium. Arch Hydrobiol 102: 461–471.

3. Vellinger C, Parant M, Rousselle P, Immel F, Wagner P, et al. (2012) Comparison of arsenate and cadmium toxicity in a freshwater amphipod (*Gammarus pulex*). Environ Pollut 160: 66–73.

4. Wallace WG, Hoexum-Brouwer TM, Brouwer M, Lopez GR (2000) Alterations in prey capture and induction of metallothioneins in grass shrimp fed cadmium-contaminated prey. Environ Toxicol Chem 19: 962–971.

5. Roast SD, Widdows J, Jones MB (2001) Impairment of mysid (*Neomysis integer*) swimming ability: an environmentally realistic assessment of the impact of cadmium exposure. Aquat Toxicol 52: 217–227.

6. Honda RT, Fernandes-de-Castilho M, Val AL (2008) Cadmium-induced disruption of environmental exploration and chemical communication in matrinxã, *Brycon amazonicus*. Aquat Toxicol 89: 204–206.

7. Kusch RC, Krone PH, Chivers DP (2008) Chronic exposure to low concentrations of waterborne cadmium during embryonic and larval de-

velopment results in the long-term hindrance of antipredator behavior in zebrafish. Environ Toxicol Chem 27: 705–710.

8. Sullivan BK, Buskey E, Miller DC, Ritacco PJ (1983) Effects of copper and cadmium on growth, swimming and predator avoidance in *Eurytemora affinis* (Copepoda). Mar Biol 77: 299–306.

9. Devin S, Beisel JN, Bachmann V, Moreteau JC (2001) *Dikerogammarus villosus* (Amphipoda: Gammaridae): another invasive species newly established in the Moselle River and French hydrosystems. Ann Limnol 37: 21–27.

10. Pöckl M (2007) Strategies of a successful new invader in European fresh waters: fecundity and reproductive potential of the Ponto-Caspian amphipod *Dikerogammarus villosus* in the Austrian Danube, compared with the indigenous *Gammarus fossarum* and *G. roeseli*. Freshw Biol 52: 50–63.

11. Pöckl M (2009) Success of the invasive Ponto-Caspian amphipod *Dikerogammarus villosus* by life history traits and reproductive capacity. Biol Invasions 11: 2021–2041.

12. Bij de Vaate A, Jazdzewski K, Ketelaars HAM, Gollasch S, Van der Velde G (2002) Geographical patterns in range extension of Ponto-Caspian macro-invertebrate species in Europe. Can J Fish Aquat Sci 59: 1159–1174.

13. Guerlet E, Ledy K, Giambérini L (2008) Is the freshwater gammarid, *Dikerogammarus villosus*, a suitable sentinel species for the implementation of histochemical biomarkers? Chemosphere 72: 697–702.

14. Gonzalo C, Camargo JA, Masiero L, Casellato S (2010) Fluoride toxicity and bioaccumulation in the invasive amphipod *Dikerogammarus villosus* (Sowinsky, 1894): a laboratory study. Bull Environ Contam Toxicol 85: 472–475.

15. Maazouzi C, Masson G, Izquierdo MS, Pihan JC (2008) Chronic copper exposure and fatty acid composition of the amphipod *Dikerogammarus villosus*: results from a field study. Environ Pollut 156: 221–226.

16. Hellou J (2011) Behavioural ecotoxicology, an "early warning" signal to assess environmental quality. Environ Sci Pollut Res 18: 1–11.

17. Gerhardt A (1995) Monitoring behavioural responses to metals in *Gammarus pulex* (L.) (Crustacea) with impedance conversion. Environ Sci Pollut Res 2: 15–23.

18. MacNeil C, Elwood RW, Dick JTA (1999) Predator-prey interaction between brown trout *Salmo trutta* and native and introduced amphipods; their implication for fish diets. Ecography 22: 686–696.

19. Kelleher B, Bergers PJM, Van den Brink FWB, Giller PS, Van der Velde G, et al. (1998) Effects of exotic amphipod invasions on fish diet in the Lower Rhine. Arch Hydrobiol 143: 363–382.

20. Eckmann R, Mörtl M, Baumgärtner D, Berron C, Fischer P, et al. (2008) Consumption of amphipods by littoral fish after the replacement of native *Gammarus roeseli* by invasive *Dikerogammarus villosus* in Lake Constance. Aquat Invasions 3: 187–191.

21. Mathis A, Hoback WW (1997) The influence of chemical stimuli from predators on precopulatory pairing by the amphipod, *G. pulex pseudolimnaeus*. Ethology 103: 33–40.

22. Brönmark C, Hansson LA (2000) Chemical communication in aquatic systems: an introduction. Oikos 88: 103–109.

23. De Meester L (1993) Genotype, fish mediated chemicals, and phototactic behaviour in *Daphnia magna*. Ecology 74: 1467–1474.

24. Kalmijn AJ (1988) Hydrodynamic and acoustic field detection. In: Atema J, Fay RR, Popper AN, Tavolga WN, editors. Sensory biology of aquatic animals. New York: Springer-Verlag. 83–130.

25. Dahl J, Nilsson PA, Pettersson LB (1998) Against the flow: chemical detection of downstream predators in running waters. Proc Biol Sci 265: 1339–1344.

26. Chivers DP, Wisenden BD, Smith RJF (1996) Damselfly larvae learn to recognize predators from chemical cues in the predator's diet. Anim Behav 52: 315–320.

27. Chivers DP, Smith RJF (1998) Chemical alarm signalling in aquatic predator-prey systems: a review and prospectus. Ecoscience 5: 338–352.

28. Sih A (1987) Predators and prey lifestyles: an evolutionary and ecological overview. In: Kerfoot WC, Sih A, editors. Predation: direct and indirect impacts on aquatic communities. Hanover, New Hampshire: University Press of New England. 203–224.

29. Pennuto C, Keppler D (2008) Short-term predator avoidance behaviours by invasive and native amphipods in the Great Lakes. Aquat Ecol 42: 629–641.

30. Wisenden BD, Cline A, Sparkes TC (1999) Survival benefit to antipredator behavior in the amphipod *Gammarus minus* (Crustacea: Amphipoda) in response to injury-released chemical cues from conspecifics and heterospecifics. Ethology 105: 407–414.

31. Wisenden BD, Pohlman SG, Watkin EE (2001) Avoidance of conspecific injury-released chemical cues by free-ranging *Gammarus lacustris* (Crustacea: Amphipoda). J Chem Ecol 27: 1249–1258.

32. Endler JA (1991) Interactions between predators and prey. In: Krebs JR, Davies NB, editors. Behavioural ecology: an evolutionary approach. Oxford: Blackwell Scientific Publications. 169–196.

33. Campbell JI, Meadows PS (1974) Gregarious behaviour in a benthic marine amphipod (*Corophium volutator*). Cell Mol Life Sci 30: 1396–1397.

34. Sornom P, Felten V, Médoc V, Sroda S, Rousselle P, et al. (2010) Effect of gender on physiological and behavioural responses of *Gammarus roeseli* (Crustacea Amphipoda) to salinity and temperature. Environ Pollut 158: 1288–1295.

35. McCahon CP, Pascoe D (1988) Increased sensitivity to cadmium of the freshwater amphipod *Gammarus pulex* (L.) during the reproductive period. Aquat Toxicol 13: 183–194.

36. Kaldonski N, Lagrue C, Motreuil S, Rigaud T, Bollache L (2008) Habitat segregation mediates predation by the benthic fish *Cottus gobio* on the exotic amphipod species *Gammarus roeseli*. Naturwissenschaften 95: 839–844.

37. Perrot-Minnot MJ, Kaldonski N, Cézilly F (2007) Increased susceptibility to predation and altered anti-predator behaviour in an acanthocephalan-infected amphipod. Int J Parasitol 37: 645–651.

38. Médoc V, Rigaud T, Bollache L, Beisel JN (2009) A manipulative parasite increasing an antipredator response decreases its vulnerability to a nonhost predator. Anim Behav 77: 1235–1241.

39. Åbjörnsson K, Dahl J, Nyström P, Brönmark C (2000) Influence of predator and dietary chemical cues on the behaviour and shredding efficiency of *Gammarus pulex*. Aquat Ecol 34: 379–387.

40. Gismondi E, Rigaud T, Beisel JN, Cossu-Leguille C (2012) Microsporidia parasites disrupt the responses to cadmium exposure in a gammarid. Environ Pollut 160: 17–23.

41. Plaistow SJ, Troussard JP, Cézilly F (2001) The effect of the acanthocephalan parasite *Pomphorhynchus laevis* on the lipid and glycogen content of its intermediate host *Gammarus pulex*. Int J Parasitol 31: 346–351.

42. Behrens W, Madère R (1991) Malonaldehyde determination in tissues and biological fluids by ion-pairing high-performance liquid chromatography. Lipids 22: 232–236.

43. Khan FR, Bury NR, Hogstrand C (2010) Cadmium bound to metal rich granules and exoskeleton from *Gammarus pulex* causes increased gut lipid peroxidation in zebrafish following single dietary exposure. Aquat Toxicol 96: 124–129.

44. Geret F, Serafim A, Barreira L, Bebianno MJ (2002) Effect of cadmium on antioxidant enzyme activities and lipid peroxidation in the gills of the clam *Ruditapes decussatus*. Biomarkers 7: 242–256.

45. Geret F, Jouan A, Turpin V, Bebianno MJ, Cosson RP (2002) Influence of metal exposure on metallothionein synthesis and lipid peroxidation in two bivalve mollusks: the oyster (*Crassostrea gigas*) and the mussel (*Mytilus edulis*). Aquat Living Resour 15: 61–66.

46. Del Rio D, Stewart AJ, Pellegrini N (2005) A review of recent studies on malondialdehyde as toxic molecule and biological marker of oxidative stress. Nutr Metab Cardiovas 15: 316–328.

47. Waisberg M, Joseph P, Hale B, Beyersmann D (2003) Target for toxicity and death due to exposure to cadmium chloride. Toxicology 192: 95–117.

48. Felten V, Charmantier G, Mons R, Geffard A, Rousselle P, et al. (2008) Physiological and behavioural responses of *Gammarus pulex* (Crustacea: Amphipoda) exposed to cadmium. Aquat Toxicol 86: 413–425.

49. Wallace WG, Estephan A (2004) Differential susceptibility of horizontal and vertical swimming activity to cadmium exposure in a gammaridean amphipod (*Gammarus lawrencianus*). Aquat Toxicol 69: 289–297.

50. Stuhlbacher A, Maltby L (1992) Cadmium resistance in *Gammarus pulex* (l.). Arch Environ Con Tox 22: 319–324.

51. Howell R (1985) Effect of zinc on cadmium toxicity to the amphipod *Gammarus pulex*. Hydrobiologia 123: 245–249.

52. Cargill AS, Cummins KW, Hanson BJ, Lowry RR (1985) The role of lipids as feeding stimulants for shredding aquatic insects. Freshwater Biol 15: 455–464.

53. Sparkes TC, Keogh DP, Pary RA (1996) Energetic costs of mate guarding behavior in male stream-dwelling isopods. Oecologia 106: 166–171.

54. Dutra BK, Fernandes F, Lauffer A, Oliveira G (2009) Carbofuran-induced alterations in the energy metabolism and reproductive behaviors of *Hyalella castroi* (Crustacea, Amphipoda). Comp Biochem Phys C 149: 640–646.

55. Barata C, Varo I, Navarro JC, Arun S, Porte C (2005) Antioxidant enzyme activities and lipid peroxidation in the freshwater cladoceran *Daphnia magna* exposed to redox cycling compounds. Comp Biochem Phys C 140: 175–186.

56. Sancho E, Ferrando MD, Fernandez C, Andreu E (1998) Liver energy metabolism of *Anguilla anguilla* after exposure to fenitrothion. Ecotoxicol Environ Saf 41: 168–175.

57. Durou C, Mouneyrac C, Pellerin J, Péry A (2008) Conséquences des perturbations du métabolisme énergétique. In: Amiard JC, Amiard-Triquet C, editors. Les biomarqueurs dans l'évaluation de l'état écologique des milieux aquatiques. Paris: Lavoisier. 273–289.

58. Baumgärtner D, Jungbluth AD, Koch U, Von Elert E (2002) Effects of infochemicals on microhabitat choice by the freshwater amphipod *Gammarus roeseli*. Arch Hydrobiol 155: 353–367.

59. Williams DD, Moore KA (1985) The role of semiochemicals in benthic community relationships of the lotic amphipod *Gammarus pseudolimnaeus*: a laboratory analysis. Oikos 44: 280–286.

60. Åbjörnsson K, Hansson LA, Brönmark C (2004) Responses of prey from habitats with different predator regimes: local adaptation and heritability. Ecology 85: 1859–1866.

61. Kinzler W, Kley A, Mayer G, Waloszek D, Maier G (2009) Mutual predation between and cannibalism within several freshwater gammarids: *Dikerogammarus villosus* versus one native and three invasives. Aquat Ecol 43: 457–464.

62. Platvoet D, Dick JTA, MacNeil C, Van Riel MC, Van der Velde G (2009) Invader-invader interactions in relation to environmental heterogeneity leads to zonation of two invasive amphipods, *Dikerogammarus villosus* (Sowinsky) and *Gammarus tigrinus* Sexton: amphipod pilot species project (AMPIS) report 6. Biol Invasions 11: 2085–2093.

63. Lürling M, Scheffer M (2007) Info-disruption: pollution and the transfer of chemical information between organisms. Trends Ecol Evol 22: 374–379.

64. Hanazato T (1999) Anthropogenic chemicals (insecticides) disturb natural organic chemical communication in the plankton community. Environ Pollut 105: 137–142.

65. Scott GR, Sloman KA, Rouleau C, Wood CM (2003) Cadmium disrupts behavioural and physiological responses to alarm substance in juvenile rainbow trout (*Oncorhynchus mykiss*). J Exp Biol 206: 1779–1790.

66. Lefcort H, Amman E, Eiger SM (2000) Antipredator behavior as an index of heavy metal pollution? A test using snails and caddisflies. Arch Environ Contam Toxicol 38: 311–316.

67. Lefcort H, Meguire RA, Wilson LH, Ettinger WF (1998) Heavy metals alter the survival, growth, metamorphosis, and antipredator behavior of Columbia spotted frog (*Rana luteiventris*) tadpoles. Arch Environ Contam Toxicol 35: 447–456.

68. Hunter K, Pyle G (2004) Morphological responses of *Daphnia pulex* to *Chaoborus americanus* kairomone in the presence and absence of metals. Environ Toxicol Chem 23: 1311–1316.

12

Species Distribution Models and Impact Factor Growth in Environmental Journals: Methodological Fashion or the Attraction of Global Change Science

Lluís Brotons[1,2]*

1 CEMFOR-CTFC, Forest Sciences Center of Catalonia, Solsona, Catalonia, Spain, **2** CREAF, Centre for Ecological Research and Forestry Applications, Cerdanyola del Vallès, Catalonia, Spain

Abstract

In this work, I evaluate the impact of species distribution models (SDMs) on the current status of environmental and ecological journals by asking the question to which degree development of SDMs in the literature is related to recent changes in the impact factors of ecological journals. The hypothesis evaluated states that research fronts are likely to attract research attention and potentially drive citation patterns, with journals concentrating papers related to the research front receiving more attention and benefiting from faster increases in their impact on the ecological literature. My results indicate a positive relationship between the number of SDM related articles published in a journal and its impact factor (IF) growth during the period 2000–09. However, the percentage of SDM related papers in a journal was strongly and positively associated with the percentage of papers on climate change and statistical issues. The results support the hypothesis that global change science has been critical in the development of SDMs and that interest in climate change research in particular, rather than the usage of SDM per se, appears as an important factor behind journal IF increases in ecology and environmental sciences. Finally, our results on SDM application in global change science support the view that scientific interest rather than methodological fashion appears to be the major driver of research attraction in the scientific literature.

Editor: Diego Fontaneto, Consiglio Nazionale delle Ricerche (CNR), Italy

Funding: This study has received financial support from the projects CGL2011-29539/BOS and Consolider Montes CSD2008-00040 granted by the Spanish Ministry of Education and Science (MEC) and it is also a contribution to the FP7-226852 project SCALES. Additional funding to LB was received from the Catalan government through project 2010 BE 272. The funders had no role in study design, data collection and analysis, decision to publish, or preparation of the manuscript.

Competing Interests: The author has declared that no competing interests exist.

* Email: lluis.brotons@ctfc.cat

Introduction

Science is under continuous change and the appearance and development of new methodologies and approaches often has profound impact on the research panorama [1]. Species distribution models (SDM) exploring the association of environmental and species location data have rapidly developed over the last 15 years and appear to have had a great influence on environmental sciences and ecology in particular. SDM applications to climate change have been identified as the broadest research front in ecology and environment from Thomson ISI according to the clustering of the co-citing highly cited papers on this topic [2].

The popularity of SDM may be rooted in a range of different factors. Since understanding species distributions is a fundamental goal of ecology, the appearance of SDMs may have provided an efficient methodological approach to estimate species distributions and allowed the use of model outputs in a wide range of ecological applications (from species-energy relationships to niche conservationism [3]. The great availability of location data sets and environmental information in digital format (GIS) and the rapid development of statistical methods allowing efficient use of available information may have influenced the successful adoption of these techniques and their rapid spread in the ecological literature [3]. Being easy to implement using widely available GIS

and distributional data coming from existing databases, SDMs may have benefited from a combination of fashion and ease of implementation. Alternatively, the popularity of SDMs may be related to the application of these techniques to expanding new ecological disciplines derived from an increasing interest in the effects of global change on biodiversity. SDMs allow a rapid estimation of the spatially explicit effects of drivers such as climate change on biodiversity at large spatial scales. Some seminal applications using SDMs have been instrumental in setting baselines of potential future impacts of climate change on a range of species [4].

In this work, I want to evaluate the impact of SDMs on the current status of ecological journals by asking the question to which degree of the use of SDMs in the literature is related to recent changes in the impact factors of ecological journals. The hypothesis evaluated derives from the idea that research fronts are likely to attract attention and drive research developments in a given discipline. Therefore, journals with a stronger focus on the research front should concentrate higher attention and receive more citations, thus benefiting from faster increases in their impact on the ecological literature. If this holds true, we predict that journals publishing more SDM-related articles should have benefited from the interest of this prolific field and show stronger increases in their citation rates and impact factors. However, if

SDM usage is related to increases in citations rates, two main mechanisms may be identified as potential explanations to the observed patterns. First, SDM-related articles may be associated with studies on global change impacts on biodiversity (climate changes, land use changes and the impact of invasive species), and therefore, one should expect that the number of papers on these topics and not on SDM per se should better explain journal citation patterns. Alternatively, SDM may have influenced journal citations rates through of their intrinsic attraction as methodological novelty allowing the easy estimation of species distributions. In this case, I expect the number of papers on SDM to be associated to changes in the journal impact factor independently of the range of global change topics included in environmental and ecology journals.

Methods

I used data from ISI web of science and test the prediction that the proportion of articles in a journal containing a larger number of SDMs related articles is related to the journal changes in impact factor during the period 2000–09. I used an objective method to select journals publishing a minimum number of articles related to SDM. This method included a general search for SDM related articles and the selection of a subset of articles included in non-multidisciplinary journals with more than 5 SDM articles published in the 2000–2009 period. Multidisciplinary journals were discarded because they included a much broader number of topics than thematic journals thus leading to potential biases in our blibliometric estimators. First, I searched the ISI web of science for articles containing the words "predictive species distribution model" "niche model" or "habitat suitability model" or a combination of these [3]. I identified a total of 2.118 articles leading to a total of 37.854 citations. Second, I selected a subset of articles published in currently active, non-multidisciplinary journals (according to the ISI categories, Thompson Scientific) with more than 5 SDM articles published during the period 2000–09 from ISI categories accounting for at least 2% of the total SDM references. These articles accounted for 1305 of the articles above. Although this subset, which accounts for over 60% of the SDM related articles included in our search, may not represent a comprehensive compilation of articles in the literature dealing with SDMs, I believe that due to the wide range of journals included, is representative of their distribution in the ecology, environmental sciences and biodiversity journals panorama. For this subset of journals mostly within the three subject areas mentioned above, I estimated impact factor trajectories and compiled the number of articles published per year during the period 2000–09. With the information derived from the databases, I was able to derive for, each of the 56 journals selected (Table S1), a measure of SDM relevance, *SDMr*, as the proportion of SDM related articles from the total number of articles published by the journal during the study period (range 0.01 to 12%). For each of the journals in this subset, I also obtained the number of articles published on different topics related to global change by searching for different combination of key words ("climate change*" (1505 articles), "land use change* or fragmentation" (603 articles) and "invasive species*" (1128 articles) in biology and ecology ("biolog* and ecology*")) and calculated the proportion of articles for each topic in each journal (Table S1). Finally, I also used two additional different controls searches to account for general patterns in general ecological studies searching for the words "population and species" in biology and ecology ("biolog* and ecology*") (2511 articles) and methodological biases searching for the word

"statistics" in biology and ecology ("biolog* and ecology*") (373 articles).

Impact factor (IF) is generally recognised as the primary measure of journal "quality" [6], but see [7]. Changes in the impact of the articles published in each journal (absolute increase) were quantified by calculating the slope of the regression of the journal's impact factor [5] and the respective year with positive slopes for journals with increasing impact factors and negative slopes for journals with decreasing impact. Finally, I also included, for each journal, the number of published articles during the study period and the year of the journal first issue (journal age) to account for general differences in article production and antiquity between journals [8]. I tested the role of journal descriptors on IF change and SDMr by means of linear models and forward variable selection using information theory based criteria (Bayesian Information Criteria, BIC) in R (package "MASS", [9]). Both variables, IF change and SDMr, were log transformed to ensure normality. To deal with collinearity problems, I also used an analytical method named hierarchical partitioning (HP hereafter). HP reduces collinearity problems by determining the independent contribution of each explanatory variable to the response variable (I) and separates it from the joint contribution (J), resulting from correlation with other variables (for a detailed explanation of how HP works, see [10]. This allows ranking the importance of the covariates in explaining the response variable independently of the others covariates. Given its usefulness for complementing multiple regression analysis, I applied HP using the "hier.part package" in R [11].

Results

The number of articles on SDM in the literature has rapidly increased during the period 2000–09 (Fig. 1) with the number of citations these papers are receiving also increasing rapidly with one third of the total amount of citations received in 2010. Ecology journals with a higher percentage of SDM related articles showed higher increases in their IFs between 2000 and 2009 (Fig. 2, Table S2). The total number of papers published by a journal, its IF at the beginning of the study period or the journal age were not significant factors behind changes in IF for the set of journals analysed and were thus discarded from the final model. The three ecology journals with values of SDMr larger than 5% (Diversity & Distributions, Global Ecology & Biogeography and Ecography) showed increases during the study period larger than 200% in their IF (Table S1). The relationship between changes in IF and SDMr was stronger and accounted for up to 16% of the variability when the journal Ecology Letters (experiencing a spectacular increase in IF during this period) was excluded from the analyses ($\beta = 11.05$, $t = 3.13$ $d.f. = 51$, p<0.005).

SDMr was highly predictable from the combination of topics analysed and contained in a given journal. SDMr was in particular strongly and positively related to the number of articles on global change topics (climate change, $\beta = 5.53$, $t = 4.74$, $d.f. = 51$, p< 0.0001, Table S2) and to the percentage of articles on statistics published by a journal ($\beta = 29.69$, $t = 4.62$, $d.f. = 51$, p<0.0001, Table S2). I also found a minor tendency for journals with lower IF in the year 2000 ($\beta = -0.00044$, $t = 2.44$, $d.f. = 51$, p<0.05) and a lower number of total articles published ($\beta = -0.0042$, $t = 2.13$, $d.f. = 51$, p<0.0001) to include a largest percentage of SDM related articles. The final model predicted 60% of variability in journal SDMr.

The effect of SDMr on IF change disappeared after the inclusion of variables accounting for the thematic scope of the journals. IF change was strongly and positively related to the

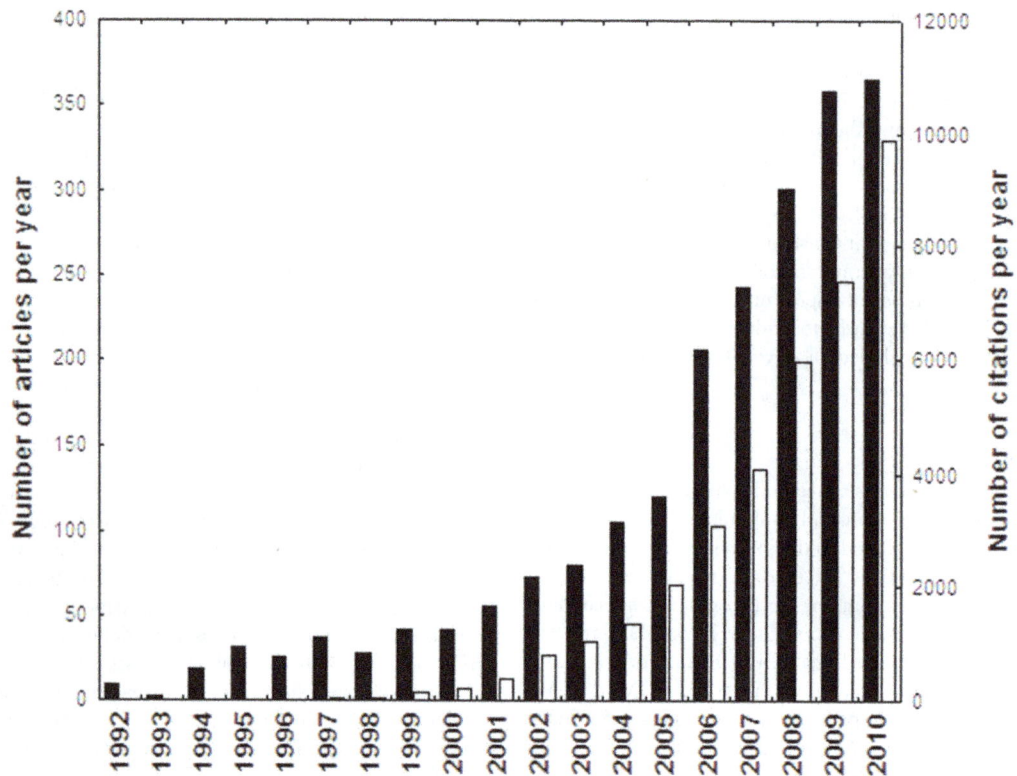

Figure 1. Number of published SDM related articles in the ecological literature (black bars) and number of citations received by these articles per year (white bars) during the period 1992–2010.

number of climate change papers published (CLIr) in a journal during the study period (Fig. 2, Table S2). In fact, CLIr showed a much higher both, independent and joint explanatory power than any other variable included in the assessments, suggesting that it is a much stronger candidate to drive IF changes than SDMr or the others bibliometric descriptors used (Table S2).

Discussion

My results indicate a positive relationship between the numbers of SDM related articles published in a journal and its IF increases during the period 2000–09. However, given the strong association between SDMr and the number of global change articles published in a journal, the role of SDM on IFs is likely to be an indirect effect of the increases in the journal IF being associated with a larger number of climate change articles published. The results support the hypothesis that global change science has been critical in the development of SDM and that climate change research in particular appears as an important factor behind increases of IF in ecological and environmental journals devoting larger attention to this topic.

Recent studies have found a positive trend in the number of articles cited by ecological journals in recent years leading to a potential for general increase in citation rates and thus impact factors [12,13]. However, it seems that increases in IF are not evenly distributed with some journals getting a disproportionally larger share of the IF growth leading to changes in the potential impact of these journals [13,14]. Our results indicate that journals with an overall higher percentage of SDMr, but specially those with more articles on climate change topics have grown at

relatively faster rates than others [15]. However, SDM use in the journals included in the analyses did not directly drive IF increases in the set of journals analysed. The application of SDMs has been described as a one of the biggest emerging fronts in ecology in the last years [2], with a large number of highly impacting articles, and appears to be rapidly growing. SDM articles often use already existing and readily available environmental data and therefore are less constrained than more traditional ecology works based on field data thus opening the way to faster and more widespread publication on different issues of general interest to ecologists. Furthermore, available software makes SDM applications very easy and potentially articles using these methodologies may be easier to write than in other ecology areas. Other studies have indeed described increases in citations rates of ecological journals associated to the number of articles published on SDMs (i.e. invasion biology, [16]). However, our results do not support the view that journals publishing more SDM related articles receive more citations per se.

Rather, the role of SDM on IF trends disappeared when climate change was included in the analyses. This indicates that journals with a larger numbers of climate change related papers have indeed grown larger IFs in the 10-year period of our study. Journals with a higher proportion of SDMs also published more articles on hotter topics such as climate change or invasion biology [17] suggesting that the increase in impact factor is not a direct consequence of the number of SDM articles published. Overall, these results show that specific topics disproportionally drive changes in research attention and appeared to influence journal citation patterns [17]. The finding that climate change research contributes to the variability in recent IF increases of environ-

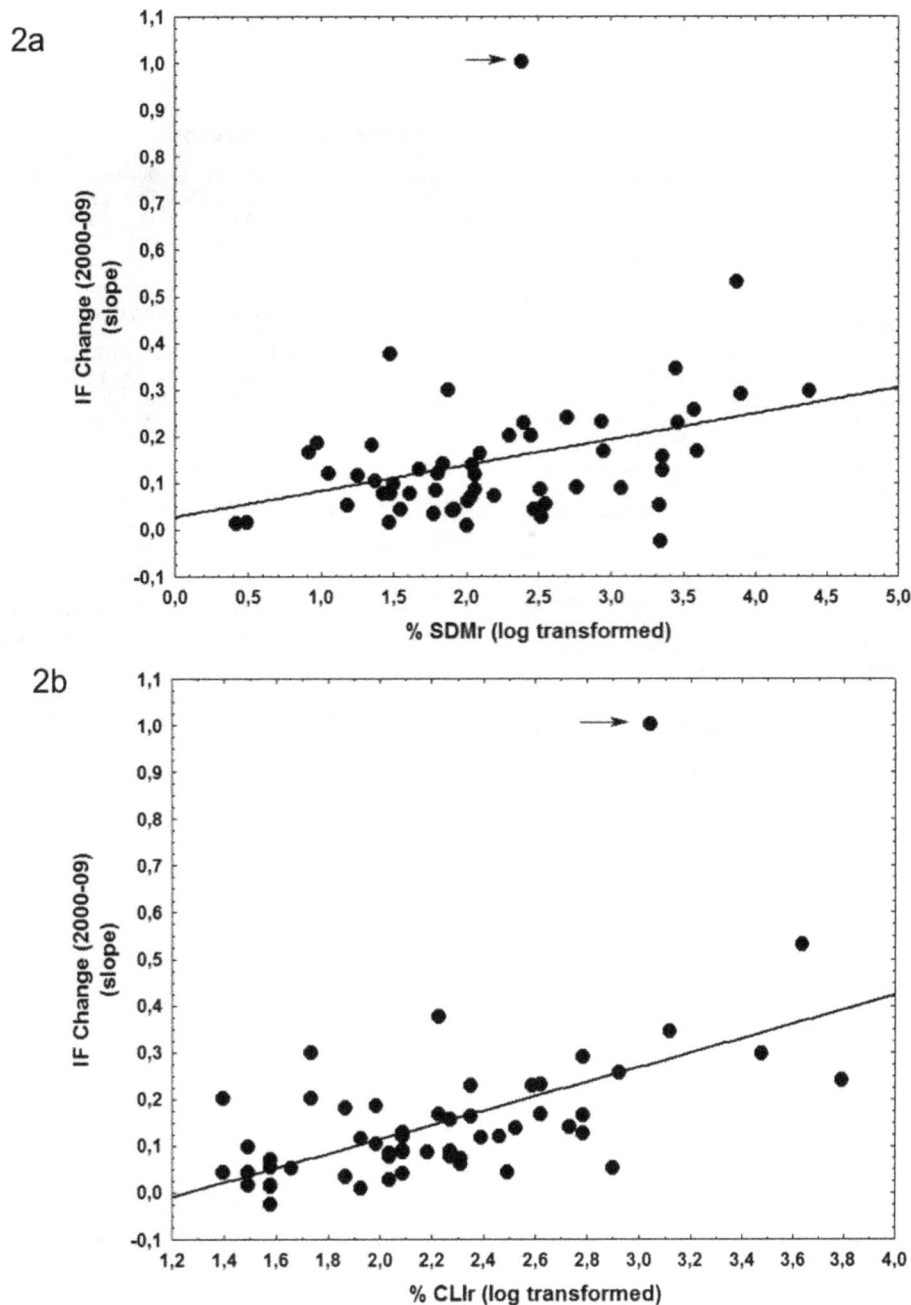

Figure 2. Increases in IF factors between 2000 and 2009 in relation to the percentage of SDM papers (SDMr, $R^2=0.12$, $\beta=10.68$, $t=2.74$, $d.f.=54$, p<0.01) (a), and in relation to the percentage of climate change papers published in each journal (CLIr, $R^2=0.31$, $\beta=17.82$, $t=4.98$, $d.f.=54$, p<0.001) (b). The arrow identifies the journal Ecology Letters.

mental and ecology journals supports the view that scientific interest and not methodological fashion appears to be a major driver of research attraction [18].

SDM usage appeared therefore related to changes in journal IF most likely because their development has been largely driven by applications in climate change science. SDM development has been instrumental in moving global change science forward due to the capability of the models to be used for a large number of species over large spatial scales [4,16]. The real impact of SDM in ecology may be therefore be better interpreted as one of the foundation stones of global change science applications in ecology.

SDMs have allowed the environmental research community to efficiently integrate the extensive availability of large-scale biological data, appropriate tools and environmental data sets into the growing needs of spatially explicit biodiversity assessments. SDMs may continue to play a significant role in the future panorama of ecology and environmental sciences as long as they remain as key methodological approaches in global change science. Spatial models allowing the projection of species distributions to future environmental conditions such as climate change are still required and tend to progressively become more complex to overcome the limitations of the correlative nature of

SDMs [19]. However, I think that the challenges faced by model building in global change science will require flexible, integrative approaches allowing the use of extensively available data, and SDMs are likely to continue playing a significant role in this context.

Acknowledgments

I thank Jane Elith and Miguel Clavero for fruitful discussions on impact factors in ecology.

Author Contributions

Conceived and designed the experiments: LB. Performed the experiments: LB. Analyzed the data: LB. Contributed reagents/materials/analysis tools: LB. Wrote the paper: LB.

References

1. Neff MW, Corley E (2009) 35 years and 160,000 articles: A bibliometric exploration of the evolution of ecology. Scientometrics 80: 657–682.
2. Institute for Scientific Information® (2005) Thomson ISI Essential Science IndicatorsSM Special Topics: Research Front Methodology [online]. Philadelphia: ISI. (http://www.esi-topics.com/RFmethodology.html).
3. Elith J, Leathwick J (2009) Species distribution models: ecological explanation and prediction across space and time. Ann Rev Ecol Sys 40: 677–697.
4. Thomas C, Cameron A, Green RE, Bakkenes M, Beaumont LJ, et al. (2004) Extinction risk from climate change. Nature 427: 145–148.
5. Garfield E (2006) The history and meaning of the journal impact factor. J Am Med Asso 295: 90–93.
6. Olden JD (2007) How do ecological journals stack-up? Ranking of scientific quality according to the h index. Ecoscience 14: 370–376.
7. Leimu R, Koricheva J (2005) What determines the citation frequency of ecological papers. TREE 20:28:32.
8. Wilson AE (2007) Journal Impact Factors are inflated. Bioscience 57: 550–551.
9. Venables WN, Ripley BD (2002) Modern Applied Statistics with S. Fourth edition. Springer.
10. Mac Nally R (2002) Multiple regression and inference in ecology and conservation biology: further comments on identifying important predictor variables. Biodivers Conserv 11: 1397–1401.
11. Walsh C, Mac Nally R (2005) "The hier.part Package" version 1.0–1. Hierarchical Partitioning. Documentation for R: A language and environment for statistical computing. R Foundation for statistical Computing, Vienna, Austria
12. Neff BD, Olden JD (2010) Not so fast: Inflation in Impact factors contributes to apparent improvents in journal quality. Bioscience 60: 455–459.
13. Althouse BM, West JD, Bergstrom CR, Bergstrom T (2009) Differences in impact factor across fields and over time. J Am Soc Inf Sci Tech 60: 27–34 http://dx.doi.org/10.1002/asi.20936.
14. Groesser SN (2012) Dynamics of journal impact factors. Sys Res Behav Sci 29: 624–644
15. Ioannidis JPA (2006) Concentration of the most-cited papers in the scientific literature: analyses of journal ecosystems. PLOS ONE 1: art. no.-e5.
16. Barbosa FG, Schneck F, Melo AS (2012) Use of ecological niche models to predict the distribution of invasive species: a scientometric analysis. Braz J Biol 72: 821–829.
17. Pysek P, Richardson DM, Vojtech J (2006) Who cites who in the invasion zoo: insights from an analysis of the most highly cited papers in invasion ecology. Preslia 78: 437–468
18. Lortie CJ, Aarssen LW, Budden AE, Leimu R (2013) Do citations and impact factors relate to the real numbers in publications? A case study of citation rates, impact, and effect sizes in ecology and evolutionary biology. Scientometrics 94: 675–682
19. Dormann CF, Schymanski SJ, Cabral J, Chuine I, Graham C, et al. (2012). Correlation and process in species distribution models: bridging a dichotomy. J. Biogeogr, 39: 2119–2131.

13

Contribution for the Derivation of a Soil Screening Value (SSV) for Uranium, Using a Natural Reference Soil

Ana Luisa Caetano[1,2]*, Catarina R. Marques[1,2], Ana Gavina[6,2], Fernando Carvalho[3], Fernando Gonçalves[1,2], Eduardo Ferreira da Silva[4], Ruth Pereira[5,6]

1 Department of Biology, University of Aveiro, Campus Universitário de Santiago, Aveiro, Portugal, **2** CESAM, University of Aveiro, Campus Universitário de Santiago, Aveiro, Portugal, **3** Nuclear and Technological Institute (ITN) Department of Radiological Protection and Nuclear Safety, Sacavém, Portugal, **4** Department of Geosciences, University of Aveiro, GeoBioTec Research Center, Campus Universitário de Santiago, Aveiro, Portugal, **5** Department of Biology, Faculty of Sciences of the University of Porto, Porto, Portugal, **6** Interdisciplinary Centre of Marine and Environmental Research (CIIMAR/CIMAR), University of Porto, Porto, Portugal

Abstract

In order to regulate the management of contaminated land, many countries have been deriving soil screening values (SSV). However, the ecotoxicological data available for uranium is still insufficient and incapable to generate SSVs for European soils. In this sense, and so as to make up for this shortcoming, a battery of ecotoxicological assays focusing on soil functions and organisms, and a wide range of endpoints was carried out, using a natural soil artificially spiked with uranium. In terrestrial ecotoxicology, it is widely recognized that soils have different properties that can influence the bioavailability and the toxicity of chemicals. In this context, SSVs derived for artificial soils or for other types of natural soils, may lead to unfeasible environmental risk assessment. Hence, the use of natural regional representative soils is of great importance in the derivation of SSVs. A Portuguese natural reference soil PTRS1, from a granitic region, was thereby applied as test substrate. This study allowed the determination of NOEC, LOEC, EC_{20} and EC_{50} values for uranium. Dehydrogenase and urease enzymes displayed the lowest values (34.9 and <134.5 mg U Kg, respectively). *Eisenia andrei* and *Enchytraeus crypticus* revealed to be more sensitive to uranium than *Folsomia candida*. EC_{50} values of 631.00, 518.65 and 851.64 mg U Kg were recorded for the three species, respectively. Concerning plants, only *Lactuca sativa* was affected by U at concentrations up to 1000 mg U kg^{1}. The outcomes of the study may in part be constrained by physical and chemical characteristics of soils, hence contributing to the discrepancy between the toxicity data generated in this study and that available in the literature. Following the assessment factor method, a predicted no effect concentration (PNEC) value of 15.5 mg kg^{-1}_{dw} was obtained for U. This PNEC value is proposed as a SSV for soils similar to the PTRS1.

Editor: Stephen J. Johnson, University of Kansas, United States of America

Funding: ALC was supported by a PhD grant from Fundação para a Ciência e Tecnologia (FCT) (http://www.fct.pt/). The funders had no role in study design, data collection and analysis, decision to publish, or preparation of the manuscript.

Competing Interests: The authors have declared that no competing interests exist.

* Email: ana.caetano@ua.pt

Introduction

Uranium (U) is a natural soil component, being originated from rocks in the Earth's crust, where it mainly occurs in the form of oxides. Natural processes acting on rocks and soils, such as wind, water erosion, dissolution, precipitation and volcanic activity contribute for U dispersal in the environment [1]. The use of U as fuel in nuclear power plants has driven to its large-scale exploration worldwide. The U exploration became significantly important in the world during the Second World War, and later on during the Cold War, in both cases to supply military needs of the greatest potencies. Recently, the World Nuclear Association estimated worldwide reserves of U at 5.4 million tons in 2009, of which Australia had about 31%, followed by Kazakhstan (12%), Canada and Russia with 9% (http://www.world-nuclear.org/info/inf75.html). The remarkable energy crisis that is currently faced worldwide due to the exhaustion of carbon based energy resources is demanding further extraction of U, as nuclear energy arises as a potential solution. Hence, it is expected that the mining and milling of U will increase in the next decades, contributing for its widespread in the environment [2].

During the last century, Portugal has actively explored radioactive ores and was for some time ranked as one of the main U producers. The extraction of U ore in Portugal started in 1908, first driven by the interest in radium (being U a by-product) and then by the interest in its military applications, till 2001 [3,4]. Most of the old U mines were located in the granitic regions of the Iberian Meseta, in the centre-north of Portugal (Beiras), [5]. Nowadays, although the mining activities ceased, like in several other places in the world, the old U mines represent a serious environmental problem, due to waste accumulation (mainly tailings and sludge) and improper disposal of radioactive material, composed by U and its daughter radionuclides [1,5–16]. Soils and water are the two major environmental matrices affected by U contamination.

U has a long half-life, persisting in nature as different isotopes, with different chemical and radiological characteristics [17]. The toxic effects induced by this metal are caused by both properties.

However, since U isotopes mainly emit alpha particles, with little penetration capacity, the main radiation hazards only occur after ingestion or inhalation of these isotopes and daughter radionuclides [17]. Once in the soil, U interacts with all the components of this matrix, such as clay minerals, aluminum and iron oxides, organic matter and microorganism, in a very complex system, where pH and organic matter seem to have the major role in controlling U mobility (pH 6) and leaching (pH<6) [18]. The high mobility/availability of U will in turn increase the ecological risks posed to soil and water compartments [19–27].

The soil has been recognized as an important compartment that provides crucial ecosystem services (e.g., filtering of contaminants, reservoir of carbon and a bank of genes) and is the support of agro-sylvo-pastoral production [28,29] and of several other human activities. The soil compartment offers raw materials (e.g., peat, clay, ore) and contributes for climate regulation and biodiversity conservation, as well as other cultural services [30,31]. The recognition of the importance of maintaining the provision of such services has increased the necessity to create appropriate legal tools to correctly and effectively protect and manage this resource. In this sense, the Soil Framework Directive proposed by the Commission of the European Communities (CEC), aims to establish a common strategy for the protection and sustainable use of soils [32]. For that end, this proposal defines measures for the identification of the main problems faced by soils, the adoption of strategies to prevent their degradation, as well as for the rehabilitation of contaminated or degraded soils [33]. The Soil Framework Directive will fill in the gap regarding soil protection, since this compartment has never been a target of specific protection policies at the European Community level [32]. Many countries, committed in regulating the management of contaminated land, have adopted generic quality standards, the soil screening values (SSVs) [34]. SSVs are concentration thresholds above which, more site-specific evaluations are required to assess the risks posed by soil contamination [35]. The SSVs should provide a level of protection to terrestrial species and ecological functions of the soil [35–37]. SSVs are particularly useful for the first tier of Ecological Risk Assessment (ERA) processes applied to contaminated sites, supporting the decision-making at this initial stage of assessment [38], which at the end is aimed in setting priorities for remediation and risk reduction measures [39]. In the case of Portugal, SSVs for soils have never been established for metals or organics. Only threshold concentrations of metals on sewage sludge were legally established to regulate the application of this solid waste on agricultural soils [40]. However, they are not appropriate for soil ERA purposes.

The use of natural reference soils in ecotoxicological tests has been recommended by several authors [41–43]. This is because the properties of the OECD artificial soil are not representative of the great majority of natural soils [44]. Different levels of toxicity, for each contaminant, can be expected in soils with different properties [45–48], hence it is important each country derives their own SSVs using natural reference soils representing the main types of soils within their territories. In this context, the main aim of this work was to obtain ecotoxicological data for U, performing soil enzymes activity tests, invertebrates and plant tests, using for that a Portuguese natural reference soil (PTRS1), that represents one of the dominant types of soil from a granitic region (cambisol) of the country [49]. As a result, enough data are gatheredas to make the first proposal of a SSV for this metal.

Materials and Methods

The present study used a natural soil that was collected in a non-protected area, requiring no specific permission for its collection. Further, no work with endangered species was performed, and no vertebrate species were used in the ecotoxicological assays. Only tests with invertebrates and plants were performed. The invertebrates were obtained from laboratorial cultures maintained by the authors of this manuscript and plant seeds were obtained from a local supplier.

1. Test soil

The natural soil (PTRS1) used as test substrate in this study was collected in Ervas Tenras [Pinhel, Guarda, Portugal center; geographical coordinates: 40°44′4.27″N and 7°10′54.3″W), at 655 m altitude, in a granitic region.

A composite soil sample was collected and immediately brought to the laboratory where it was air dried. Another portion of the soil, was immediately sieved through a 2 mm mesh size and the sieved fraction (<2 mm) was stored in polyethylene bags, at −20°C, until further analysis of soil microbial parameters, which were performed within the period of one month. For the tests with soil organisms and plants, the soil was passed through a 4 mm mesh sieve and the sieved fraction (<4 mm) was defaunated through two freeze–thawing cycles (48 h at −20°C followed by 48 h at 25°C), before the beginning of the assays.

The physical and chemical properties (including total metal contents) of the PTRS1 soil were presented in a preliminary study by Caetano et al. [49], aimed in characterizing this soil as a reference substrate for ecotoxicological purposes. The main properties of the PTRS1 are also described in Tables 1 and 2. Briefly, soil-KCl 1 M and soil-deionized water suspensions (1:5 m/v) were used for pH (KCl, 1 M) and pH-H_2O measurements, respectively, according to ISO 17512–1 [50]- After 15 min of magnetic stirring and 1 h resting period, the pH of the suspension was measured using a WTW 330/SET-2 pH meter. A soil water suspension (1:5 w/v) was used for the measurement of soil conductivity [51] Ten grams of PTRS1 were mechanically shaken in polypropylene flasks with 50 ml with deionized water filtered in a Milli-Q equipment (hereinafter referred as deionized water), water for 15 min. The mixture was left to rest overnight for soil bulk settling [51]. The conductivity of the resulting suspension was measured using an LF 330/SET conductivity meter. Soil water content was determined from the loss of weight after drying at 105°C, for 24 h. Organic matter (OM) content was determined by loss of ignition of dried soil samples at 450°C during 8 h [52]. For determination of water holding capacity (WHC) polypropylene flasks were prepared with a filter paper-replaced bottom, which after being filled up with soil samples, were immersed in water for 3 h. After this period, samples were left for water drainage during 2 h and the WHC was determined accounting to the loss of weight after drying at 105°C until weight stabilization [50].

2. Test substance

For all the test organisms, the natural soil was spiked with a stock solution of uranyl nitrate 6-hydrate, $UO_2(NO_3)_26H_2O$ (98%, PANREAC) prepared with deionized water in order to obtain a range of concentrations, which were ascertained by range finding tests performed with the different test species.

For soil enzyme tests, the PTRS1 soil was spiked with the following concentrations: 0.0, 134.6, 161.5, 193.8, 232.5, 279.0, 334.8, 401.8, 482.2, 578.7, 694.4, 833.3, 1000 mg U kg^{-1}_{dw}. To obtain these concentrations, the stock solution of uranyl nitrate

Table 1. Physical and chemical properties of PTRS1 soil (retrieved from Caetano et al. [49]).

| | pH (H$_2$O) | pH (KCl, 1 M) | Conductivity (mS cm^{-1}) | OM (%) | WHC (%) | Particle-size distribution (%) | | | | [U] (mg Cu kg^{-1} soil$_{dw}$) |
						Clay (<4 µm)	Silt (4-63 µm)	Sand (63 µm-2 mm)	Gravel (>2 mm)	
PTRS1	5.9±0.09	4.3±0.02	4.8±0.23	6.5±0.004	23.9±1.84	3.3	22.8	46.9	23.9	9.0

pH (H$_2$O), pH (KCl, 1 M), OM (organic matter), and WHC (water holding capacity) are represented as average ± STDEV.

was diluted in the volume of deionized water required to adjust the soil moisture at 80% of its maximum water holding capacity (WHC$_{max}$).

The following U concentrations were used to expose the earthworms in the reproduction tests: 0.0, 113.1, 124.4, 136.9, 150.5, 165.6, 231.9, 324.6, 454.5, 500.0, 550.0, 605.0, 665.5 mg U kg^{-1}$_{dw}$. For potworms, collembolans and terrestrial plant assays the same range of concentrations was tested: 0.0, 167.4, 192.5, 221.4, 254.6, 292.7, 336.6, 420.8, 526.0, 657.5, 756.1, 1000 mg U kg^{-1}$_{dw}$.

The volume of deionized water required to adjust the WHC of the soil to a given percentage of its maximum value was used to dilute the stock solution for these tests. After spiking the soil was left to rest for equilibration for 48 h before testing.

3. Ecotoxicological assessment

3.1 Soil microbial activity. For testing the effect of increasing concentrations of U on soil microbial parameters, a 30-day exposure was firstly conducted. Ten grams of sieved PTRS1 soil per replicate and concentration were spiked with different U concentrations, a total of three replicates were used per treatment. Six replicates with the same amount of soil only moistened with deionized water were also prepared for the control. The soil was incubated at $20 \pm 2°C$ and a photoperiod of 16 h^L:8 h^D. During the incubation period, the soil moisture was weekly monitored by weighing the pots, and whenever needed it was adjusted to 80% of its WHC$_{max}$ by adding deionized water. At the end, 1 g of each replicate from the control and concentrations tested was stored in individual falcon tubes at $-20°C$ for approximately one month. Thereby, a total of 9 sub-replicates were made for each concentration. The soil was thawed at 4°C before analysis.

The biochemical parameters analyzed were: the activity of arylsulphatase, dehydrogenase, urease, and cellulase enzymes and changes in the nitrogen mineralization (N mineralization) and potential nitrification.

For the determination of arylsulphatase activity, the method proposed by Tabatabai and Bremner [53] and Schinner et al. [54] was followed. After addition of 1 mL of p-nitrophenylsulfate (0.02 M), soil sub-samples were incubated for one hour, at 37°C. The nitrophenyl liberated by the activity of arylsulphatase was extracted and colored with a 4 mL of sodium hydroxide (0.5 M) and determined photometrically at 420 nm. The results were expressed as µg p-nitrophenylsulfate (p-NP) g^{-1} soil $_{dw}$ h^{-1}.

The method proposed by Öhlinger [55] was used to assess the dehydrogenase activity. The samples were suspended in 1 mL of trifeniltetrazol chloride (TTC) (3.5 g L^{-1}) and incubated at 40°C for 24 h. The triphenylformazan (TPF) produced was extracted with acetone and measured spectrophotometrically at 546 nm. The results were expressed as µg TPF g^{-1} soil$_{dw}$ h^{-1}.

The cellulase activity was tested according to the method proposed by Schinner et al. [54] and Schinner and von Mersi [56]. The reducing sugars produced during the incubation period, after addition of 1.5 mL of acetate buffer (2 M), caused the reduction of hexacyanoferrate (III) potassium to hexacyanoferrate (II) potassium in an alkaline solution. This last compound reacts with ferric ammonium sulfate in acid solution to form a ferric complex of hexacyanoferrate (II), of blue colour, which is colorimetrically measured at 690 nm and expressed as µg glucose g^{-1} soil$_{dw}$ 24 h^{-1}.

N mineralization activity was measured according to Schinner et al. [54]. For this purpose the soil samples were incubated for 7 days at 40°C. During this period, the organic forms of N were converted to inorganic forms (mainly ammonium ion, NH$_4$$^+$),

Table 2. Pseudo-total concentrations (mg/kg) of metals recorded in PTRS1 soil (average ± standard deviation) extracted with aqua régia, (retrieved from Caetano et al.[49]).

Metal	PTRS1
Ag	0.1±0.0
Al	25628.5±5130.0
B	2.2±0.8
Ba	45.8±8.0
Be	1.2±0.2
Cd	0.1±0.1
Co	5.6±1.1
Cr	10.8±2.1
Cu	9.0±1.8
Fe	24921.4±4534.4
Li	124.4±22.9
Hg	5253.5±1025.5
Mn	386.8±77.9
Mo	0.9±0.2
Na	78.1±14.9
Ni	4.6±0.9
Pb	12.5±2.2
Sb	0.2±0.0
Sn	10.4±1.9
U	7.8±1.7
V	37.8±14.1
Zn	57.1±8.9

which were determined by a modification of the Berthelot reaction, after extraction with 3 mL of potassium chloride (2 M). The reaction of ammonia with sodium salicylate in the presence of sodium dichloroisocyanurate formed a green colored complex in alkaline pH that was measured at 690 nm and expressed as μg N g^{-1} soil$_{dw}$ d^{-1}.

The urease activity was assayed according to the method proposed by Kandeler and Gerber [57] and, Schinner et al. [54]. The samples were incubated for 2 h at 37°C after the addition of 4 mL of a buffered urea solution (720 mM). The ammonia released was extracted with 6 mL of potassium chloride (2 M) and determined by the modified Berthelot reaction. The quantification was based on the reaction of sodium salicylate with ammonia in the presence of chlorinated water. UR was detected at 690 nm and expressed as μg N g^{-1} soil$_{dw}$ 2 h^{-1}.

The quantification of potential nitrification was determined by the method of Kandeler [58], which is a modification of the technique proposed by Berg and Rosswall [59]. The ammonium sulphate (4 mL, 10 mM) was used as substrate, and soil samples were incubated for 5 h, at 25°C. Nitrate released during the incubation period was extracted with 1 mL of potassium chloride (2 mM) and determined colorimetrically at 520 nm. This reaction was expressed as μg nitrite (N) g^{-1} soil$_{dw}$ h^{-1}.

3.2. Invertebrate and plant tests. Test organisms and culture conditions:
The earthworm *Eisenia andrei* (Oligochaeta: Lumbricidae), the potworm *Enchytraeus crypticus* (Oligochaeta: Enchytraeidae) and the springtail *Folsomia candida* (Collembola: Isotomidae) were used as invertebrate test organisms.

All organisms were obtained from laboratorial cultures, kept under controlled environmental conditions (temperature: 20±2°C; photoperiod: 16 h^L:8 h^D). The earthworms (*E. andrei*) are maintained in plastic boxes (10 to 50 L) containing a substrate composed by peat, dry and defaunated horse manure (through two freeze–thawing cycles (48 h at −20°C followed by 48 h at 65°C), and deionized water. The pH of the culture medium is adjusted to 6.0–7.0 with CaCO$_3$. The organisms are fed, every 2 weeks, with six table spoons of oatmeal previously hydrated with deionized water and cooked for 5 min. The potworms (*E. crypticus*) are cultured in plastic containers (25.5 cm length; 17.4 cm width; 6.5 cm height), which are filled with pot soil moistened to the nearest 60% of its WHC$_{max}$ and with pH adjusted to 6.0±0.5. The organisms are fed twice a week with a tea spoon of macerated oat. The collembolans (*F. candida*) are maintained in plastic containers filled with culture medium composed by moistened Plaster of Paris mixed with activated charcoal 8:1 (w:w). They are fed with half of a tea spoon of granulated dry yeast, twice a week. The food is added in small amounts to avoid spoilage by fungi.

Seeds from four plant species (two dicotyledonous and two monocotyledonous), purchased from a local supplier, were used for seed germination and growth tests: *Avena sativa*, *Zea mays*, *Lacuta sativa* and *Lycopersicon esculentum*.

Reproduction tests with invertebrates: Previous studies from our team, at least with earthworms from the same laboratorial cultures, have proved that these organisms were not exposed to meaningful levels of metals (especially U, in laboratorial culture conditions) [60]. The accomplishment of validity criteria, by all the controls of the assays (herein described) with the three invertebrate species, also confirmed that the test animals were not previously exposed to toxic levels of metals through test containers, substrates or food. The reproduction tests with *E. andrei*, *E. albidus* and *F. candida* were carried out according to the ISO guidelines 11268-2 [61], 16387 [62] and 11267 [63], respectively. Each replicate of the invertebrate tests contained 10 individuals in a certain developmental stage: the earthworms had a fully developed clitellum and an individual fresh weight between 250 and 600 mg; the potworms were 12-mm size; and the springtails were 10–12 days old. Five hundred grams of dry soil were weighted per test vessel for earthworms. For the tests with potworms and collembolans 20 g and 30 g of soil were weighted per replicate, respectively. Following an ECx sampling design, which considers more concentrations and less number of replicates, two replicates per concentration and five replicates for the control were prepared in the reproduction tests with *E. andrei*. Adult earthworms were removed from the test containers after 28 days. The produced cocoons persisted in the soil until 56 days have been completed. After this period, the juveniles from each test container were counted. During the test, organisms were fed once a week, with 5 g per box of defaunated horse manure (using the same procedure above described), and the soil moisture content was weekly monitored (following the procedures outlined in ISO guideline 11268-2 [61]).

The *E. albidus* reproduction test was held for 28 days and the adults were left in the vessels until the end of the test. About 2 mg of rolled oats were placed on the soil surface, weekly to feed the animals. At the end of the test, the potworms were killed with alcohol, colored with Bengal red and counted according to the Ludox Flotation Method, as described in ISO 16387 [62]. The reproduction tests with *F. candida* took four weeks to be completed. The collembolans were fed with granulated dry yeast, obtained from a commercial supplier, being weekly added (about 2 mg of yeast per test vessel) to the soil surface. At the end of the test, the containers were filled with water and the juveniles were

Table 3. Toxicity data obtained for copper (mg U kg^{-1} soil$_{dw}$) in PTRS1 soil on soil microbial processes, invertebrates and plants.

Biota	Endpoint	NOEC	LOEC	EC$_{20}$	EC$_{50}$
Microbial parameters					
Arylsulphatase		232.5	279	155.3 (84.76–255.87)	295.6 (216.09–375.17)
Dehydrogenase		<134.5	≤134.5	34.9 (20.52–59.35)	110.3 (83.25–137.47)
Nitrogen mineralization	Enzyme activity	694.4	833.3	152.2 (46.66–257.79)	347.0 (211.25–482.91)
Celulase		≤134.5	≥134.5	n.d.	n.d.
Urease		<134.5	≤134.5	<134.5	<134.5
Potencial nitrification		<134.5	≤134.5	429.5 (229.53–629.46)	610.0 (459.07–761.11)
Invertebrates					
Eisenia andrei	Rep. (56 days)	500.0	550.0	474.8 (391.47–558.04)	631.0 (532.78–699.21)
Enchytraeus crypticus	Rep. (28 days)	420.8	526.0	469.7 (355.47–584.04)	518.6 (480.40–556.90)
Folsomia candida	Rep. (28 days)	675.5	756.1	343.4 (172.23–514.60)	851.64 (606.10–1097.18)
Plants					
Avena sativa	Germination	≥1000	>1000	n.d.	n.d.
Zea mays	Germination	≥1000	>1000	n.d.	n.d.
Lactuca sativa	Germination	≥1000	>1000	n.d.	n.d.
Lycopersicon esculentum	Germination	≥1000	>1000	n.d.	n.d.
Avena sativa	Fresh mass	≥1000	>1000	n.d	n.d.
Zea mays	Fresh mass	≥1000	>1000	n.d	n.d.
Lactuca sativa	Fresh mass	≥1000	>1000	n.d	n.d.
Lycopersicon esculentum	Fresh mass	≥1000	>1000	n.d	n.d.
Avena sativa	Dry mass	≥1000	>1000	n.d.	n.d.
Zea mays	Dry mass	≥1000	>1000	n.d.	n.d.
Lactuca sativa	Dry mass	<167.4	≤167.4	n.d.	n.d.
Lycopersicon esculentum	Dry mass	≥1000	>1000	n.d.	n.d.

For ECx point estimates the 95% confidence limits are presented in brackets. n.d.- not determined; Rep. – reproduction.

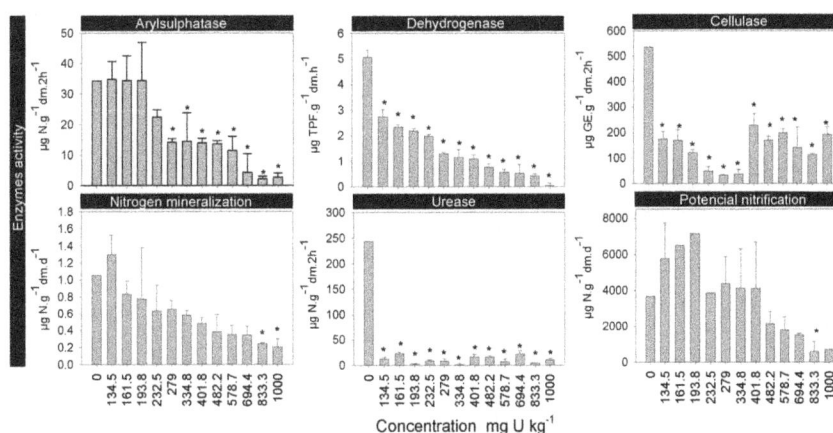

Figure 1. Soil enzyme activities, N mineralization and potential nitrification. Response of the arylsulphatase, dehydrogenase, cellulase urease, activity, N mineralization and potential nitrification to soils spiked with a range of uranium concentrations. The error bars indicate the standard deviation. The asterisks point out significantly differences from the control (P<0.05).

Table 4. Toxicity of copper (mg U kg^{-1} soil$_{dw}$) reported in the literature for the reproduction of soil invertebrates using different soil types with different physical and chemical characteristics.

Species	Endpoint	Soil type	Physical-chemical parameters			Point estimates (mg U kg^{-1} soil$_{dw}$)				Reference
			pH	OM (%)	Clay (%)	NOEC	LOEC	EC$_{20}$	EC$_{50}$	
Eisenia fetida	Rep. (56 days)	Canadian soil	6.2	1.0	2.0	n.d.	n.d.	>1000	n.d.	Sheppard and Stephenson [86]
		Canadian soil	6.2	1.0	2.0	n.d.	n.d.	>1120	n.d.	Sheppard and Stephenson [86]
		Canadian soil	7.5	2.2		>838	n.d.	n.d.	n.d.	Sheppard and sheppard [98]
		Canadian soil	7.5	18.4		>994	n.d.	n.d.	n.d.	Sheppard and sheppard [98]
Folsomia Candida	Rep. (28 days)	Canadian soil	7.5	2.2	24.0	n.d.	n.d.	840.0	n.d.	Sheppard and Stephenson [86]
		Canadian soil	7.5	n.d.	n.d.	n.d.	n.d.	>720	n.d.	Sheppard and Sheppard [98]
Elymus lanceolatus	Germination	Canadian soil	6.2	1	2	n.d.	>1000	n.d.	n.d.	Sheppard and Stephenson [86]
Elymus lanceolatus	Germination	Canadian soil	7.5	2.2	24	n.d.	>1001	n.d.	n.d.	Sheppard and Stephenson [86]
Zea mays	Dry mass	European soil	5.2	2.5	n.d.	n.d.	>100	n.d.	n.d.	Stojanović et al., [108]

OM - organic matter, Rep. - reproduction, n.d. - not determined, germ.- germination.

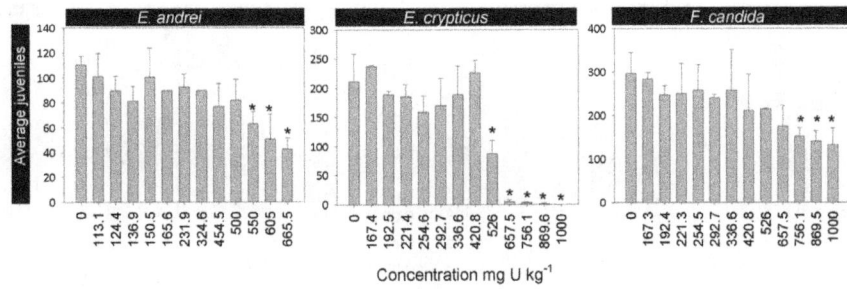

Figure 2. Reproduction of invertebrates. Results obtained exposing *Eisena andrei*, *Enchytraeus crypticus* and *Folsomia candida*, to natural PTRS1 soil, contaminated with different concentrations of U. The error bars indicate the standard deviation. The asterisks point out significantly differences from the control (P<0.05).

counted after flotation. The addition of a few dark ink drops provided a higher contrast between the white individuals and the black background. The organisms were then counted through the use of the ImageJ software (online available: http://rsb.info.nih.gov/ij/download.html). The exposure was carried out at $20\pm2°C$ and a photoperiod of $16^L:8^D$. For both species five replicates of uncontaminated natural PTRS1 soil were prepared for the control. The same ECx sampling design applied for earthworms was followed. However, in order to reduce the variability of the results, three replicates were prepared per test concentration (instead of two for the earthworms).

Seed germination and plant growth tests: Germination and growth tests with terrestrial plants were performed following standard procedures described by the ISO guideline 11269-2 [64]. For this purpose, 200 g_{dw} of the spiked soil with the concentrations

described above was placed in plastic pots (11.7 cm diameter, 6.2 cm height) and tested. In this case, the amount of water required to adjust the WHC_{max} of the soil to 45% was used to dilute the stock solution and to moist the soil at the beginning of the test. The soil was placed in the plastic pots (11.7 cm diameter, 6.2 cm height). In the bottom of each plastic pot a hole was previously made to let a rope passing through, hence allowing communication with the pot below that was filled with deionized water. After soil spiking and soil saturation with water twenty seeds were added to each pot and gently covered with the spiked soil. The level of water in the lower recipient was adjusted whenever needed, as to guarantee the necessary conditions of moisture according to, the recommendations specified in [64]. Five replicates of uncontaminated natural PTRS1 soil were prepared for the control, while three replicates were tested per

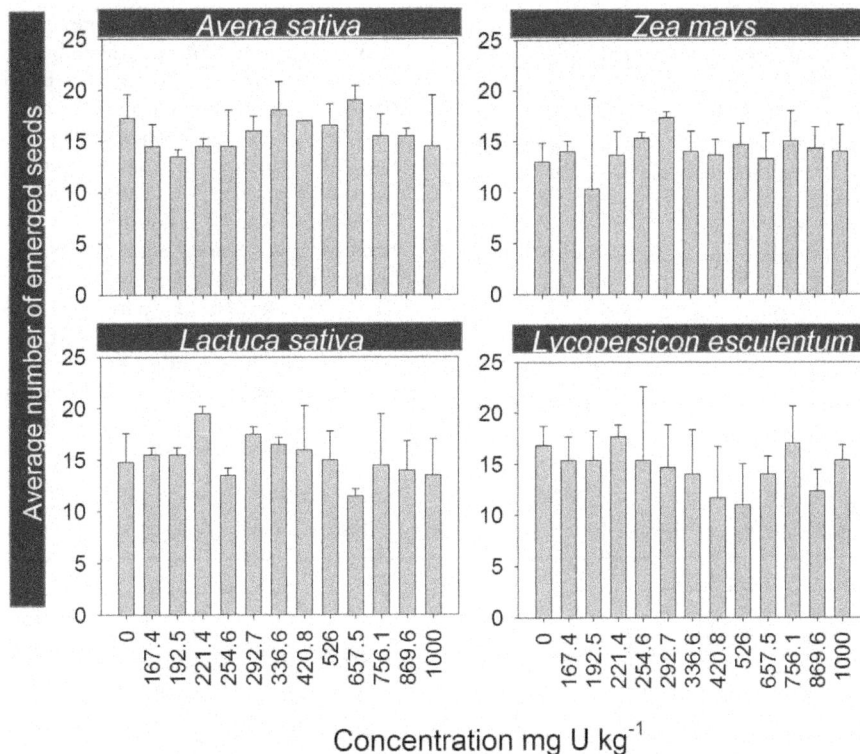

Figure 3. Seed germination of plants. Average number of emerged seeds in monocotyledonous, *Avena sativa and Zea mays* and in dicotyledonous species, *Lycopersicon esculentum and Lactuca sativa*, grown in PTRS1 soil contaminated with U. The error bars indicate the standard deviation.

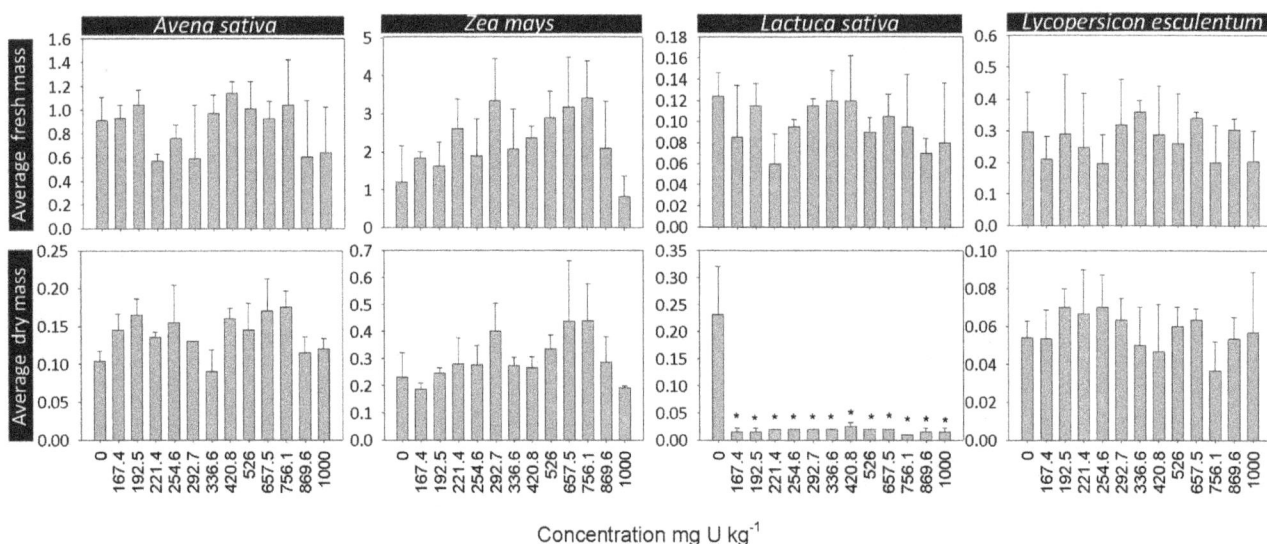

Figure 4. Growth of plants. Average values of fresh mass and dry mass in monocotyledonous, *Avena sativa and Zea mays* and in dicotyledonous species, *Lycopersicon esculentum and Lactuca sativa* grown in PTRS1 soil, contaminated with U. The error bars indicate the standard deviation. The asterisks point out significantly differences from the control (P<0.05).

concentration, in order to minimize the variability of the results, and to follow the ECx sampling design, similarly used for the invertebrate tests.

At the beginning of the test, nutrients (Substral - Plants fertilizer using 1 bottle cap for 2 L of water proportion according to the manufacturer recommendation; Fertilizer NPK: 6-3-6; nitrogen (N): 6%; phosphate (P_2O_5): 3%; potassium (K_2O): 6%; iron (Fe): 0,03%; trace elements: Cu, Mn, Mo and Zn), were added in each lower recipient containing the water. Pots were maintained at constant conditions of temperature ($20\pm2°C$), photoperiod (16 hL: 8 hD) and light intensity (25.000 lux). Daily observations were carried out to record the number of emerged seeds. Only the first five emerged seeds were left to grow, the remaining ones were counted and harvested. Fourteen days later, the assay was finished and the fresh and dry biomass above soil was assessed for each test species at the end of the exposure period.

The endpoints seed germination, and fresh and dry biomass, above soil, were assessed for each species at the end of the exposures according to the methods outlined in ISO guideline 11269-2 [64].

For this work, a battery of enzymes involved in different biogeochemical cycles S (sulfur cycle), N (Nitrogen cycle), C (Carbon cycle), as well as enzymes more indicative of the good physiological conditions of the whole microbial community (e.g.

dehydrogenases) were selected. The species of invertebrates and plants were selected based on the availability of standard protocols. Since we aimed to obtain data for the derivation of SSVs, for regulatory purposes, this procedure is recommended.

Statistical Analysis

A one-way analysis of variance (one-way ANOVA) was performed to test significant differences between the uranium concentrations tested for each endpoint analyzed: the activity of enzymes, the number of juveniles produced by potworms and collembolans, the number of emerged seeds, and the fresh and dry mass of the plants. The Kolmogorov-Smirnov test was applied to check data normality, whereas homoscedasticity of variances was checked by the Levene's test. When these two assumptions of the one-way ANOVAs were not met, a Kruskal-Wallis analysis was performed. The statistical analysis was run in the SigmaPlot 11.0 software for Windows. When statistical significant differences were recorded, the Dunnett's (for parametric one-way ANOVA) or the Dunn's test (for non-parametric ANOVA) was carried out to perceive which concentrations were significantly different from the respective control. Based on the outcomes of the multiple comparison tests the NOEC (no-observed-effect-concentration) and LOEC (low-observed-effect-concentration) values were deter-

Table 5. Soil quality guideline values derived for copper in Portugal, USA and Canada (mg U Kg^{-1} soildw).

Portugal				Canada	Other reference
Backgound concentrations	PNEC		Proposed SSV[b]	SQG$_E$[c]	
	NOEC	EC$_{20}$			
7.8[a]	23.3	15.5	15.5	23	100[d]

[a]Caetano et al.[49];
[b]SSV - soil screening value;
[c]Canadian Soil Quality Guidelines for environmental health (SQGe), Scott-Fordsmand and Pedersen [116].;
[d]Sheppard and Sheppard [98].

mined. The EC_{20} and EC_{50} values for each endpoint were calculated whenever possible, after fitting the data to a log-logistic model using the STATISTICA 7.0 software.

PNEC-Based SSV Derivation

Following the approach suggested by the Technical Guidance Document published by the European Commission [65], a predicted no effect concentration (PNEC) for U in the PTRS1 soil was determined, based on the assessment factor method For that, it was by used the lowest point estimate (i.e., NOEC and EC_{20} values) and applied the appropriate assessment factor based on the criteria of the Guidance Document [65]. The lowest point estimate calculated was for arylsulphatase activity. Considering that more than three NOEC values were obtained in this study, for at least three species, an assessment factor of 10 was applied, The PNEC value was calculated through the application of the following equation:

$$PNEC = \frac{\text{lowest point estimate}}{10}$$

Results and Discussion

1. Soil microbial activity

As far as authors are aware, this study is one of the few studies gathering data regarding the ecotoxicity of spiked soils with U on soil microbial parameters. Only a study from Sheppard et al. [66] has analyzed the effect of U on soil phosphatase activity in eleven different Canadian soils (including an agricultural, a boreal forest and a garden soil). This study recorded a significantly depressed activity only at the highest concentration tested (1000 mg U kg $soil_{dw}{}^{-1}$) for all the soils. These results suggested that probably, soil phosphatase activity was one of the less sensitive soil microbial parameters to U. In fact, Pereira et al. [7] also reported the low sensitivity of this parameter in mine soils contaminated with metals.

The variation in soil enzyme activities, N mineralization and potential nitrification in the PTRS1 soil, spiked with different U concentrations, is shown in Figure 1, and the Table 3 summarizes the toxicity values obtained for each biochemical parameter.

U had a clear inhibitory effect in almost all functional parameters tested. Overall, dehydrogenase and urease were the most affected soil enzymes by U, being their activity significantly inhibited at concentrations equal or lower than 134.5 mg U kg $soil_{dw}{}^{-1}$ (Table 3). Dehydrogenases have a relevant role in the oxidation of soil organic matter (SOM), being a good indicator of the active microbial biomass in the soil compartment [67]. As such, U (in the form of uranyl) strongly affected the normal microbial activity in PTRS1 soil. Meyer et al. [68] also observed a significant reduction in respiration rates of a soil exposed to depleted uranium (DU), but only for concentration equal and higher than 500 mg U kg $soil_{dw}{}^{-1}$. Indeed, the inhibition of urease activities indicates that U had a deleterious effect on soil N-cycle (Figure 1, Table 3). The reduction in the activity of this enzyme may have been caused by a negative effect of U on the overall microbial biomass, which in turn was translated into a reduction in the oxidation rate of organic N into ammonium [58,69]. Arylsulphatase is regularly involved in the S-cycle by catalyzing hydrolysis reactions in the biogeochemical transformation of S [67]. This parameter was significantly affected by U at a LOEC of 279.0 mg U kg $soil_{dw}{}^{-1}$. On its turn, the cellulase activity was

significantly inhibited at intermediate U concentrations. However in the highest concentrations the tendency was reversed and the activity increased, but not for levels significantly different from the control (Figure 1). Thereby, we can conclude that the C-metabolism associated with the degradation of SOM and catalyzed by these extracellular enzymes [70] was constrained by U. N mineralization and potential nitrification are indicators of the functioning of the N-cycle, hence providing an overview of the activity of specific microbial groups (nitrifying bacteria) directly involved in both processes [71]. The general pattern of response observed for these two parameters corresponded to stimulation at the lower U concentrations and inhibition under the highest ones (Figure 1), leading to EC_{50} values of 347.0 and 610.0 mg U kg $soil_{dw}{}^{-1}$ (Table 3), respectively. It has been stated that N mineralization is normally less sensitive than potential nitrification, since the former is carried out by a wider diversity of microorganisms [71]. However, our data showed the opposite (Figure 1). Meyer et al [68] did not observe effects on nitrogen mineralization of the test soil for U concentrations up to 25000 mg kg $soil_{dw}{}^{-1}$, however the form of U tested by these authors (schoepite $UO2(OH_2).H_2O$) was less soluble than the one tested in this soil.

The sensitivity of soil microbial parameters to metals has already been demonstrated by several authors, either in metal-polluted or in artificially spiked soils (e.g.,[4,72–77]). Dehydroge-nase and urease had generally been referred as the most affected enzymes for different metals (e.g., Cu, Pb, Zn, Cd, Fe, Cr, Ni), (e.g.,[72,73,78,79]). Arylsulphatase and cellulase, however, have shown contradictory responses in different studies. Some authors observed negative correlations between arylsulphatase and cellu-lase activities and Zn [75] and Cu concentrations, respectively [80,81]; while others observed positive correlations between arylsulphatase and Cd [81], and no changes on cellulase activities in the presence of metals in urban soils was observed [82]. Usually, potential nitrification is negatively influenced by the presence of metals and metalloids such as Pb, Cu and As [7,81]; and the inhibitory effect of some metals (like Zn, Cd and Pb) on N mineralization was also reported by Dai et al. [83]. Antunes et al [81] found negative correlations (based on the Spearman coefficient) between U levels in soils from an abandoned U mine (presenting a mixture of metals) and the activities of urease and cellulase enzymes. For dehydrogenase, potential nitrification and arylsulphatase no significant correlations were detected. Never-theless, this study analyzed mine contaminated soils, where the mixture of metals, may cause either synergistic or antagonistic effects, and where a well adapted and functional microbial community was likely established.

The inhibition of soil enzyme activities recorded could have been caused by toxicological effects of metals on soil microorgan-isms with subsequent decrease in their abundance and/or biomass; and/or by the direct inactivation of extracellular enzymes by metals [84]. Notwithstanding, the levels of metals may be not the sole effect on soil microbial activity. Soil properties (e.g., pH, organic matter content, nutrients and soil texture) may also interfere and modulate the bioavailability and toxicityof metals on soil enzymes [74,85]. Clays can retain and protect extracellular hydrolases, namely urease [73]. But the low clay content of PTRS1 soil (3.32%) (Table 1) might have increased U bioavail-ability, leading to the impairment of soil microbial community through cytotoxic effects, hence reducing their metabolic activity [81]. Additionally, the low pH of PTRS1 soil (Table 1) might have contributed for U availability and impacts on enzyme processes, potential nitrification and N mineralization, particularly at higher

U concentrations, as previously observed by Coppolecchia et al. [75] for arylsulphatase in the presence of Zn and low pH.

The above results illustrated well the effects of U in the performance of soil enzymes, reinforcing the importance of these parameters as bioindicators of soil quality. Indeed, the EC_{20} values calculated for dehydrogenase (34.9 mg Ukg $soil_{dw}^{-1}$), urease ($<$ 135.5 mg Ukg $soil_{dw}^{-1}$), N mineralization (152.2 mg Ukg $soil_{dw}^{-1}$) and arylsulphatase (155.3 mg Ukg $soil_{dw}^{-1}$) are within the environmental concentrations quantified in soils from an abandoned U mine, following extractions with *aqua regia* or with rainwater [8]. In this sense, the data herein generated represent a great asset for the derivation of SSVs, since they have a great ecological representativeness.

2. Uranium toxicity to the reproduction of soil invertebrates

The reproduction tests with the three invertebrate species revealed that *E. andrei*, *E. crypticus* and *F. candida* were quite sensitive to U in the PTRS1 soil. Tests fulfilled the validity criteria established by the standard guidelines for control replicates [61–63]. The resulting NOEC, LOEC, EC_{20} and EC_{50} values obtained in this study and toxicity data available in the literature are summarized in the Table 3.

The effects of U in the reproduction of *E. andrei* were evident, since statistical significant differences were found between the control and the highest tested concentrations of U for this organism ($F = 5.218$, d.f. $= 23$, $p = 0.002$) (Figure 2). The tested metal did not significantly affect the reproduction of *E. andrei* at concentrations up to 500.0 mg U kg $soil_{dw}^{-1}$ (NOEC) but compromised this endpoint for concentrations above 550.0 mg U kg $soil_{dw}^{-1}$ (LOEC). EC_{20} and EC_{50} values of U for *E. andrei* reproduction were 474.83 mg U kg $soil_{dw}^{-1}$ and 631.00 mg U kg $soil_{dw}^{-1}$, respectively (Table 3). The results obtained in our study, did not support those of Sheppard and Stephenson [86] (Table 4) that did not record toxic effects for *E. andrei* below 1000 mg U $soil_{dw}^{-1}$ (soils (carbonated): pH 7.5, 18% organic matter, 18% clay). However, they found an inhibition of juveniles production in two soils spiked with U, presenting low organic matter (2.2% and 1%) and a pH of 7.5 and 6.2, respectively (Table 4). According to the literature, the adsorption of metals to soil components is dependent on its physical and chemical properties, therefore influencing their toxicity to soil organisms [41,47,48]. Chelinho et al. [87], observed that soils with an organic matter content below 4% reduced or completely inhibited earthworms reproduction. However, the PTRS1 natural soil, had a high organic matter content, 6.2% (according to the classification provided by Murphy et al. [88]). Besides, as previously checked, the intrinsic properties of this soil did not compromise the performance of earthworms [49]. A high organic matter content of soils is usually related with a decrease in the toxicity of the contaminants for the organisms [41,43,89]. However, this was not the case in the study. In fact, Lourenço [60,90] exposed *E. andrei* to a U mine soil with 215.72±8.50 mg U kg $soil_{dw}^{-1}$, a pH of 7.79±0.01, and 7.71±0.60% of organic matter and observed that the bioaccumulation of U and daughter radionuclides was in tandem with loss of DNA integrity of coelomocyte cells, changes in the frequency of cells of immune system and also with histopathological changes (especially of the epidermis and chloragogenous tissue and intestinal epithelium). In fact, some other authors [91] had also suggested that the direct dermal exposure of earthworms to metals in the soil pore water, the ingestion of water, and/or soil particles may strongly favor the bioaccumulation of metals. Since pH is variable in the different compartments of gastrointestinal tract of

earthworms, it can increase the mobilization of contaminants from soil after its ingestion [92,93].

Although, other metals were present in the contaminated soil tested by Lourenço et al. [60,90] U likely had a crucial role in the toxic effects observed, because it persisted in the whole body till 56 days. These authors suggested that the changes observed in DNA integrity were likely early warning indicators of effects on the growth and reproduction of the organisms. And in fact, effects on reproduction were observed in our study. Further, Giovanetti et al. [94]. exposed *E. fetida* natural U- and DU-contaminated soil (no information on soil type) for 7 and 28 days. Regarding natural U, no mortality or significant changes in weight were observed for both exposure periods up to 600 mg U kg$^{-1}_{dw}$. The chloragogeneous tissue, the main storage tissue of U, presented meaningful changes after 7 days of exposure for 300 mg U Kg^{-1}, while DNA strand breaks increased in a dose dependent manner above 150 mg U kg $soil_{dw}^{-1}$.

Regarding *E. crypticus* reproduction, it was significantly not reduced above 526.0 mg U kg $soil_{dw}^{-1}$ (LOEC) (Table 3) ($F = 31.05$, d.f. $= 12$, $p<0.05$). The EC_{20} and EC_{50} values estimated were respectively 469.7 and 518.6 mg U kg $soil_{dw}^{-1}$. Although no toxicity values are reported for the lowest concentrations tested, enchytraeids showed considerable sensitivity to U, since the number of juveniles was minimal or no juveniles were produced by *E. crypticus* at concentrations above 657.5 mg U kg $soil_{dw}^{-1}$ (Figure 2). Despite enchytraeids are commonly used in standardized toxicity tests, to the best of our knowledge, no data are available in the literature regarding the effects of U on the reproduction of this species. The available information concerns only the toxic effects caused by other metals or by natural soil properties in the reproduction of this species [43,48,95–97]. Thus, taking into account this literature review, pH and CEC were the most important parameters controlling the high sensitivity of enchytraeids to metals. Additionally, and according to Kuperman et al. [43], adults survival and juveniles production by *E. crypticus* can be maximized in natural soils with properties within the following ranges: 4.4–8.2 pH; 1.2–42% OM; 1–29% clay. The PTRS1 natural soil used as test substrate fell into in these ranges (Table 3), and similarly to *E. andrei*, the reproduction of this species was not compromised during the validation of the PTRS1 natural soil as a reference soil [49], meaning that the soil properties did not limit the performance of *E. crypticus*.

Concerning *F. candida*, U affected the production of juveniles, as shown by a significant decrease of this endpoint along the concentrations tested ($F = 11.6$, d.f. $= 12$, $p<0.05$) (Figure 2). The number of juveniles was not significantly affected up to 675.50 mg U kg $soil_{dw}^{-1}$ (NOEC), but it was significantly decreased for U concentrations equal to or greater than 756.10 mg U Kg^{-1} (LOEC) (Table 3). The EC_{20} value estimated for reproduction was 343.41 mg U kg $soil_{dw}^{-1}$ which is considerably lower than the toxicity data reported by Shepard et al. [98], $EC_{20}>710$ mg U kg $soil_{dw}^{-1}$, in two loam soils with pH 7.5 (Table 4). The low sensitivity of *F. candida* to U was also observed by Sheppard and Stephenson [86] which tested 3 soils amended with a range of U concentrations and aged for 10 years before testing. In this study, the lowest EC_{20} value obtained was 840 mg U kg $soil_{dw}^{-1}$ in a loam soil (pH 7.5, 24% clay, 2.2% OM) (Table 4). Despite this, *F. candida* was more sensitive in the study of Sheppard and Stephenson (since their EC_{20} value was similar to the EC_{50} recorded in our study 851.64 mg U kg $soil_{dw}^{-1}$). When considering the number of juveniles produced, U was less toxic to *F. candida* comparatively to *E. andrei* and *E. crypticus*. The lower sensitivity of *F. candida* is also consistent with other studies, when the effects of other metals in the reproduction of the three species was investigated [97,99], or even when other species of collembolans were analyzed [86]. The exposure of *F. candida* to chemicals in soil is apparently lower than for earthworms, which are exposed both by ingestion of contaminated soil (mineral particles, organic matter and chemicals in the soil solution) and also through direct dermal contact [100]. Despite the widely known influence of soil parameters on the bioavailability of chemicals and their influence on the reproduction of soil organisms, less is known about the intrinsic effects of physico-chemical parameters of the soils in the reproduction of *F. candida*. In general, several authors had reported a high tolerance of *F. candida* reproduction to a wide range of soil textural classes, organic matter contents and soil pH [48,101,102]. Once again the performance of this species was not compromised by the intrinsic properties of the PTRS1 soil. Hence, the effects observed can undoubtedly be attributed to U exposure.

3. Phytotoxicity of uranium

Relatively to terrestrial plants tests, all the validity criteria as described by the standard guidelines were attained [64]. Data obtained showed no significant effects on seeds emergence for all species tested (p>0.05). In fact, it was observed a relatively high rate of germination, either in monocotyledonous and dicotyledonous species (Figure 3). This outcome was somewhat expected, based on previous studies (e.g.,[22]). Seed coats form a barrier which protects embryos from a wide range of contaminants, especially metals. Thus, the germination relies almost exclusively on the seed reserves making it a less sensitive endpoint to the toxicity of soil pollutants [103].

An apparent hormetic effect was recorded for the other endpoints measured for almost all plant species. Such occurrence was recorded by other authors and it was attributed to the use of U as uranyl nitrate, which corresponds to a supplementary dose of N given to plants [98].

With regard to production of fresh- and dry-mass, it was possible to perceive that the tested plants displayed different sensitivities to this metal. However, no significant differences were generally observed comparatively to the control, exception for *L. sativa* dry mass (H = 22.8, d.f. = 12, p = 0.029). Thus, and according to Figure 4, *L. sativa* was the most sensitive terrestrial plant to U. The high sensitivity of *L. sativa* was also found by Hubálek et al. [104] and Soudek [105]. This was probably caused by the high capacity of this species to bioaccumulate high concentrations of metals, including U [22].

The exposure of plants to metals, was already extensively studied, showing that these contaminants can induce biological effects on germination, growth and development, as well as, alterations in the nutrient profile of plants [22,106]. However, only some studies (e.g., [66,107] and others reviewed [98]) have assessed the ecotoxicological effects of U on terrestrial plant species.

Based on our study, once again was proved the diverse ecotoxicological outcomes for U effects on plant species, since no effects were observed, in the range of tested concentrations for the three evaluated endpoints (in three out of four species), in PTRS1 soil. Similar results were obtained by Sheppard and Sheppard [86] in acidic soils (Table 4), when testing the emergence and growth of wheatgrass *Elymus lanceolatus*. Like in our study, these authors did not observe any effect on this species (up to 1000 mg U kg $soil_{dw}^{-1}$). In opposition, Sheppard and Sheppard [81] revised data on U toxicity to terrestrial plants and reported EC_{25} values ranging from 300 to 500 mg U kg $soil_{dw}^{-1}$, considering only the most reliable studies. Stojanović et al. [108] also reported phytotoxic effects of U on *Zea mays* exposed, in different soil types, to 250, 500 and 1000 mg U kg $soil_{dw}^{-1}$, but especially at the highest concentration tested and in the most acidic soil. However, no statistical analysis of the data was performed in this study.

Soil properties are also the factors that most strongly affect U uptake and phytotoxic effects, [18,109–111]. The bivalent uranyl ion (UO_2^{2+}) is sorbed to the negatively charged surfaces of clay minerals and organic compounds. In acidic soils subjected to pH increase, more negatively charged binding sites are available on mineral surfaces due to the progressive reduction of protons occupying these sites. However, pH values close to 6, like the one of PTRS1, favors U availability, since the concentrations of carbonates tends to increase, and U is released to the soil solution in the form of U-carbonate complexes [18]. The natural soil PTRS1, besides being acidic, has a lower clay content, which means lower binding sites for the bivalent uranyl ion (UO_2^{2+}), hence constraining U bioavailability. other soil properties and plant mechanisms may explain the reduced sensitivity of the plants in comparison with soil microbial parameters and invertebrates. Viehweger and Geipel [112] reported an increased U absorption by *Arabidopsis halleri* attributed to Fe deficiency in the medium of hydroponically grown plants. With respect to this metal, in the natural PTRS1 soil, the analyses done by Caetano et al. [49] showed that Fe surpassed the soil benchmark values proposed by two EPA regions (http://rais.ornl.gov/tools/eco_search.php http://rais.ornl.gov/tools/eco_search.php). In this sense, it is hypothesized that the high Fe content of the PTRS1 soil, may have also contributed for reducing the absorption of U by plants. As far as plant mechanisms are considered, in several studies reviewed by Mitchell et al. [113] the transport of U within plants was reduced and higher concentrations were consistently found in the roots. Using X-ray absorption spectroscopy (XAS) and transmission electron microscopy (TEM), Laurette et al. [114] observed that when plants are exposed to U and phosphates, needle-like U-phosphates are formed and precipitate, both outside and inside the cells, or persist in the subsurface of root tissues. The precipitation of U-phosphate complexes acts as a protective mechanism preventing U translocation to the shoots and leaves. This can also occur when the culture medium of the plants has no phosphate, since some plants are able to exudate phosphates. Further, U may be also absorbed like UO_2^{2+} and linked to endogenous organophospate groups [114]. In opposition, when translocation occurs within plants, U has mainly formed U-carboxylated complexes. Plants can also exudate organic acids to the rhizosphere environment or UO_2^{2+} may form complexes with endogenous compounds like malic, citric, oxalic and acetic acid [114]. In summary, the different resistance mechanisms described above could explain the lack of toxic effects observed for *A. sativa*, *Z. mays* and *L. esculentum*, in opposition to *L. sativa*. Most concerning is the fact that the majority of studies testing the phytotoxicity of U, including those performed by us, were made with the addition of nutrients solution, which increased the availability of phosphates to the test soil, likely decreasing the sensitivity of plants to U. Hence, to enhance the protection level of SSVs derived for plants, more assays with different plant species should be performed and the addition of nutrients should be prevented, or at least the tests may include replicates with and without nutrients.

Derivation of a Soil Screening Value (SSV) for Uranium Applying Assessment Factors

The PNEC values obtained for U were based in EC_{20} and NOEC values varied between 15.5 and 23.3 mg kg $soil_{dw}^{-1}$, respectively (Table 5). These values were six to four times lower than the PNEC value suggested by Sheppard and Sheppard [86] (Table 5), which was 100 mg Ukg soil$_{dw}$$^{-1}$. In opposition, they are close to the lowest Canadian Soil Quality Guideline for both environment and human health (23 mg U kg soil$_{dw}$$^{-1}$). Thereby, while more ecotoxicological data is being obtained or other methods are being applied to derive soil screening values (SSVs) we prefer to be precautious by proposing a PNEC of 15.5 mg Kg^{-1} soil$_{dw}$ as a SSV for U, in soils similar to the PTRS1. This SSV value is near the background value found in non-contaminated soils [8,48], but not in some areas with naturally occurring U anomalies in soils, where concentrations ranging between 13–724 mg U kg soil$_{dw}$$^{-1}$ can be found [115].

Conclusion

With the present study it was possible to generate a set of important ecotoxicological data for the derivation of a SSV for U, using a Portuguese natural soil representative of a granitic region, where this type of mine exploration occurred.

Soil enzyme activities were clearly inhibited by U. The obtained results depended not only on the concentrations of U but also on the properties of soil, which were likely responsible for the bioavailability of U and subsequent impairments on soil microbial population and, consequently, in their activity. Dehydrogenase and urease were particularly sensitive to U. Further, and comparatively to the remaining effect concentrations obtained/estimated for invertebrates and plants, the soil microbial parameters were more affected by U contamination[1].

The toxic effects of U in soil invertebrates were also confirmed, but the tested species showed a variable sensitivity to this metal.

The increasing order of species sensitivity to U based on EC$_{50}$ values for reproduction was *E. crypticus* > *E. andrei* > *F. candida*. However, if EC$_{20}$ values are considered *F. candida* is the most sensitive invertebrate, since its EC$_{20}$ value was 343.41 mg U kg soil$_{dw}$$^{-1}$, compared to 474.83 mg U kg soil$_{dw}$$^{-1}$ and 469.76 mg U kg soil$_{dw}$$^{-1}$ EC$_{20}$ values estimated for *E. andrei* and *E. crypticus*, respectively. The EC$_{20}$ values estimated were lower than the NOEC values for *E. andrei* and *F. candida*. Thus, the EC$_{20}$ point estimate should be selected for the derivation of more protective SSVs. Relatively to the plants, the tested species showed no adverse effects caused by U in PTRS1, with the exception of *L. sativa* dry mass yield. Considering the results obtained, it was possible to verify a great variability between the EC$_x$ values estimated in this study and those reported in the scientific literature. Multiple factors can contribute to this discordance, but probably at least for some species, soils physical and chemical properties were the main factors responsible for such differences. Although, this reinforces, at least in part, the importance of using natural soils representatives of the main types of soil from each region in ecotoxicological evaluations and for the derivation of SSVs, the data generated suggests that the SSV (15.5 mg Kg^{-1} soil$_{dw}$) derived for U, was six times lower than the PNEC value proposed by other authors. Nevertheless, as mentioned previously, more data should be obtained following standard protocols.

Author Contributions

Conceived and designed the experiments: ALC CRM RP. Performed the experiments: ALC AG. Analyzed the data: ALC CRM RP. Contributed reagents/materials/analysis tools: FC FG EFS. Contributed to the writing of the manuscript: ALC CRM RP.

References

1. Gavrilescu M, Pavel L, Cretescu I (2009) Characterization and remediation of soils contaminated with uranium. J Hazard Mater 163: 475–510. doi:10.1016/j.jhazmat.2008.07.103.

2. Malyshkina N, Niemeier D (2010) Future sustainability forecasting by exchange markets: basic theory and an application. Environ Sci Technol 44: 9134–9142. Available: http://www.ncbi.nlm.nih.gov/pubmed/21058697.

3. Carvalho F, Oliveira J, Faria I (2009) Alpha emitting radionuclides in drainage from Quinta do Bispo and Cunha Baixa uranium mines (Portugal) and associated radiotoxicological risk. Bull Environ Contam Toxicol 83: 668–673. Available: http://www.ncbi.nlm.nih.gov/pubmed/19590808. Accessed 19 November 2012.

4. Pereira R, Barbosa S, Carvalho FP (2014) Uranium mining in Portugal: a review of the environmental legacies of the largest mines and environmental and human health impacts. Environ Geochem Health 36: 285–301. Available: http://www.ncbi.nlm.nih.gov/pubmed/24030454. Accessed 7 March 2014.

5. Carvalho F, Madruga M, Reis M, Alves J, Oliveira J, et al. (2007) Radioactivity in the environment around past radium and uranium mining sites of Portugal. J Environ Radioact 96: 39–46. Available: http://www.ncbi.nlm.nih.gov/pubmed/17433852. Accessed 25 October 2012.

6. Vandenhove H, Sweeck L, Mallants D, Vanmarcke H, Aitkulov A, et al. (2006) Assessment of radiation exposure in the uranium mining and milling area of Mailuu Suu, Kyrgyzstan. J Environ Radioact 88: 118–139. Available: http://www.ncbi.nlm.nih.gov/pubmed/16581165.

7. Pereira R, Sousa J, Ribeiro R, Gonçalves F (2006) Microbial Indicators in Mine Soils (S. Domingos Mine, Portugal). Soil Sediment Contam 15: 147–167. Available: http://www.informaworld.com/openurl?genre=article&doi=10%2e1080%2f15320380500506813&magic=crossref%7c%7cD404A21C5BB053405B1A640AFFD44AE3. Accessed 5 October 2013.

8. Pereira R, Antunes S, Marques S, Gonçalves F (2008) Contribution for tier 1 of the ecological risk assessment of Cunha Baixa uranium mine (Central Portugal): I soil chemical characterization. Sci Total Environ 390: 377–386. Available: http://www.ncbi.nlm.nih.gov/pubmed/17919686. Accessed 6 September 2011.

9. Arogunjo A, Höllriegl V, Giussani A, Leopold K, Gerstmann U, et al. (2009) Uranium and thorium in soils, mineral sands, water and food samples in a tin mining area in Nigeria with elevated activity. J Environ Radioact 100: 232–240. Available: http://www.ncbi.nlm.nih.gov/pubmed/19147259. Accessed 14 October 2013.

10. Momčilović M, Kovačević J, Dragović S (2010) Population doses from terrestrial exposure in the vicinity of abandoned uranium mines in Serbia.

Radiat Meas 45: 225–230. Available: http://linkinghub.elsevier.com/retrieve/pii/S1350448710000363. Accessed 25 October 2012.

11. Figueiredo M, Silva T, Batista M, Leote J, Ferreira M, et al. (2011) Uranium in surface soils: An easy-and-quick assay combining X-ray diffraction and X-ray fluorescence qualitative data. J Geochemical Explor 109: 134–138. Available: http://linkinghub.elsevier.com/retrieve/pii/S0375674210001366. Accessed 23 October 2012.

12. Patra A, Sumesh C, Mohapatra S, Sahoo S, Tripathi R, et al. (2011) Long-term leaching of uranium from different waste matrices. J Environ Manage 92: 919–925. Available: http://www.ncbi.nlm.nih.gov/pubmed/21084148. Accessed 14 October 2013.

13. Scheele F (2011) Uranium from Africa: Mitigation of Uranium Mining Impacts on Society and Environment by Industry and Governments. SSRN Electron J. Available: http://papers.ssrn.com/abstract=1892775. Accessed 14 October 2013.

14. Niemeyer J, Moreira-Santos M, Nogueira M, Carvalho G, Ribeiro R, et al. (2010) Environmental risk assessment of a metal-contaminated area in the Tropics. Tier I: screening phase. J Soils Sediments 10: 1557–1571. Available: http://www.springerlink.com/index/10.1007/s11368-010-0255-x. Accessed 29 October 2012.

15. Carvalho F (2011) Environmental Radioactive Impact Associated to Uranium Production Nuclear and Technological Institute Department of Radiological Protection and Nuclear Safety,. Journal, Am Sci Environ Publ Sci 7: 547–553.

16. Wang J, Lu A, Ding A (2007) Effect of cadmium alone and in combination with butachlor on soil enzymes. Heal (San Fr: 395–403. doi:10.1007/s10653-007-9084-2.

17. ASTDR (2011) Agency for Toxic Substances & Disease Registry-Toxicological profile for uranium. Draft. U.S. Department of Health and Human Services. Public Health Service and Agency for Toxic Substances and Disease Registry. 452 + annexes.

18. Vandenhove H, Van Hees M, Wouters K, Wannijn J (2007) Can we predict uranium bioavailability based on soil parameters? Part 1: effect of soil parameters on soil solution uranium concentration. Environ Pollut 145: 587–595. Available: http://www.ncbi.nlm.nih.gov/pubmed/16781802. Accessed 19 November 2012.

19. Gongalsky K (2003) Impact of pollution caused by uranium production on soil macrofauna. Environ Monit Assess 89: 197–219. Available: http://www.ncbi.nlm.nih.gov/pubmed/14632090. Accessed 15 October 2013.

20. Geras'kin S, Evseeva T, Belykh E, Majstrenko T, Michalik B, et al. (2007) Effects on non-human species inhabiting areas with enhanced level of natural

radioactivity in the north of Russia: a review. J Environ Radioact 94: 151–182. Available: http://www.ncbi.nlm.nih.gov/pubmed/17360083. Accessed 26 October 2012.

21. Joner E, Munier-Lamy C, Gouget B (2007) Bioavailability and microbial adaptation to elevated levels of uranium in an acid, organic topsoil forming on an old mine spoil. Environ Toxicol Chem 26: 1644–1648. Available: http://www.ncbi.nlm.nih.gov/pubmed/17702337.

22. Pereira R, Marques CR, Ferreira MJS, Neves MFJV, Caetano AL, et al. (2009) Phytotoxicity and genotoxicity of soils from an abandoned uranium mine area. Appl Soil Ecol 42: 209–220. Available: http://linkinghub.elsevier.com/retrieve/pii/S0929139309000778. Accessed 6 September 2011.

23. Kenarova A, Radeva G, Danova I, Boteva S, Dimitrova I (2010) Soil bacterial abundance and diversity of uranium impacted. Second Balkan conference on Biology: 5–9.

24. Islam E, Sar P (2011) Molecular assessment on impact of uranium ore contamination in soil bacterial diversity. Int Biodeterior Biodegradation 65: 1043–1051. Available: http://linkinghub.elsevier.com/retrieve/pii/S0964830511001697. Accessed 29 October 2012.

25. Geng F, Hu N, Zheng J-F, Wang C-L, Chen X, et al. (2011) Evaluation of the toxic effect on zebrafish (*Danio rerio*) exposed to uranium mill tailings leaching solution. J Radioanal Nucl Chem 292: 453–463. Available: http://link.springer.com/10.1007/s10967-011-1451-x. Accessed 15 October 2013.

26. Lourenço J, Pereira R, Silva A, Carvalho F, Oliveira J, et al. (2012) Evaluation of the sensitivity of genotoxicity and cytotoxicity endpoints in earthworms exposed in situ to uranium mining wastes. Ecotoxicol Environ Saf 75: 46–54. Available: http://www.sciencedirect.com/science/article/pii/S014765131100 2685. Accessed 14 October 2013.

27. Islam E, Sar P (2011) Molecular assessment on impact of uranium ore contamination in soil bacterial diversity. Int Biodeterior Biodegradation 65: 1043–1051. Available: http://www.sciencedirect.com/science/article/pii/S0964830511001697. Accessed 25 October 2013.

28. Lavelle P, Decaens T, Aubert M, Barot S, Blouin M, et al. (2006) Soil invertebrates and ecosystem services. Eur J Soil Biol 42: S3–S15. Available: http://linkinghub.elsevier.com/retrieve/pii/S1164556306001038. Accessed 19 July 2011.

29. O'Halloran K (2006) Toxicological Considerations of Contaminants in the Terrestrial Environment for Ecological Risk Assessment. Hum Ecol Risk Assess An Int J 12: 74–83. Available: http://www.tandfonline.com/doi/abs/10.1080/10807030500428603. Accessed 7 September 2011.

30. Barrios E (2007) Soil biota, ecosystem services and land productivity. Ecol Econ 64: 269–285. Available: http://linkinghub.elsevier.com/retrieve/pii/S0921800907001693. Accessed 4 October 2012.

31. Dominati E, Patterson M, Mackay A (2010) A framework for classifying and quantifying the natural capital and ecosystem services of soils. Ecol Econ 69: 1858–1868. Available: http://www.sciencedirect.com/science/article/pii/S0921800910001928. Accessed 14 October 2013.

32. CEC (2006) Comission of the European Cominities, Directive 2004/35/EC. Commissiona of the European Communities, Brussels.

33. Bone J, Head M, Barraclough D, Archer M, Scheib C, et al. (2010) Soil quality assessment under emerging regulatory requirements. Environ Int 36: 609–622. Available: http://www.ncbi.nlm.nih.gov/pubmed/20483160. Accessed 6 October 2012.

34. Jensen J, Mesman M (2006) Ecological Risk Assessment of Contaminated Land. Decision support for site specific investigations. ISBN 90-6960-138-9 978-90-6960-138-0.

35. Fishwick S (2004) Soil screening values for use in UK ecological risk assessment. Soil Quality & Protection, Air Land and Water Group. Science Environment Agency's Project Manager.

36. USEPA (2003) United States Environmental Protection Agency. Guidance for Developing Ecological Soil Screening Levels (Eco-SSLs). Attachment 1–2: Assessment of Whether to Develop Ecological Soil Screening Levels for Microbes and Microbial Processes, OSWER Directi. Environ Prot.

37. Carlon C (2007) Derivation methods of soil screening values in europe. a review and evaluation of national procedures towards harmonisation. European Commission, Joint Research Centre, Ispra, EUR 22805-EN, 306.

38. Provoost J, Reijnders L, Swartjes F, Bronders J, Carlon C, et al. (2008) Parameters causing variation between soil screening values and the effect of harmonization. J Soils Sediments 8: 298–311. Available: http://www.springerlink.com/index/10.1007/s11368-008-0026-0. Accessed 29 October 2012.

39. Van Gestel C (2012) Soil ecotoxicology: state of the art and future directions. Zookeys 296: 275–296. Available: http://www.pubmedcentral.nih.gov/articlerender.fcgi?artid=3335420&tool=pmcentrez&rendertype=abstract. Accessed 29 October 2012.

40. MAOTDR (2006) Ministerio do Ambiente do Ordenamento do Territorio e do Desenvolvimento Regional. Diário da República, 1. a série – N. o 208–27 de Outubro.

41. Römbke J, Jänsch S, Junker T, Pohl B, Scheffczyk A, et al. (2006) Improvement of the applicability of ecotoxicological tests with earthworms, springtails, and plants for the assessment of metals in natural soils. Environ Toxicol Chem 25: 776–787. Available: http://www.ncbi.nlm.nih.gov/pubmed/16566163.

42. Van Assche F, Alonso JL, Kapustka L, Petrie R, Stephenson GL, et al. (2002) Terrestrial plant toxicity tests. In: Fairbrother A., Glazebrock P.W., Van Straalen N.M., Tarazona J.V. (eds) Test methods to determine hazards of

sparingly soluble metal compounds in soils. SETAC, Pensacola, Forida, 37–57. Available: https://www.setac.org/store/view_product.asp?id=1038018.

43. Kuperman R, Amorim M, Römbke J, Lanno R, Checkai R, et al. (2006) Adaptation of the enchytraeid toxicity test for use with natural soil types. Eur J Soil Biol 42: S234–S243. Available: http://www.sciencedirect.com/science/article/pii/S1164556306000719. Accessed 14 October 2013.

44. Hofman J, Hovorková I, Machát J (2009) Comparison and Characterization of OECD Artificial Soils. Ecotoxicological Characterization of Waste 2009, 223–229. Available: http://www.rivm.nl/bibliotheek/rapporten/711701047.html.

45. Song J, Zhao F, McGrath S, Luo Y (2006) Influence of soil properties and aging on arsenic phytotoxicity. Environ Toxicol Chem 25: 1663–1670. Available: http://www.ncbi.nlm.nih.gov/pubmed/16764487. Accessed 14 October 2013.

46. Rooney C, Zhao F, McGrath S (2007) Phytotoxicity of nickel in a range of European soils: influence of soil properties, Ni solubility and speciation. Environ Pollut 145: 596–605. Available: http://www.ncbi.nlm.nih.gov/pubmed/16733077. Accessed 14 October 2013.

47. Van Gestel C, Borgman E, Verweij R, Ortiz M (2011) The influence of soil properties on the toxicity of molybdenum to three species of soil invertebrates. Ecotoxicol Environ Saf 74: 1–9. Available: http://www.ncbi.nlm.nih.gov/pubmed/20951431. Accessed 14 October 2013.

48. Domene X, Chelinho S, Campana P, Natal-da-Luz T, Alcañiz JM, et al. (2011) Influence of soil properties on the performance of Folsomia candida: implications for its use in soil ecotoxicology testing. Environ Toxicol Chem 30: 1497–1505. Available: http://www.ncbi.nlm.nih.gov/pubmed/21437938. Accessed 13 March 2013.

49. Caetano A, Gonçalves F, Sousa J, Cachada A, Pereira E, et al. (2012) Characterization and validation of a Portuguese natural reference soil to be used as substrate for ecotoxicological purposes. J Environ Monit 14: 925–936. Available: http://www.ncbi.nlm.nih.gov/pubmed/22297688. Accessed 31 May 2013.

50. ISO 17512-1 (2008) International Organization for Standardization 17512-1: 2008.Soil Quality: Avoidance Test for Testing the Quality of Soils and the Toxicity of Chemicals-Test with Earthworms (*Eisenia Fetida*). Geneva, Switzerland. International Organization for Standard. Geneva, Switzerland.

51. FAOUN (1984) Food and agriculture organization of the United Nations – physical and chemical methods of soil and water analysis. Soils Bull. 10, 1–275.

52. SPAC (2000) Soil and Plant Analysis Council – Handbook of Reference Methods. CRC Press, Boca Raton, Florida.

53. Tabatabai M, Bremner J (1970) Arylsulfatase activity in soils. Soil Science Society of America, 34: 225–9.

54. Schinner F, Kandeler E, Öhlinger R, Margesin R (1996) Methods in soil biology. Germany: Springer-Verlag.

55. Öhlinger R (1996) Soil sampling and sample preparation. In: Schinner F, Öhlinger R, Kandeler E, Margesin R, editors. Methods in Soil Biology. Springer-Verlag.

56. Schinner F, von Mersi W (1990) Xylanase, CM-cellulase and invertase activity in soil, an improved method. Soil Biol Biochem 22: 511–5.

57. Kandeler E, Gerber H (1988) Short-term assay of soil urease activity using colorimetric determination of ammonium. biol Fert Soils 6: 68–72.

58. Kandeler E (1996) Potential nitrification. In: Schinner F, Öhlinger R, Kandeler E, Margesin R, editors. Methods in soil biology. Germany: Springer-Verlag-Berlin-Heidelberg, 146–9.

59. Berg P, Rosswall T (1985) Ammonium oxidizer numbers, potential and actual oxidation rates en two Swedish arable soils. Biology Fertility Soils 1: 131–40.

60. Lourenço JI, Pereira RO, Silva AC, Morgado JM, Carvalho FP, et al. (2011 a) Genotoxic endpoints in the earthworms sub-lethal assay to evaluate natural soils contaminated by metals and radionuclides. J Hazard Mater 186: 788–795. Available: http://www.ncbi.nlm.nih.gov/pubmed/21146299. Accessed 21 September 2013.

61. ISO 11268-2 (1998) International Organization for Standardization ISO 11268-2. Soil quality: effects of pollutants on earthworms (*Eisenia fetida*) - Part 2: Determination of effects on reproduction. ISO 11268-2. Geneva, Switzerland: International Organization for Standardiza. Geneva, Switzerla.

62. ISO 16387 (2004) International Organization for Standardization ISO 16387. Soil quality: effects of pollutants on Enchytraeidae (*Enchytraeus sp.*)-Determination of effects on reproduction and survival. ISO16387. Geneva, Switzerland: International Organization for Standard. Geneva, Switz.

63. ISO 11267 (1999) International Organization for Standardization ISO 11267. Soil quality: inhibition of reproduction of Collembola (*Folsomia candida*) by soil pollutants. ISO 11267. Geneva, Switzerland: International Organization for Standardization. Geneva, Switz.

64. ISO 11269-2 (2005) International Organization for Standardization ISO 11269-2. Soil quality: determination of the effects of pollutants on soil flora - Part 2: Effects of chemicals on the emergence and growth of higher plants. ISO 11269-2. Geneve, Switzerlan: International O. Geneva, Switz.

65. EC. Commission European (2003) Commission European. Technical Guidance Document on Risk Assessment. in support of Commission Directive 93/67/EEC on Risk Assessment for new notified substances Commission Regulation (EC) No 1488/94 on Risk Assessment for existing substances Directive 98/.

66. Sheppard S, Evenden W (1992) Bioavailability Indices for Uranium: Effect of Concentration in Eleven Soils. Arch Environ Contain Toxicol 23, 117–124.

67. Taylor J, Wilson B, Mills M, Burns R (2002) Comparison of microbial numbers and enzymatic activities in surface soils and subsoils using various techniques. 34.

68. Meyer MC, Paschke MW, McLendon T, Price D (1998) Decreases in Soil Microbial Function and Functional Diversity in Response to Depleted Uranium. J Environ Qual 27: 1306. Available: https://www.agronomy.org/publications/jeq/abstracts/27/6/JEQ0270061306. Accessed 31 March 2014.

69. Wang M, Markert B, Shen W, Chen W, Peng C, et al. (2011) Microbial biomass carbon and enzyme activities of urban soils in Beijing. Environ Sci Pollut Res Int 18: 958–967. Available: http://www.ncbi.nlm.nih.gov/pubmed/21287285. Accessed 25 September 2013.

70. Alvarenga P, Palma P, Gonçalves AP, Baião N, Fernandes RM, et al. (2008) Assessment of chemical, biochemical and ecotoxicological aspects in a mine soil amended with sludge of either urban or industrial origin. Chemosphere 72: 1774–1781. Available: http://www.ncbi.nlm.nih.gov/pubmed/18547605. Accessed 13 May 2014.

71. Winding A, Hund-Rinke K, Rutgers M (2005) The use of microorganisms in ecological soil classification and assessment concepts. Ecotoxicol Environ Saf 62: 230–248. Available: http://www.ncbi.nlm.nih.gov/pubmed/15925407.

72. Khan S, Cao Q, Hesham AE-L, Xia Y, He J-Z (2007) Soil enzymatic activities and microbial community structure with different application rates of Cd and Pb. J Environ Sci (China) 19: 834–840. Available: http://www.ncbi.nlm.nih.gov/pubmed/17966871. Accessed 15 October 2013.

73. Lee S, Kim E, Hyun S, Kim J (2009) Metal availability in heavy metal-contaminated open burning and open detonation soil: assessment using soil enzymes, earthworms, and chemical extractions. J Hazard Mater 170: 382–388. Available: http://www.ncbi.nlm.nih.gov/pubmed/19540045. Accessed 5 April 2013.

74. Papa S, Bartoli G, Pellegrino A, Fioretto A (2010) Microbial activities and trace element contents in an urban soil. Environ Monit Assess 165: 193–203. Available: http://www.ncbi.nlm.nih.gov/pubmed/19444636. Accessed 13 May 2013.

75. Coppolecchia D, Puglisi E, Vasileiadis S, Suciu N, Hamon R, et al. (2011) Soil Biology & Biochemistry Relative sensitivity of different soil biological properties to zinc. Soil Biol Biochem 43: 1798–1807. Available: http://dx.doi.org/10.1016/j.soilbio.2010.06.018.

76. Lee S-H, Park H, Koo N, Hyun S, Hwang A (2011) Evaluation of the effectiveness of various amendments on trace metals stabilization by chemical and biological methods. J Hazard Mater 188: 44–51. Available: http://www.ncbi.nlm.nih.gov/pubmed/21333442. Accessed 17 August 2013.

77. Hu B, Liang D, Liu J, Xie J (2013) Ecotoxicological effects of copper and selenium combined pollution on soil enzyme activities in planted and unplanted soils. Environ Toxicol Chem 32: 1109–1116. Available: http://www.ncbi.nlm.nih.gov/pubmed/23401089. Accessed 8 October 2013.

78. Gülser F, Erdoğan E (2008) The effects of heavy metal pollution on enzyme activities and basal soil respiration of roadside soils. Environ Monit Assess 145: 127–133. Available: http://www.ncbi.nlm.nih.gov/pubmed/18027096. Accessed 17 August 2013.

79. Thavamani P, Malik S, Beer M, Megharaj M, Naidu R (2012) Microbial activity and diversity in long-term mixed contaminated soils with respect to polyaromatic hydrocarbons and heavy metals. J Environ Manage 99: 10–17. Available: http://dx.doi.org/10.1016/j.jenvman.2011.12.030.

80. Alvarenga P, Palma P, de Varennes A, Cunha-Queda AC (2012) A contribution towards the risk assessment of soils from the São Domingos Mine (Portugal): chemical, microbial and ecotoxicological indicators. Environ Pollut 161: 50–56. Available: http://www.ncbi.nlm.nih.gov/pubmed/22230067. Accessed 5 October 2013.

81. Antunes S, Pereira R, Marques S, Castro B, Gonçalves F (2011) Impaired microbial activity caused by metal pollution A field study in a deactivated. Sci Total Environmen 410–411.

82. Sivakumar S, Nityanandi D, Barathi S, Prabha D, Rajeshwari S, et al. (2012) Selected enzyme activities of urban heavy metal-polluted soils in the presence and absence of an oligochaete, Lampito mauritii (Kinberg). J Hazard Mater 227–228: 179–184. Available: http://www.ncbi.nlm.nih.gov/pubmed/22658212. Accessed 17 August 2013.

83. Dai J, Becquer T, Rouiller JH, Reversat G, Bernhard-Reversat F, et al. (2004) Influence of heavy metals on C and N mineralisation and microbial biomass in Zn-, Pb-, Cu-, and Cd-contaminated soils. Appl Soil Ecol 25: 99–109. Available: http://www.sciencedirect.com/science/article/pii/S0929139303001355. Accessed 15 October 2013.

84. Kızılkaya R, Bayraklı B (2005) Effects of N-enriched sewage sludge on soil enzyme activities. Appl Soil Ecol 30: 192–202. Available: http://www.sciencedirect.com/science/article/pii/S0929139305000594. Accessed 15 October 2013.

85. Turner BL, Hopkins DW, Haygarth PM, Ostle N (2002) β-Glucosidase activity in pasture soils. Available: http://www.sciencedirect.com/science/article/pii/S0929139302000203. Accessed 15 October 2013.

86. Sheppard SC, Stephenson GL (2012) Ecotoxicity of aged uranium in soil using plant, earthworm and microarthropod toxicity tests. Bull Env Contam Toxicol 8843–47: 43–47. doi:10.1007/s00128-011-0442-5.

87. Chelinho S, Domene X, Campana P, Natal-da-Luz T, Scheffczyk A, et al. (2011) Improving ecological risk assessment in the Mediterranean area: selection of reference soils and evaluating the influence of soil properties on avoidance and reproduction of two oligochaete species. Environ Toxicol Chem

30: 1050–1058. Available: http://www.ncbi.nlm.nih.gov/pubmed/21305581. Accessed 17 October 2013.

88. Murphy S, Giménez D, Muldowney LS, Heckman JR (2012) Soil Organic Matter Level and Interpretation How is the Organic Matter Content of Soils Determined? How Is Organic Matter Level: 1–3.

89. Natal-da-Luz T, Ojeda G, Pratas J, Van Gestel CA, Sousa JP (2011) Toxicity to Eisenia andrei and Folsomia candida of a metal mixture applied to soil directly or via an organic matrix. Ecotoxicol Environ Saf 74: 1715–1720. Available: http://www.ncbi.nlm.nih.gov/pubmed/21683441. Accessed 13 March 2013.

90. Lourenço J, Silva A, Carvalho F, Oliveira J, Malta M, et al. (2011) b) Histopathological changes in the earthworm Eisenia andrei associated with the exposure to metals and radionuclides. Chemosphere 85: 1630–1634. Available: http://www.ncbi.nlm.nih.gov/pubmed/21911243. Accessed 4 October 2013.

91. Hobbelen PHF, Koolhaas JE, van Gestel CAM (2006) Bioaccumulation of heavy metals in the earthworms Lumbricus rubellus and Aporrectodea caliginosa in relation to total and available metal concentrations in field soils. Environ Pollut 144: 639–646. Available: http://www.ncbi.nlm.nih.gov/pubmed/16530310. Accessed 15 October 2013.

92. Li L, Wu J, Tian G, Xu Z (2009) Effect of the transit through the gut of earthworm (Eisenia fetida) on fractionation of Cu and Zn in pig manure. J Hazard Mater 167: 634–640. Available: http://www.ncbi.nlm.nih.gov/pubmed/19232822. Accessed 19 September 2013.

93. Peijnenburg W, Jager T (2003) Monitoring approaches to assess bioaccessibility and bioavailability of metals: matrix issues. Ecotoxicol Environ Saf 56: 63–77. Available: http://www.ncbi.nlm.nih.gov/pubmed/12915141. Accessed 15 October 2013.

94. Giovanetti A, Fesenko S., Cozzella ML, Asencio LD, Sansone U (2010) Bioaccumulation and biological effects in the earthworm Eisenia fetida exposed to natural and depleted uranium. J Environ Radioact 101: 509–516. Available: http://www.ncbi.nlm.nih.gov/pubmed/20362371. Accessed 15 October 2013.

95. Peijnenburg W, Baerselman R, de Groot C, Jager T, Posthuma L, et al. (1999) Relating environmental availability to bioavailability: soil-type-dependent metal accumulation in the oligochaete Eisenia andrei. Ecotoxicol Environ Saf 44: 294–310. Available: http://www.ncbi.nlm.nih.gov/pubmed/10581124.

96. Amorim M, Römbke J, Scheffczyk A, Soares AM (2005) Effect of different soil types on the enchytraeids Enchytraeus albidus and Enchytraeus luxuriosus using the herbicide Phenmedipham. Chemosphere 61: 1102–1114. Available: http://www.ncbi.nlm.nih.gov/pubmed/16263380.

97. Kuperman R (2004) Manganese toxicity in soil for Eisenia fetida, Enchytraeus crypticus (Oligochaeta), and Folsomia candida (Collembola). Ecotoxicol Environ Saf 57: 48–53. Available: http://linkinghub.elsevier.com/retrieve/pii/S0147651303001544. Accessed 22 June 2011.

98. Sheppard S, Sheppard M (2005) Derivation of ecotoxicity thresholds for uranium. J Environ 79: 55–83. Available: http://www.ncbi.nlm.nih.gov/pubmed/15571876. Accessed 19 November 2012.

99. Lock K, Janssen CR (2001) Cadmium Toxicity for Terrestrial Invertebrates: Taking Soil Parameters Affecting Bioavailability into Account. Environ Toxicol: 315–322.

100. Layinka T, Idowu B, Dedeke A, Akinloye A, Ademolu O, et al. (2011) Earthworm as bio-indicator of heavy metal pollution around Lafarge, Wapco Cement Factory, Ewekoro, Nigeria. Proceedings of the Environmental Management Conference, Federal University of Agriculture, Abeokuta, Nigeria: 489–496.

101. Amorim M, Rçmbke J, Scheffczyk A, Nogueira A, Soares A (2005) Effects of Different Soil Types on the Collembolans Folsomia candida and Hypogastrura assimilis using the Herbicide Phenmedipham. 352: 343–352. doi:10.1007/s00244-004-0220-z.

102. Jänsch S, Amorim MJ, Römbke J (2011) Identification of the ecological requirements of important terrestrial ecotoxicological test species. Environmental Reviews, 13(2): 51–83. Available: http://www.nrcresearchpress.com/doi/abs/10.1139/a05-007#.U10AxlDENPB. Accessed 15 October 2013.

103. Liu X, Zhang S, Shan X-Q, Christie P (2007) Combined toxicity of cadmium and arsenate to wheat seedlings and plant uptake and antioxidative enzyme responses to cadmium and arsenate co-contamination. Ecotoxicol Environ Saf 68: 305–313. Available: http://www.ncbi.nlm.nih.gov/pubmed/17239437. Accessed 15 October 2013.

104. Hubálek T, Vosáhlová S, Matejů V, Kovácová N, Novotný C (2007) Ecotoxicity monitoring of hydrocarbon-contaminated soil during bioremediation: a case study. Arch Environ Contam Toxicol 52: 1–7. Available: http://www.ncbi.nlm.nih.gov/pubmed/17106791. Accessed 13 August 2013.

105. Soudek P, Petrová S, Benešová D, Dvořáková M, Vaněk T (2011) Uranium uptake by hydroponically cultivated crop plants. J Environ Radioact 102: 598–604. Available: http://www.ncbi.nlm.nih.gov/pubmed/21486682. Accessed 13 September 2013.

106. Gopal R, Rizvi A (2008) Excess lead alters growth, metabolism and translocation of certain nutrients in radish. Chemosphere 70: 1539–1544. Available: http://www.ncbi.nlm.nih.gov/pubmed/17923149. Accessed 15 October 2013.

107. Sheppard S, Evenden W, Anderson A (1992) Multiple assays of uranium toxicity in soil. Environ. Toxicol. Water Qual., 7: 275–294.

108. Stojanović M, Stevanović D, Milojković J, Grubišić M, Ileš D (2009) Phytotoxic Effect of the Uranium on the Growing Up and Development the Plant of Corn.

Water, Air, Soil Pollut 209: 401–410. Available: http://www.springerlink.com/index/10.1007/s11270-009-0208-4. Accessed 25 September 2011.

109. Bednar A, Medina V, Ulmer-Scholle D, Frey B, Johnson B, et al. (2007) Effects of organic matter on the distribution of uranium in soil and plant matrices. Chemosphere 70: 237–247. Available: http://www.sciencedirect.com/science/article/pii/S0045653507008168. Accessed 15 October 2013.

110. Tunney H, Stojanovic M, Mrdakovic Popic J, McGrath D, Zhang C (2009) Relationship of soil phosphorus with uranium in grassland mineral soils in Ireland using soils from a long-term phosphorus experiment and a National Soil Database. J Plant Nutr Soil Sci 172: 346–352. Available: http://doi.wiley.com/10.1002/jpln.200800069. Accessed 13 September 2013.

111. Soudek P, Petrová Š, Benešová D, Vaněk T (2011) Uranium uptake and stress responses of in vitro cultivated hairy root culture of Armoracia rusticana. Agrochimical 1: 15–28.

112. Viehweger K, Geipel G (2010) Uranium accumulation and tolerance in Arabidopsis halleri under native versus hydroponic conditions. Environ Exp Bot 69: 39–46. Available: http://dx.doi.org/10.1016/j.envexpbot.2010.03.001.

113. Mitchell N, Pérez-Sánchez D, Thorne MC (2013) A review of the behaviour of U-238 series radionuclides in soils and plants. J Radiol Prot 33: R17–48. Available: http://www.ncbi.nlm.nih.gov/pubmed/23612607. Accessed 18 September 2013.

114. Laurette J, Larue C, Llorens I, Jaillard D, Jouneau P-H, et al. (2012) Speciation of uranium in plants upon root accumulation and root-to-shoot translocation: A XAS and TEM study. Environ Exp Bot 77: 87–95. Available: http://www.sciencedirect.com/science/article/pii/S0098847211002814. Accessed 15 October 2013.

115. Pereira A, Neves L (2012) Estimation of the radiological background and dose assessment in areas with naturally occurring uranium geochemical anomalies–a case study in the Iberian Massif (Central Portugal). J Environ Radioact 112: 96–107. Available: http://www.ncbi.nlm.nih.gov/pubmed/22694913. Accessed 14 September 2013.

116. Scott-Fordsmand JJ, Pedersen MB (1995) Soil quality criteria for selected inorganic compounds. Danish Environmental Protection Agency, Working Report No. 48, 200.

14

Comparative DNA Damage and Repair in Echinoderm Coelomocytes Exposed to Genotoxicants

Ameena H. El-Bibany, Andrea G. Bodnar, Helena C. Reinardy*

Molecular Discovery Laboratory, Bermuda Institute of Ocean Sciences, St. George's, Bermuda

Abstract

The capacity to withstand and repair DNA damage differs among species and plays a role in determining an organism's resistance to genotoxicity, life history, and susceptibility to disease. Environmental stressors that affect organisms at the genetic level are of particular concern in ecotoxicology due to the potential for chronic effects and trans-generational impacts on populations. Echinoderms are valuable organisms to study the relationship between DNA repair and resistance to genotoxic stress due to their history and use as ecotoxicological models, little evidence of senescence, and few reported cases of neoplasia. Coelomocytes (immune cells) have been proposed to serve as sensitive bioindicators of environmental stress and are often used to assess genotoxicity; however, little is known about how coelomocytes from different echinoderm species respond to genotoxic stress. In this study, DNA damage was assessed (by Fast Micromethod) in coelomocytes of four echinoderm species (sea urchins *Lytechinus variegatus*, *Echinometra lucunter lucunter*, and *Tripneustes ventricosus*, and a sea cucumber *Isostichopus badionotus*) after acute exposure to H_2O_2 (0–100 mM) and UV-C (0–9999 J/m^2), and DNA repair was analyzed over a 24-hour period of recovery. Results show that coelomocytes from all four echinoderm species have the capacity to repair both UV-C and H_2O_2-induced DNA damage; however, there were differences in repair capacity between species. At 24 hours following exposure to the highest concentration of H_2O_2 (100 mM) and highest dose of UV-C (9999 J/m^2) cell viability remained high ($>94.6\pm1.2\%$) but DNA repair ranged from $18.2\pm9.2\%$ to $70.8\pm16.0\%$ for H_2O_2 and $8.4\pm3.2\%$ to $79.8\pm9.0\%$ for UV-C exposure. Species-specific differences in genotoxic susceptibility and capacity for DNA repair are important to consider when evaluating ecogenotoxicological model organisms and assessing overall impacts of genotoxicants in the environment.

Editor: Sam Dupont, University of Gothenburg, Sweden

Funding: The authors would like to thank a Bermuda Charitable Trust for the financial support to conduct this study. AHEB was supported by the National Science Foundation's Research Experiences for Undergraduates (REU, Grant No. 1266880). The funders had no role in study design, data collection and analysis, decision to publish, or preparation of the manuscript.

Competing Interests: The authors have declared that no competing interests exist.

* Email: helena.reinardy@bios.edu

Introduction

There has been much interest to integrate assessment of genetic effects into environmental studies to broaden the understanding of ecotoxicological impacts on organisms and populations [1–4]. Maintenance of DNA integrity is essential for proper cellular and organismal function, and the capacity to withstand genotoxic challenge is important to avoid long-term genetic instability and population vulnerability [5]. Unrepaired DNA damage can lead to mutations, cellular senescence, apoptosis, progression of cancer [6], and the process of aging [7]. Of particular concern in ecotoxicology is the potential for chronic effects and trans-generational impacts on populations by transfer of damaged DNA to offspring [8]. To minimize the harmful consequences of DNA damage, organisms are equipped with a variety of cellular defense and DNA repair mechanisms.

DNA is constantly damaged by both endogenous and exogenous sources, and genotoxicity can be considered as an imbalance between DNA damage and DNA repair mechanisms. Two major model genotoxicants are ultraviolet (UV) radiation and hydrogen peroxide (H_2O_2), which each induce different forms of DNA lesions. UV-C (<280 nm) is absorbed by the ozone in the earth's atmosphere and UV-B is the main component of UV radiation of environmental concern [9]; however, both UV-B and UV-C induce formation of cyclobutane-pyrimidine dimers (CPDs) and 6-4 photoproducts (6-4PPs) [10], in addition to DNA strand breaks [11]. UV-C induces high levels of DNA damage [12] and is commonly used as a model genotoxicant to investigate biological effects of UV irradiation [13,14]. H_2O_2 is produced as a byproduct of metabolic processes and cellular defense mechanisms [15], and is an important reactive oxygen species (ROS) involved in exogenously-induced oxidative DNA damage [16]. Antioxidant activity can restrict oxidative DNA lesions to several hundred per day, but excess ROS or a deficiency in antioxidants can lead to increased base oxidation and DNA strand breaks [17]. UV- and H_2O_2-induced DNA damage are primarily repaired by nucleotide excision repair (NER) and base excision repair (BER), respectively [18,14]. Investigation of DNA damage and repair after exposure to these two genotoxicants can inform on susceptibility to both oxidative damage and UV-induced DNA lesions, in addition to the capacity for both BER and NER.

Marine invertebrates have been extensively studied as bioindicators of environmental stress [19], and the sea urchin embryo test has served as a sensitive indicator of pollutant genotoxicology, embryo-toxicology, and teratogenicity [20–23]. Activation of DNA damage checkpoints, DNA repair, and apoptosis in sea urchin embryos have been demonstrated in response to genotoxicants such as methyl methanesulfonate, bleomycin, and exposure to ultraviolet radiation [24–26]. Despite the fact that sea urchin embryos are frequently used in toxicity testing, little is known of the effects of genotoxicants on the cells of adult sea urchins. Information about the cellular response of adult sea urchins to environmental stress is valuable for ecotoxicological studies and would increase understanding of the life history traits of these animals. Life history studies show that different species of sea urchins exhibit a very large range of reported lifespans (from approximately 3 to more than 100 years) [27–30], there is little evidence of senescence [31], and few reported cases of neoplasia [32,33]. Investigating DNA damage and DNA repair in cells of different sea urchin species would provide valuable information on selection of appropriate bioindicator species, allow assessments of environmental stress on different species, and shed light on mechanisms underlying life history traits of these animals.

The open circulatory system of echinoderms is comprised of coelomic fluid containing different cells types, collectively termed 'coelomocytes'. Coelomocytes fall into one of three categories: phagocytes, spherule cells (red and colorless), and vibratile cells, with further sub-categories within each cell type [34]. Coelomocytes play an integral role in immune cell functions such as fighting microbial infections and wound healing [34]. Damage to coelomocytes can compromise these essential functions, directly affecting the health of organisms and stability of populations. Coelomocytes (or circulating cells) from a variety of terrestrial and aquatic organisms (e.g. earthworms, bivalves, fish) have been useful bioindicators of environmental stress and are frequently used to assess genotoxicity [35–40]. Changes in the number and/or composition of coelomocytes have been reported in sea urchins from contaminated environments and those exposed to elevated $pCO2$ or increased temperature, suggesting that sea urchin coelomocytes may also serve as sensitive indicators of environmental stress [9,41–44]. However, another study showed that DNA from coelomocytes of the sea urchin *Lytechinus variegatus* is relatively resistant to genotoxicants [45]. Understanding susceptibility to DNA damage and DNA repair capacity of coelomocytes from different echinoderm species would be useful in assessing the value of coelomocytes as bioindicator cells and understanding the overall impacts of genotoxicants on these organisms. Persistent genotoxic damage is dependent on the balance between repair and replacement of damaged cells. Studies on echinoderms indicate a low level of cell turnover in the coelomocyte population (<1.5% BrdU incorporation in 3 hours [46] or 16 hours [47] in star fish) and low levels of apoptosis following acute exposures to UV-B [9], UV-C, hydrogen peroxide, methylmethane sulfonate and benzo[a]pyrene [45]; however, the DNA repair capacity of coelomocytes from different echinoderm species has not been investigated.

The objectives of this study are to assess the capacity to which cells from different echinoderm species are able to repair different types of DNA damage after exposure to two model genotoxicants, UV-C and H_2O_2. The specific aims are to comparatively evaluate the DNA damage and DNA repair capabilities in coelomocytes of four echinoderm species (sea urchins *L. variegatus*, *Echinometra lucunter lucunter*, *Tripneustes ventricosus*, and sea cucumber *Isostichopus badionotus*). We hypothesize that echinoderm coelomocytes will be able to repair some level of DNA damage, and the extent of genotoxicity sensitivity and DNA repair capacity will differ among species.

Materials and Methods

Animal collection and maintenance

All animals were collected and maintained in strict accordance with the Collecting and Experimental Ethics Policy (CEEP) of the Bermuda Institute of Ocean Sciences. All experiments complied with the ethical policy of the CEEP committee and did not require specific approval. All experiments were carried out on coelomocytes extracted from animals with minimal impact, except for a single small *E. l. lucunter* which was sacrificed in order to collect sufficient coelomic fluid for the experiment, and all efforts were made to minimize suffering. Except as mentioned above, all animals showed no adverse behavioral effects of the coelomocyte sampling procedure, all animals survived the procedure, and all animals were returned to their collection location.

Collection of animals complied with the collection policy of CEEP, no species were endangered, and no animals were collected from protected locations. Collection numbers of *L. variegatus*, *I. badionotus*, and *T. ventricosus* were within the CEEP collection limits and no specific collection permission was required. Collection of *E. l. lucunter* was carried out under a Department of Environmental Protection special permit (permit no, 131002, Bermuda Government), approved by the Director of Environmental Protection. All species were collected from the shallow sublittoral zone (less than 2 m depth at low tide), September–October, 2013, in Bermuda. *L. variegatus* and *I. badionotus* were collected from Harrington Sound (32°19.4′N, 64°43.6′W), *T. ventricosus* were collected from Fort St. Catherine beach (32°23.3′N, 64°40.3′W), and *E. l. lucunter* were collected from Castle Harbor (32°21.2′N, 64°39.8′W) and Gravelly Bay (32°19.1′N, 64°42.8′W). Animal husbandry and maintenance complied with CEEP policy. Sea urchins were maintained in flow-through aquaria with ambient temperature and light, and were left to acclimate for a minimum of 1 week after collection. *I. badionotus* were maintained in an outdoor flow-through aquarium with a layer of sediment on the bottom, and were left to acclimate for 1 week. Sea urchins were fed weekly with fresh sea grass, and sediment was replenished fortnightly in the *I. badionotus* aquarium.

Coelomocyte collection and treatment

Unless otherwise specified, all chemicals were sourced from Sigma-Aldrich (Sigma-Aldrich Co., St. Louis, MO, USA). Sea urchin test diameter was measured with calipers, and 2–6 ml coelomic fluid was extracted by syringe with an 18-guage needle inserted through the peristomial membrane surrounding the Aristotle's lantern. Sea cucumber size was estimated by weight, width, and length measurements, and 6–10 ml coelomic fluid was extracted by syringe with a 21-guage needle inserted laterally in the mid-body region. The experiments were designed to include a single coelomocyte collection per animal, division of the coelomic fluid for UV-C or H_2O_2 treatment, and proceeding concurrently with exposure and recovery period of both sets of treatment samples. Cell concentration, cell viability, and differential cell counts (red and other coelomocytes) were calculated after 1:1 dilution with trypan blue [0.8% trypan blue in calcium-magnesium-free seawater (CMFSW: 460 mM NaCl, 10 mM KCL, 7 mM Na_2SO_4, 2.4 mM $NaHCO_3$, pH 7.4) containing 30 mM EDTA] using a haemocytometer (Neubauer Bright Line haemocytometer). The volume for 50,000 cells per assay reaction (in triplicate or quadruplicate) was estimated and aliquoted into microcentrifuge tubes for each exposure. From the species selected for this study, cell aggregation was not a considerable factor in the experimental set-up. Coelomocytes from *L. variegatus*, *T. ventricosus*, and *I. badionotus* did not exhibit a strong agglutination reaction and could easily be dissociated to single cell suspensions by gently pipetting or vortexing. *E. l. lucunter* coelomocytes did exhibit some aggregation but clumps of cells were avoided when sample aliquots were taken. Differential cell counts and cell viability were estimated on all control and highest-exposed (9999 J/m^2 and 100 mM for UV-C and H$_2$O$_2$, respectively) samples after 24 hours recovery.

For the UV-C (254 nm) treatment, coelomocyte samples (25–132 µl volume) were irradiated (0, 250, 1000, 3000, or 9999 J/m^2) in 0.5 ml open microcentrifuge tubes in a Stratalinker UV Crosslinker 1800 (Stratagene, La Jolla, CA, USA). The recovery period was timed to begin immediately after dose delivery, and samples were left to recover for 0, 1, 3, 6, and 24 hours in the dark at room temperature. At each recovery timepoint, samples were placed on ice to halt DNA repair and processed for the Fast Micromethod assay.

For the H$_2$O$_2$ treatment, coelomocyte samples were exposed in 1.5 ml microcentrifuge tubes. H$_2$O$_2$ stock dilutions were prepared in CMFSW and added to coelomocyte samples to give the following final concentrations: 0, 0.1, 1, 10, or 100 mM H$_2$O$_2$. Samples were left in the dark for 10 min followed by 5 min centrifugation (8000 g) at room temperature. H$_2$O$_2$ exposure was halted by removal of supernatant after centrifugation, and cells were re-suspended in cell-free coelomic fluid (CFCF, prepared by collection of supernatant after centrifugation, 13000 g for 5 min, of coelomic fluid to remove cells) and the recovery period was started. At each recovery timepoint, samples were placed on ice to halt DNA repair, and processed for the Fast Micromethod assay.

Fast Micromethod for estimation of DNA damage

The method for fluorescent detection of alkaline DNA unwinding was carried out as described by Schröder et al. [48], with minor modifications. In brief, samples were assayed after respective periods of recovery and coelomocyte sample volume was adjusted with CFCF to make up to 50,000 cells per reaction. Samples were assayed in triplicate or quadruplicate by loading 20 µl sample to each replicate well on a black-walled 96-well microplate (USA Scientific, Inc., Ocala, FL, USA), and 20 µl of suitable blank (CMFSW or CFCF) were added to control wells. In some instances for *L. variegatus*, fewer cells were used per reaction when the cell concentration in coelomic fluid was low. Cells were lysed by adding 20 µl of lysing solution (9.0 M urea, 0.1% SDS, 0.2 M EDTA) containing 1:49 PicoGreen (Life Technologies, Grand Island, NY, USA, P7581), and left in the dark on ice for 40 min. DNA unwinding solution (20 mM EDTA, 1 M NaOH) was added (200 µl) to initiate alkaline unwinding (pH 12.4±0.02), fluorescence was detected (kinetic mode, excitation 480 nm, emission 520 nm, SpectraMax M2 Microplate Reader, Molecular Devices, CA, USA), and relative fluorescent units (RFU) was recorded every 5 min for a 30-min period. DNA unwinding was carried out at room temperature.

DNA damage was calculated according to the strand scission factor (SSF) equation [48]: SSF = log (% dsDNA$_{sample}$/% dsDNA$_{control}$)×(−1), where dsDNA$_{sample}$ are the treated samples and dsDNA$_{control}$ are the unexposed samples, and percentages are calculated from RFU after 20-min unwinding compared with initial (0 min unwinding) RFU, after subtracting respective blank RFU (CMFSW or CFCF). Due to high background fluorescence in CFCF from *I. badionotus*, RFU for that species were blanked

with CMFSW RFU, but other species' RFU were blanked with individual CFCF RFU.

Analyses

Both treatments (UV-C and H$_2$O$_2$) were conducted concurrently on a single coelomocyte sample per animal, and different animals (*T. ventricosus* n = 5, *L. variegatus* n = 12, *E. l. lucunter* n = 8, and *I. badionotus* n = 8) were considered biological replicates in all analyses. DNA damage estimation by Fast Micromethod included technical replicates (n = 3–4) of each sample to give an overall SSF per sample for each animal, and all biological replicates were combined for analyses of coelomocyte parameters, initial dose/concentration response, and DNA damage profiles over the 24-hour period of recovery.

Statistical analyses were performed in Statgraphics Centurion XVI.I (StatPoint Technologies, Inc., VA, USA). Intraspecific effects of size on DNA damage (SSF) during the 24-hour recovery period was tested by general linear model (GLM) with test diameter (average length for *I. badionotus*), dose/concentration, and recovery time as quantitative variables. To investigate intraspecific effects of concentration/dose and time, all individuals within a species were combined and DNA damage (SSF) was tested by GLM with concentration/dose and time as quantitative independent variables; dose/concentration differences from controls after 24 hours recovery were tested by one-way ANOVA or Kruskal-Wallis (for normally distributed or non-normally distributed data, respectively), with *post-hoc* Fisher's least significant difference (LSD) test at the 95% confidence level. Differences in DNA repair between species were tested by GLM, with species as a categorical factor, and concentration/dose and time as quantitative independent variables; species differences were established by *post-hoc* multiple range tests. Additionally, DNA repair was estimated as the percentage of DNA damage after 24 hours recovery compared with initial (0-hours recovery) DNA damage for each individual and for each exposure level, following the equation: % DNA repair = 100−((T$_{24}$ SSF/T$_0$ SSF)×100), where T$_{24}$ SSF is SSF after 24 hours recovery, and T$_0$ SSF is the initial (0-hours recovery) SSF; negative DNA repair values indicated no DNA repair and were set to zero. DNA repair (%) data was arcsine transformed to test for intraspecific differences in repair capacity (ANOVA, *post-hoc* multiple range tests). DNA repair capacity was categorized as follows: low (<25% DNA repair), moderate (25–50% DNA repair), high (50–75% DNA repair), or very high (>75% DNA repair).

Results

No anti-coagulant was used for collection of coelomic fluid and there was minimal or no cell aggregation in coelomocytes from *L. variegatus*, *T. ventricosus*, and *I. badionotus*. A proportion of coelomocytes from *E. l. lucunter* aggregated within the first few minutes after collection, clumps were disaggregated by gently pipetting before analysis, and persistent clumps were avoided. Coelomocytes isolated from the different species were evaluated for cell concentration, proportion of white to red cells, and cell viability. *E. l. lucunter* and *I. badionotus* had significantly higher total coelomocyte concentrations compared with the other species (Kruskal-Wallis and multiple range test, p<0.05), and no red coelomocytes were observed in any sample from *I. badionotus* (Table 1). There was no significant cell death in any of the coelomocyte samples over the course of the study, with cell viability >94% 24 hours after exposure to UV-C or H$_2$O$_2$ (Table 1). A slight reduction in overall coelomocyte size was

Table 1. Number of individuals, size ranges, coelomocyte characterization, and cell viability after 24-hours recovery from highest levels exposures to UV-C and H_2O_2 of all echinoderms tested.

Species	n	Test diameter range (mm)	Pre-treatment		Viability 24 hours recovery after treatment (% of total cells)	
			Coelomocyte concentration (cells/μl)	Red coelomocytes (% of total cells)	UV-C (9999 J/m²)	H_2O_2 (100 mM)
T. ventricosus	5	101–115	1855±280 A	7.6±2.4	99.8±0.2 C	94.6±1.2
L. variegatus	12	47–85	1957±220 A	9.1±1.6	98.5±0.4 C	98.4±0.6
E. l. lucunter	8	27–71	4565±745 B	8.0±3.1	99.4±0.3 C	99.7±0.1
I. badionotus	8	87–258*	4386±839 B	0	99.2±0.3 D	98.1±0.5

*Length (mm, average of several measurements during sampling) was measured for I. badionotus. Different letters (A/B/C/D) denote significant interspecific difference in means (multiple range tests, p<0.05). Data are mean ± s.e.m.

observed after 24 hours recovery from the highest levels of UV-C and H_2O_2.

Coelomocytes from all species showed an increase in DNA damage with increasing concentration or dose of genotoxicant (Figure 1). Patterns of dose responses indicated higher sensitivity in *T. ventricosus* and lower sensitivity in *E. l. lucunter* coelomocytes exposed to H_2O_2. *I. badionotus* had a lower magnitude of DNA damage after both genotoxicant treatments compared with the sea urchin species, and there was considerable inter-individual variation. The different sea urchin species responses to UV-C exposure were similar, with a slight indication of higher DNA damage at the highest doses in *T. ventricosus*.

Individuals of each species varied in size but there was no significant size effect over the 24-hour period of recovery after exposure to either H_2O_2 or UV-C in *T. ventricosus*, *L. variegatus*, *E. l. lucunter* (H_2O_2 only) or *I. badionutus* (GLM, p>0.05). There was a significant effect of size of DNA damage in *E. l. lucunter* after exposure to UV-C; however, the sample size was small and only 3 large individuals were collected therefore the biological significance is unknown and all individuals were grouped for further analyses.

Each species had a different response in reduction in DNA damage over a 24-hour period of recovery after exposure to UV-C, however *L. variegatus* and *E. l. lucunter* were not different from each other after exposure to H_2O_2 (Figure 2, GLM p<0.05, *post-hoc* multiple range test). The temporal pattern of DNA damage over time was consistent among species, with clear DNA repair for most treatment levels for both exposures only evident after 6–24 hours recovery, and *I. badionotus* had greater inter-individual variation compared with the sea urchin species (Figure 2). None of the sea urchin species showed very high repair of DNA damage in the highest two exposures (10 and 100 mM H_2O_2, and 3000 and 9999 J/m² UV-C) after 24 hours recovery, however *I. badionotus* showed high (>55%) or very high (>75%) repair of DNA damage at all exposure levels after 24 hours recovery (Figure 2, *post-hoc* Fisher's LSD, p<0.05, Table 2). *T. ventricosus* had highest DNA repair 24 hours after exposure to 0.1 mM H_2O_2 (59%) and 250 J/m² UV-C (20%), compared with controls, but *L. variegatus* and *E. l. lucunter* had high (>65%) DNA repair up to 10 mM H_2O_2, and *E. l. lucunter* had moderate (38%) DNA repair at 3000 J/m² UV-C.

There was a trend in overall DNA repair capacity (% DNA repair) between species: *T. ventricosus*<*L. variegatus*<*E. l. lucunter*<*I. badionotus* (Table 2). *E. l. lucunter* and *I. badionotus* had moderate (42%) and high (71%) repair of DNA damage, respectively, 24 hours following exposure to the highest concentration of H_2O_2 (100 mM), and high (53%) and very high (80%) repair of DNA damage, respectively, 24 hours following exposure to the highest dose of UV-C (9999 J/m²); these values contrast with low (<25%) repair in *L. variegatus* and *T. ventricosus* for the highest levels of both UV-C and H_2O_2. *I. badionotus* had high or very high DNA repair at all levels of exposure, and *E. l. lucunter* had high or very high levels of DNA repair after exposure to concentrations of H_2O_2 up to 10 mM. *T. ventricosus* had moderate or low DNA repair at all levels of exposure, except 0.1 mM H_2O_2 (59%), and both *T. ventricosus* and *L. variegatus* had reduced DNA repair at high concentrations or doses. There was an indication among all species for higher DNA repair capacity for H_2O_2-induced DNA damage, compared with UV-C-induced DNA damage.

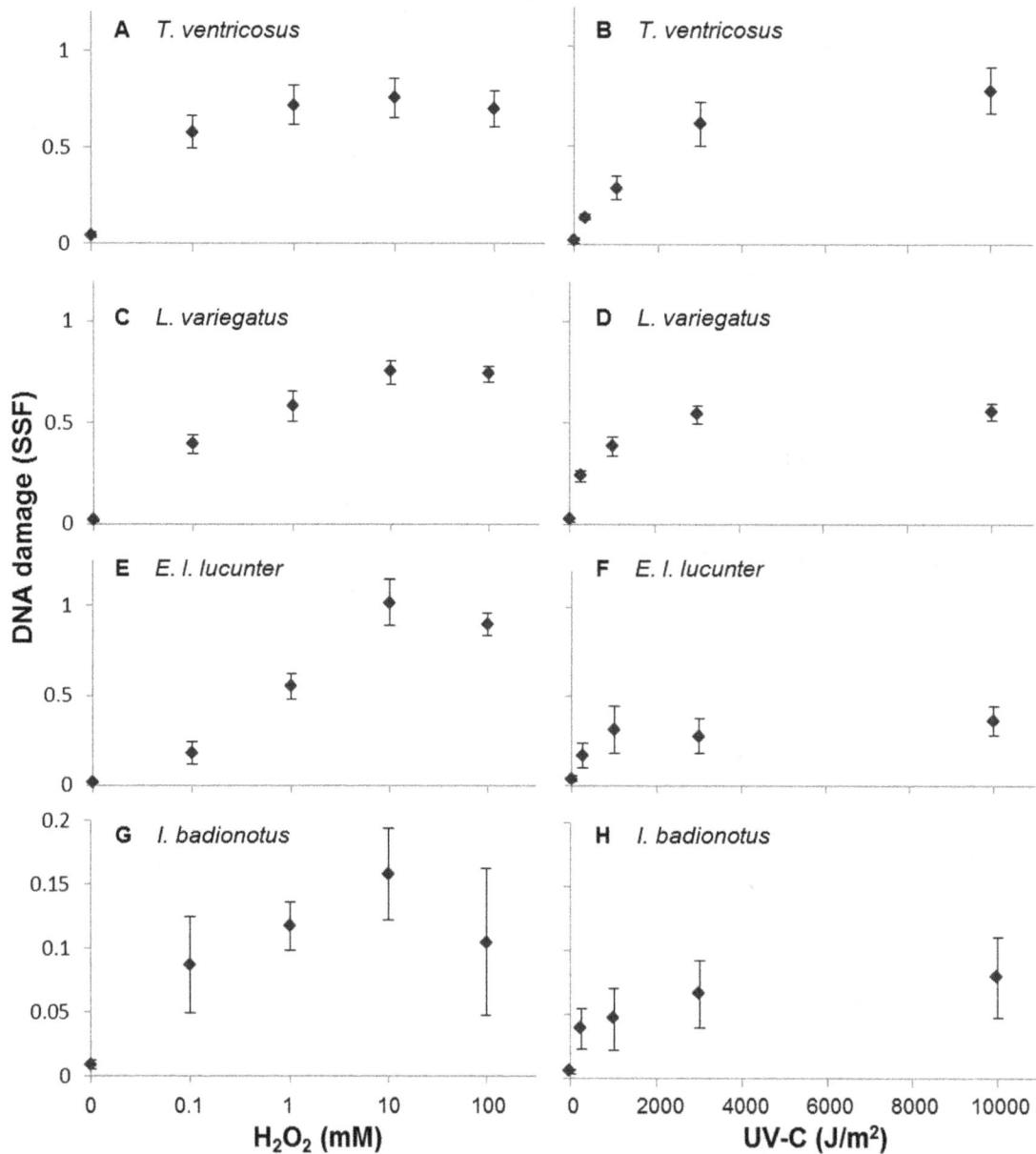

Figure 1. Dose/concentration response in echinoderm coelomocytes. Increase in DNA damage (strand scission factor, SSF, Fast Micromethod) with increasing concentration of H_2O_2 (**A**, **C**, **E**, and **G**) or dose of UV-C (**B**, **D**, **F**, and **H**) after acute exposure of coelomocytes from *T. ventricosus* (**A** and **B**, n = 5), *L. variegatus* (**C** and **D**, n = 11–12), *E. l. lucunter* (**E** and **F**, n = 6–7), and *I. badionotus* (**G** and **H**, n = 8). Data are means ± s.e.m.

Discussion

The objective of this study was to comparatively evaluate DNA damage and DNA repair capabilities of coelomocytes from four echinoderm species (*L. variegatus*, *E. l. lucunter*, *I. badionotus*, and *T. ventricosus*). Investigating DNA damage and DNA repair in cells of these different species can provide information on the value of coelomocytes as bioindicator cells and increase understanding of the overall impacts of genotoxicants on these organisms. Coelomocytes were chosen to evaluate the response to DNA damaging agents because they are well characterized cells involved in immunity and wound healing that have been proposed to be sensitive indicator cells for environmental stress [9,35–38,43–44], yet little is known of their response to genotoxicants. The

coelomocyte populations differed between species with *E. l. lucunter* and *I. badionotus* having higher cell concentrations than *L. variegatus* and *T. ventricosus*. There were no differences in the percentage of red spherule cells in the coelomocytes of the three sea urchin species; however, no red spherule cells were identified in the coelomic fluid of *I. badionotus*. This is consistent with a study on the sea cucumber *Apostichopus japonicus* which identified six cell types, none of which were red spherule cells [49]. Because little is known about the DNA repair capacity of various coelomocyte types, it is unknown whether differences in composition of coelomic fluid among species play a role in the ability for coelomocytes to repair damaged DNA. In addition, differences in coelomocyte composition between individuals may be a potential source of inter-individual variations observed in both treatment

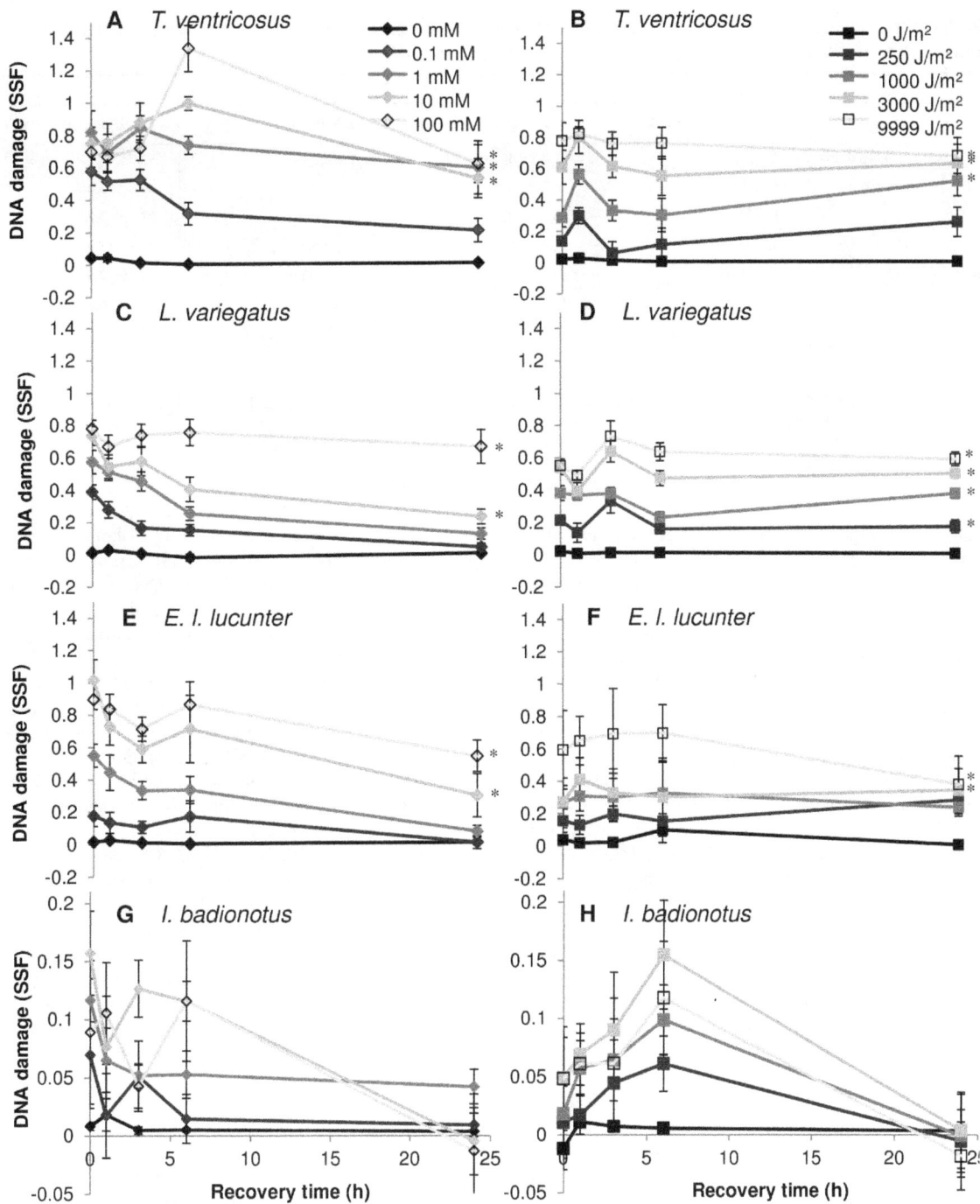

Figure 2. DNA repair in echinoderm coelomocytes. DNA repair [reduction in DNA damage (SSF)] over a 24-hour period of recovery after acute exposure to H_2O_2 (**A**, **C**, **E**, and **G**) or UV-C (**B**, **D**, **F**, and **H**) in coelomocytes from *T. ventricosus* (**A** and **B**, n = 5), *L. variegatus* (**C** and **D**, n = 12), *E. I. lucunter* (**E** and **F**, n = 8), and *I. badionotus* (**G** and **H**, n = 8). Data are means ± s.e.m. *Significantly higher than controls, indicating incomplete repair (within 24-hour timepoint, Fisher's LSD, p<0.05).

groups, in particular for *I. badionotus*. Apoptosis has been reported in sea urchin embryos exposed to UV radiation [26], but high coelomocyte viability 24 hours post-exposure over the course of this study suggests that apoptosis was not a factor contributing to the levels of DNA damage. This is consistent with the report of low levels of apoptosis in coelomocytes of the sea urchin *P. lividus* exposed to up to 2000 J/m² of UV-B (312 nm) [9]. Despite little cell death, 24 hours after exposure to the highest

levels of H_2O_2 and UV-C, coelomocytes were observed to be smaller in size. A study on cultured mouse myotubes found that 24 hours of chronic exposure to H_2O_2 significantly reduced myotube diameter *in vitro* [50]; however, it is unknown whether this decrease in cell size may have an impact on DNA repair activity in the nucleus.

In this study, DNA damage was detected by the Fast Micromethod, as recommended for high-throughput genotoxic

Table 2. Percent DNA repair (DNA damage at 24 hours recovery compared with initial DNA damage[†]) in echinoderm coelomocytes after 24 hours recovery from acute exposure to H_2O_2 or UV-C.

		T. ventricosus (n = 5)	L. variegatus (n = 12)	E. l. lucunter (n = 7–8)	I. badionotus (n = 8)
H_2O_2 (mM)	0.1	58.6±12.2	81.8±5.4	59.6±17.6	57.8±17.0
	1	33.4±19.8	73.2±6.8	79.8±5.2	59.0±13.0
	10	32.6±8.2	65.0±7.2	71.2±7.8	83.4±12.2
	100	18.2±9.2	24.8±6.6*	41.8±8.4	70.8±16.0
UV-C (J/m^2)	250	20.0±20.0	35.2±10.4	23.6±11.8	54.8±17.6
	1000	11.0±11.0	13.2±4.4	27.0±15.8	67.0±16.2
	3000	15.6±6.6	13.6±3.8	38.0±18.0	61.2±17.0
	9999	16.2±15.8	8.4±3.2	53.2±15.8	79.8±9.0

[†]% DNA repair = 100 − ((T_{24} SSF/T_0 SSF)×100), where T_{24} SSF is strand scission factor (SSF) after 24 hours recovery and T_0 SSF is the initial (0-hour) recovery SSF; negative % DNA repair values indicated no DNA repair and were set to zero.
*Significant reduction in DNA repair within a species (arcsine transformed, ANOVA, p<0.05, *post-hoc* multiple range test).
Data are means ± s.e.m. from individually calculated % DNA repair.

analyses [13] and comparable with the comet assay for DNA strand break detection and sensitivity [48]. There was a clear concentration- and dose-dependent increase in DNA damage for all echinoderm species tested. DNA damage levels in coelomocytes from *I. badionotus* appeared to be much lower than those for the sea urchin species; however, CMFSW blanks (not CFCF blanks) were subtracted from *I. badionotus* samples due to high relative fluorescent units in the CFCF from this species, which may underestimate the amount of DNA damage. Further investigation is needed to determine whether differences in the overall magnitude of SSF values of *I. badionotus* reflect high genotoxicity resistance in this species, and interspecific comparisons of overall levels of DNA damage with this species are carried out with caution. Based on the response over a similar concentration range of H_2O_2, the sensitivities of echinoderm coelomocytes are similar to that reported for zebrafish larvae exposed to H_2O_2 *in vivo*, where DNA damage (as estimated by comet assay) reached a plateau in the response curve between 100–200 mM H_2O_2 [51]. Other marine invertebrates such as shrimp (embryo and larvae exposures) and mussels (*in vitro* haemocyte exposures) have high levels of reported DNA damage at concentrations of H_2O_2 below 1 mM [37,52–53]. These interspecific differences highlight the need for consideration of suitable genotoxic bioindicator species. Genotoxic exposure of HeLa cells, mouse lymphoma cells, and peripheral blood mononuclear cells resulted in SSF values in a similar range to the levels of initial damage induced in coelomocytes of sea urchins [48]. However, comparable treatments of HeLa cells exposed to 1000 J/m^2 UV-C resulted in a SSF of 1.196 [48], considerably higher than the SSFs values of 0.28, 0.38, and 0.26 from coelomocytes of *T. ventricosus*, *L. variegatus and E. l. lucunter*, respectively, exposed to the same dose. This is consistent with the observation that LD$_{50}$ values for sea urchin coelomocytes (*L. variegatus*) exposed to H_2O_2 and UV-C are much higher than those of mammalian cells [45,54–56] and suggests that echinoderm coelomocytes are generally more resistant to genotoxicity than mammalian cells.

Comparisons of SSF values and DNA repair capacity revealed clear differences between the four species after exposure to UV-C and H_2O_2. It is thought that shallow coastal marine species may be readily exposed to genotoxicants and therefore evolutionarily well-adapted to repair DNA damage [57]. It is clear from the present results that coelomocytes from all species were able to repair some level of DNA damage from both genotoxic treatments, resulting in

reduction in DNA damage levels within 24 hours. The time profile and temporal delay in reduction of SSF within the first 6 hours of recovery could be indicative of direct DNA repair activity as both NER and BER pathways involve removal of a nucleotide or base which temporarily produces a single-strand break in the DNA [11,14]. The lower levels of DNA damage and pattern of a peak in DNA damage 1–6 hours after acute exposure to UV-C might indicate the relative lack of direct DNA strand breaks induced initially by UV-C exposure, and NER-induced strand breaks during repair [11,58–59]. Clear indication of DNA repair in coelomocytes indicates that these cells are active in the DNA damage response system of echinoderms and supports the need for further studies of the biology of these cells.

Variability after 24 hours recovery between the four species in the present study highlights important differences in DNA repair capacity even among species that share similar habitats and presumably similar exposure to genotoxicants. Sediment-dwelling species including sea cucumbers are thought to be more susceptible to genotoxicant exposure due to direct contact with the sediment [57]; however, the results of *I. badionotus* coelomocytes indicated the species was the most effective of the selected species in DNA repair, with very high repair of DNA damage after 24 hours of recovery. Phylogenetic relationships among the echinoderms reveal *T. ventricosus* and *L. variegatus* to belong to the family Toxopneustidea, whereas, *E. l. lucunter* belongs to the family Echinometridea; both families belong to the class Echinoidea [60]. All four echinoderm species belong to the same subphylum Echinozoa. Because *T. ventricosus* and *L. variegatus* are more closely related to each other than to *E. l. lucunter*, and even less so to *I. badionotus*, it is striking that there are differences in DNA repair capacity between the two, suggesting factors more significant than ancestry are involved in determining repair capacity. One determining factor may be the lifespan of the species, and the four echinoderm species included in this study vary in their natural lifespan. Life history data indicate that *T. ventricosus* and *L. variegatus* are relatively short-lived species (<4 years) [27,30] while *E. l. lucunter* is a longer-lived species with an estimated maximum lifespan of approximately 50 years [61]. There are very few studies of life history traits of sea cucumbers and no specific information is available for *I. badionotus* growth, survival, and longevity. However, growth data of other sea cucumber species suggest that sea cucumbers are slow-growing and long-lived. It is estimated that *Cucumaria frondosa* may take more than 25 years to reach a harvestable size [62] and modeled growth of *Holothuria nobilis* suggests that it may live for several decades [63]. DNA repair capacity (% DNA repair) after H_2O_2 exposure was greater in *E. l. lucunter* and *L. variegatus* than in the shorter-lived *T. ventricosus*. Additionally, percentage repair of UV-C-induced DNA damage indicated greater repair in the longer-lived *E. l. lucunter* group than in both other shorter-lived sea urchin species. A link between longevity and resistance to genotoxic stress has also been shown in bivalves with varying natural lifespans [64–65], and a greater repair capacity in longer-lived sea urchin species supports the idea that longer-lived species invest greater energy in cellular maintenance and repair [66–67]. Lack of lifespan information for *I. badionotus* restricts comparison between the species with regards to lifespan, but their highly efficient DNA repair capacity supports the speculation that they may be relatively long-lived in concordance with other sea cucumber species [62–63].

In conclusion, coelomocytes from different echinoderm species showed distinct differences in their sensitivity to DNA-damaging agents and their ability to repair damaged DNA over a 24-hour recovery period, therefore the choice of a single 'sensitive' species for ecotoxicological studies must be made with caution and consideration of differences within and between species. It is clear that coelomocytes from all species tested show some capacity for DNA repair, indicating involvement of these cells in the DNA damage response system of echinoderms; these results warrant further investigation into the biology of the DNA damage response and immune cell system in echinoderms. There was a trend for longer-lived echinoderms to have a greater DNA repair capacity

compared with shorter-lived species, and it would be interesting to investigate this further with more species over a great range of natural life spans. Complete DNA repair after 24 hours recovery from exposure to both H_2O_2 and UV-C was evident for *I. badionotus*, while *T. ventricosus* (with the shortest estimated lifespan) had the lowest overall capacity for DNA repair. Interspecific variability in echinoderms, however, must be taken into account when considering suitable model organisms for ecotoxicological investigations, and life history characteristics such as longevity may be important determinants for species vulnerability to environmental genotoxicity.

Acknowledgments

The authors would like to thank Thomas Ebert at Oregon State University for assistance with life history information of various echinoderm species and many thanks to the editor and reviewers for helpful suggestions to improve this manuscript.

Author Contributions

Conceived and designed the experiments: HCR AGB. Performed the experiments: AHEB. Analyzed the data: HCR AHEB. Contributed reagents/materials/analysis tools: AGB HCR. Wrote the paper: AHEB HCR AGG.

References

1. Belfiore NM, Anderson SL (2001) Effects of contaminants on genetic patterns in aquatic organisms: a review. Mutat Res 489: 97–122.
2. Kleinjans JCS, van Schooten FJ (2002) Ecogenotoxicology: the evolving field. Environ Tox Pharm 11: 173–179.
3. Klerks PL, Xie L, Levinton JS (2011) Quantitative genetics approaches to study evolutionary processes in ecotoxicology; a perspective from research on the evolution of resistance. Ecotoxicology 20: 513–523.
4. Ribeiro R, Lopes I (2013) Contaminant driven genetic erosion and associated hypotheses on alleles loss, reduced population growth rate and increases susceptibility to future stressors: an essay. Ecotoxicology 22:889–899.
5. Wurgler FE, Kramers PGN (1992) Environmental effects of genotoxins (ecogenotoxicology). Mutagenesis 7:321–327.
6. Enoch T, Norbury C (1995) Cellular responses to DNA damage: cell-cycle checkpoints, apoptosis and the roles of p53 and ATM. Trends Biochem Sci 20:426–430.
7. Cooke MS, Evans MD, Dizdaroglu M, Lunec J (2003) Oxidative DNA damage: mechanisms, mutation, and disease. FASEB J 17:1195–1214.
8. Dubrova YE (2003) Radiation-induced transgenerational instability. Oncogene 22:7087–7093.
9. Matranga V, Pinsino A, Celi M, Di Bella G, Natoli A (2006) Impacts of UV-B radiation on short-term cultures of sea urchin coelomocytes. J Mar Biol 149:25–34.
10. Sinha RP, Hader DP (2002) UV-induced DNA damage and repair: a review. Photochem Photobiol Sci 1:225–236.
11. Rastogi RP, Richa, Kumar A, Tyagi MB, Sinha RP (2010) Molecular mechanisms of ultraviolet radiation-induced DNA damage and repair. J Nucleic Acids 2010.
12. Misovic M, Milenkovic D, Martinovic T, Ciric D, Bumbasirevic V, et al. (2013) Short-term exposure to UV-A, UV-B, and UV-C irradiation induces alteration in cytoskeleton and autophagy in human keratinocytes. Ultrastruct Pathol 37:241–248.
13. Bihari N, Batel R, Jaksic Z, Muller WEG, Waldmann P, et al. (2002) Comparison between the comet assay and fast micromethod for measuring DNA damage in HeLa cells. Croat Chem Acta 75:793–804.
14. Ramos-Espinosa P, Rojas E, Valverde M (2012) Differential DNA damage response to UV and hydrogen peroxide depending of differentiation stage in a neuroblastoma model. Neurotoxicology 33:1086–1095.
15. Henle ES, Linn S (1997) Formation, prevention, and repair of DNA damage by iron/hydrogen peroxide. J Biol Chem 272:19095–19098.
16. Valavanidis A, Vlahogianni T, Dassenakis M, Scoullos M (2006) Molecular biomarkers of oxidative stress in aquatic organisms in relation to toxic environmental pollutants. Ecotoxicol Environ Saf 64:178–189.

17. Azqueta A, Shaposhnikov S, Collins AR (2009) DNA oxidation: investigating its key role in environmental mutagenesis with the comet assay. Mutat Res 674:101–108.
18. Friedberg EC (2003) DNA damage and repair. Nat 421:436–440.
19. Jha AN (2004) Genotoxicological studies in aquatic organisms: an overview. Mut Res 552:1–17.
20. Anderson SL, Wild GC (1994) Linking genotoxic responses and reproductive success in ecotoxicology. Environ Health Perspect 102:9–12.
21. Bay S, Burgess R, Nacci D (1993) Status and applications of echinoid (Phylum Echinodermata) toxicity test methods. In: Landis WG, Hughes JS, Lewis MA, editors. Environmental Toxicology and Risk Assessment, pp. 281–302.
22. Hose JE (1985) Potential uses of sea-urchin embryos for identifying toxic chemicals - description of a bioassay incoporating cytologic, cytogenetic and embryologic endpoints. J Appl Toxicol 5:245–254.
23. Saco-Alvarez L, Duran I, Ignacio Lorenzo J, Beiras R (2010) Methodological basis for the optimization of a marine sea-urchin embryo test (SET) for the ecological assessment of coastal water quality. Ecotoxicol Environ Saf 73:491–499.
24. Lamare MD, Barker MF, Lesser MP, Marshall C (2006) DNA photorepair in echinoid embryos: effects of temperature on repair rate in Antarctic and non-Antarctic species. J Exp Biol 209:5017–5028.
25. Le Bouffant R, Cormier P, Cueff A, Belle R, Mulner-Lorillon O (2007) Sea urchin embryo as a model for analysis of the signaling pathways linking DNA damage checkpoint, DNA repair and apoptosis. Cell Mol Life Sci 64:1723–1734.
26. Lesser MP, Kruse VA, Barry TM (2003) Exposure to ultraviolet radiation causes apoptosis in developing sea urchin embryos. J Exp Biol 206:4097–4103.
27. Beddingfield SD, McClintock JB (2000) Demographic characteristics of *Lytechinus variegatus* (Echinoidea: Echinodermata) from three habitats in a North Florida Bay, Gulf of Mexico. Mar Ecol 21:17–40.
28. Ebert TA, Southon JR (2003) Red sea urchins (*Strongylocentrotus franciscanus*) can live over 100 years: confirmation with A-bomb (14)carbon. Fishery Bulletin 101:915–922.
29. Moore HB, Jutare T, Bauer JC, Jones JA (1963) The biology of *Lytechinus variegatus*. Bull Mar Sci 23–53.
30. Pena MH, Oxenford HA, Parker C, Johnson A (2010) Biology and fishery management of the white sea urchin, *Tripneustes ventricosus*, in the eastern Caribbean. Rome: Food and Agriculture Organization of the United Nations. FAO Fisheries and Aquaculture Circular No 1056. p 43.
31. Ebert TA (2008) Longevity and lack of senescence in the red sea urchin *Strongylocentrotus franciscanus*. Exp Gerontol 43:734–738.

32. Jangoux M (1987) Diseases of Echinodermata. 4. Structural abnormalities and general considerations on biotic diseases. Dis Aquat Organ 3:221–229.

33. Robert J (2010) Comparative study of tumorigenesis and tumor immunity in invertebrates and nonmammalian vertebrates. Dev Comp Immunol 34:915–925.

34. Smith LC, Ghosh J, Buckley KM, Clow KA, Dheilly NM, et al. (2010) Echinoderm Immunity. In: Söderhäll K, editor. Invertebrate Immunology. Springer Science+Business Media, LLC, Landes Bioschience pp. 260–301.

35. Bolognesi C, Hayashi M (2011) Micronucleus assay in aquatic animals. Mutagenesis 26:205–213.

36. Canty MN, Hutchinson TH, Brown RJ, Jones MB, Jha AN (2009) Linking genotoxic responses with cytotoxic and behavioural or physiological consequences: differential sensitivity of echinoderms (Asterias rubens) and marine molluscs (Mytilus edulis). Aquat Toxicol 94:68–76.

37. Dallas LJ, Bean TP, Turner A, Lyons BP, Jha AN (2013) Oxidative DNA damage may not mediate Ni-induced genotoxicity in marine mussels: assessment of genotoxic biomarkers and transcriptional responses of key stress genes. Mutat Res Genet Toxicol Environ Mutagen 754:22–31.

38. Kolarevic S, Knezevic-Vukcevic J, Paunovic M, Kracun M, Vasiljevic B, et al. (2013) Monitoring DNA damage in haemocytes of freshwater mussel Sinanodonta woodiana sampled from the Velika Morava River in Serbia with the comet assay. Chemosphere 93:243–251.

39. Muangphra P, Gooneratne R (2011) Comparative genotoxicity of cadmium and lead in earthworm coelomocytes. Applied and Environmental Soil Science 2011:1–7.

40. Reinecke SA, Reinecke AJ (2004) The comet assay as biomarker of heavy metal genotoxicity in earthworms. Arch Environ Contam Toxicol 46:208–215.

41. Branco PC, Borges JCS, Santos MF, Jensch BE, da Silva JRMC (2013) The impact of rising sea temperature on innate immune parameters in the tropical subtidal sea urchin Lytechinus variegatus and the intertidal sea urchin Echinometra lucunter. Mar Environ Res 92:95–101.

42. Dupont S, Thorndyke M (2012) Relationship between CO2-driven changes in extracellular acid-base balance and cellular immune response in two polar echinoderm species. J Exp Mar Bio Ecol 424:32–37.

43. Pinsino A, Della Torre C, Sammarini V, Bonaventura R, Amato E, et al. (2008) Sea urchin coelomocytes as a novel cellular biosensor of environmental stress: a field study in the Tremiti Island Marine Protected Area, Southern Adriatic Sea, Italy. Cell Biol Toxicol 24:541–552.

44. Matranga V, Toia G, Bonaventura R, Muller WE (2000) Cellular and biochemical responses to environmental and experimentally induced stress in sea urchin coelomocytes. Cell Stress Chaperones 5:113–120.

45. Loram J, Raudonis R, Chapman J, Lortie M, Bodnar A (2012) Sea urchin coelomocytes are resistant to a variety of DNA damaging agents. Aquat Toxicol 124–125:133–138.

46. Holm K, Dupont S, Sköld H, Stenius A, Thorndyke M, et al (2008) Induced cell proliferation in putative haematopoietic tissues of the sea star, Asterias rubens (L.). J Exp Biol 211:2551–2558.

47. Hernroth B, Farahani F, Brunborg G, Dupont S, Dejmek A, et al (2010) Possibility of mixed progenitor cells in sea star arm regeneration. J Exp Zool B Mol Dev Evol 341B:457–468.

48. Schröder HC, Batel R, Schwertner H, Boreiko O, Müller WEG (2006) Fast micromethod DNA single-strand-break assay. In: Henderson DS, editor. Methods in Molecular Biology: DNA Repair Protocols: Mammalian Systems, 2nd ed. Humana Press Inc, Totowa, NJ, pp 287–305.

49. Xing K, Yang HS, Chen MY (2008) Morphological and ultrastructural characterization of the coelomocytes in Apostichopus japonicus. Aquat Biol 2:85–92.

50. McClung JM, Judge AR, Talbert EE, Powers SK (2009) Calpain-1 is required for hydrogen peroxide-induced myotube atrophy. Am J Physiol Cell Physiol 296:C363–371.

51. Reinardy HC, Dharamshi J, Jha AN, Henry TB (2013) Changes in expression profiles of genes associated with DNA repair following induction of DNA damage in larval zebrafish Danio rerio. Mutagenesis 28:601–608.

52. Cheung VV, Depledge MH, Jha AN (2006) An evaluation of the relative sensitivity of two marine bivalve mollusc species using the Comet assay. Mar Environ Res 62:S301–S305.

53. Hook SE, Lee RF (2004) Genotoxicant induced DNA damage and repair in early and late developmental stages of the grass shrimp Paleomonetes pugio embryo as measured by the comet assay. Aquat Toxicol 66:1–14.

54. Long AC, Colitz CMH, Bomser JA (2004) Apoptotic and necrotic mechanisms of stress-induced human lens epithelial cell death. Exp Biol Med 229:1072–1080.

55. Murakami S, Salmon A, Miller RA (2003) Multiplex stress resistance in cells from long-lived dwarf mice. FASEB J 17:1565–1566.

56. Salmon AB, Akha AAS, Buffenstein R, Miller RA (2008) Fibroblasts from naked mole-rats are resistant to multiple forms of cell injury, but sensitive to peroxide, ultraviolet light, and endoplasmic reticulum stress. J Gerontol A Biol Sci Med Sci 63:232–241.

57. Depledge MH (1998) The ecotoxicological significance of genotoxicity in marine invertebrates. Mut Res 399:109–122.

58. Collins AR (2014) Measuring oxidative damage to DNA and its repair with the comet assay. Biochim Biophys Acta 1840:794–800.

59. Azqueta A, Langie SAS, Slyskova J, Collins AR (2013) Measurement of DNA base and nucleotide excision repair activities in mammalian cells and tissues using the comet assay – a methodological overview. DNA Repair 12:1007–1010.

60. WoRMS (2013) Echinozoa, World Register of Marine Species. Available: http://www.marinespecies.org/aphia.php?p=taxdetails&id=148744.

61. Ebert TA, Russell MP, Gamba G, Bodnar A (2008) Growth, survival, and longevity estimates for the rock-boring sea urchin Echinometra lucunter lucunter (Echinodermata, Echinoidea) in Bermuda. Bull Mar Sci 82:381–403.

62. So JJ, Hamel JF, Mercier A (2010) Habitat utilisation, growth and predation of Cucumaria frondosa: implications for an emerging sea cucumber fishery. Fish Manag Ecol 17:473–484.

63. Uthicke S, Welch D, Benzie JAH (2004) Slow growth and lack of recovery in overfished holothurians on the Great Barrier Reef: evidence from DNA fingerprints and repeated large-scale surveys. Conservation Biol 18:1395–1404.

64. Ungvari Z, Ridgeway I, Philipp EER, Campbell CM, McQuary P, et al. (2011) Extreme longevity is associated with increased resistance to oxidative stress in Arctica islandica, the longest-living non-colonial animal. J Gerontol A Biol Sci Med Sci 66:741–750.

65. Ungvari Z, Sosnowska D, Mason JB, Gruber H, Lee SW, et al. (2013) Resistance to genotoxic stress in Arctica islandica, the longest living noncolonial animal: is extreme longevity associated with a multistress resistance phenotype? J Gerontol A Biol Sci Med Sci 68:521–529.

66. Kirkwood TBL (2005) Understanding the odd science of aging. Cell 120:437–447.

67. Bodnar AG (2009) Marine invertebrates as models for aging research. Exp Gerontol 44:477–484.

15

Predicting Vulnerabilities of North American Shorebirds to Climate Change

Hector Galbraith[1,4], David W. DesRochers[2], Stephen Brown[1], J. Michael Reed[3]*

1 Manomet Center for Conservation Sciences, Manomet, Massachusetts, United States of America, **2** Department of Natural Sciences, Dalton State College, Dalton, Georgia, United States of America, **3** Department of Biology, Tufts University, Medford, Massachusetts, United States of America, **4** National Wildlife Federation, Springfield, Massachusetts, United States of America

Abstract

Despite an increase in conservation efforts for shorebirds, there are widespread declines of many species of North American shorebirds. We wanted to know whether these declines would be exacerbated by climate change, and whether relatively secure species might become at–risk species. Virtually all of the shorebird species breeding in the USA and Canada are migratory, which means climate change could affect extinction risk via changes on the breeding, wintering, and/or migratory refueling grounds, and that ecological synchronicities could be disrupted at multiple sites. To predict the effects of climate change on shorebird extinction risks, we created a categorical risk model complementary to that used by Partners–in–Flight and the U.S. Shorebird Conservation Plan. The model is based on anticipated changes in breeding, migration, and wintering habitat, degree of dependence on ecological synchronicities, migration distance, and degree of specialization on breeding, migration, or wintering habitat. We evaluated 49 species, and for 3 species we evaluated 2 distinct populations each, and found that 47 (90%) taxa are predicted to experience an increase in risk of extinction. No species was reclassified into a lower–risk category, although 6 species had at least one risk factor decrease in association with climate change. The number of species that changed risk categories in our assessment is sensitive to how much of an effect of climate change is required to cause the shift, but even at its least sensitive, 20 species were at the highest risk category for extinction. Based on our results it appears that shorebirds are likely to be highly vulnerable to climate change. Finally, we discuss both how our approach can be integrated with existing risk assessments and potential future directions for predicting change in extinction risk due to climate change.

Editor: Grant Ballard, Point Blue Conservation Science, United States of America

Funding: This study was internally funded. Each of the authors were supported primarily by their employers; Galbraith's time was supported by Manomet Center for Conservation Sciences and by the National Wildlife Federation; DesRochers was supported by Dalton State College; Brown was supported by Manomet Center for Conservation Sciences; and Reed was supported by Tufts University and by Manomet Center for Conservation Sciences. The funders had no role in study design, data collection and analysis, decision to publish, or preparation of the manuscript.

Competing Interests: The authors have declared that no competing interests exist.

* Email: michael.reed@tufts.edu

Introduction

Shorebirds are important components of the ecosystems in which they live, they are valued by the general public, can exhibit extremely large and impressive aggregations during migration, and they can act as sentinels of global environmental change [1–3]. There also is a growing demand to move beyond evaluating climate change impacts on single species or habitats and to evaluate expected broad scale ecological impacts on communities and ecosystems [4]. Consequently, we are concerned about the current documented widespread declines of many species of North American shorebirds [1,5–8], particularly the recent steep declines in Atlantic populations of Red Knots (scientific names of North American shorebirds are given below) [9–10] and Semipalmated Sandpipers [11–12].

The U.S. Fish and Wildlife Service currently lists three North American shorebirds as Threatened or Endangered [13]. IUCN lists five shorebird species in North America as Near Threatened or at higher risk, and four additional species in these categories for the Western Hemisphere [14]. The causes of these declines are not well understood but most likely include loss of breeding, migration, and wintering habitats, and disturbance and exploitation [1,15–17]. It should be recognized, however, that the factors causing such changes could be global, since population reductions have been seen in virtually all shorebird flyways from North and South America, to East Africa, to Asia and Australia, e.g., [18–19].

Global climate change is an anthropogenic stressor that could adversely affect shorebird populations across species' ranges. Shorebirds that breed and/or winter at high latitudes may be among the most sensitive of bird species to this stressor because this is where climate change is expected to be most severe [20]. They also have several additional risk factors, including lengthy, energetically expensive migrations where they may be vulnerable to changes in wind patterns, dependence upon coastal migration stopover sites that are vulnerable to sea level rise, and dependence upon ecological synchronicities that may be disrupted by a changing climate [16,21–23]. Small–Lorenz et al. [24] point out that assessments of vulnerability to climate change often ignore problems associated with a migratory life–history, causing them to underestimate vulnerabilities. Shorebirds are already in a vulnerable condition and climate change may exacerbate this.

If we are to understand what may happen to shorebirds in the near future and initiate appropriate conservation measures it is essential that we be able to predict the likely vulnerabilities of shorebird species to various aspects of the changing climate, cf. [25]. To be useful for conservation, predictive frameworks should be based on the ecologies and life histories of the species, should incorporate what we know about how the planet's climate will alter, and should generate at least qualitative estimates of species vulnerabilities, e.g. [26–27].

Categorizing vulnerability to extinction based on a suite of characteristics, such as population size and rate of decline, is used widely e.g. [28–31]. The best known models are those of Partners in Flight and IUCN (also used by BirdLife International) [32]. Their categorization approach to vulnerability also can be used to evaluate species' changes in vulnerability as ecosystems change over time, e.g. [33]. Partners in Flight (PIF) uses a model to assess vulnerability based on population trend, relative abundance, threats during the breeding and non–breeding seasons, and breeding and non–breeding range sizes. For each category, each species receives a score of 1 to 5, with 5 associated with greatest risk. These scores are summed using several different formulas, each of which is used to determine species of conservation concern for particular reasons. A similar system was developed based on the same set of basic variables for the U.S. Shorebird Conservation Plan [1], although the resulting risk categories are defined somewhat differently. None of these systems includes risk due to climate change. In this paper, it was our overarching goal to determine the degree to which climate change will alter the extinction risk level assigned to shorebird species in the U.S. Shorebird Conservation Plan, and for this method to be compatible with the PIF ranking system.

We approached our reconsideration of risks under climate change by developing an assessment framework, and then used it to evaluate the vulnerabilities to climate change of North American (north of Mexico) shorebird species, whose life histories extend across wide ecological and behavioral spectra. Specifically, we (1) identified risk factors, (2) created a framework for quantifying the change in risk due to climate change for each of the factors, including the possibility of decreased extinction risk due to climate change, (3) identified the effects climate change would have on the risk factors, (4) reviewed the literature on each shorebird species we assessed to determine species–specific risk for each factor, and (5) assigned species to their new extinction–risk categories. We also (6) did a sensitivity analysis to determine how the results were affected by different decision rules for changing PIF risk categories.

Methods

We included 49 species in our assessment. For three species (Willet, Piping and Snowy Plover) we evaluated two distinct populations each, so in all 52 taxa were evaluated. We excluded Eskimo Curlew (*Numenius borealis*) from our analysis because it is likely extinct [34]. Our assessments are compatible with both the U.S. Shorebird Conservation Plan and PIF frameworks, although because of the increased risk to some species already at the highest risk categories, we needed to add a new risk category – critical – to distinguish species at greatly increased risk.

To achieve the goal of creating a framework that could be integrated with both PIF and the U.S. Shorebird Conservation Plan, we first evaluated other existing approaches. The State of the

Birds [35] developed a framework to assess changes in risk due to climate change, with the goal of applying it to all bird species. They included migration distance and timing as bivariate factors (birds that migrate long distances and use daylight cues = 1; else 0); degree of breeding habitat obligation (high = 1 vs. not = 0); dispersal ability (1 vs. 0); niche specificity (1, 0); reproductive potential (lays one egg per year = 1, else 0); and habitat susceptibility (divided into 3 levels, 2, 1, 0, from highest to lowest susceptibility). Scores were summed to assess overall risk. This approach apparently ignores risks associated with migration and wintering habitat obligation, does not allow for extinction risk to decrease due to climate change, and there is heavy weighting of reproductive potential, which is evaluated on a narrow scale that distinguishes only between one–egg clutches and all other clutch sizes. Also, while reproductive potential may be important for population size recovery following sudden decline, it may be less important with respect to gradual climate change. This approach is applicable to other species included in the PIF prioritization system.

We included six factors in our risk framework, each of which had 3–5 risk levels. Factors included: expected losses or gains in (1) breeding, (2) migration, and (3) non–breeding habitat (4) degree of dependence on ecological synchronicities; (5) migration distance; and (6) degree of habitat specialization (on breeding, migration, and non–breeding grounds). All risk factors were given equal weight in the assessment, and each factor is described in detail below.

Expected Losses or Gains in Breeding, Migration, and Non–breeding Habitat (1–3)

We accepted that the atmospheric concentrations of greenhouse gases will approximately double (over pre–industrial levels) by the middle to end of the century [36–37]. We then summarized, based on current understanding reported in the literature, the effects climatic change should have on habitats used by our focal shorebirds in the western hemisphere. What follows is our assessment of these changes (designated B1–B5), and brief statements about our confidence in these changes. These approximate confidence levels of >70%, 30–70%, and <30% are modified from the 5–category scale developed by [38] for the Intergovernmental Panel on Climate Change Third Assessment Report. We reduced the number of categories because we did not think the implied precision of 5 levels of confidence was defensible.

(B1) Northern hemispheric boreal and arctic areas. Tundra habitat will be reduced in extent as the tree line moves poleward; areas that persist as tundra will become less dominated by graminoids and other low–growth species and will become increasingly dominated by more shrubby species, reducing the habitat value for breeding shorebirds [39–47]. Also, the boreal forest will extend its range northward as it replaces tundra, but its southern distribution will contract northwards [37–44]. Although it is true that new areas of bare ground are likely to be created by ice cap and glacial recession in high tundra areas, we do not believe that this will result in more habitat for most breeding shorebirds since it will persist as gravel or bouldery moraine for a long period until vegetated and soil–forming processes can occur. Confidence = medium.

Changes in precipitation and evapotranspiration are also likely, but the aggregate effects on tundra hydrology are difficult to predict [48]. Drier overall conditions may be likely, and may reduce food availability during the breeding season [48]. It is unclear how climate change will affect the water balance on tundra breeding habitats due to the complex interaction of several factors, including amounts and timing of precipitation events, timing and extent of spring thaw, depth of the active layer, and erosion events [48–53]. While annual rainfall is predicted to increase throughout the breeding range, evapotranspiration is also expected to increase enough to more than offset the effect of increased precipitation [36]. The result may be a loss of some wetland breeding habitat to dryer conditions, but this is unclear. Confidence = medium.

(B2) **North American Great Plains.** Much of the climate modeling that has been performed indicates that these interior grassland regions will become hotter and drier [44,54–57]. This is likely to result in adverse impacts to shorebird species that depend on seasonally or permanently flooded wetlands for their migration stopovers. Confidence = medium

(B3) **Coastal habitats.** Based on IPCC [36] and more recent modeling [58–60] we assume that sea levels will rise globally by between 1 and 2 meters, resulting in the loss of coastal shorebird habitats. This applies to North, Central and South America [61], and will be worst in areas with, for example, high tidal amplitudes in shallow lagoons and broad estuaries [62–65]. Consequently, we anticipate major loss of coastal wintering habitat for shorebirds, particularly in areas where the land surface is subsiding or accretion rates of intertidal habitats are low (e.g., most Gulf Coast sites) [66–67]. If coastal habitats are able to move inland in response to sea level rise, it could offset losses, but at many sites this will be precluded by human infrastructure and interventions [21,65,68–69]. Confidence = high

(B4) **Interior South America.** Ecological modeling based on climate change models indicates that increased aridification in South America will have the following effects: first it is likely to result in the replacement of currently forested areas in the Amazon by savanna habitat and seasonal forests [70–71]. Experimental droughts in the eastern Brazilian Amazon resulted in increased tree mortality, which also supports the expectation of declining rainforest habitat [72]. This is unlikely to benefit shorebirds as few use the existing savanna habitats in central South America. Second, the existing grassland areas in central and southern South America will become drier [36,73], but the effect on the grassland habitats on which North American shorebirds currently winter is uncertain. Confidence = low.

(B5) **Eastern North American forests.** The only North American shorebird species that primarily uses temperate forest habitat for breeding is the American Woodcock. The species prefers young forest with openings, and the species tolerates a wide a range of tree species [74]. In much of the woodcock's range afforestation is occurring due to ecological succession resulting from abandonment of historical agricultural areas [75]. As a result, young forests adjacent to fields or containing areas of open habitat are declining, resulting in loss of required breeding habitat. Additionally, climate change likely will result in increased vegetation growth at higher latitudes in North America [38–41]. This will result in the establishment of more woody vegetation and a subsequent increase in young forest habitat in the north. It is unclear if northward expansion of the woodcock's range is occurring, so changes in forest landscape may outpace range expansion. Another potential concern for forest breeding habitat is climate change's impact on tree mortality. There is growing evidence that drought resulting from climate change leads to increased tree mortality [76–78]. This may open breeding areas for American Woodcock locally, but widespread forest loss could result in loss of breeding habitat. Confidence = Medium.

(B6) **Ocean.** One of the primary mechanisms through which climate change could impact oceanic habitats is through acidification [79–80]. This likely will reduce the quality of marine habitats, but the extent to which this might affect pelagic non-

breeding shorebirds is uncertain [36,81–82]. One hypothesis is that ocean acidification could reduce the fitness of many plankton species by reducing calcification and other physiological processes [83–84]. If ocean acidification does negatively impact marine plankton food resources, the decrease could be offset, however, by increased ocean upwelling which could function to increase food resources [85]. Confidence = low.

Ecological Synchronicities (4)

We recognize two types of ecological synchronicities important to shorebirds that we think could be affected by climate change.

(ES1) **Breeding season food resources.** Arctic temperatures are rising and are projected to further increase in the future, resulting in earlier spring thaws and ice melts [36]. This likely will result in earlier invertebrate hatches because arctic invertebrate emergence is temperature dependent [86]. Long–term field observations and recent experimental warming studies of arctic plots support this hypothesis [87–88]. If birds are unable to alter migration timing, then arctic nesting shorebirds may have insufficient food resources for young.

(ES2) **Migration food resources.** Some migrants depend on highly seasonal food sources during migration [89]. For example, shorebirds such as Ruddy Turnstones, Red Knots, Sanderlings, and Semipalmated Sandpipers are highly reliant on American Horseshoe Crab (*Limulus polyphemus*) eggs for refueling during northward migration stopovers [90–91]. If climate change affects timing of horseshoe crab breeding, this would disrupt synchronicity between horseshoe crab egg laying and spring migration.

Migration Distance (5)

We treat migration distance as others have, as a surrogate for things that can go wrong that have not been captured by other factors [31,35,92]. The assumption is that the farther a species has to migrate, the more ecological disruption can occur [92–94]. In the context of climate change, for example, migratory connectivity interacts with habitat loss from sea level rise [95–96] and species may encounter more severe weather during migration [97–98]. Our separation of species into distance categories was done by looking for natural breaks in the migration distance data, resulting in distances being divided into 5 categories (Fig. 1). Migration distances were calculated from the approximate center of each species' breeding range to the approximate center of each species' wintering range using data from NatureServe [99]. The two exceptions were Bristle–thighed Curlew and Bar–tailed Godwit, which are not covered by this database. Known migration distances placed these species in the greatest–distance category.

Degree of Habitat Specialization (6)

This variable refers to degree of specialization to a certain habitat type, rather than the vulnerability of the habitat type. We assert that being specialized increases your extinction risk to climate change because of reduced response capability. If a species specializes on a habitat type at any time in its life cycle (breeding, migration, non–breeding), it was considered to be specialized. We divided this risk factor into three categories (Table 1).

Assessment Framework Development

Each risk factor was assessed for each species using information from the literature regarding the natural history of the species and anticipated changes due to climate change. A summary of each species' risk level associated with climate change for each risk factor narrative, as well as confidence scores can be found in Appendices S1 and S2. For each risk factor, for each species, we

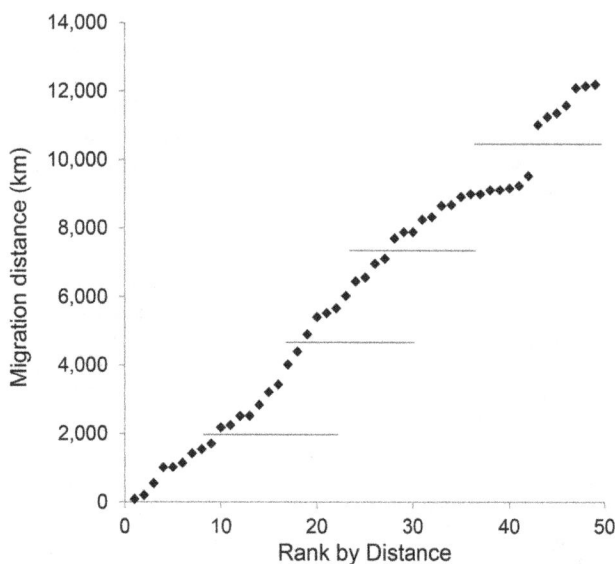

Figure 1. One–way migration distances calculated as mid–point to mid–point of their summer and winter geographic ranges. Ranges were downloaded from the NatureServe database. Horizontal lines separate dispersal distances as ranked in Table 1, with the shortest distances associated with rank 1 and the greatest distances with rank 5. The exceptions are the Bristle–thighed Curlew and Bar–tailed Godwit, which do not overwinter in the New World so they are not covered by the database. They fall into the greatest migration distance category, and are represented arbitrarily in the figure by the 2 points showing the greatest migration distances.

also included a subjective confidence score (1 = low to 5 = high confidence). We recognize that a species might have increased extinction risk due to climate change, but it might not increase enough to change risk categories.

We described changes in risk using two systems: a numeric scoring system that had maximum values for each factor of 5, and a graphical depiction of the change in risk using arrows because we thought they were more intuitive for rapid visual assessment of changes and patterns. Risk factors were scaled from 0 to 5 to match PIF scaling. For the three habitat factors (1–3), we allowed for the possibility of improved conditions due to climate change. Improvement resulted in negative scores (or down–arrows) to show reduced risk. The factors, and their subdivision and scoring, are shown in Table 1.

For our purposes, we decided that an increase in risk score of 10 (equivalent to 4 ↑s; the arrows indicate the direction and degree of effect) was sufficient to increase by a single risk category because a score of 10 would mean that a species is at extreme risk in two of the six categories. This assignment is a first approximation based on best professional judgment cf. [32], but should be revisited as more information about shorebird ecology and vulnerability to habitat changes becomes available.

To investigate the importance of our decision for how much change in risk is sufficient to cause a change in risk category, we did a sensitivity analysis. Specifically, we assessed the sensitivity of our results – which species were placed into which risk category – to the amount of change in extinction risk that was required for a species to change risk categories. We did this by making the criterion for changing categories more sensitive, requiring the accumulation of only 3 arrows to make the transition between risk categories. We also evaluated the effect of making the criterion less sensitive, evaluating the effects of requiring 5, 6 and 7 arrows to

allow a species to change risk categories. If our method is insensitive to this criterion, we would expect little change in categorization with changing criteria.

Results

Each species' account and changes in risk level are found in Appendices S1 and S2, but we briefly go through the account for the Semipalmated Sandpiper to demonstrate the procedure. (1) We anticipate moderate loss of breeding habitat (score 3; 1 ↑). Our reasoning is based largely on the expectation that tundra breeding habitat will be reduced over the longer term by the increase of woody vegetation, which will invade current areas of tundra [43]. Additional impacts may also occur from changes in precipitation, but it is unclear how climate change will affect the water balance on tundra breeding habitats due to the complex interaction of several factors, including amounts and timing of precipitation, timing of spring thaw, and depth of the active layer [48]. While annual rainfall is predicted to increase throughout the breeding range, evapotranspiration is also expected to increase enough to more than offset the effect of increased precipitation. The result may be a loss of some wetland breeding habitat to dryer conditions, but this is unclear. Our confidence in the assessment of the overall score for moderate loss of breeding habitat is low. (2) We anticipate major loss of wintering habitat (score 5; 2 ↑s) because winter range includes almost exclusively coastal shoreline habitat, so sea level rise (SLR), storm surges, and changing fresh–salt water mixes pose a large threat. Since the species uses estuaries with large tidal amplitudes in Brazil, this may buffer against the SLR impacts, at least locally. Our confidence in this estimate is high. (3) We anticipate moderate loss of migration habitat (score 3; 1 ↑) because SLR likely will cause the loss of some coastal migratory areas. Expected decrease in rainfall in southern areas of North America will cause a decrease in spring migration habitat. In contrast, rainfall is expected to increase in northern portions of North America during spring migration, likely resulting in increased habitat in the interior. Our confidence in this estimate is high. (4) This species has a high degree of dependence on ecological synchronicities (score 5; 2 ↑s). Arctic temperatures are expected to increase, resulting in earlier spring thaws and ice melts. This, in turn, will likely result in earlier invertebrate emergence. If birds are unable to alter migration timing, then arctic nesting shorebirds may have insufficient food resources to support reproduction. Our confidence in this estimate is high. (5) Migration distance is 7886 km (score 4; 2 ↑s). (6) We categorize this species as being moderately specialized in its habitat use (score 4; 2 ↑s). It has fairly specific wintering habitat requirements, including shorelines with wide intertidal mudflats, near shallow lagoons, and wide estuaries with large tidal amplitudes. Our confidence in this estimate is high. This assessment generates a total score of 24 (9 ↑s), which is enough in our protocol to push the species up two risk categories from its place in the current U.S. Shorebird Conservation Plan, from a species of Moderate Concern to Highly Imperiled.

Of the 52 taxa we evaluated, 45 (87%) are predicted to qualitatively increase their risks of extinction as a result of climate change; 33 by one level in the U.S. Shorebird Conservation Plan, and 12 by 2 levels (Table 2, Fig. 2). Only three species had risk factors that we predict will lower a species' extinction risk due to climate change: Solitary Sandpiper, due to the creation of more breeding habitat; Bristle–thighed Curlew, due to the expansion of breeding and wintering habitat; and White–rumped Sandpiper, due to more wintering habitat. The U.S. Shorebird Conservation Plan currently lists 29 species at risk levels of High Concern or

Table 1. List of risk factors evaluated for species sensitivity to climate change.

1) Loss/gain in breeding habitat under climate change:	Score	Arrow
Major losses (>50%)	5	↑↑
Moderate losses (10–50%)	3	↑
Limited or no losses (−10–10%)	0	0
Moderate increase (10–50%)	−3	↓
Major increase (>50%)	−5	↓↓
2) Loss/gain in wintering habitat under climate change:		
Major losses (>50%)	5	↑↑
Moderate losses (10–50%)	3	↑
Limited or no losses (−10–10%)	0	0
Moderate increase (10–50%)	−3	↓
Major increase (>50%)	−5	↓↓
3) Loss/gain in migration habitat under climate change:		
Major losses (>50%)	5	↑↑
Moderate losses (10–50%)	3	↑
Limited or no losses (−10–10%)	0	0
Moderate increas (10–50%)	−3	↓
Major increase (>50%)	−5	↓↓
4) Degree of dependence on ecological synchronicities:		
High	5	↑↑
Moderate	3	↑
Low	0	0
5) Migration distance (=surrogate for a suite of issues):		
see figure 1 for distance categories	5	↑↑
	4	↑
	3	↑
	2	0
	1	0
6) Degree breeding, wintering or migration habitat specialization		
Highly specialized	5	↑↑
Specialized	4	↑↑
Somewhat specialized	3	↑
Not specialized	0	0

Values are given by scores (similar to the PIF approach) and by arrows. Note that negative scores/down arrows indicate a decreased extinction risk due to climate change. Our current assessment is based on 4 arrows in the same direction (up or down) being sufficient to shift a species to the next risk category.

higher, and no species are considered Not at Risk. Based on our assessments, we categorize 43 taxa (species+races, hereafter 'species' or 'taxa') as High Concern or higher due to increased risks resulting from climate change, with 15 of these being in the newly created Critical category (Table 2).

Of the 52 taxa assessed, 38 (73%) showed increased vulnerabilities due to effects of climate change on breeding habitat, 36 (69%) due to effects on wintering habitat, and 34 (65%) due to migration habitat (Table 2). More taxa also exhibited maximal negative responses (criteria in Table 1) to climate change on the breeding grounds than to winter or migration habitat (24 taxa vs. 19 and 16, respectively). That is, more taxa exhibited increased risk due to climate change on the breeding grounds than for the wintering and migration grounds, and the risks were higher. The number of taxa predicted to have no response or a positive response to climate change was similar across breeding, winter, and migration habitat (13, 15, and 18 taxa respectively). Ecological

synchronicity and migration distance, by comparison, had less of an effect on extinction risk due to climate change, with 17 (33%) and 14 (27%) species, respectively, showing no negative effect due to climate change. The greatest risk factor of those assessed, however, was degree of habitat specialization, with 47 (90%) of the taxa showing a negative response to climate change (Table 2).

A natural potential comparison of our results is with those of the State of the Birds [35]. This is a somewhat difficult comparison to make, however, because we used different scales for our risk categories. However, there appears to be general, qualitative concordance for many species. For example, of the 12 species where they predict no (0 score out of 5) or a low (1 score) increase of extinction risk due to climate change, we predict no or low effects on all of them; i.e., our results leave the species in the same risk category or increase by one category (Table 2). However, we predict an increase of only a single risk category on an additional 19 species where State of the Birds predicts greater impacts of

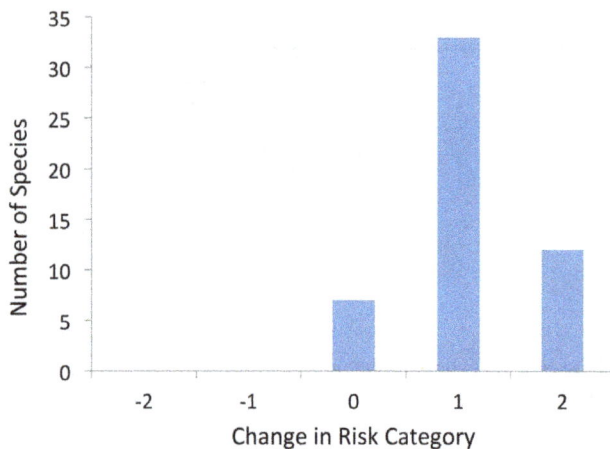

Figure 2. Number of species that we predict will not change U.S. Shorebird Conservation Plan Risk Categories due to climate change (0), and the number that will have increased risk of extinction (positive values); we predicted no species to have reduced risk (negative values). Data are summarized from Table 2 (differences between last two columns).

climate change (scores of ≥ 2). Although there is a lot of variability, our results are generally, but not closely, consistent with those of the State of the Birds ($r^2 = 0.27$). Our biggest difference occurs for the Purple Sandpiper, where we predict no change in risk category due to climate change, while State of the Birds predicts a strong response (score = 4). Although not to the same degree, we also predict substantially lower increases in extinction risk due to climate change for Black Oystercatcher, Wandering Tattler, Bristle–thighed Curlew, Hudsonian Godwit, Surfbird, Western Sandpiper, and Rock Sandpiper (Table 2).

The number of species that change risk categories in our assessment was sensitive to how much of an effect of climate change is required to cause the shift (Table 3; Appendices S3 and S4). When we make it easier to shift categories (3 arrows to change), we are left with only five species in the moderate or lower concern categories and 22 species in the highest (newly created 'critical') risk category, compared to 9 and 15, respectively, when 4 arrows are required to change categories. There is less sensitivity in the other direction. Even when we require 7 arrows to change risk category, we still have 20 species in the highly imperiled or critical risk categories, compared to only 6 when climate change is not considered (Table 3; Appendices S3 and S4). Consequently, one might argue about the most appropriate degree of increased risk required to change risk categories; however, regardless of the threshold used, we conclude that there is an important shift in the numbers of North American shorebirds species at risk of extinction due to climate change.

Discussion

Many species of shorebirds are the focus of conservation efforts aimed at reversing population declines e.g. [9,100], so there is a need to prioritize conservation actions that can have the largest impact on the species most in need. The system currently in use for prioritizing shorebird conservation efforts in the United States was developed in 1999–2001 [1], and did not explicitly include vulnerability to the impacts of a changing climate, e.g. [101]. Many studies have shown that climate change poses risks to populations of plants and animals and that impacts to vulnerable species are already occurring, e.g. [102–104]. It is expected that

such adverse impacts will become more severe and widespread in the future as the climate continues to change. One major application of the system developed in the present study would be to revise the priority scores given to shorebird species by updating the threat scores with the information presented here regarding vulnerability to climate change. We recommend that the U.S. Fish and Wildlife Service revise shorebird priority scores as suggested here, so that the impacts of a changing climate can be more fully integrated into efforts to conserve shorebirds. In addition to applying this information to shorebird species, the same approach could also be applied to other birds. The Partners–in–Flight prioritization system also could be updated to include the approach presented here, if the information on relative risks were collected for other species. This would allow a similar update to reflect vulnerability to climate change across a wide range of bird taxa. We do note that the species assessments and criteria assigned in this manuscript should be considered as first approximations, and will undoubtedly be revised with further discussion by a wider audience. Our primary goal was to establish a system for evaluating the increased risk to species from climate change with respect to existing threat assessments, and to start a discussion about the appropriate values for various species.

Shorebird populations and flyways across the planet are currently being affected by other stressors, many of them unknown, in addition to climate change, e.g., [5]. These impacts are resulting in severe population reductions [1,6–8]. Based on our analyses, adding the stresses and risks imposed by a changing climate to this already threatened baseline renders shorebirds even more vulnerable to extinction. If we are correctly to understand the risks to which shorebirds are exposed, and to identify and implement effective conservation strategies and actions, it is important that we understand these vulnerabilities, particularly those that will occur due to climate change. The purpose of this study was to assess the climate change risks to shorebirds and incorporate these into existing vulnerability evaluations so that we gain a better understanding of the entire panoply of risk factors to which these species are exposed, and their resulting overall vulnerabilities.

Based on our results it appears that shorebirds, as a group, are likely to be highly vulnerable to the changing climate. These vulnerabilities are due to a number of factors. First, many species breed, migrate through, or winter in areas that are likely to be severely impacted by climate change (particularly arctic tundra, coastal breeding, and wintering, and migration stopover sites). Second, the extensive migrations that many of them undertake expose them to risks of changing weather patterns (increased frequencies and intensities of hurricanes, for example) [98]. Shorebirds that require particular staging areas might be more vulnerable to climate change than are those species using stopover sites [95,105]. Lastly, the ecological synchronicities that many shorebirds depend on (e.g., the complementary timing of the arctic snowmelt and invertebrate prey availability) might suffer disruptions [16,21–24]. Our results reflect these vulnerabilities.

Of the 52 shorebird taxa (49 species, 3 split into 2 populations) that breed in North America and that we evaluated, 45 (87%) were predicted to exhibit an increased extinction risk when the risks posed by climate change were added to their current vulnerabilities as estimated in the U.S. Shorebird Conservation Plan [1]. No species was reclassified into a lower–risk category, although prior to the analysis it had been a possibility. The factors responsible for these increased vulnerabilities were risks of: loss of breeding habitat (particularly for arctic– and coastal–breeders); loss of coastal and inland migration stopover habitats due to sea level rise and drought; and loss of coastal wintering habitat due to sea level

Table 2. Results of predicted change in extinction risk to shorebird species based on climate change.

Common name	Scientific name	Breeding habitat	Wintering habitat	Migration habitat	Ecological synchronicity	Migration distance	Degree of habitat specialization	Sum of change (arrows)	State of the Birds 2010[1]	Current risk category	Revised risk category[2]
Black-necked Stilt	*Himantopus mexicanus*	↑↑	↑	↑	0	0	↑↑	6	0	2	3
American Avocet	*Recurvirostra americana*	↑	↑↑	↑↑	0	0	↑	7	1	3	4
American Oystercatcher	*Haematopus palliatus*	↑↑	↑↑	↑↑	0	0	↑↑	8	4	4	6
Black Oystercatcher	*Haematopus bachmani*	↑	↑	↑	0	0	↑↑	5	4	4	5
Black-bellied Plover	*Pluvialis squatarola*	↑↑	↑	↑	↑	↑↑	↑↑	8	3	3	5
American Golden-Plover	*Pluvialis dominica*	↑↑	↑	↑	↑	↑↑	↑	8	2	4	6
Pacific Golden-Plover	*Pluvialis fulva*	↑↑	0	0	↑	0	↑↑	5	3	4	5
Snowy Plover – coastal	*Charadrius nivosus*	↑↑	↑↑	↑↑	0	0	↑↑	8	2	5	6
Snowy Plover – inland		↑	↑	↑↑	0	0	↑↑	6	2	5	6
Wilson's Plover	*Charadrius wilsonia*	↑↑	↑↑	↑↑	0	0	↑↑	8	3	4	6
Semipalmated Plover	*Charadrius semipalmatus*	↑	↑	↑	0	↑	↑↑	7	1	2	3
Piping Plover – coastal	*Charadrius melodus*	↑↑	↑↑	↑↑	0	0	↑↑	8	3	5	6
Piping Plover – inland		↑↑	↑↑	↑	0	0	↑↑	7	3	5	6
Killdeer	*Charadrius vociferous*	0	0	0	0	0	0	0	0	3	3
Mountain Plover	*Charadrius montanus*	↓	↑	↑	0	0	↑↑	3	1	5	5
Spotted Sandpiper	*Actitis macularius*	0	0	0	0	↑	0	1	1	2	2
Solitary Sandpiper	*Tringa solitaria*	↓	↑↑	↑↑	↑	↑	↑↑	7	2	4	5
Wandering Tattler	*Tringa incana*	0	↑	0	↑	↑	↑	4	4	3	4
Greater Yellowlegs	*Tringa melanoleuca*	0	0	0	↑	↑	↑↑	4	1	3	4
Willet – eastern	*Tringa semipalmata*	↑↑	↑↑	↑↑	0	0	↑	7	2	3	4
Willet – western		↑	↑	↑↑	0	0	↑	5	2	3	4
Lesser Yellowlegs	*Tringa flavipes*	↓	↑	↑	↑	↑	↑	4	1	3	4
Upland Sandpiper	*Bartramia longicauda*	0	0	0	0	↑↑	↑	3	2	4	4
Whimbrel	*Numenius phaeopus*	↑↑	↑↑	↑↑	↑	↑	↑	9	3	4	6
Bristle-thighed Curlew	*Numenius tahitiensis*	↓	↑↑	↑↑	↑	↑↑	↑	6	4	4	5
Long-billed Curlew	*Numenius americanus*	↑↑	↑	0	0	0	↑↑	5	2	5	6
Hudsonian Godwit	*Limosa haemastica*	0	↑	0	↑↑	↑↑	0	4	4	4	5
Bar-tailed Godwit	*Limosa lapponica*	↑↑	↑↑	↑↑	↑	↑↑	↑↑	11	4	4	6
Marbled Godwit	*Limosa fedoa*	↑	↑	↑	0	0	↑↑	5	1	4	5
Ruddy Turnstone	*Arenaria interpres*	↑	↑↑	↑↑	↑↑	↑	↑↑	10	3	4	6
Black Turnstone	*Arenaria melanocephala*	↑	↑	↑	↑	0	↑↑	6	4	4	5
Red Knot	*Calidris canutus*	↑	↑↑	↑↑	↑↑	↑	↑↑	10	4	4	6
Surfbird	*Calidris virgata*	↑	↑↑	↑↑	↑↑	↑	↑↑	9	5	4	6
Stilt Sandpiper	*Calidris himantopus*	↑↑	0	0	↑	↑	↑↑	6	3	3	4

Table 2. Cont.

Common name	Scientific name	Breeding habitat	Wintering habitat	Migration habitat	Ecological synchronicity	Migration distance	Degree of habitat specialization	Sum of change (arrows)	State of the Birds 2010[1]	Current risk category	Revised risk category[2]
Sanderling	*Calidris alba*	↑↑	↑↑	↑↑	↑↑	↑	↑↑	11	3	4	6
Dunlin	*Calidris alpina*	↑↑	↑↑	↑	↑	0	↑↑	8	3	3	5
Rock Sandpiper	*Calidris ptilocnemis*	↑	0	0	↑	0	↑↑	4	4	3	4
Purple Sandpiper	*Calidris maritima*	↑	0	0	0	0	↑↑	3	4	2	2
Baird's Sandpiper	*Calidris bairdii*	0	0	0	↑	↑↑	↑↑	5	3	2	3
Least Sandpiper	*Calidris minutilla*	0	↑	↑	↑	↑		5	3	3	4
White–rumped Sandpiper	*Calidris fuscicollis*	↑↑	→	↑	↑	↑↑	↑	6	3	2	3
Buff–breasted Sandpiper	*Calidris subruficollis*	↑↑	0	0	↑	↑↑	↑↑	7	3	4	5
Pectoral Sandpiper	*Calidris melanotos*	↑↑	0	0	↑	↑↑	↑↑	6	3	2	3
Semipalmated Sandpiper	*Calidris pusilla*	↑↑	↑↑	↑	↑↑	↑	↑↑	10	4	3	5
Western Sandpiper	*Calidris mauri*	↑	↑↑	↑	↑	↑	↑	7	4	4	5
Short–billed Dowitcher	*Limnodromus griseus*	0	↑↑	↑↑	↑	↑		7	4	4	5
Long–billed Dowitcher	*Limnodromus scolopaceus*	↑↑	↑	↑	↑	↑		7	3	4	5
Wilson's Snipe	*Gallinago delicata*	0	0	0	0	↑	↑	2	0	3	3
American Woodcock	*Scolopax minor*	↑	→	→	0	0	↑↑	1	1	4	4
Wilson's Phalarope	*Phalaropus tricolor*	↑	↑	↑	0	↑	0	4	3	4	5
Red–necked Phalarope	*Phalaropus lobatus*	↑↑	0	0	↑	↑	↑↑	6	1	3	4
Red Phalarope	*Phalaropus fulicarius*	↑↑	0	0	↑	↑	↑↑	6	3	3	4

Arrows depict extent and direction of change in risk associated with climate change. See Table 1 for description of risk factors and scoring, and see Appendices S1 and S2 for species specific discussion. Also included in current U.S. Shorebird Conservation Plan (USSCP) Risk Categories and State of the Birds vulnerability score and our proposed revised USSCP risk categories based on the added effects of climate change. Our Assessment of change in risk due to climate change is assessed by adding arrows across rows (1 up arrow and 1 down arrow result in a net of 0 arrows), and using the decision rule of 4 arrows (net, up or down) to shift risk categories.
[1]The lower scores, which were not published as part of the State of the Birds, were provided by the U.S. NABCI Committee.
[2]Based on our risk analysis, this would be the new U.S. Shorebird Conservation Plan risk category; we added a new risk category – 6 (Critical, Table 3) – to account for qualitatively greater risk than the current U.S. Shorebird Conservation Plan assessment allows.

Table 3. Results of sensitivity analysis of risk categorization for shorebird species.

Risk Category	Current	Projected under with climate change, from most (7) to least (3) conservative transition criterion				
		7 ↑	6 ↑	5 ↑	4 ↑	3 ↑
Not at risk	0	0	0	0	0	0
Low concern	7	6	3	2	2	1
Moderate concern	16	12	12	11	7	4
High concern	23	14	13	13	13	11
Highly imperiled	6	17	18	17	15	14
Critical	–[1]	3	6	9	15	22

What is shown in the first column of results is the current distribution of taxa across risk categories by the U.S. Shorebird Conservation Plan (USSCP). The columns that follow are the predicted distributions under different criteria for changing risk category. In Table 2 we assume that the accumulation of 4 arrows across risk factors is sufficient for a species to change risk category; this table shows the sensitivity of this result using more liberal (3 arrows) and more conservative (5, 6, and 7 arrows) criteria for changing risk category. We added a new risk category to those used by Partners-in-Flight (PIF) and the USSCP, Critical, to account for species being at categorically greater risk than previously considered. (See Table 2 and Appendices S3 and S4 for species–specific assessments and summaries.)
[1]Category does not exist in current PIF framework.

rise. Of particular note, for high-Arctic breeders, there is minimal latitude and land for northward range expansion. Extreme weather events were also projected to increase vulnerabilities due to negative effects on habitat, migration mortality, and disruption of ecological synchronicities, e.g. [94].

The increased vulnerabilities of 10 species could not be accommodated using the existing PIF scoring system and we had to create an even higher level of risk than is currently available. These Critical species (including coastal Snowy, Piping, and Wilson's plovers, and inland species such as Mountain Plover and Long–billed Curlew) are already at a high risk level due to other stressors (particularly anthropogenic habitat destruction) and their populations are already declining and jeopardized [32]. The addition of climate change to their risk factors raises them to an even higher level of vulnerability, which may pose even higher threats to their continued existence.

Also of concern is that the addition of climate change to the vulnerability calculations elevates another 18 species to the highest U.S. Shorebird Conservation Plan risk category. Thus, a total of 28 of 49 species are now at the highest risk category under the U.S. Shorebird Conservation Plan, or they exceeded this risk level and had to have an additional category created. The degree to which species changed risk categories was sensitive to our rules of category change. To some degree, as with population viability analyses using stochastic simulation models, which rule we use for category change is a value judgment [106]. Regardless of what rules are used, however, our analysis suggests that shorebirds will have increased vulnerability under climate change, perhaps to a large extent.

Our assessment of extinction risk might be criticized because it does not allow for adaptive capacity in shorebird populations. That is, shorebirds might modify their breeding, migratory, and/or wintering habitat use, foraging, and/or timing to accommodate the changing climate. We already know that some shorebirds in Western Europe have apparently truncated their fall migrations to winter in the Baltic, rather than in oceanic Atlantic countries, such as the UK [104]. Previously, the winter conditions in the Baltic were so harsh that birds had to move farther to exploit the milder conditions of the UK, Holland, etc. Thus, the ameliorating winter conditions in the Baltic have encouraged changes in migration distance [104]. Similarly, In North America, some migratory populations of Hudsonian Godwits have advanced their timing of migration during warm periods, which allows their breeding to

synchronize with peak food abundance, while other populations have not [107]. There also is some evidence that Semipalmated and Pectoral sandpipers and Red–necked and Red phalaropes have been observed breeding earlier during warm years [108]. As another example of adapting to changing conditions, Dunlin nestlings can exhibit accelerated growth during periods of low food availability during warm conditions [109]. However, it would be unwise of us to assume that such adaptive capacities were likely to apply across all shorebird species because there is evidence that high Arctic shorebird species may have little capacity for adaptation due to low genetic variability resulting from bottleneck events from previous climate shifts [110]. Time constraints can also cause conflicts among competing life–history requirements, as has been reported in Pied Flycatchers *Ficedula hypoleuca* [111]. Clearly more research needs to be done to determine the degree to which climate adaptation might occur in shorebirds.

What would it take to accurately and precisely predict change in extinction risk due to climate change for migratory shorebirds, or for any species, rather than taking the relatively coarse approach we did in this paper? Certainly there have been detailed assessments of expected regional changes in shorebird populations in response to climate change [112–113], and one could create models to link species to landscapes via simulation. But what would be required for accurate, reliable predictions? Strictly speaking, to build a convincing case for an accurate prediction, the first thing we would need is accurate models of climate change. Although there are many models of climate change, and they agree in general with climate trends, there is still a great deal of uncertainty in the exact amount of changes in expected temperature and precipitation, e.g. [114–115], particularly at the fine geographic scale that would be needed to understand biotic responses, including the effects of changes in wind patterns [116]. Because hydrological models are complex (i.e., non–linear, with feedback and chaotic dynamics), more accurate data are unlikely to improve model predictions [117]. In addition, accurate regional and local downscaling of global climate models might not be possible [118–120].

The next requirement is accurate models linking climate change to hydrologic responses, so we could accurately determine changes in hydrology, amount of sea level rise, the degree to which plant communities will change in response to climate change, in both inland and coastal regions,. Accurate models that allow these

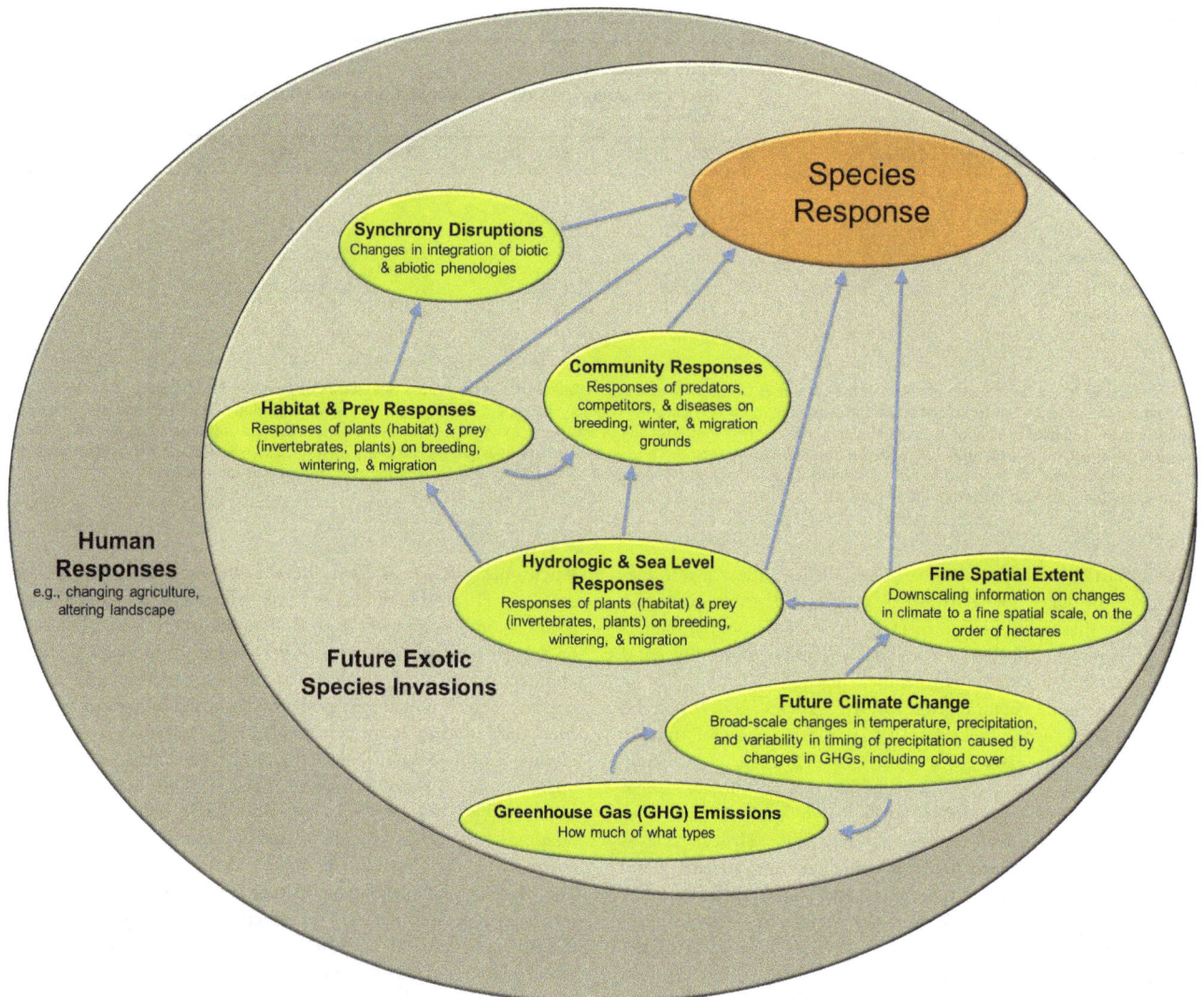

Figure 3. Digraph showing relationships (arrows) for which we need accurate information in order to accurately predict species-specific shorebird responses to climate change. By accurate, we mean variation explained between nodes is >90% or near that, not merely determining statistically significant relationships. Subheadings specify the relationships, and 'species response' includes adaptive responses as well as non-adaptive responses. 'Fine spatial extent' refers to downscaling climate change estimates to the spatial scale at which species respond; factors at this scale affect species' responses directly and indirectly. The digraph is nested within the contexts of future introductions of exotic, invasive species, and human responses to climate change to indicate that all of the relationships from the digraph can be affected by these particular occurrences or responses.

predictions do not exist [117,121–123]. Even if we had accurately developed models, we also would need accurate assessments of species' ranges as well as niche–based models for each species we want to evaluate that accurately predicts, with a very high level of variability in distribution explained, the distribution of species, cf. [106]. We do not yet have these, e.g. [124–127], and it is not clear to what extent or rate different bird might respond behaviorally to climate change [128]. Finally, we need models that accurately depict community–wide biotic responses to climate change, including accurate anticipation of inter–specific interactions, how local species invasions and extinctions will affect resource availability, how they might change as niches shift [129–133]. We do not have these either, and we might be unlikely to accurately anticipate shifting realized niches for a variety of practical reasons [134–136]. These challenges are exacerbated by migration because the relationships must be known in breeding,

non–breeding, and migration habitats [137]. These relationships we just described are depicted in Fig. 3. Even the highly restrictive requirements we just presented might ultimately be insufficient, because they do not take into account human responses to climate change. For example, what will be the human responses in changes to agricultural practices, relocation away from coastal areas, and so–called adaptive response measures, e.g. [138–140], and how will they affect the capacity for ecosystems and shorebirds to respond?

Consequently, we suspect that detailed regional and local biological forecasting of the effects of climate change, even if the correct (but currently unknown) IPCC scenario is selected, is likely to be only generally accurate. Therefore, we think that the relatively coarse assessment of changes in extinction risk that we present here is a useful level of assessment for species at a continental scale; see [31] for another example of a categorical risk assessment at a smaller geographic scale. We stress that the somewhat bleak picture we paint regarding prediction accuracy at small spatial scales should not be used as an excuse to not make models or predictions, or to avoid planning for climate change. Rather, we encourage model development and testing, followed by model revision as more data become available. As with all models, we suggest treating the structure, parameter values, and predictions as hypotheses to test. We also support alternative modeling approaches that might be effective at accommodating model uncertainty, such as robust decision-making [141].

Supporting Information

Appendix S1 Vulnerability scores and associated confidence levels for 49 North American breeding shorebird species

Appendix S2 Degree of habitat specialization described for each of the taxa

Appendix S3 Sensitivity analysis of risk category in which shorebirds are placed

Appendix S4 Species in each of the risk categories under the current system, and revised based on climate change.

Acknowledgments

We thank B. Harrington and N. Warnock for many discussions about shorebird ecology and for commenting on an earlier draft of this manuscript; B. Tavernia for his ArcGIS work to determine migration distances; A. Madden for presentation advice on figure 3; B. Andres for sharing U.S. NABCI Committee unpublished data; and three anonymous reviewers provided useful input on an earlier draft of this manuscript. Data provided by NatureServe in collaboration with Robert R., J. Zook, The Nature Conservancy – Migratory Bird Program, Conservation International – CABS, World Wildlife Fund – US, and Environment Canada – WILDSPACE.

Author Contributions

Conceived and designed the experiments: HG DWD SB JMR. Performed the experiments: HG DWD SB JMR. Analyzed the data: HG DWD SB JMR. Contributed reagents/materials/analysis tools: HG DWD SB JMR. Wrote the paper: HG DWD SB JMR.

References

1. Brown S, Hickey C, Harrington B, Gill R (2001) United States shorebird conservation plan, 2nd ed. Manomet: Manomet Center for Conservation Sciences. 61 p.
2. Warnock N, Elphick C, Rubega MA (2002). Shorebirds in the marine environment. In: Schreiber EA, Burger J, editors. Biology of marine birds. Boca Raton: CRC Press. pp. 581–615.
3. Piersma T, Lindström Å (2004) Migrating shorebirds as integrative sentinels of global environmental change. Ibis 146(S), 61–69.
4. Russel BD, Harley CDG, Wernberg T, Mieszkowska N, Widdicombe S, et al. (2012) Predicting ecosystem approaches that integrate the effects of climate change across entire systems. Biology Letters 8: 164–166.
5. International Wader Study Group (2003) Waders are declining worldwide. Conclusions from the 2003 International Wader Study Group Conference, Cadiz, Spain. Wader Study Group Bulletin 101/102: 8–12.
6. Morrison RIG, McCaffery BJ, Gill RE, Skagen SK, Jones SL, et al. (2006) Population estimates of North American Shorebirds, 2006. Wader Study Group Bulletin 111: 67–85.
7. Thomas GH, Lanctot RB, Székely T (2006) Can intrinsic factors explain population declines in North American breeding shorebirds? A comparative analysis. Animal Conservation 9: 252–258.
8. Bart J, Brown S, Harrington B, Morrison RIG (2007) Survey trends of North American shorebirds: population declines or shifting distributions? Journal of Avian Biology 38: 73–82.
9. Morrison RIG, Ross PK, Niles LJ (2005) Declines in wintering populations of Red Knots in southern America. Condor 106: 60–70.
10. Federal Register (2006) Vol. 71, No. 176. Tuesday, September 12, Proposed Rules.
11. Jehl J (2007) Disappearance of breeding Semipalmated Sandpipers from Churchill, Manitoba: More than a local phenomenon. Condor 109: 351–360.
12. Morrison RIG, Mizrahi DS, Ross RK, Ottema OH, De Pracontal N, et al. (2012) Dramatic declines of Semipalmated Sandpipers on their major wintering areas in the Guianas, northern South America. Waterbirds 35: 120–134.
13. US Fish and Wildlife Service (2011) Available: http://ecos.fws.gov/tess_public/SpeciesReport.do?groups=B&listingType=L&mapstatus=1, Accessed 2011 Jun 22.
14. International Union for the Conservation of Nature (2013) IUCN Red List of Threatened Species. Version 2012.2. Available: http://www.iucnredlist.org. Accessed 2013 Jan 13.
15. Howe MA, Geissler PH, Harrington BA (1989) Population trends of North American shorebirds based on the International Shorebird Survey. Biological Conservation 49: 185–199.
16. Gratto-Trevor CL (1997) Climate change: proposed effects on shorebird habitat, prey, and numbers in the Outer Mackenzie Delta. Mackenzie Basin Impact Study Final Report, pp. 205–210.
17. Austin GE, Rehfisch MM (2003) The likely impact of sea level rise on waders (Charadrii) wintering on estuaries. Journal of Nature Conservation 11: 43–58.
18. Milton D (2003) Threatened shorebird species of the East Asian–Australasian Flyway: significance for Australian wader study groups. Wader Study Group Bulletin 100: 105–110.

19. Stroud DA, Davidson NC, West R, Scott DA, Haanstra L, et al, compilers(2004) Status of migratory wader populations in Africa and Western Eurasia in the 1990s. International Wader Studies 15: 1–259.
20. Parry ML, Canziani OF, Palutikof JP, van der Linden PJ, Hanson CE, editors(2007) Contribution of Working Group II to the Fourth Assessment Report of the Intergovernmental Panel on Climate Change, 2007. Cambridge University Press, Cambridge, United Kingdom and New York, NY, US.
21. Galbraith H, Jones R, Park R, Clough J, Herrod–Julius S et al. (2002) Global climate change and sea level rise: potential losses of intertidal habitat for shorebirds. Waterbirds 25: 173–183.
22. Hedenström A, Barta Z, Helm B, Houston AI, McNamara JM, et al. (2007) Migration speed and scheduling of annual events by migrating birds in relation to climate change. Climate Res 35: 79–91.
23. Colwell MA (2010) Shorebird Ecology, Conservation, and Management. Berkeley: Univ. of California Press. 344 p.
24. Small–Lorenz SL, Culp LA, Ryder TB, Will TC, Marra PP (2013) A blind spot in climate change vulnerability assessments. Nature Climate Change 3: 91–93.
25. Lee TM, Jetz W (2008) Future battlegrounds for conservation under global change. Proceedings of the Royal Society, B 275: 1261–1270.
26. Galbraith H, Price J (2009) A Framework for Categorizing the Relative Vulnerability of Threatened and Endangered Species to Climate Change. U.S. Environmental Protection Agency. EPA/600/R–09/011. Available: http://cfpub.epa.gov/ncea/cfm/recordisplay.cfm?deid=203743# Accessed 2011 Jun 20.
27. Lee TM, Jetz W (2011) Unravelling the structure of species extinction risk for predictive conservation science. Proceedings of the Royal Society, B 278: 1329–1338.
28. Millsap BA, Gore JA, Runde DR, Cerulean CI (1990) Setting priorities for the conservation of fish and wildlife species in Florida. Wildlife Monographs 111.
29. Herman TB, Scott FW (1994) Protected areas and global climate change: assessing the regional or local vulnerability of vertebrate species. In: Pernetta JC, Leemans R, Elder D, Humphrey S, editors Impacts of climate change on ecosystems and species: implications for protected areas. Gland, Switzerland: IUCN. pp. 13–27.
30. Reed JM (1995) Relative vulnerability to extirpation of montane breeding birds in the Great Basin. Great Basin Naturalist 55: 342–351.
31. Gardali T, Seavy NE, DiGaudio RT, Comrack LA (2012) A climate change vulnerability assessment of California's at–risk birds. PLoS ONE 7(3): e29507. doi:10.1371/journal.pone.0029507
32. Panjabi AO, Dunn EH, Blancher PJ, Hunter WC, Altman B, et al. (2005) The Partners in Flight handbook on species assessment. Version 2005. Partners in Flight Technical Series No. 3. Rocky Mountain Bird Observatory website, Available: http://www.rmbo.org/pubs/downloads/Handbook2005.pdf
33. Higdon JW, MacLean DA, Hagan JM, Reed JM (2006) Risk of extirpation for vertebrate species on an industrial forest in New Brunswick, Canada: 1945, 2002, and 2027. Canadian Journal of Forest Research 36: 467–481.
34. Elphick CS, Roberts DL, Reed JM (2010) Estimated dates of recent extinctions for North American and Hawaiian birds. Biological Conservation 143: 617–624.

35. North American Bird Conservation Initiative, U.S. Committee (2010) The State of the Birds 2010 Report on Climate Change, United States of America. Washington, DC: U.S. Department of the Interior.

36. Intergovernmental Panel on Climate Change (2007) Climate Change 2007: Mitigation. Contribution of Working Group III to the Fourth Assessment Report of the Intergovernmental Panel on Climate Change. Metz B, Davidson OR, Bosch PR, Dave R, Meyer, LA, editors. Cambridge, UK: Cambridge University Press.

37. Chapin FS III, Shaver GR, Giblin AE, Nadelhoffer KG, Laundre JA (1995) Response of arctic tundra to experimental and observed changes in climate. Ecology 76: 694–711.

38. Moss RH, Schneider SH (2001) Towards Consistent Assessment and Reporting of Uncertainties in the IPCC TAR. In: Pachauri R, Taniguchi T, editors. Cross–Cutting Issues in the IPCC Third Assessment Report. Global Industrial and Social Progress Research Institute (for IPCC). Tokyo.

39. Keeling CD, Chin JFS, Whorf TP (1996) Increased activity of northern vegetation inferred from atmospheric CO_2 measurements. Nature 382: 146–149.

40. Zhou L, Tucker CJ, Kaufmann RK, Slayback D, Shabanov NV, et al. (2001) Variation in northern vegetation activity inferred from satellite data of vegetation index during 1981–1999. Journal Geophysical Research–Atmospheres 106(D17), 20069–20083.

41. Lucht W, Prentice IC, Myneni RB, Sitch S, Friedlingstein P, et al. (2002) Climatic control of the high–latitude vegetation greening trend and Pinatubo effect. Science 296: 1687–1689.

42. Arctic Climate Impact Assessment (2004) Impacts of a Warming Arctic. Arctic Climate Impact Assessment. Cambridge: Cambridge University Press. 144 p.

43. Sturm M, Schimel J, Michaelson G, Welker JM, Oberbauer SF, et al. (2005) Winter biological processes could help convert arctic tundra to shrubland. Bioscience 55: 17–26.

44. Karl T, Melillo M, Peterson T (2009) Global Climate Change Impacts in the United States. Cambridge University Press. 192 p.

45. Bhatt US, Walker DA, Raynolds MK, Comiso JC, Epstein HE, et al. (2010) Circumpolar Arctic tundra vegetation change is linked to sea ice decline. Earth Interactions 14: 1–20.

46. Elmendorf SC, Henry GHR, Hollister RD, Björk RG, Boulanger–Lapointe N, et al. (2012) Plot–scale evidence of tundra vegetation change and links to recent summer warming. Nature Climate Change 2: 453–457.

47. Raynolds MK, Walker DA, Epstein HE, Pinzon JE, Tucker CJ (2012) A new estimate of tundra–biome phytomass from trans–Arctic field data and AVHRR NDVI. Remote Sensing Letters 3: 403–411.

48. Martin PD, Jenkins JL, Adams FJ, Jorgenson MT, Matz AC, et al. (2009) Wildlife response to environmental arctic change: predicting future habitats of arctic Alaska. Report of the Wildlife Response to Environmental Arctic Change (WildREACH): Predicting Future Habitats of Arctic Alaska Workshop, 17–18 November 2008. Fairbanks, Alaska: U.S. Fish and Wildlife Service. 138 p.

49. Karlsson JM, Bring A, Peterson GD, Gordon LJ, Destouni G (2011) Opportunities and limitations to detect climate–related regime shifts in inland Arctic ecosystems through eco–hydrological monitoring. Environmental Research Letters 6: 1–9.

50. MacDonald LA, Turner KW, Balasubramaniam AM, Wolfe BB, Hall RI, et al. (2012) Tracking hydrological responses of a thermokarst lake in the Old Crow Flats (Yukon Territory, Canada) to recent climate variability using aerial photographs and paleolimnological methods. Hydrological Processes 26: 117–129.

51. Assini J, Young KL (2012) Snow cover and snowmelt of an extensive High Arctic wetland: spatial and temporal seasonal patterns. Hydrological Sciences Journal 57: 738–755.

52. Prowse T (2012) Lake and river ice in Canada. In: French H, Slaymaker O, editors. Changing cold environments: a Canadian perspective. Sussex: Wiley–Blackwell. pp. 163–181

53. Woo M–K, Pomeroy J (2012) Snow and runoff: Processes, sensitivity and vulnerability. In: French H, Slaymaker O, editors. Changing cold environments: a Canadian perspective. Sussex: Wiley–Blackwell. pp. 105–125

54. Sheffield J, Wood WF (2008) Projected changes in drought occurrence under future global warming from multi–model, multi–scenario, IPCC AR4 simulations. Climate Dynamics 31: 79–105.

55. Johnson WC, Werner B, Guntenspergen GR, Voldseth RA, Millett B, et al. (2010) Prairie wetland complexes as landscape functional units in a changing climate. BioScience 60: 128–140.

56. Strzepek K, Yohe G, Neumann J, Boehlert B (2010) Characterizing changes in drought risk for the United States from climate change. Environmental Research Letters 5: 1–9.

57. Dai A (2011) Drought under global warming: a review. Wiley Interdisciplinary Reviews: Climate Change 2: 45–65.

58. Rahmstorf S (2007) A Semi–Empirical Approach to Projecting Future Sea–Level Rise. Science 315: 368–370.

59. Pfeffer WT, Harper JT, O'Neel SO (2008) Kinematic Constraints on Glacier Contributions to 21st–Century Sea–Level Rise. Science 321: 1340–1343.

60. Mcleod E, Poulter B, Hinkel J, Reyes E, Salm R (2010) Sea–level rise impact models and environmental conservation: A review of models and their applications. Ocean & Coastal Management 53: 507–517.

61. Menon S, Soberón J, Li X, Peterson AT (2010) Preliminary global assessment of biodiversity consequences of sea level rise mediated by climate change. Biodiversity and Conservation 19: 1599–1609.

62. Shaw J, Taylor RB, Forbes DL, Ruz MH, Solomon S (1998). Sensitivity of the coasts of Canada to sea–level rise. Ottawa: Geological Survey of Canada, Report No. 505. 79 p.

63. Boruff BJ, Emrich C, Cutter SL (2005) Erosion hazard vulnerability of U.S. coastal counties. Journal of Coastal Research 21: 932–942.

64. Nicholls RJ, Wong PP, Burkett VR, Codignotto JO, Hay JE, et al. (2007) Coastal systems and low–lying areas. In: Parry ML, Canziani OF, Palutikof JP, van der Linden PJ, Hanson CE, editors. Climate change 2007: impacts, adaptation and vulnerability. Contribution of Working Group II to the Fourth Assessment Report of the Intergovernmental Panel on Climate Change. Cambridge: Cambridge University Press. pp. 315–356.

65. Strauss BJ, Ziemlinski R, Weiss JL, Overpeck JT (2012) Tidally adjusted estimates of topographic vulnerability to sea level rise and flooding for the contiguous United States. Environmental Research Letters 7: 014033.

66. Chu–Agor ML, Muñoz–Carpena R, Kiker G, Emanuelsson A, Linkov I (2011) Exploring vulnerability of coastal habitats to sea level rise through global sensitivity and uncertainty analyses. Environmental Modelling & Software 26: 593–604.

67. Stralberg D, Brennan M, Callaway JC, Wood JK, Schile LM, et al. (2011) Evaluating tidal marsh sustainability in the face of sea–level rise: A hybrid modeling approach applied to San Francisco Bay. PLoS One 6(11): e27388. doi:10.1371/journal.pone.0027388.

68. Titus JG, Hudgens DE, Trescott DL, Craghan M, Nuckols WH, et al. (2009) State and local governments plan for development of most land vulnerable to rising sea level along the US Atlantic coast. Environmental Research Letters 4: 044008.

69. Seavey JR, Gilmer B, McGarigal KM (2011) Effect of sea–level rise on piping plover (Charadrius melodus) breeding habitat. Biological Conservation 144: 393–401.

70. Magrin G, Gay García C, Cruz Choque D, Giménez JC, Moreno AR, et al. (2007) Latin America. In: Parry ML, Canziani OF, Palutikof JP, van der Linden PJ, Hanson CE, editors. Climate change 2007: impacts, adaptation and vulnerability. Contribution of Working Group II to the Fourth Assessment Report of the Intergovernmental Panel on Climate Change. Cambridge: Cambridge University Press. pp. 581–615.

71. Malhi Y, Aragão LEOC, Galbraith D, Huntingford C, Fisher R, et al. (2009) Exploring the likelihood and mechanism of a climate–change–induced dieback of the Amazon rainforest. Proceedings of the National Academy of Sciences USA 106: 20610–20615.

72. Lola da Costa AC, Galbraith D, Almeida S, Portela BTT, da Costa M, et al. (2010) Effect of 7 yr of experimental drought on vegetation dynamics and biomass storage of an eastern Amazonian rainforest. New Phytologist 187: 579–591.

73. Hirota M, Nobre C, Oyama MD, Bustamante MMC (2010) The climatic sensitivity of the forest, savanna and forest–savanna transition in tropical South America. New Phytologist 187: 707–719.

74. Keppie D, Whiting JRM (1994) American Woodcock (Scolopax minor). The Birds of North America Online, Poole A, editor. Ithaca: Cornell Lab of Ornithology; Retrieved from the Birds of North America Online: http://bna.birds.cornell.edu/bna/species/100.

75. Flinn K, Vellend M (2005) Recovery of forest plant communities in post–agricultural landscapes. Frontiers in Ecology and the Environment 3: 243–250.

76. Adams HD, Guardiola–Claramonte M, Barron–Gafford GA, Villegas JC, Breshears DD, et al. (2009) Temperature sensitivity of drought–induced tree mortality: implications for regional die–off under global–change type drought. Proceedings of the National Academy of Sciences USA 106: 7063–7066.

77. Allen CD, Macalady AK, Chenchouni H, Bachelet D, McDowell N, et al. (2010) A global overview of drought and heat–induced tree mortality reveals emerging climate change risks for forests. Forest Ecology and Management 259: 660–684.

78. Jactel H, Petit J, Desprez–Loustau M–L, Delzon S, Piou D, et al. (2012) Drought effects on damage by forest insects and pathogens: a meta–analysis. Global Change Biology 18: 267–276.

79. Raven J, Caldeira K, Elderfield H, Hoegh–Guldberg O, Liss P, et al. (2005) Ocean Acidification due to Increasing Atmospheric Carbon dioxide. Policy Document 12/05. London: Royal Society. 60 p.

80. Flynn KJ, Blackford JC, Baird ME, Raven JA, Clark DR, et al. (2012) Changes in pH at the exterior surface of plankton with ocean acidification. Nature Climate Change 2: 510–513.

81. Blackford JC (2010) Predicting the impacts of ocean acidification: Challenges from an ecosystem perspective. Journal of Marine Systems 81: 12–18.

82. Milligan AJ (2012) Plankton in an acidified ocean. Nature Climate Change 2: 489–490.

83. Fabry VJ, Seibel BA, Feely RA, Orr JC (2008) Impacts of ocean acidification on marine fauna and ecosystem processes. ICES Journal of Marine Science 65: 414–432.

84. Taylor AR, Chrachri A, Wheeler G, Goddard H, Brownlee C (2011) A voltage–gated H^+ channel underlying pH homeostasis in calcifying coccolithophores. PLoS Biology 9e1001085. doi:10.1371/journal.pbio.1001085.

85. Snyder MA, Sloan LC, Diffenbaugh NS, Bell JL (2003) Future climate change and upwelling in the California Current. Geophysical Research Letters 30: 8-1-8-4.
86. Sweeney BW, Jackson JK, Newbold JD, Funk DH (1992) Climate change and the life histories and biogeography of aquatic insects in eastern North America. In: Firth P, Fisher SG, editors. Global change and freshwater ecosystems. Springer. pp. 143–176
87. Hodkinson ID, Webb NR, Bale JS, Block W, Coulson SJ, et al. (1998) Global change and Arctic ecosystems: Conclusions and predictions from experiments with terrestrial invertebrates on Spitsbergen. Arctic and Alpine Research 30: 306–313.
88. Tulp I, Schekkerman H (2008) Has prey availability for Arctic birds advanced with climate change? Hindcasting the abundance of tundra arthropods using weather and seasonal variation. Arctic 61: 48–60.
89. Castro G, Myers JP (1993) Shorebird predation on eggs of horseshoe crabs during spring stopover on Delaware Bay. The Auk 110: 927–930.
90. Baker AJ, Gonzalez PM, Piersma T, Niles LJ, De Serrano Do Nascimento I (2004) Rapid population decline in Red Knots: fitness consequences of decreased refuelling rates and late arrival in Delaware Bay. Proceedings of the Royal Society of London B 271: 875–882.
91. Mizrahi DS, Peters KA (2009) Relationships between sandpipers and horseshoe crab in Delaware Bay: A synthesis. In: Tanacredi JT, et al. editors. Biology and Conservation of Horseshoe Crabs, DOI 10.1007/978-0-387-89959-6_4, Springer.
92. Jones T, Cresswell W (2010) The phenology mismatch hypothesis: are declines of migrant birds linked to uneven global climate change? Journal of Animal Ecology 79: 98–108.
93. Both C, Van Turnhout CAM, Bijlsma RG, Siepel H, Van Strien AJ, et al. (2010) Avian population consequences of climate change are most severe for long-distance migrants in seasonal habitats. Proceedings of the Royal Society B 1685: 1259–1266.
94. Saino N, Ambrosini R, Rubolini D, von Hardenberg J, Provenzale A, et al. (2011) Climate warming, ecological mismatch at arrival and population decline in migratory birds. Proceedings of the Royal Society B 278: 835–842.
95. Iwamura T, Possingham HP, Chadès I, Minton C, Murray NJ, et al. (2013) Migratory connectivity magnifies the consequences of habitat loss from sea-level rise for shorebird populations. Proceedings of the Royal Society B 280: 20130325.
96. Convertino M, Bockelie A, Kiker GA, Muñoz-Carpena R, Linkov I (2012) Shorebird patches as fingerprints of fractal coastline fluctuations due to climate change. Ecological Processes 1:9.
97. Robinson RA, Crick HQP, Learmoth JA, Maclean IMD, Thomas CD (2009) Travelling through a warming world: climate change and migratory species. Endangered Species Research 7: 87–99.
98. Klaassen M, Hoye BJ, Nolet BA, Buttemer WA (2012) Ecophysiology of avian migration in the face of current global hazards. Philosophical Transactions of the Royal Society B 367: 1719–1732.
99. Ridgely RS, Allnutt TF, Brooks T, McNicol DK, Mehlman DW, et al. (2007) Digital distribution maps of the birds of the western hemisphere, version 3.0. Arlington: NatureServe.
100. Nebel S, Porter JL, Kingsford RT (2008) Long-term trends of shorebird populations in eastern Australia and impacts of freshwater extraction. Biological Conservation 141: 971–980.
101. Godet L, Jaffré M, Devictor V (2011) Waders in winter: long-term changes of migratory bird assemblages facing climate change. Biology Letters 7: 714–717.
102. Parmesan C, Galbraith H (2004) Observed ecological impacts of climate change in North America. Arlington: Pew Center for Global Climate Change. Available: http://www.c2es.org/publications/observed-impacts-climate-change-united-states.
103. Parmesan C (2006) Ecological and evolutionary responses to recent climate change. Annual Review of Ecology and Systematics 37: 637–669.
104. MacLean IMD, Austin GE, Rehfisch MM, Blew J, Crowe O, et al. (2008) Climate change causes rapid changes in the distribution and site abundance of birds in winter. Global Change Biology 14: 2489–2500.
105. Warnock N (2010) Stopover vs. staging: the difference between a hop and a jump. Journal of Avian Biology 41: 621–626.
106. Reed JM, Akçakaya HR, Burgman M, Bender D, Beissinger S., et al. (2006) Critical habitat. In: Scott JM, Goble DD, Davis FW, editors. The Endangered Species Act at thirty: conserving biodiversity in human-dominated landscapes. Washington D.C.: Island Press. pp. 164–177
107. Senner NR (2012) One species but two patterns: Populations of the Hudsonian Godwit (Limosa haemastica) differ in spring migration timing. The Auk 129: 670–682.
108. Liebezeit JR, Gurney KEB, Budde M, Zack S, Ward D (2014) Phenological advancement in arctic bird species: relative importance of snow melt and ecological factors. Polar Biology in press, 10.1007/s00300-014-1522-x.
109. McKinnon L, Nol E, Juillet C (2013) Arctic-nesting birds find physiological relief in the face of trophic constraints. Scientific Reports 3:1816 DOI: 10.1038/srep01816
110. Meltofte H, Piersma T, Boyd H, McCaffery B, Ganter B, et al. (2007) Effects of climate variation on the breeding ecology of Arctic shore birds. Danish Polar Center, Meddelelser om Grønland Bioscience 59, Copenhagen, 1–48.
111. Both C, Visser ME (2001) Adjustment to climate change is constrained by arrival date in a long-distance migrant bird. Nature 411:296–298.
112. Rehfisch MM, Austin GE, Freeman SN, Armitage MJS, Burton NHK (2004) The possible impact of climate change on the future distributions and numbers of waders on Britain's non-estuarine coast. Ibis 146 (Suppl. 1): 70–81.
113. Aiello-Lammens ME, Chu-Agor ML, Convertino M, Fischer RA, Linkov I, et al. (2011) The impact of sea-level rise on Snowy Plovers in Florida: integrating geomorphological, habitat, and metapopulation models. Global Change Biology 17: 3644–3654.
114. Araújo MB, Whittaker RJ, Ladle RJ, Erhard M (2005) Reducing uncertainty in projections of extinction risk from climate change. Global Ecology and Biogeography 14: 529–538.
115. Davidson EA, Janssens IA (2006) Temperature sensitivity of soil carbon decomposition and feedbacks to climate change. Nature 440: 165–173.
116. Gill R Jr, Tibbitts T, Douglas D, Handel C, Mulcahy D, et al. (2009) Extreme endurance flights by landbirds crossing the Pacific: ecological corridor rather than barrier? Proceeding of the Royal Society B 276: 447–457.
117. Vano JA, Udall B, Cayan DR, Overpeck JT, Brekke LD, et al. (2014) Understanding uncertainties in future Colorado River streamflow. Bulletin of the American Meteorological Society January, 59–78.
118. Anagnostopolos GG, Koutsoyiannis D, Christofides A, Efstratiadis A, Mamassis N (2010) A comparison of local and aggregated climate model outputs with observed data. Hydrological Sciences Journal 55:1094–1110.
119. Kundzewicz ZW, Stakhiv EZ (2010) Are climate models "ready for prime time" in water resources management applications, or is more research needed? Hydrological Sciences Journal 55:1085–1089.
120. Wilby RL (2010) Evaluating climate model outputs for hydrological applications. Hydrological Sciences Journal 55:1090–1093.
121. He Y, Wetterhall F, Bao H, Cloke H, Li Z, et al. (2010) Ensemble forecasting using TIGGE for the July–September 2008 floods in the Upper Huai catchment: a case study. Atmospheric Science Letters 11: 132–138.
122. Convertino M, Welle P, Muñoz-Carpena R, Kiker GA, Chu-Agor MaL., et al. (2012) Epistemic uncertainty in predicting shorebird biogeography affected by sea-level rise. Ecological Modelling 240: 1–15.
123. O'Gorman PA (2012) Sensitivity of tropical precipitation extremes to climate change. Nature Geoscience 5: 697–700.
124. Pearson RG, Dawson TP (2003) Predicting the impacts of climate change on the distribution of species: are bioclimate envelope models useful? Global Ecology & Biogeography 12: 361–371.
125. Araújo MB, Pearson RG, Thuillers W, Erhard M (2005) Validation of species-climate impact models under climate change. Global Change Biology 11: 1504–1513.
126. Rotenberry JT, Wiens JA (2009) Habitat relations of shrubsteppe birds: A 20-year retrospective. Condor 111: 401–413.
127. Essl F, Rabitsch W, Dullinge S, Moser D, Milasowszky N (2013) How well do we know species richness in a well-known continent? Temporal patterns of endemic and widespread species descriptions in the European fauna. Global Ecology and Biogeography 22: 29–39.
128. Devictor V, Julliard R, Couvet D, Jiguet F (2008) Birds are tracking climate warming, but not fast enough. Proceedings of the Royal Society B 275: 2743–2748.
129. Smith PA, Gilchrist G, Forbes MR, Martin J-L, Allard K (2010) Inter-annual variation in the breeding chronology of arctic shorebirds: effects of weather, snow melt and predators. Journal of Avian Biology 41: 292–304.
130. Pe'er G, Henle K, Dislich C, Frank K (2011) Breaking functional connectivity into components: a novel approach using an individual-based model, and first outcomes. PLoS ONE 6. doi:10.1371/journal.pone.0022355
131. Schmidt NM, Ims RA, Høye TT, Gilg O, Hansen LH, et al. (2012) Response of an arctic predator guild to collapsing lemming cycles. Proceedings of the Royal Society, B 279: 4417–4422.
132. Cahill AE, Aiello-Lammens ME, Fisher-Reid MC, Hua X, Karanewsky CJ, et al. (2013) How does climate change cause extinction? Proceedings of the Royal Society, B 280, doi: 10.1098/rspb.2012.1890.
133. Brantley S, Ford CR, Vose JM (2013) Future species composition will affect forest water use after loss if eastern hemlock from southern Appalachian forests. Ecological Applications 23: 777–790.
134. Williams JW, Jackson ST (2007) Novel climates, no-analog communities, and ecological surprises. Frontiers in Ecology and the Environment 5: 475–482.
135. Correa-Metrio A, Bush MB, Cabrera KR, Sully S, Brenner M, et al. (2012) Rapid climate change and no-analog vegetation in lowland Central America during the last 86,000 years. Quaternary Science Reviews 38: 63–75.
136. Veloz SD, Williams JW, Blois JL, He F, Otto-Bliesner B, et al. (2012) No-analog climates and shifting realized niches during the late quaternary: implications for 21st-century predictions by species distribution models. Global Change Biology 18: 1698–1713.
137. Knudsen E, Lindén A, Both C, Jonzén N, Pulido F, et al. (2011) Challenging claims in the study of migratory birds and climate change. Biological Reviews 86: 928–946.
138. Olesen JE, Trnka M, Kersebaum KC, Skjelvag AO, Seguin B, et al. (2011) Impacts and adaptation of European crop production systems to climate change. European Journal of Agronomy 34: 96–112.
139. Khailani DK, Perera R (2013) Mainstreaming disaster resilience attributes in local development plans for the adaptation to climate change induced flooding: A study based on the local plan of Shah Alam City, Malaysia. Land Use Policy 30: 615–627.

140. Lloyd MG, Peel D, Duck RW (2013) Towards a social–ecological resilience framework for coastal planning. Land Use Policy 30: 925–933.

141. Veloz SD, Nur N, Salas L, Jongsoomijt D, Wood J, et al. (2013) Modeling climate change impacts on tidal marsh birds: restoration and conservation planning in the face of uncertainty. Ecosphere 4(4):49. http://dx.doi.org/10.1890/ES12-00341.1.

16

The Lesser Known Challenge of Climate Change: Thermal Variance and Sex-Reversal in Vertebrates with Temperature-Dependent Sex Determination

Jennifer L. Neuwald, Nicole Valenzuela*

Department of Ecology, Evolution and Organismal Biology, Iowa State University, Iowa, United States of America

Abstract

Climate change is expected to disrupt biological systems. Particularly susceptible are species with temperature-dependent sex determination (TSD), as in many reptiles. While the potentially devastating effect of rising mean temperatures on sex ratios in TSD species is appreciated, the consequences of increased thermal variance predicted to accompany climate change remain obscure. Surprisingly, no study has tested if the effect of thermal variance around high-temperatures (which are particularly relevant given climate change predictions) has the same or opposite effects as around lower temperatures. Here we show that sex ratios of the painted turtle (*Chrysemys picta*) were reversed as fluctuations increased around low *and* high unisexual mean-temperatures. Unexpectedly, the developmental and sexual responses around female-producing temperatures were decoupled in a more complex manner than around male-producing values. Our novel observations are not fully explained by existing ecological models of development and sex determination, and provide strong evidence that thermal fluctuations are critical for shaping the biological outcomes of climate change.

Editor: Renee Reijo Pera, Stanford University, United States of America

Funding: This study was supported in part by the research grant IOS0743284 from the United States National Science Foundation and associated supplements IOS-0826664, IOS-0824550, IOS-0924290, and IOS-0925486 to Nicole Valenzuela. No additional external funding was received for this study. The funders had no role in study design, data collection and analysis, decision to publish, or preparation of the manuscript.

Competing Interests: The authors have declared that no competing interests exist.

* E-mail: nvalenzu@iastate.edu

Introduction

Climate helps determine many fundamental traits of organisms, from geographic distributions to life history patterns (e.g. [1]). Modifications of global and local biological patterns can thus be expected in response to climate change. Documented climatic-induced alterations of biological systems (e.g. [2,3]) stress the urgency of understanding the effect of current and future climatic variation.

Specifically, changes in environmental temperature can profoundly alter the sex ratio of temperature-dependent sex determination (TSD) species, many of which are endangered. While most concern has focused on rising mean temperatures (e.g. [4,5]), research on TSD reptiles indicates that natural sex ratios produced under daily temperature fluctuations may differ from those produced at constant incubation (e.g. [6,7,8], and references therein). Yet, the proximate TSD thermal mechanism remains unresolved. Notably, larger thermal fluctuations are predicted to accompany rising mean temperatures under climate change among years and decades [9], as well as seasonally [10], the scale at which sexual development of many TSD vertebrates occurs. Thus, to fully unravel the consequences of the complex thermal inputs experienced by TSD species, the full spectrum of ecologically-relevant temperatures and variation requires investigation. Previous work found that increasing the variance around low (male-producing) or intermediate (mixed-sex) temperatures feminized TSD turtle sex ratios [6,11,12,13]. However, whether a similar variance experienced around high (female-producing)

temperature induces females, males, or is lethal remains untested experimentally. Thus, it is unclear if enhancing the thermal variance around both unisexual means has the same or opposite effects on sex ratios. Here we address this question, which is critical for understanding the impact of climate change as the frequency of higher temperatures and the variance around those values increases, using the emerging model TSD turtle, *Chrysemys picta* [14].

Ecological models of sex determination

A persistent challenge to understanding sex ratio evolution in TSD species is the difficulty of predicting sex ratios from natural nests where temperature fluctuates daily and often unpredictably [7,15]. Many models have been proposed to address this issue, the simplest of which is the use of the mean incubation temperature as the sole predictor of nest sex ratio (e.g. [7,16,17]), or the combined use of the mean and variance of nest temperature in a bivariate plot under a sum-of-squares criterion to best fit a line separating male- from female-biased sex ratios [18,19]. Other models have taken into account the cumulative effect of temperature on sex ratio by using as a predictor the number of hours at or above the temperature that produces 1:1 sex ratios during the thermo-sensitive period [16,18,19,20,21,22]. These models proved to be poor predictors of sex ratio or only suitable to few species. More recent ecological models have been devised to account for cumulative and differential effects of lower and higher temperatures on developmental rate and sex determination [6,7,23,24,25]. The most corroborated Constant Temperature Equivalent (CTE) model measures the proportion of development occurring above the threshold temperature, and predicts that fluctuations with constant variance about a stationary mean produce equal sex ratios as a constant temperature (i.e. CTE value) [6]. Second, a Cumulative Temperature Units (CTU) model accounts for temperature fluctuations by measuring the integrated time that the embryos spend above a biological threshold in a manner akin to a degree-day model [7]. However, the applicability of both models is restricted to temperatures within the range of values that have a linear effect on development [23], whereas temperatures experienced in natural nests and used in this study include values above and below this optimal temperature range (OTR) [8,12,26,27,28]. Finally, the latest is a variable degree model [25] developed for our study species, *Chrysemys picta*, which predicts the sex ratio of the nest as 100% male if the highest cumulative development over the thermosensitive period occurred within 22–28°C, 100% female if it occurred exclusively below 22°C or above 28°C, or 50% each sex if it occurred within 21–23°C or within 27.5–28.5°C. Thus, this model assumes that *C. picta* exhibits a low female-male threshold, an early conclusion that has been refuted by empirical data [29]. Additionally, this model's predictions are limited to trimodal sex ratios (0, 50, and 100% females) rather than accounting for the continuous sex ratios observed in nature.

To bypass the caveats described above and to account for the effect of temperatures outside the OTR, here we fit a non-linear model of development by temperature [23,30] to incubation data from our study combined with data for *Chrysemys picta* from the literature [12,13,31,32,33,34]. Developmental rate expressed as percentage per day (r_a) was thus calculated as

$$r_a = b_1 10^{-v^2\left(1-b_5+b_5 v^2\right)}, \text{ where} \quad (1)$$

$$v = (u + e^{b_4 u})/c_2, \quad (2)$$

$$u = (T - b_3)/(b_3 - b_2) - c_1, \quad (3)$$

$$c_1 = 1/(1 + 0.28 b_4 + 0.72 \ln(1 + b_4), \text{ and} \quad (4)$$

$$c = 1 + [b_4/(1 + 1.5 b_4 + 0.39 b_4^2)], \quad (5)$$

where T represents the incubation temperature [23,30]. The parameters describe the maximal developmental rate (b_1) and the temperature (b_3) at which it occurs, and b_2 corresponds to the temperature at which developmental rate is $b_1/10$ [23,30]. We then used this resulting developmental function (r_a) to calculate the constant temperature predicted to induce a developmental rate equal to that observed in our fluctuating experiments ("*non-linear*" CTE values or nl-CTE), akin to the method used by the CTE model [6]. Implementation details are described in the methods section.

Our experimental approach reveals a more complex effect on developmental rate and sex ratio for the higher temperature regimes than previously anticipated, opening challenging questions about the potential response of TSD systems under predicted climate change.

Methods

Ethics Statement

All animal procedures were approved by Iowa State University IACUC under protocol # 5-05-5902-J.

Freshly-laid eggs of *Chrysemys picta* were incubated in moistened-sand [35] distributed among incubators set to fluctuate ±3°C and ±5°C around male−(26°C) and female−(31°C) producing means (Fig. 1, Table 1) [36]. Incubation temperatures included values that are experienced by natural nests [8,26,27,28], the higher of which are predicted to become more frequent as mean temperature and variance increase with climate change [9,10]. Daily temperature ranges used in our study are also within those experienced by nests in the field [37,38]. Treatments minimized ramping time between temperatures as part of a parallel gene expression study to better disentangle the effect of mean versus thermal variance, and of high versus low temperature on sex ratios (Fig. 1). Twenty out of 93–123 eggs incubated per treatment were

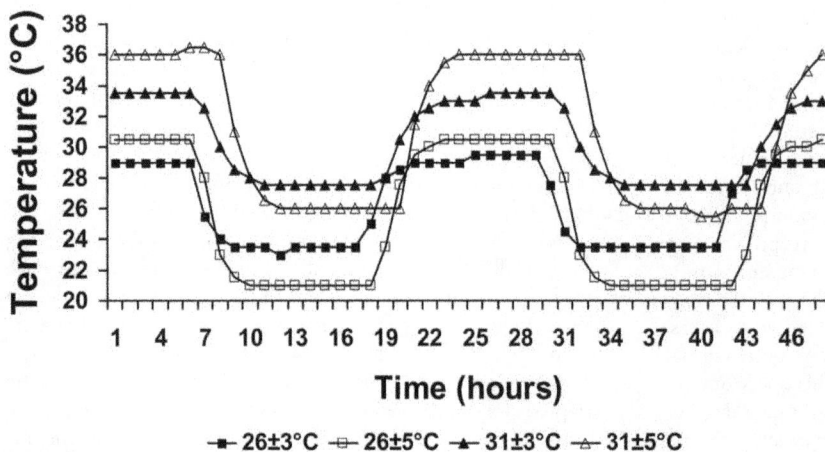

Figure 1. Two-day trace of the thermal regimes used in this study to incubate *Chrysemys picta* eggs.

Table 1. Incubation experimental design used in this study of *Chrysemys picta*, and incubation parameters calculated from a non-linear model of development by temperature [23,30], linear models [CTE (Georges et al. 1994) and CTU (Valenzuela 2001) models], and a variable degree model [25] of sex determination as described in the text.

	Incubation Treatment			
	26±3°C	26±5°C	31±3°C	31±5°C
Minimum temperature (°C)	23	21	28	26
Maximum temperature (°C)	29	31	34	36
Mean temperature (°C)	26	26	31	31
nl-CTE (i.e. constant temperature value predicted to produce an equivalent developmental rate as the fluctuating profiles, determined from the non-linear model)	26.2	25.6	28.9	27.2
CTE (i.e. constant temperature value predicted to produce an equivalent development per treatment as the fluctuating profiles)	26.7	28.0	31.6	32.5
Daily CTU from thermal traces (integral of hourly temperature records above developmental threshold of 14°C)	595	559	784	814
VDM temperature (i.e. temperature at which the highest cumulative development over the thermosensitive period occurred)	29	31	≥28	26

CTE and CTU units were calculated using 14°C as the developmental zero [33]. VDM temperature was calculated as 29°C and 31°C, respectively, for the 26±3 and 26±5°C treatments, and as 26°C for the 31±5°C treatment, since the minimum temperature in the former two treatments retards development compared to the maximum values while the opposite is true for the 31±5°C treatment. VDM temperature was calculated as ≥28°C for the 31±3°C treatment since embryos were exposed to temperatures between 28 and 34°C.

targeted for hatching for this study. Individuals were sexed by gonadal inspection 2.5–3.5 months post-hatching [36].

Sex ratios were compared to those from constant temperatures (i.e. ±0°C) to elucidate the effect of increasing amplitude of thermal fluctuations with respect to the effect of mean temperature [6,11,12,13,23,33]. Deviations from expected sex ratios across treatments (100% males and 100% females from a 26°C and 31°C mean, respectively [36]) were evaluated using chi-square tests. The expected value for female-producing temperatures (frequency = 0), was replaced by frequency = 1 to avoid division by zero.

As mentioned before, natural nest temperatures and those used in our incubation experiments include values above and below the optimal temperature range (OTR) [8,12,26,27,28]. The non-linear model of development by temperature [23,30] described by equations 1–5 above was fitted to *Chrysemys picta* incubation data for from our study and others [12,13,31,32,33,34] using DEVARA software [30]. Temperature input data for the model included thermal profiles with constant or fluctuating temperatures and their corresponding observed incubation period (days) (this study and [12,13,31,32,33,34]). Three of the constant parameters of the model (b_1–b_3) were obtained from the literature, while the remaining two parameters (b_4 and b_5) were determined by iterative fitting as implemented in DEVARA using existing developmental data from our study and the literature. The parameters' values were: $b_1 = 2.2$ (maximal developmental rate), which occurs at $b_3 = 32°C$ [32], and $b_2 = 15.5$ (temperature at which developmental rate is $b_1/10$), which was calculated from the linear relationship of constant temperature and developmental rate for reported values within the OTR [12,13,31,32,33,34]. The constant temperature predicted to induce a developmental rate equal to that observed in our fluctuating experiments ("*non-linear*" CTE values or nl-CTE) (Table 1) was calculated from r_a. Additionally, we compared the results from this analysis with the results from the simpler CTE and CTU models [6,7], as well as the VDM [25] to examine the discrepancies among model predictions.

Embryonic mortality was measured as the percent of eggs that died during incubation from the total number of eggs in each treatment. G-tests of independence were used to test for a temperature treatment effect in embryonic mortality. Incubation time was measured as days to hatching. The effect of temperature treatment on incubation length was tested using a Tukey-Kramer HDS test.

Results

The experiment was replicated in 2008 and 2009 and produced consistent results (Fig. 2; $\chi^2 = 5.8$, df = 3, P>0.12; Table 2). Augmenting the variance around each unisexual mean reversed the sex ratios from those yielded by constant temperatures (Fig. 2). Sex ratios deviated significantly from expectation ($\chi^2 = 337.5$, 66.1, and 724.4 for 2008, 2009, and across years, respectively; df = 1; P<0.0001). Results were robust to removing cells with expected values of zero ($\chi^2 = 13.5$, 17.1, and 47.4 for 2008, 2009, and across years, respectively; df = 1; P<0.0002). Additionally, all sex ratios differed from parity (P<0.0001). Thus, while greater variance around low and intermediate mean temperature had a feminizing effect in this and other studies [6,11,12,13], larger fluctuations around the high female-producing mean induced male differentiation. Mortality was significantly-higher under 31±5°C than under other treatments on both years (Table 3). Final sample sizes varied among treatments (Fig. 1) due to mortality or because eggs were diverted to, or added from, the concurrent gene expression study.

Not surprisingly, existing linear models of fluctuating temperature effects on development and sex determination [6,7,25] do not account for our results (Table 2). Indeed, the non-linear model fitted to the combined data from our study and from constant and fluctuating temperature experiments reported in the literature confirmed that 21°C (the minimum temperature in the 26±5°C treatment), 34°C and 36°C (the maximum temperatures in the 31±3°C and 31±5°C treatments, respectively) fall outside the OTR for *C. picta* of ~22–32°C [12], and thus, have a retarding effect on developmental rate (Fig. 3). Accordingly, although incubation time was shorter overall in the 31°C-mean experiments than in the 26°C-mean experiments, development was slower

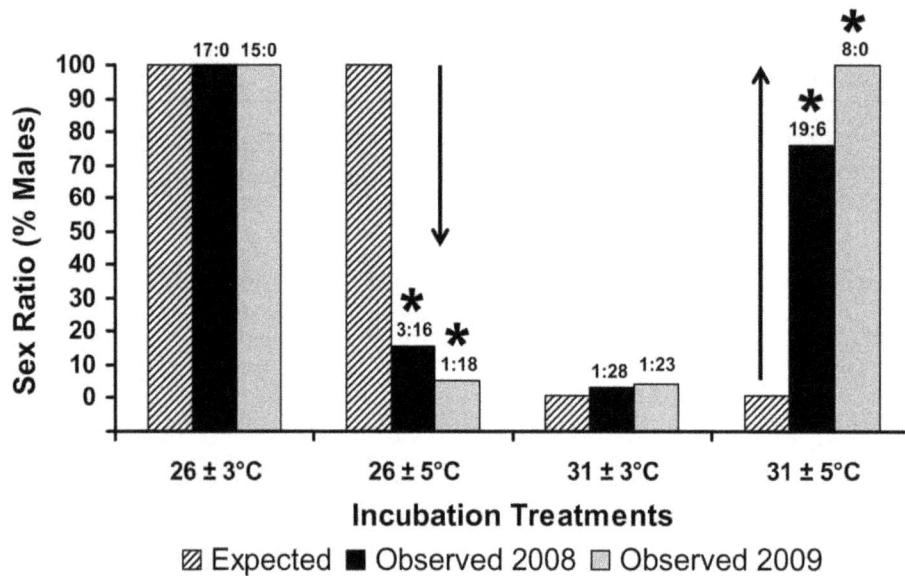

Figure 2. Reversing effect of increasing thermal variance on sex ratios of *Chrysemys picta* **turtles.** Observed males and observed females (m:f) per treatment and year are indicated above columns. Asterisks indicate statistically significant deviations from sex ratios expected by the mean temperature. Arrows indicate direction of change from expectation.

under the ±5°C treatments as compared to the ±3°C treatments for each mean temperature (Table 3).

Notably, our results indicate that while increasing the variance around the male producing mean (26°C) had little effect in developmental rate, sex ratio was decoupled from the thermal effect on development under the largest fluctuations, whereas around the female producing mean (31°C) both the developmental rate and the sex ratios were affected by the thermal variance experienced. This unexpected observation reveals that the effect of increasing thermal fluctuations on sex determination depends upon the region of the temperature range where they fall. For instance, both the 26±3°C and 26±5°C treatments exhibited a developmental rate similar to that predicted by the non-linear model for a 26°C constant-temperature-equivalent

(Table 2; Fig. 3), which should produce 100% males as observed for the 26±3°C experiment. The 26±°C treatment however, yielded 9% males instead of the predicted 100% males (Table 2). On the other hand, the 31±3°C treatment was expected to have produced ≥50% females as would a constant 28.9°C equivalent rather than the 6% males obtained. Further, the non-linear model predicted 31±5°C to have produced 100% males as would a constant 27.2°C rather than the 82% males obtained (Table 2).

Discussion

We found that greater thermal variance around both low and high unisexual mean temperatures reversed the sex ratios of the painted turtle from those expected by the mean alone. Impor-

Table 2. Observed sex ratios of *Chrysemys picta*, and predictions from a non-linear model of development by temperature [23,30], linear models [CTE (Georges et al. 1994) and CTU (Valenzuela 2001) models], and a variable degree model [25] of sex determination as described in the text.

	Incubation Treatment			
	26±3°C	**26±5°C**	**31±3°C**	**31±5°C**
Mean temperature (°C)	26	26	31	31
Observed sex ratio (% male) 2008	100§	15.8*†§	3.4†	76*†§
Observed sex ratio (% male) 2009	100§	5.3*†§	4.2†	100*†§
Overall observed % male	100§	8.8*†§	5.9†	81.8*†§
Expected sex ratio given the mean temperature	100	100	0	0
Sex ratio (% male) predicted by the CTE values	100	≈50	0	0
Sex ratio (% male) predicted by the nl-CTE values	100	100	≤50	100
Sex ratio (%male) predicted by the VDM values	0	0	0	100
CTU-predicted (*and observed*) sex ratio order from 100% male [1] to 100% female [4]	2 (1)	1 (3)	3 (4)	4 (2)

* = observed sex ratios unexplained by the mean temperature,
§ = observed sex ratios unexplained by the linear and the VDM models,
† = observed sex ratios unexplained by the non-linear model.

Table 3. Observed developmental rate and mortality of *Chrysemys picta* in this study.

	Incubation Treatment			
	26±3°C	26±5°C	31±3°C	31±5°C
Mean *(StDev)* **incubation time 2008‡** (days to hatching). Incubation time differed significantly among treatments with different letter superscript (Tukey-Kramer test, experimentwise α = 0.05).	70[A] (2.2)	70[A] (1.9)	56[B] (1.5)	63[C] (3.5)
Mean *(StDev)* **incubation time 2009** (days to hatching). Incubation time differed significantly among all treatments (Tukey-Kramer test, experimentwise α = 0.05).	64[A] (2.9)	68[B] (4.1)	51[C] (2.7)	58[D] (1.5)
Embryonic mortality (%) 2008‡. [§]Mortality statistically higher than other treatments ($\chi^2 = 10.1$, df = 3, P = 0.02 overall, versus $\chi^2 = 2.8$, df = 2, P = 0.24 when excluding 31±5°C).	7.53	10.64	14.95	22.22[§]
Embryonic mortality (%) 2009. [§]Mortality statistically higher than other treatments ($\chi^2 = 16.8$, df = 3, P = 0.0008 overall, versus $\chi^2 = 1.1$, df = 2, P = 0.6 when excluding 31±5°C).	7.14	3.88	4.92	18.52[§]

Tukey-Kramer HSD significant differences in incubation time are indicated by the lettered superscripts.
‡ = 2008 values of mortality and incubation time are higher than in 2009 due to logistical problems with data-recording and thus are presented only to illustrate the relative effect of thermal treatments which is consistent across years, but not their absolute magnitude.

tantly, our results reveal that all levels of thermal fluctuation decouple developmental rate and sex ratios if experienced around the female-producing mean while this decoupling was only observed under the largest fluctuation around the male-producing mean temperature. Such decoupling means that the effect of temperature on sex ratio is not perfectly predicted by current models based solely on the effect that temperature has on development, revealing that the potency that temperature has to influence developmental rate differs somewhat from its potency to induce sex determination as described below.

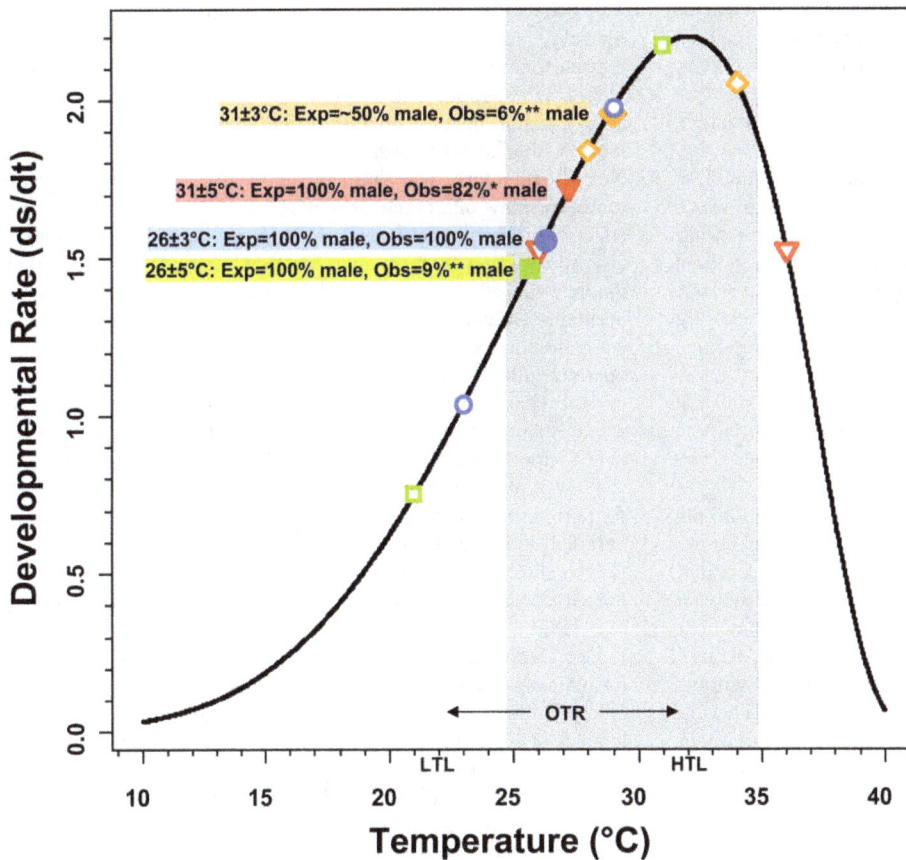

Figure 3. Developmental rate of *Chrysemys picta* embryos as a function of temperature. Solid symbols denote the constant temperature predicted by the non-linear model to exhibit a developmental rate equal to that observed at each fluctuating experiment conducted in this study (nl-CTE, see text for full description). Open symbols denote the developmental rate predicted for the minimum and maximum temperatures used in each fluctuating experiment. Symbols of the same shape and color correspond to a single fluctuating experiment as described in the color legends. Exp = sex ratio expected by the nl-CTE values. Obs = observed sex ratio. Asterisks denote deviations from expectation. OTR = Optimal Thermal Range (gray area), LTL = Low Thermal Limit, HTL = High Thermal Limit [12].

Thermal variance, development and sex determination

Despite our findings not matching predictions from the linear and non-linear ecological models, these patterns may nonetheless be explained by heat accumulation theory [6,39]. Indeed, the high-variance treatments included temperatures that fall outside the OTR for *C. picta* [12] but which sustained development, which is consistent with theoretical expectations from and empirical observations in other TSD taxa [39]. Namely, embryos under 26±5°C (a treatment which produced sex ratios counter to expected) cycled between a 21°C minimum, a value below the lower thermal limit for this species [12] that slows down development (Fig. 3), and a 31°C maximum, a high female-producing value in the optimal range [12] (Fig. 3), such that greater embryonic development likely occurred under the female-producing temperatures despite the male-producing mean (Fig. 1). Similar to the 26±5°C treatment, embryos under 31±5°C cycled between 26°C (male-producing) and 36°C, a value above the upper thermal limit which retards growth (Fig. 3), such that development occurred mostly under the male-producing temperatures despite the female-producing mean. Consistently, incubation time was shorter for ±3°C than for ±5°C treatments (Table 3). Additionally, while mortality rates (3.5–22.2%) were commensurate with wild nests [40], 31±5°C induced significantly higher mortality than other treatments (Table 3). Interestingly, embryos under 31±3°C cycled between 28°C (a value near the pivotal temperature that produces 1:1 sex ratios) and 34°C, a value above the OTR which appeared to retard growth as expected (Fig. 3) while at the same time, exhibiting a greater potency than the lower temperature in inducing females (the sex expected by mean temperature). Thus, a highly female-biased sex ratio not significantly different from that predicted by the mean temperature alone was produced (Fig. 2).

The alternative that 21°C may induce female-differentiation in *C. picta* [21,41] as low temperatures do in TSDII species (in which females are produced at both low and high temperatures, while males are produced at intermediate temperatures), was ruled out by extensive experimental data [29]. Furthermore, the variable degree model which assumes a TSDII model for *C. picta* [25] predicted that all treatments should produce 100% females except 31±5°C which should produce 100% males (Tables 1 and 2), an expectation far from our observed sex ratios. It is worth noting that while moisture was kept constant in our study and it may vary in the field, previous research has demonstrated that moisture levels do not affect sex determination in *C. picta* [42].

Our results emphasize that a general model accounting for both mean and variance across the full range of viable temperatures remains overdue to explain sex determination and accurately forecast sex ratios under climate change. Importantly, we show for the first time that the amplitude of thermal fluctuations mediate the sex ratio response to mean temperature around female-producing values in a more complex way than it does around male-producing values ([6,11,12,13,33], and this study). Thermal values and fluctuations used in this experiment are within the range experienced by nests of this species in the field [8,26,27,28,37,38]. Thus, while the 31°C-mean treatments had a higher mean than the averages recorded in natural populations, our design permitted testing the tolerance of this species to long exposure to high temperatures that are already experienced in the wild and which may be encountered more frequently in the future if mean and variance increase due to climate change as currently predicted [10]. Therefore, our experimental design helped place the female-producing temperature used broadly in laboratory studies (e.g. [36,43,44]) in the same context as the better-studied male-producing and intermediate temperatures [6,11,12,13,33]).

Our novel observations reveal that the effect of increasing thermal fluctuations on sex determination depends upon the region of the temperature range where they fall, consistent with reports for other phenotypes in TSD and GSD taxa [12,45,46]. Furthermore, our results open the question about whether the effects of temperature mean and variance on multiple traits in other biological systems may be decoupled as observed here in ways that have not been previously anticipated.

Climate change and TSD evolution

Our results strengthen the concern about the fate of TSD systems facing chronic environmental disturbances (e.g. [6,7,8], and references therein). Variance in continental temperature is expected to increase during the summer [10] when air temperatures influence sex ratios of wild *C. picta* [4]. Although potential negative effects of climate change might be lessened by compensatory plastic or rapid evolutionary responses [6], these may be constrained in endangered TSD taxa under low population sizes or disturbed habitats. At first glance, our observations would misleadingly suggest that if thermal variance increases during the reproductive season as current climate change models predict [10], the sex-reversing effect of greater fluctuations would be beneficial by helping buffer against the effect of changes in the temperature mean alone. Furthermore, a recent study in *C. picta* showed that increasing the thermal variance around an intermediate temperature had no effect in hatchling morphological, behavioral or immunological phenotypes, nor in embryonic mortality [13] thus, ruling out several potential negative effect of thermal variance at lower temperatures. However, whether phenotypic responses differ under larger thermal fluctuations around higher mean temperatures remains untested, and our data indicate that at least mortality is higher under those conditions. Notably therefore, our findings suggest that higher embryonic mortality may offset any benefit accrued by the masculinizing effect of higher variance around female-producing mean temperatures, and may consequently interact with other factors that mediate the effect of a thermally-changing world. For instance, nesting behavior has been proposed as another potential compensatory response to climate change given that canopy openness affects the daily temperature range (a proxy of thermal variance) and consequently the sex ratios in TSD species [47]. The sex-reversing effect of thermal variance observed in our study would superficially suggest that shallow nesting TSD species, such as *C. picta*, may be more able to respond to climate change than deeper nesting taxa, such as sea turtles [6]. However, larger variation resulting from a compensatory nesting response in *C. picta* might expose the shallower eggs to lethally- or suboptimaly-high temperatures, as occurred in our study and in other species (e.g. [48]).

Fig. 4 summarizes our findings in the context of climate change for *Chrysemys picta*, a turtle exhibiting a TSDIa pattern of sex determination (where males are produced at low temperatures and females at high)[49], but the same logic can be extended in the opposite direction for TSDIb taxa (where females are produced at low temperatures and males at high) or TSDII taxa (where males are produced at intermediate temperatures and females above and below these values) since global warming will affect them in a similar way as for TSDIa taxa. Under limited thermal variance, increases in mean temperatures below the optimal temperature range (OTR) accelerate development and improve survival as values move closer to the OTR, although these values are exclusively male-producing (Fig. 4). Within the OTR, higher mean temperatures also accelerate development but mortality is unaffected since this range is optimal, and the proportion of

		Below OTR	OTR Optimal Temperature Range	Above OTR
Effect of Increased Mean T°	Developmental rate	Faster	Faster	Slower
	Embryonic mortality	Lower	Unaffected	Higher
	Sex ratio	Unaffected: (100% males)	Feminizing: (Higher female-bias)	Unaffected: (100% females)
Effect of Increased Thermal Variance	Developmental rate	Slower	Faster	Slower
	Embryonic mortality	Unaffected	Unaffected	Higher
	Sex ratio	Feminizing (Higher female-bias)	Feminizing: Higher female-bias	Masculinizing: (Higher male-bias)

Figure 4. Observed effects of increased mean temperature and increased thermal variance on life history parameters of the TSDIa turtle, *Chrysemys picta*, and implications in the context of climate change predictions. Effects are divided into three thermal ranges: optimal temperatures (OTR), colder temperatures below the OTR, and warmer temperatures above the OTR. Inner cells correspond to neutral effects (gray), beneficial effects (green), and detrimental effects (pink) on developmental rate, embryonic survival, and sex ratio, as described in the text. Listed effects correspond to those of increased mean temperature alone under low or no variance scenarios, and to those of increased thermal variance when compared to mean temperature effects.

females increases as a logistic function of the mean temperature from all-male- to all-female- producing (Fig. 4)[49]. One detrimental effect of climate change derives from causing TSDIa taxa to produce excessively female-biased sex ratio that endangered population persistence due to male-limitation (or female limitation due to excessively male-biased sex ratio in the case of TSDIb taxa). Under these low thermal variance conditions above the OTR, higher temperatures reach values that are detrimental for development and survival but remain all-female-producing (Fig. 4). The combination of extreme biased sex ratios as described earlier and increased mortality represent the dangers of global warming at these values. In contrast, as thermal variance increases below the OTR, embryos are exposed to lower suboptimal temperatures that slow down development but do not affect mortality (if they remain mainly above lethal values), such that most development occurs at the higher temperatures which thus tend to produce higher female-biases compare to the mean temperature (Fig. 4). This feminizing effect will be beneficial for a TSDIa population suffering from excessively male-biased sex ratios. Within the OTR, increasing the thermal variance exposes the embryos to higher optimal values that accelerate development and have a feminizing effect, and which exhibit more potency to stimulate development than lower values to slow down development and affect sex ratio. These temperatures within the OTR do not affect mortality (Fig. 4). Thus, within the OTR, the potential danger of increased variance due to climate change would be the

production of excessively female-biased sex ratios in TSDIa taxa. Above the OTR, increasing the thermal variance exposes embryos to excessively high temperatures that inhibit development and cause higher mortality, such that development occurs mostly at the lower temperatures resulting in a net masculinizing effect (Fig. 4). This masculinizing effect would be beneficial to counter the female-biases induced by higher mean temperatures under climate change, but these benefits may be offset by the higher mortality suffered which may lead to severe population bottlenecks.

Further research is warranted to explore the effect of increased variance around actual natural thermal profiles on sex and fitness to test more realistic scenarios of sex determination and environmental perturbations. Additionally, studies to reveal the molecular mechanism responsible for our observations are urgently needed. Both research studies are ongoing. Whether greater fluctuations in the wild have the same effect as those observed here and around intermediate temperatures [13] requires investigating the interaction between thermal variance and other biological traits, including incubation length, nesting timing, generation time, and TSD pattern, among others. For instance, turtles and lizards with contrasting TSD patterns may respond differently to such changes [49]. Importantly, if TSD is an adaptive trait (*sensu* [50]), an increased thermal variance may decouple the environmental variables that confer sex-specific fitness at different temperatures throughout the reproductive season or geographic locations, potentially breaking down the adaptiveness of this sex-determining mechanism, and perhaps even inducing a transition in sex determining mechanisms. Interestingly, such transitions between TSD and GSD systems over 200 my of turtle evolution are associated with dramatic genomic rearrangements and appear to coincide with climate change events [51]. Alternatively, the untested hypothesis that thermal variance itself induces differential fitness, or is correlated with such a variable [50], and therefore underlies the evolution or maintenance of TSD [15], should be considered. A recent initial study in C. picta suggested that sex and other phenotypic responses to thermal variance are decoupled at least for embryonic and hatchling life stages [13], but research on other systems demonstrates that responses to incubation conditions may be delayed to the reproductive life stages [52].

References

1. Sarma SSS, Nandini S, Gulati RD (2005) Life history strategies of cladocerans: comparisons of tropical and temperate taxa. Hydrobiologia 542: 315–333.
2. Umina PA, Weeks AR, Kearney MR, McKechnie SW, Hoffmann AA (2005) A rapid shift in a classic clinal pattern in Drosophila reflecting climate change. Science 308: 691–693.
3. Kausrud KL, Mysterud A, Steen H, Vik JO, Ostbye E, et al. (2008) Linking climate change to lemming cycles. Nature 456: 93–97.
4. Janzen FJ (1994) Climate change and temperature-dependent sex determination in reptiles. Proceedings of the National Academy of Sciences of the United States of America 91: 7487–7490.
5. Hays GC, Broderick AC, Glen F, Godley BJ (2003) Climate change and sea turtles: a 150-year reconstruction of incubation temperatures at a major marine turtle rookery. Global Change Biology 9: 642–646.
6. Georges A, Limpus C, Stoutjesdijk R (1994) Hatchling sex in the marine turtle Caretta caretta is determined by proportion of development at a temperature, not daily duration of exposure. J Exp Zool 270: 432–444.
7. Valenzuela N (2001) Constant, shift and natural temperature effects on sex determination in Podocnemis expansa turtles. Ecology 82: 3010–3024.
8. Sternadel LL, Packard GC, Packard MJ (2006) Influence of the nest environment on bone mineral content in hatchling painted turtles (Chrysemys picta). Physiological and Biochemical Zoology 79: 1069–1081.
9. Boer GJ (2009) Changes in interannual variability and decadal potential predictability under global warming. Journal of Climate 22: 3098–3109.
10. Stouffer RJ, Wetherald RT (2007) Changes of variability in response to increasing greenhouse gases. Part I: Temperature. Journal of Climate 20: 5455–5467.
11. Du WG, Shen JW, Wang L (2009) Embryonic development rate and hatchling phenotypes in the Chinese three-keeled pond turtle (Chinemys reevesii): The influence of fluctuating temperature versus constant temperature. Journal of Thermal Biology 34: 250–255.
12. Les HL, Paitz RT, Bowden RM (2009) Living at extremes: development at the edges of viable temperature under constant and fluctuating conditions. Physiological and Biochemical Zoology 82: 105–112.
13. Paitz RT, Clairardin SG, Griffin AM, Holgersson MCN, Bowden RM (2010) Temperature fluctuations affect offspring sex but not morphological, behavioral, or immunological traits in the Northern Painted Turtle (Chrysemys picta). Canadian Journal of Zoology-Revue Canadienne De Zoologie 88: 479–486.
14. Valenzuela N (2009) The painted turtle, Chrysemys picta: A model system for vertebrate evolution, ecology, and human health. Cold Spring Harbor Protocols; DOI:10.1101/pdb.emo124.
15. Valenzuela N (2004) Evolution and maintenance of temperature-dependent sex determination. In: Valenzuela N, Lance VA, eds. Temperature dependent sex determination in vertebrates. Washington, DC: Smithsonian Books. pp 131–147.
16. Pieau C (1982) Modalities of the action of temperature on sexual differentiation in field-developing embryos of the European pond turtle Emys orbicularis (Emydidae). J Exp Zool 220: 353–360.
17. Schwarzkopf L, Brooks RJ (1987) Nest-site selection and offspring sex ratio in painted turtles, Chrysemys picta. Copeia. pp 53–61.
18. Bull JJ (1985) Sex ratio and nest temperature in turtles comparing field and laboratory data. Ecology 66: 1115–1122.
19. Souza RRd, Vogt RC (1994) Incubation temperature influences sex and hatchling size in the neotropical turtle Podocnemis unifilis. J Herpetol 28: 453–464.
20. Wilhoft DC, Hotaling E, Franks P (1983) Effects of temperature on sex determination in embryos of the snapping turtle, Chelydra serpentina. J Herpetol 17: 38–42.
21. Schwarzkopf L, Brooks RJ (1985) Sex determination in northern painted turtles: effect of incubation at constant and fluctuating temperatures. Canadian Journal of Zoology 63: 2543–2547.
22. Mrosovsky N, Provancha J (1992) Sex ratio of hatchling loggerhead sea turtles: data and estimates from a 5-year study. Canadian Journal of Zoology 70: 530–538.
23. Georges A, Beggs K, Young JE, Doody JS (2005) Modelling development of reptile embryos under fluctuating temperature regimes. Physiological and Biochemical Zoology 78: 18–30.
24. Georges A, Doody S, Beggs K, Young J (2004) Thermal models of TSD under laboratory and field conditions. In: Valenzuela N, Lance VA, eds. Temperature dependent sex determination in vertebrates. Washington, DC: Smithsonian Books. pp 79–89.
25. Parrott A, Logan JD (2010) Effects of temperature variation on TSD in turtle (C. picta) populations. Ecological Modelling 221: 1378–1393.
26. Cagle KD, Packard GC, Miller K, Packard MJ (1993) Effects of the microclimate in natural nests on development of embryonic painted turtles, Chrysemys picta. Functional Ecology 7: 653–660.
27. Morjan CL (2003) Variation in nesting patterns affecting nest temperatures in two populations of painted turtles (Chrysemys picta) with temperature-dependent sex determination. Behavioral Ecology and Sociobiology 53: 254–261.
28. Valenzuela N, Neuwald JL (unpublished data).
29. Etchberger CR, Ewert MA, Raper BA, Nelson CE (1992) Do low incubation temperatures yield females in painted turtles? Canadian Journal of Zoology 70: 391–394.
30. Dallwitz MJ, Higgins JP (1992) User's guide to DEVAR, a computer program for estimating development rate as a function of temperature. CISRO Division of Entomology Report No.2: 1–23.
31. Ewert M (1985) Embryology of Turtles. In: Gans C, ed. Biology of the Reptilia, Development A. New York: John Wiley and Sons, Inc.
32. Gutzke WHN, Packard GC, Packard MJ, Boardman TJ (1987) Influence of the hydric and thermal environments on eggs and hatchlings of painted turtles (Chrysemys picta). Herpetologica 43: 393–404.
33. Les HL, Paitz RT, Bowden RM (2007) Experimental test of the effects of fluctuating incubation temperatures on hatchling phenotype. Journal of Experimental Zoology Part a-Ecological Genetics and Physiology 307A: 274–280.
34. Mahmoud IY, Hess GL, Klicka J (1973) Normal embryonic stages of western painted turtle, Chrysemys picta bellii. Journal of Morphology 141: 269–279.
35. Valenzuela N (2009) Egg incubation and collection of painted turtle embryos. Cold Spring Harbor Protocols; DOI:10.1101/pdb.prot5238.
36. Ewert MA, Nelson CE (1991) Sex determination in turtles: diverse patterns and some possible adaptive significance. Copeia. pp 50–69.
37. Janzen F (1994) Vegetational cover predicts the sex ratio of hatchling turtles in natural nests. Ecology 75: 1593–1599.
38. Weisrock DW, Janzen FJ (1999) Thermal and fitness-related consequences of nest location in painted turtles (Chrysemys picta). Functional Ecology 13: 94–101.
39. Deeming DC, Ferguson MWJ (1989) The mechanism of temperature dependent sex determination in Crocodilians: a hypothesis. American Zoologist 29: 973–985.
40. Valenzuela N, Janzen FJ (2001) Nest-site philopatry and the evolution of temperature-dependent sex determination. Evolutionary Ecology Research 3: 779–794.
41. Gutzke WHN, Paukstis GL (1984) A Low Threshold Temperature for Sexual-Differentiation in the Painted Turtle, Chrysemys-Picta. Copeia. pp 546–547.
42. Packard GC, Packard MJ, Benigan L (1991) Sexual differentiation, growth, and hatching success by embryonic painted turtles incubated in wet and dry environments at fluctuating temperatures. Herpetologica 47: 125–132.
43. Janzen FJ, Morjan CL (2002) Egg size, incubation temperature, and posthatching growth in painted turtles (Chrysemys picta). Journal of Herpetology 36: 308–311.
44. Valenzuela N (2010) Multivariate expression analysis of the gene network underlying sexual development in turtle embryos with temperature-dependent and genotypic sex determination Sexual Development 4: 39–49.
45. Mullins MA, Janzen FJ (2006) Phenotypic effects of thermal means and variances on smooth softshell turtle (Apalone mutica) embryos and hatchlings. Herpetologica 62: 27–36.

In summary, our study underlines the importance of investigating the role of thermal variance to understand TSD sex ratio evolution, its consequences, and its effect on other fitness-relevant phenotypes to understand the response of biodiversity to local and global disturbances at multiple time scales.

Acknowledgments

We thank D.C. Adams and D. Warner for critical comments.

Author Contributions

Conceived and designed the experiments: JLN NV. Performed the experiments: JLN NV. Analyzed the data: NV. Contributed reagents/materials/analysis tools: NV. Wrote the paper: JLN NV.

46. Patterson LD, Blouin-Demers G (2008) The effect of constant and fluctuating incubation temperatures on the phenotype of black ratsnakes (*Elaphe obsoleta*). Canadian Journal of Zoology-Revue Canadienne De Zoologie 86: 882–889.

47. Doody JS, Guarino E, Georges A, Corey B, Murray G, et al. (2006) Nest site choice compensates for climate effects on sex ratios in a lizard with environmental sex determination. Evolutionary Ecology 20: 307–330.

48. Andrewartha SJ, Mitchell NJ, Frappell PB (2010) Does incubation temperature fluctuation influence hatchling phenotypes in reptiles? A test using parthenogenetic geckos. Physiological and Biochemical Zoology 83: 597–607.

49. Valenzuela N, Lance VA, eds (2004) Temperature Dependent Sex Determination in Vertebrates. Washington, DC: Smithsonian Books.

50. Charnov EL, Bull JJ (1977) When is sex environmentally determined? Nature 266: 828–830.

51. Valenzuela N, Adams DC (2011) Chromosome number and sex determination co-evolve in turtles Evolution. *(in press)*.

52. Warner DA, Shine R (2008) The adaptive significance of temperature-dependent sex determination in a reptile. Nature 451: 566–U565.

17

Comparative Toxicity of Nanoparticulate CuO and ZnO to Soil Bacterial Communities

Johannes Rousk[1,2]*, **Kathrin Ackermann**[1], **Simon F. Curling**[3], **Davey L. Jones**[1]

1 Environment Centre Wales, Bangor University, Gwynedd, United Kingdom, **2** Section of Microbial Ecology, Department of Biology, Lund University, Lund, Sweden, **3** BioComposites Centre, Bangor University, Gwynedd, United Kingdom

Abstract

The increasing industrial application of metal oxide Engineered Nano-Particles (ENPs) is likely to increase their environmental release to soils. While the potential of metal oxide ENPs as environmental toxicants has been shown, lack of suitable control treatments have compromised the power of many previous assessments. We evaluated the ecotoxicity of ENP (nano) forms of Zn and Cu oxides in two different soils by measuring their ability to inhibit bacterial growth. We could show a direct acute toxicity of nano-CuO acting on soil bacteria while the macroparticulate (bulk) form of CuO was not toxic. In comparison, $CuSO_4$ was more toxic than either oxide form. Unlike Cu, all forms of Zn were toxic to soil bacteria, and the bulk-ZnO was more toxic than the nano-ZnO. The $ZnSO_4$ addition was not consistently more toxic than the oxide forms. Consistently, we found a tight link between the dissolved concentration of metal in solution and the inhibition of bacterial growth. The inconsistent toxicological response between soils could be explained by different resulting concentrations of metals in soil solution. Our findings suggested that the principal mechanism of toxicity was dissolution of metal oxides and sulphates into a metal ion form known to be highly toxic to bacteria, and not a direct effect of nano-sized particles acting on bacteria. We propose that integrated efforts toward directly assessing bioavailable metal concentrations are more valuable than spending resources to reassess ecotoxicology of ENPs separately from general metal toxicity.

Editor: Melanie R. Mormile, Missouri University of Science and Technology, United States of America

Funding: JR was funded by the Swedish research council (www.vr.se) grants nos 2009-7343 and 2011-5719. The funders had no role in study design, data collection and analysis, decision to publish, or preparation of the manuscript.

Competing Interests: The authors have declared that no competing interests exist.

* E-mail: johannes.rousk@biol.lu.se

Introduction

Manufactured particles with at least one dimension between 1 and 100 nm [1] have been termed Engineered nanoparticles (ENPs). Metal oxide ENPs are receiving increasing attention in material science and nano-technology based industries for a large variety of applications, including catalysts, sensors and for their incorporation into commercial products [2]. For instance, ENP CuO is used in semiconductors, catalysts and in photovoltaic cells [3], while ENP ZnO is used in personal care products as well as coatings and paints [4] due to its UV absorption efficiency and transparency to visible light that increases with smaller particle size. The increasing industrial application of ENPs is likely to increase their environmental release, especially exposing soils and freshwaters [4,5], prompting a careful evaluation of their ecotoxity in this environment. The small size of the ENP, and the greater mobility and potentially increased risk of uptake by organisms that this confers, have been proposed to increase the toxic potential of ENP substances generally, as demonstrated *in vitro* [6]. Several studies have demonstrated the potential of metal oxide ENPs as environmental toxicants [7,8]. However, lack of suitable control treatments have compromised the power of early assessments to determine and quantify the potential environmental impact in a useful context, i.e. whether the ENP property of the substance made it more toxic than it would have been in a generic (non-ENP) form. To enable progress in our understanding of the

ecotoxicology of ENPs, it is important to assess (i) whether the nano-form size that characterizes ENP *per se* increases the toxicity beyond other forms of the substance, and (ii) how the observed toxicity of the ENP compares to the well-known toxicity of ionic forms of heavy metals. That is, while many studies have been conducted to determine if heavy metal ENPs can be toxic, there is a scarcity of studies that have investigated if the ENP form of heavy metal toxicants are more toxic than non-ENP forms of the same substance.

Here, we evaluated the ecotoxicity of ENP forms of Zn and Cu oxides in two different soils. In addition to the ENP forms of the metal oxides, we also included two reference toxicant forms, (i) bulk oxide of non-nanoparticulate form, and (ii) highly soluble sulfate forms of the metals. The bulk form of the metal oxide is designed as a control to show if any observed toxicity was due to its nano-particulate form, while the soluble sulfate form acts as a control to elucidate the extent to which observed toxicity is due to metal ion solubilization in soil solution. By measuring the resulting metal concentrations in soil solution, we strengthen the connection between metals and toxicity. To provide a sensitive measure of ecotoxicity we measured the effect of the substance additions on bacterial growth using the leucine incorporation method [9,10], previously successfully used to accurately determine toxicity of environmental toxicants including metals [11,12], antibiotics [13–15], phenols [16] and salt [17].

Materials and Methods

Soils and chemicals

We used two different soils for the experiment, one mineral pasture soil (Typic Dystrochrept, organic-C = 40 mg g^{-1}, total-N = 3.3 mg g^{-1}, pH(H$_2$O) = 5.0; henceforth "mineral soil") and one organic pasture soil (Typic Fragiochrept, organic-C = 154 mg g^{-1}, total-N = 9.3 mg g^{-1}, pH(H$_2$O) = 6.6; henceforth "organic soil"). These soils are described in detail elsewhere [18]. Soils were fresh sieved (<2 mm), and then adjusted to a moisture content of 40% of water holding capacity (WHC) for the mineral soil, and 60% WHC for the organic soil, both deemed to be optimal for microbial activity based on previous work. After these preparations the soils were incubated at 20°C for one week before experimentation commenced. The CuO (particle size 40–80 nm; henceforth "nano-CuO") and ZnO (particle size 20 nm; henceforth "nano-ZnO") ENPs used in the experiment were supplied and guaranteed by IO-LI-TEC nanomaterials (Heilbronn, Germany) and had CAS reference numbers 1317-38-0 and 1314-13-2, respectively. The bulk forms of the metal oxides and sulfates of the metals were standard laboratory grade chemicals (ZnO CAS: 1314-13-2, henceforth "bulk-ZnO"; ZnSO$_4$·7H$_2$O CAS: 7446-20-0; CuO CAS: 1317-38-0, henceforth "bulk-CuO"; CuSO$_4$·5H$_2$O CAS: 7758-99-8, all supplied by Sigma Aldrich, St Louis, USA).

Experimental design

Subsamples of soil (2.0 g dry weight equivalents) were weighed into 50 ml centrifugation tubes, to which immediately were added 200 mg laboratory grade acid washed sand (40–100 μm mesh) carrying the different toxicants. Nano-ZnO, bulk-ZnO, ZnSO$_4$, nano-CuO, bulk-CuO and CuSO$_4$ were added at 8 concentrations at logarithmic intervals from 0 to 200 mmol metal g^{-1} soil. The sand-toxicant mixtures were mixed into the soil samples through vigorous shaking and stirring with a clean spatula to ensure homogenous application, and all treatments were run in independent duplicates. The samples were incubated for a period of 5–7 h to allow sufficient mixing and equilibration of the sample, yet sufficiently brief to ensure that the innate soil bacterial tolerance to the toxicant additions, rather than the induced tolerance following the selective growth of a tolerant community [12–14,16], were assessed. After this incubation, all samples were analyzed for bacterial growth using the leucine incorporation method [19] adapted for soil [9,20], importantly using short incubation periods (2 h). To estimate this, 20 ml water were added to the 2 g soil sample, followed by a homogenization/centrifugation step [9] to extract a bacterial suspension, which subsequently was used to estimate bacterial growth. Subsets of the same suspension were also analyzed for concentrations of Zn and Cu and for pH. After a filtration step (0.45 μm), soil solutions were analysed for metal concentrations using an inductively coupled plasma optical emission spectrophotometer (ICP-OES; Varian 700 – ES, Varian Inc. Scientific Instruments, Palo Alto, USA). Initially a semi-quantitative analysis was performed to determine major constituents using internal calibration to screen for any interference. The target metal ion concentrations were then determined quantitatively by calibration against a series of standard solutions derived from a commercial multi-element standard (Sigma-Aldrich, St Louis, USA) with preparation blanks used to determine background concentrations.

Particle size characterization

To characterize the particle size of the bulk form metal oxides, 10 g subsamples were added to stainless steel sieves (53 μm grid; a size fraction routinely used to differentiate between particulate and soluble organic matter [21]), that were subjected to shaking on a rotary shaker (200 rpm) overnight (16 h), with filter paper collectors placed below. The fraction of the subsample falling through the sieve was collected and weighed to characterize the proportion of fine particles in the bulk materials.

Statistical analysis and calculations

Tolerance values were expressed as the logarithm of the concentration of toxicant resulting in 50% inhibition (concentration at 50% effect, EC$_{50}$) of the short term assay for bacterial growth. A more potent toxicant inhibits the bacterial growth at a lower concentration, and thus a lower log(EC$_{50}$) indicates higher toxicity. The log(EC$_{50}$) values were determined by fitting a sigmoidal curve to model the concentration-response relationship, i.e. bacterial growth along the range of added metal concentrations, $y = c/(1+e^{b(x-a)})$, where y is the relative bacterial growth, x is the logarithm of the added concentration of metal, c is the bacterial growth in the no-addition control (at 0 mmol metal g^{-1} soil) a is the value of log(EC$_{50}$) and b is the parameter indicating the slope of the inhibition curve. The bacterial growth data was normalized to unity to present data as relative bacterial growth. KaleidaGraph 4.0 for Mac (Synergy software) was used to fit the model to the experimental data. To provide a more accurate estimate of bacterial growth in the unamended control (i.e. at 0 mmol g^{-1} added toxicant), the average of all the 0 mmol g^{-1} metal additions (i.e. nano, bulk and sulfate forms of both metals, all being the same treatment) were combined for each soil. The concentration dependence of metal concentrations in soil solution was tested using regression analysis (JMP 9.0 for Mac, SAS Inst., USA).

Results

Cu-toxicity for soil bacteria

The toxic effect of CuSO$_4$ on bacterial growth was clear in both soil types (Fig. 1C, F), with increasing concentrations of added toxicant effectively reducing bacterial growth to virtually zero, resulting in a tight fit of the sigmoidal curve equation used to establish inhibition curves (R^2>0.99 for both soils). Using the log(EC$_{50}$) as an index for toxicity of the metals, the bacterial growth was more susceptible to CuSO$_4$ in the mineral soil (log(EC$_{50}$) = 0.52±0.09; estimate ±1 SE) than in the organic soils (log(EC$_{50}$) = 1.28±0.04). This suggested that 50% of the bacterial growth was inhibited by about 3.4 mmol CuSO$_4$ g^{-1} soil in the mineral soils, and that about 19 mmol CuSO$_4$ g^{-1} resulted in a similar inhibition of bacterial growth in the organic soil. Bulk-CuO appeared inert by comparison, and did not appear to affect the bacterial growth in either soil in the studied interval (Fig. 1B, E). In contrast, the ENP form of CuO produced a pronounced inhibition of bacterial growth in the mineral soil (Fig. 1A), resulting in a clear concentration response curve that could be well-described by a sigmoidal model (R^2 = 0.98), and thus proved toxic to the bacterial community (log(EC$_{50}$) = 1.55±0.10). The toxic effect of the nano-CuO was much less pronounced in the organic soil (Fig. 1D), and we could not see clear evidence for a concentration-response relationship. However, we note that the highest concentration of nano-CuO (200 mmol g^{-1}) did appear to suppress bacterial growth somewhat in the organic soil.

Zn-toxicity for soil bacteria

The toxic effect of ZnSO$_4$ on bacterial growth was clear in both soil types (Fig. 2C, F), and increasing concentrations of added toxicant effectively suppressed bacterial growth to virtually zero,

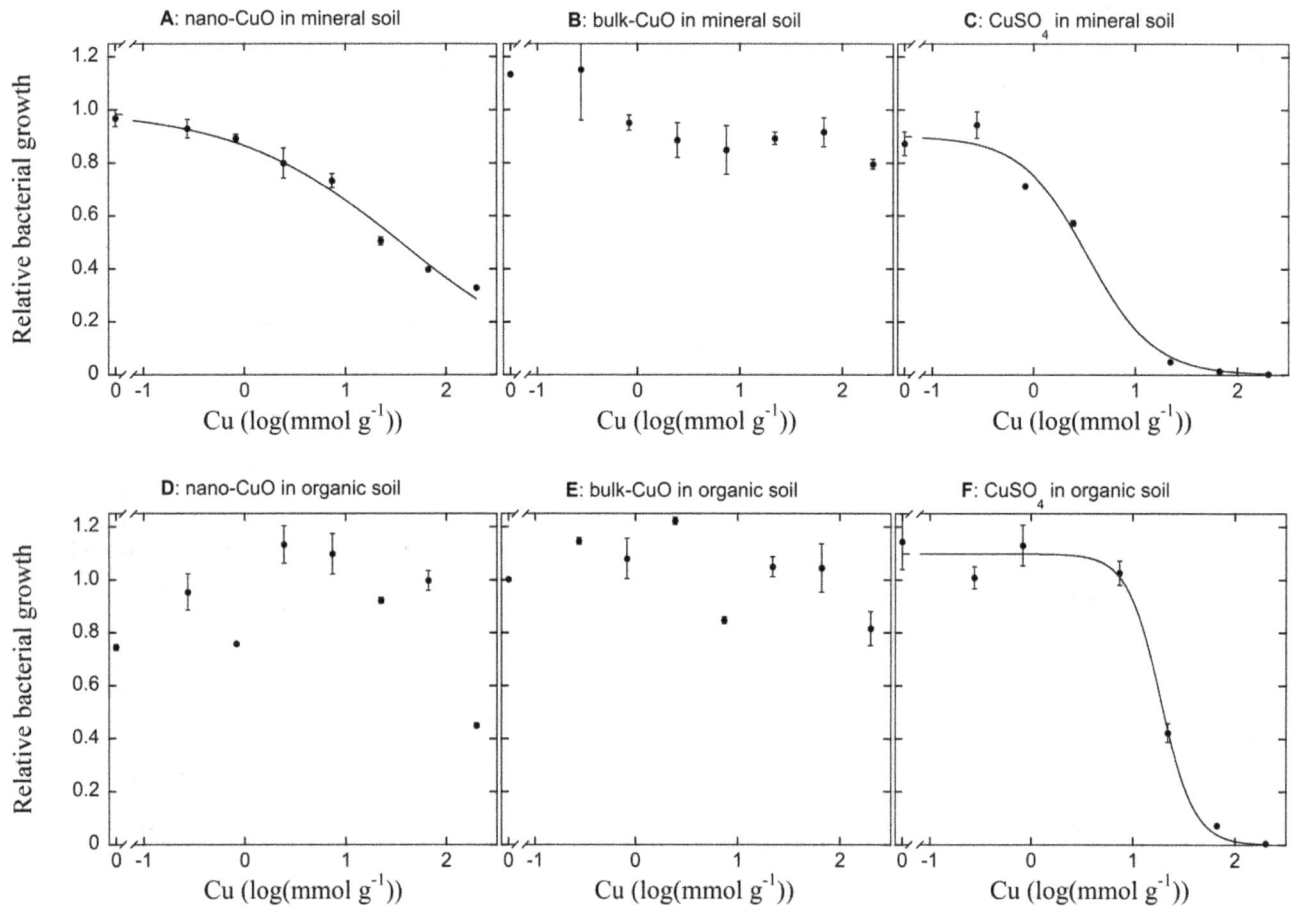

Figure 1. Cu toxicity to bacterial communities in mineral (panels A, B, C) and organic (panels D, E, F) soils. The effects of nano-sized (i.e. ENPs; panels A, D) and macroparticulate 'bulk-sized' (i.e. non-ENP) oxide (panels B, E) and as well as sulfate forms of Cu (panels C, F) on soil bacterial community growth rate are contrasted. The relationship between the relative bacterial growth (normalized relative to the bacterial growth rate in unamended soils) and rate of Cu application are described with a sigmoidal curve to establish the concentration response relationship. Only statistically significant relationships are presented as lines. Datapoints represent the mean of two independent replicates ±1 SE.

resulting in a good fit of the sigmoidal curve used to model the concentration-response relationships ($R^2 > 0.97$ for in both soils). Using the $\log(EC_{50})$ as an index for toxicity of Zn, the bacterial growth was more susceptible to $ZnSO_4$ in the mineral soil ($\log(EC_{50}) = 0.38 \pm 0.15$) than in the organic soil ($\log(EC_{50}) = 1.52 \pm 0.07$). This suggested that 50% of the bacterial growth was inhibited by about 2.4 mmol $ZnSO_4$ g^{-1} soil in the mineral soil, and that about 33 mmol $ZnSO_4$ g^{-1} resulted in a similar inhibition in the organic soil. Bulk-ZnO also effectively inhibited the bacterial growth in both soils in the studied interval (Fig. 2B, E), and clear concentration-response relationships could be estimated (both $R^2 > 0.97$; Fig. 2B, E). Using the $\log(EC_{50})$ as an index for potency of the substances in the two soils, it was evident that bulk-ZnO was more toxic to bacteria in the organic soil ($\log(EC_{50}) = 0.62 \pm 0.15$) than in the mineral soil ($\log(EC_{50}) = 1.11 \pm 0.08$). This suggested that 50% of the bacterial growth was inhibited already by 4.2 mmol bulk-ZnO g^{-1} soil in the organic soils, and that the same inhibition of bacteria only occurred by 13 mmol bulk-ZnO g^{-1} in the mineral soil. The ENP forms of ZnO also effectively reduced bacterial growth in both soils, and thus we could establish significant concentration-response relationships in both mineral ($R^2 = 0.83$; Fig. 2A), and in the organic soils ($R^2 = 0.68$; Fig. 2D). The nano-ZnO appeared to inhibit the bacterial communities more effectively in the mineral

soil ($\log(EC_{50}) = 1.81 \pm 0.10$) than in the organic soil ($\log(EC_{50}) = 2.27 \pm 0.14$). This suggested that 50% of the bacterial growth was inhibited by about 64 mmol nano-ZnO g^{-1} soil in the mineral soil, whilst the same inhibition was only reached at 185 mmol Zn g^{-1} soil in the organic soil.

Cu in soil solution

The presence of Cu in soil solution, as indicated by the ICP-OES-measurements after a filtration step, increased with higher added concentrations of $CuSO_4$ in both the mineral (P<0.001, $R^2 > 0.99$; Fig. 3C) and organic (P<0.001, Fig. 3F) soils. The incremental increases of higher added concentrations appeared to be small below 10 mmol Cu g^{-1}, after which the presence of Cu in soil solution increased at a higher rate, suggesting a threshold effect. Further, the solubility of Cu was not complete, and the presence in soil solution only reached maximal levels of around 70 μmol Cu g^{-1} or 20 Cu μmol g^{-1} in mineral and organic soils, respectively, i.e. only a fraction of 1×10^{-4}–4×10^{-4} of the added Cu was present in solution. Corresponding measurements showed that the Cu presence following the Bulk-CuO treatment appeared to only at background levels, and no concentration dependence was found in either soil (Fig. 3B, E). Increasing application rates of nano-CuO increased the Cu concentration, as indicated by the ICP-OES-measurements after a filtration step, in soil solution in

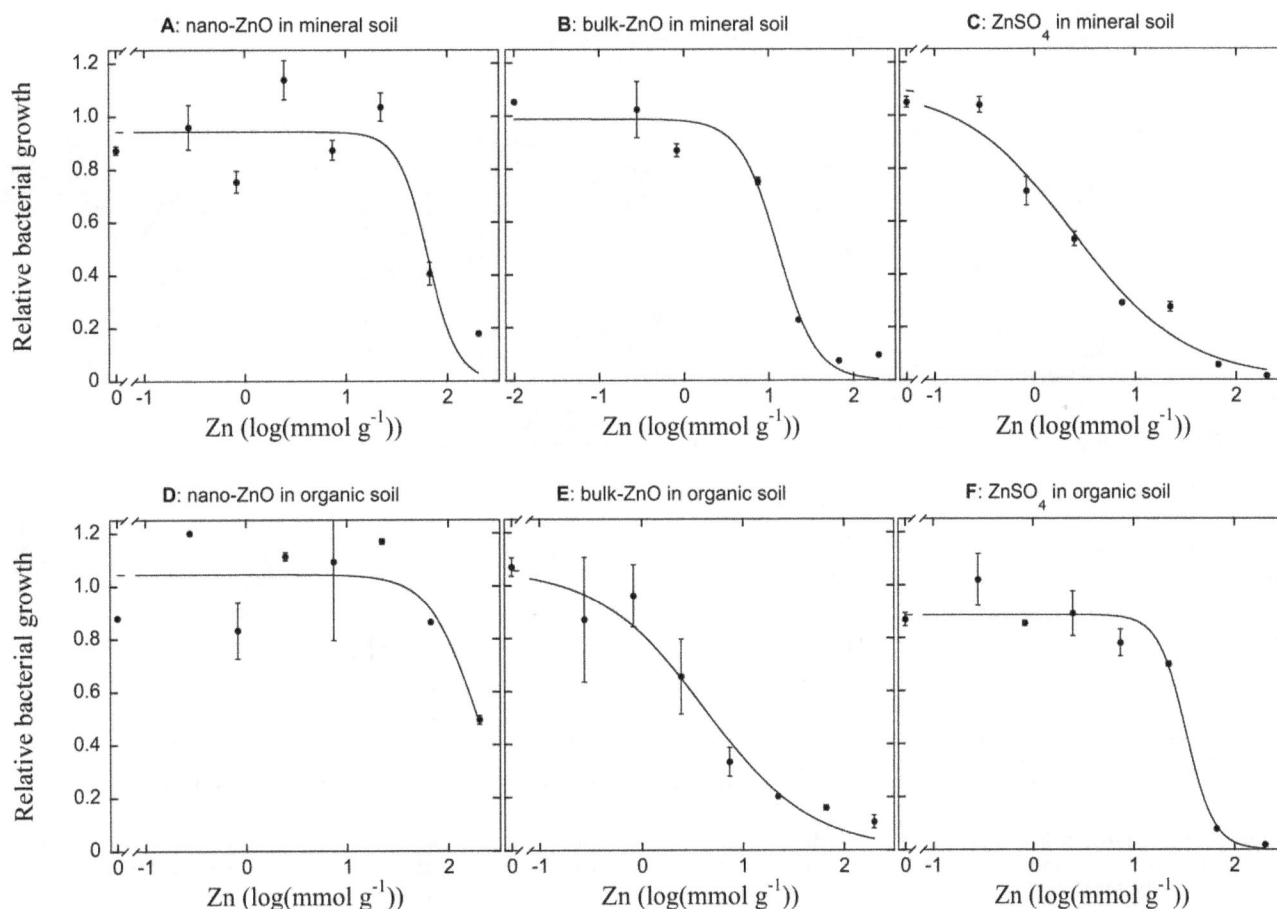

Figure 2. Zn toxicity to bacterial communities in mineral (panels A, B, C) and organic (panels D, E, F) soils. The effects of nano-sized (i.e. ENPs; panels A, D) and macroparticulate 'bulk-sized' (i.e. non-ENP) oxide (panels B, E) and as well as sulfate forms of Zn (panels C, F) on soil bacterial community growth rate are contrasted. The relationship between the relative bacterial growth (normalized relative to the bacterial growth rate in unamended soils) and rate of Zn application are described with a sigmoidal curve to establish the concentration response relationship (presented as lines). Datapoints represent the mean of two independent replicates ±1 SE. Sometimes error bars are hidden by symbols.

both soils (P<0.001, R^2>0.96 for both; Fig. 3A, D), and while the highest concentration of added nano-CuO increased Cu concentration 15 (Fig. 3A) to 20 (Fig. 3D) fold compared to the lowest concentration, to 0.15 and 0.20 µmol Cu g^{-1} in mineral and organic soils, respectively, these levels were less than 1% of those released by $CuSO_4$ at the highest concentrations.

Zn in soil solution

The presence of Zn in soil solution, as indicated by the ICP-OES-measurements after a filtration step, increased with higher application rates of $ZnSO_4$ in both the mineral (P<0.001, R^2 = 0.96; Fig. 4C) and organic (P<0.001, R^2 = 0.95; Fig. 4F) soils. Like for Cu, the incremental increases with higher application rates appeared to be small up to 10 mmol Zn g^{-1}, after which the presence in soil solution increased at a higher rate, again suggesting a threshold effect. Further, the solubility of Zn was not complete, and the presence in soil solution only reached maximal levels of around 30 µmol Zn g^{-1} in both soils, i.e. only a fraction of 2×10^{-4} was present in solution. In contrast with Cu, the bulk form of ZnO increased Zn concentrations in soil solution, as measured by the ICP-OES, as shown by clear relationships between application rates and concentration in solution for both mineral (P<0.001; R^2>0.99, Fig. 4B) and organic (P<0.001; R^2>0.99; Fig. 4E) soils. The Zn in soil solution increased 10–20

fold between lowest and highest application rates of bulk-ZnO, and reached maximal levels of 0.55 µmol Zn g^{-1} in mineral and 0.25 µmol Zn g^{-1} in organic soils. Nano-ZnO applications resulted in a clear relationship between the estimated concentration in solution and application rate in the mineral soil (P<0.01; R^2 = 0.94; Fig. 4A) whilst a tendency for the same pattern was also observed in the organic (P = 0.06; R^2 = 0.63; Fig. 4D) soil. Maximal levels of Zn in solution were just over 0.10 µmol Zn g^{-1} in the mineral soil and less than 0.05 µmol Zn g^{-1} in the organic soil.

Soil pH effects

Higher concentrations of nano-CuO gradually increased soil pH in both soils, by nearly 1 unit from pH 5.0 to nearly pH 6.0, in the mineral soil (Fig. S1A), and by about 0.3 units, from pH 6.6 to just under pH 7.0, in the organic soil (Fig. S1C). There was very little effect by bulk-CuO on pH in either soil. $CuSO_4$ drastically decreased soil pH in both soils, by nearly 2 units, from pH 5.0 to just over pH 3.0 in the mineral soil (Fig. S1A) and by more than 2 units, from pH 6.6 to less than pH 4.0 in the organic soil (Fig. S1C). There was very little influence by nano-ZnO on soil pH in both soils (Fig. S1B, D), while the bulk-ZnO increased soil pH by about 0.5 units in both soils. $ZnSO_4$ decreased soil pH by about 2 units in both soils, from pH 5 to just over pH 3 in the mineral soils

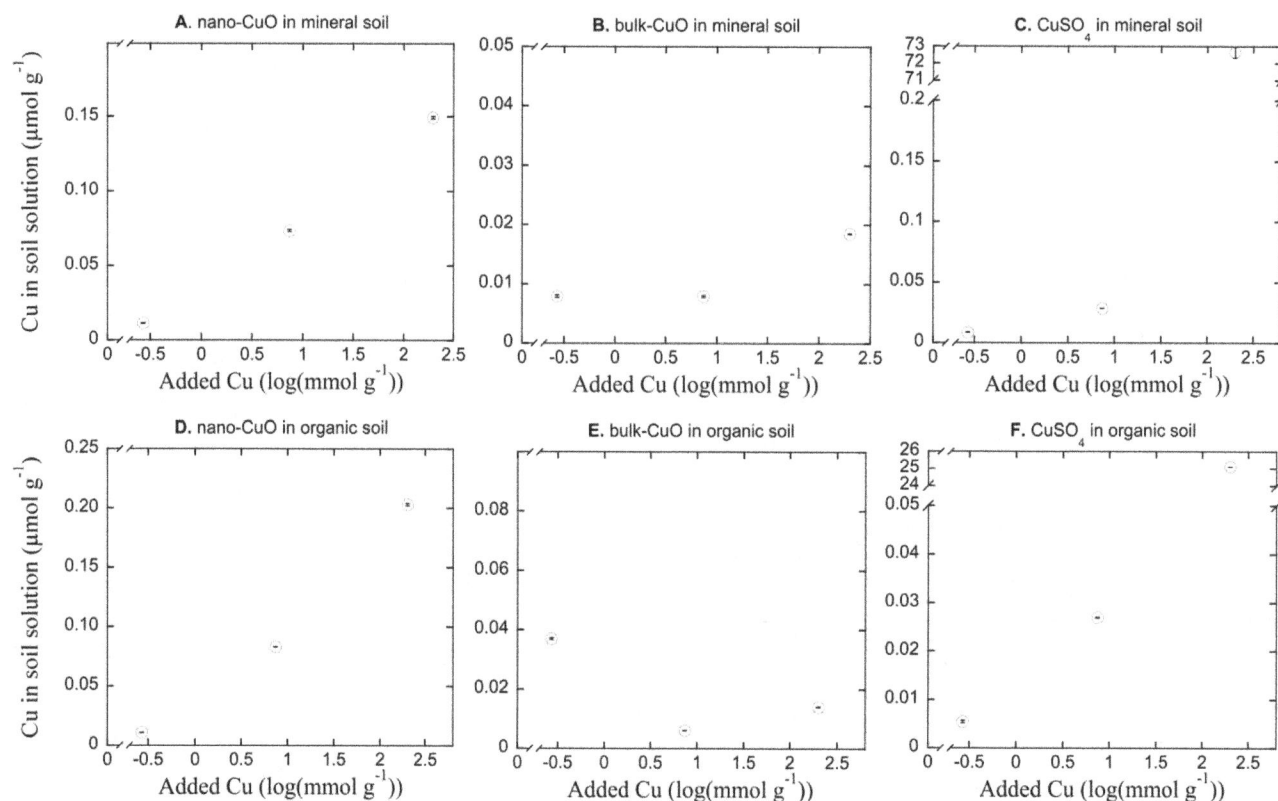

Figure 3. Free Cu in in soil solutions. The relationship between Cu concentration in soil solution and the application rate of nano-CuO (panels A, D), bulk-CuO (panels B, E) and CuSO₄ (panels C, F) in mineral (panels A, B, C) and organic (D, E, F) soils. Note the broken y-axis scales (panels C, F). Datapoints are the mean of three replicate analyses ±1 SE. Sometimes error bars are hidden by symbols.

(Fig. S1B) and from pH 6.6 to about pH 4.5 in the organic soils (Fig. S1D).

Particle size characterization

Size fractionation showed that 80.8% of the bulk-CuO and 95.5% of the ZnO was made up by particles (or aggregates) larger than 53 µm in at least one dimension.

Discussion

Comparative toxicity of nano-CuO

We established a clear dose-response relationship between higher concentrations of nano-CuO and reduced bacterial growth in the mineral soil, and a tendency for reduced growth at the highest concentration of nano-CuO in the organic soil, showing that there was a direct acute toxicity effect acting on soil bacteria. In addition, we could show that macroparticulate (i.e. non-ENP) forms of CuO, bulk-CuO, had no concentration response relationship for bacterial growth. That the ENP form of CuO, nano-CuO, rendered the compound more toxic compared with the bulk form suggested that the toxic effect was directly related to its nano-particulate form. The toxicity of nano-CuO, as well as the relative inertness of bulk-Cu, appeared to be directly related to their dissolution and presence of Cu in solution. While there was no relationship between higher application rates of bulk-CuO and Cu in solution, the dissolved Cu increased with higher application rates of nano-CuO. Comparing the nano-CuO with a soluble form, CuSO₄, showed that the presence of Cu in solution increased similarly between the compounds up to additions of

about 10 mmol Cu g⁻¹, but that the CuSO₄ contributed incrementally more to the dissolved Cu concentration at additions rates beyond this level. This also coincided with a more sharply inhibited bacterial growth in CuSO₄ treatments at rates higher than 10 mmol Cu g⁻¹, further strengthening the connection between measured bacterial toxicity and the presence of Cu in solution. In short, CuO was more toxic in an ENP form than in a macroparticulate (bulk) form, but a more soluble form, CuSO₄ was yet more toxic. It should be noted, however, that the acidifying effect of higher rates of CuSO₄ is likely to have added to acute toxicity of the compound, since an unambiguous connection between soil pH and bacterial growth has been established [22,23]. Further, it has been shown that acute reduction of soil pH by 2 units can reduce soil bacterial growth by about half [24], and additionally, a pH reduction will reduce Cu²⁺ solubility [25], thus affecting the presence of Cu in solution. Thus, it is possible that the acute toxicity of CuSO₄ is exaggerated by the change in pH in comparison to the oxide forms, while the small positive pH effects by the oxide additions (<0.5 pH-unit alterations between pH 6 and 8 in the different soils) would only be expected to affect bacterial growth negligibly [24]. The consequences of this confounding effect of metal sulphate additions for its property as a positive control are further discussed in following sections.

Comparative toxicity of nano-ZnO

Similar to the effects of Cu, we determined a clear dose-response relationship between higher concentrations of nano-ZnO and reduced bacterial growth in both soils, again indicating a direct acute toxicity response of the substance on soil bacteria. In

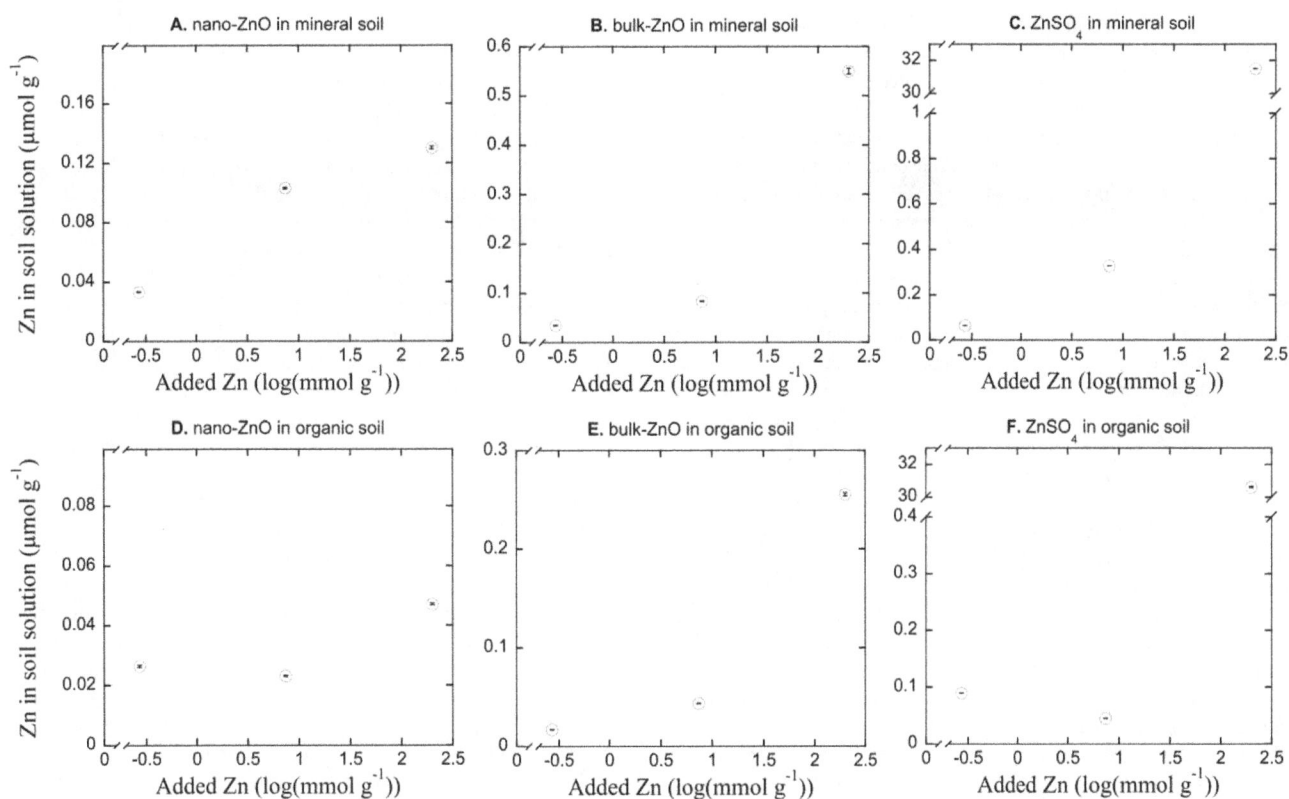

Figure 4. Free Zn in soil solutions. The relationship between Zn concentration in soil solution and the added concentration of nano-ZnO (panels A, D), bulk-ZnO (panels B, E) and ZnSO$_4$ (panels C, F) in mineral (panels A, B, C) and organic (panels D, E, F) soils. Note the broken y-axis scales (panels C, F). Datapoints are the mean of three replicate analyses ±1 SE. Sometimes error bars are hidden by symbols.

addition, we showed that macroparticulate (i.e. non-ENP) forms of ZnO, bulk-ZnO, possessed equally clear or stronger concentration response relationships for bacterial growth, and were even more toxic to bacterial growth (i.e. lower EC$_{50}$ values). As occurred for CuO, the toxicity of both forms of ZnO, appeared to be directly related to their dissolution and presence in soil solution. The dissolved Zn increased with higher application rates of nano-ZnO, but the effect size was rather small. There was an even stronger, or more pronounced, relationship between higher application rates of bulk-ZnO and dissolved Zn. Comparing the nano-ZnO with the bulk-form and the soluble form, ZnSO$_4$, showed that the presence of Zn in solution increased similarly for the three compounds up to additions of about 10 mmol Zn g^{-1} in the mineral soil, but that the bulk-ZnO and ZnSO$_4$ contributed incrementally more to the Zn concentration in solution than did nano-ZnO with higher additions beyond this rate. In the organic soil the pattern was different, and nano-ZnO or ZnSO$_4$ did not clearly increase Zn concentrations in solution with higher application rates up to 10 mmol Zn g^{-1}. The bulk-ZnO, in contrast, appeared to contribute to higher concentrations of Zn in solution also up to 10 mmol Zn g^{-1}. At higher application rates a threshold was reached for bulk-ZnO and ZnSO$_4$, and Zn concentrations in solution were significantly elevated at the highest application rate of ZnSO$_4$ compared to bulk-ZnO, and both were many-fold higher than nano-ZnO. The inhibition of bacterial growth correlated intimately with this pattern for Zn in solution, with lowest toxicity (highest EC$_{50}$ values) for nano-ZnO, intermediate for bulk ZnO and highest toxicity (lowest EC$_{50}$) for ZnSO$_4$ in mineral soils, while in the organic soil nano-ZnO had lowest toxicity (highest EC$_{50}$), with the ZnSO$_4$ being intermediate due to

its low contribution to Zn below 10 mmol Zn g^{-1} and bulk-ZnO being most toxic (lowest EC$_{50}$). The toxicity of ZnSO$_4$ also had potential to overestimate the toxicity of Zn due to the addition's soil acidifying effect (see discussion for CuSO$_4$ above).

Nano-particulate toxicity

The EC$_{50}$ values determined for the sulphate addition of the metals are well within the span previously obtained for soil organisms, validating our assessments of toxic effects by the additions in general and adding credence to our assessments of the ENP toxicity specifically. Obtained EC$_{50}$ values for the sulphate forms of Cu and Zn were slightly lower than corresponding SIR-estimated levels for soil microorganisms [26] and plants [27] but similar to other assessments using growth-based assays [11,28], which should be expected given the higher sensitivity of the growth-based assay [14,15].

The two soils used in this study had very different characteristics. In addition to very different organic matter concentrations, the soils also differed in clay content, pH and cation concentrations. It has been noted that ENPs are highly influenced by the concentration and form of organic matter in soil [29,30], influencing the ENPs tendency to form aggregates [31] and interaction with biomolecules [32]. Further, it has been suggested that one of the most influential parameters for the toxicity of metal ions, once in solution, is the effective cation exchange capacity [33], largely a product of the soil pH. While, simplistically, it can be assumed that the higher surface area of nano-form compared to bulk-form metal oxides should increase their potential to be dissolved into soil solution, the transition between oxides and free metal ions in solution is a two-stage process. First, metal oxide particles may interact with particles and organic matter constituents of the soils [34], e.g. forming aggregates, affecting their effective surface area in the soil. Second, once dissolved into soil solution, the metals will interact with other ions in solution and electronegative or charged functional groups on solids or macromolecules, forming metal complexes, which can reduce their concentration as free ions in solution, affecting the mass balance. The resulting contribution by metal additions to metal ion presence in soil solution is, consequently, hard to predict generally [35–37], and especially hard for ENP forms [29,31]. Thus, we safely can conclude that the presence of the metals in solution was differentially affected between the three different forms of Zn and Cu added in the two soils, but we are not able to assign these differences to specific factors such as organic matter content or soil pH, and more systematic comparisons of ranges of soils are required before we can start assigning these differences to mechanisms.

Using metal sulphates proved to be imperfect control treatments to evaluate the toxic effects of soluble metals. Both Cu and Zn sulphates greatly reduced the soil pH by up to 2 units in the highest metal addition rates, a well-known property of metal sulphate additions [38]. It is likely that this extensive acidification added to the toxic effects of the metal sulphates, and the presence of protons in solution is also known to modulate metal toxicity by competing with metal cations for biotic binding sites [39,40,41]. Further, higher solubility of metals at reduced pH, with acidification commencing and quickly increasing at 10 mmol metal g^{-1} and beyond, could have contributed the threshold-like effect observed for metals in solution (Fig. 3, 4). However, the comparative nature of our experiment design allows for the evaluation of the putative confounding toxicity that the acidification affected our bacterial growth based toxicity assay with. While the sulphide form of Zn consistently decreased the pH in both soil types by about 2 units, the bulk-ZnO had negligible effects on pH (supplementary fig. S1). A mechanism of toxicity shared by the substances, on the other hand, were bacterial exposure to Zn. However, the bulk-ZnO proved more toxic to bacteria, and reduced bacterial growth to levels comparable to ZnSO$_4$ at the highest application rates. This suggests that while pH in principle could have added to the metal toxicity, there is no evidence from our data to suggest that it did so to an important degree, but rather that the metal itself exerted the toxic effect. This result was also consistent for both soils types. The same argument would also be applicable to evaluating putative osmotic effect and ionic strength effects of the metal sulphate additions [42], where this also would have acted to make the salt form, ZnSO$_4$, more toxic than bulk-ZnO.

Three candidate hypotheses for the putative toxicity of ENP forms of metal oxides have been forwarded: (i) generation of reactive oxygen species that can cause lipid peroxidation and disrupt cell membranes as well as damage DNA [43,44], (ii) membrane disorganization [45] and (iii) release of metal ions [44,46] of well known ecotoxicity [33,47]. While we are not able to pinpoint the precise reasons for how the different forms of metal oxides and sulphates contributed to Cu and Zn concentrations in soil solution, we found a robust and unambiguous connection between bacterial growth inhibition and the measured metal concentration in soil solution. Our estimate of metal concentrations in soil solution is not capable of resolving in what form it is present there, as metal ions, suspended ENPs or dissolved metal complexes. All we can assess is the total metal concentrations (with a particulate size <0.45 μm due to the filtration step) carried in solution. If our additions resulted in ENP presence in soil solution after addition to soil samples, and these contributed to toxicity, detectable metal concentrations in the nano-Cu and nano-Zn treatments would be expected to expose bacteria to a higher

toxicity than similar concentrations of other forms (bulk and sulphate forms). There is no such evidence in our data, and the toxicity of metal concentrations in soil solution, irrespective of source (ENP or not), were found to inhibit bacteria to the same extent. Although we are not able to explicitly rule out that nano-particle related generation of reactive oxygen species or disorganization of membranes contributed to the toxicity of the ENPs, our results are consistent with only one of these mechanisms being influential for bacterial growth inhibition – the contribution to metal ion concentration in soil solution. Thus, the most parsimonious interpretation of our results would be that the contribution by the added metals to metal ion concentrations in soil solution was the more important mechanism for the observed toxicity to bacterial growth in soils, and that the direct influence of nano-particles on bacteria was negligible beyond this. While this conclusion is built on conjecture, direct measurements of metal ions concentrations, by means of e.g. a Cu specific electrode to measure Cu^{2+} [48,49], could be used to confirm the causality in our interpretation, and suggests a way forward.

That metal exposure can be detrimental for soil biota has been well-known for decades [42] and heavy metal toxicity is growing to be a mature subject field [33], as evidenced by the development of the biotic ligand model (BLM) to predict environmental toxicity of Cu, Zn and other metals [39,40,41]. The increasing application of ENP forms of metals have resulted in a new surge in studies of metal ENP [2,5], and assessments of ENP metal oxides in soil have been able to determine and show clear toxicity to e.g. soil bacteria [7]. However, to date, there is a shortage of careful evaluations of how the ENP form of the metal, *per se*, modulates its toxicity. We show that ENP oxides of Zn and Cu can inhibit bacterial growth in different soils. Moreover, the toxicity of the ENP metal oxide form differed from the non-ENP, but the difference was contingent on the soil studied. More soluble forms (sulphates) of Cu and Zn proved more toxic to soil bacteria than the metal oxide forms, ENP or otherwise. Emerging from these results, we find a tight connection between the presence of metal in soil solution and the resulting toxicity of the added metals, suggesting that the ENP form can be more toxic than non-ENP forms, but only when dissolution of the metal is higher in this form (this is the soil dependence). Although a framework to assess bioavailable metal concentrations has proved elusive [33,36,37], we suggest that efforts toward the synthetic goal of e.g. BLM [39,40,41] are more valuable than spending resources to reassess ecotoxicology of metal ENPs separately from general ecotoxicology of metals.

Supporting Information

Figure S1 The relationship between the pH in soil solution and the added concentration of nano-CuO, bulk-CuO and CuSO$_4$ (panels A, C) and nano-ZnO, bulk-ZnO and ZnSO$_4$ (panels B, D) in mineral (panels A, B) and organic (C, D) soils. Datapoints are the mean of two replicate analyses ± 1 SE. Sometimes error bars are hidden by symbols.

Acknowledgments

We thank Dr. Graham A. Ormondroyd for helpful comments on an earlier version of the manuscript.

Author Contributions

Conceived and designed the experiments: JR DLJ. Performed the experiments: JR. Analyzed the data: JR. Contributed reagents/materials/analysis tools: JR KA SFC DLJ. Wrote the paper: JR. Participated actively in revision of manuscript: JR KA SFC DLJ.

References

1. Auffan M, Rose J, Bottero J-Y, Lowry GV, Jolivet J-P, et al. (2009) Towards a definition of inorganic nanoparticles from an environmental, health and safety perspective. Nat Nanotechnol 4: 634–641.
2. Godwin HA, Chopra K, Bradley KA, Cohen Y, Harthorn BH, et al. (2009) The University of California center for the environmental implications of nanotechnology. Environ Sci Technol 43: 6453–6457.
3. Jiang X, Herricks T, Xia Y (2002) CuO nanowires can be synthesized by heating copper substrates in air. Nano Lett 71: 1308–1316.
4. Klaine SJ, Alvarez PJJ, Batley GE, Fernandez TF, Handy RD, et al. (2008) Nanomaterials in the environment: behaviour, fate, bioavailability, and effects. Environ Sci Technol 27: 1825–1851.
5. Colvin VL (2003) The potential environmental impact of engineered nanomaterials. Nat Biotechnol 21: 1166–1170.
6. Chithrani BD, Ghazani AA, Chan WC (2006) Determining the size and shape dependence of gold nanoparticle uptake into mammalian cells. Nano Lett 6: 662–668.
7. Ge Y, Schimel JP, Holden PA (2011) Evidence for negative effects of TiO$_2$ and ZnO nanoparticles in soil bacterial communities. Environ Sci Technol 45: 1659–1664.
8. Li M, Zhu L, Lin D (2011) Toxicity of ZnO nanoparticles to Escherichia coli: mechanism and the influence of medium components. Environ Sci Technol 45: 1977–1983.
9. Bååth E, Pettersson M, Söderberg KH (2001) Adaptation of a rapid and economical microcentrifugation method to measure thymidine and leucine incorporation by soil bacteria. Soil Biol Biochem 33: 1571–1574.
10. Rousk J, Bååth E (2011) Growth of saprotrophic fungi and bacteria in soil. FEMS Microbiol Ecol 78: 17–30.
11. Rajapaksha RNCP, Tobor-Kapłon MA, Bååth E (2004) Metal toxicity affects fungal and bacterial activities in soil differently. Appl Environ Microbiol 70: 2966–2973.
12. Nolsø Aaen K, Holm PE, Priemé A, Ngo Hung N, Brandt KK (2011) Comparison of aerobic and anaerobic [^{3}H]leucine incorporation assays for determining pollution-induced bacterial community tolerance in copper-polluted, irrigated soils. Environ Toxicol Chem 30: 588–595.
13. Aldén Demoling L, Bååth E (2008) Use of pollution-induced community tolerance of the bacterial community to detect phenol toxicity in soil. Environ Toxic Chem 27: 334–340.
14. Brandt KK, Sjöholm OR, Krogh KA, Halling-Sørensen B, Nybroe O (2009) Increased pollution-induced bacterial community tolerance to sulfadiazine in soil hotspots amended with artificial root exudates. Environ Sci Technol 43: 2963–2968.
15. Rousk J, Aldén Demoling L, Bååth E (2009) Contrasting short-term antibiotic effects on respiration and bacterial growth compromises the validity of the selective respiratory inhibition technique to distinguish fungi and bacteria. Microb Ecol 58: 75–85.
16. Aldén Demoling L, Bååth E (2008) No long-term persistence of bacterial pollution-induced community tolerance in tylosin-polluted soil. Environ Sci Technol 42: 6917–6921.
17. Rousk J, Elyaagubi FK, Jones DL, Godbold DL (2011) Bacterial salt tolerance is unrelated to soil salinity across an arid agroecosystem salinity gradient. Soil Biol Biochem 17: 1881–1887.
18. Rousk J, Jones DL (2010) Loss of low molecular weight dissolved organic carbon (DOC) and nitrogen (DON) in H$_2$O and 0.5 M K$_2$SO$_4$ soil extracts. Soil Biol Biochem 42: 2331–2335.
19. Kirchman D, K'Nees E, Hodson R (1985) Leucine incorporation and its potential as a measure of protein synthesis by bacteria in natural aquatic systems. Appl Environ Microbiol 59: 3605–3617.
20. Bååth E (1994) Measurement of protein synthesis by soil bacterial assemblages with the leucine incorporation technique. Biol Fert Soils 17: 147–153.
21. Baldock JA, Skjemstad JO (1999) Soil organic carbon/soil organic matter. Soil analysis. Collingwood, Civ, Australia, CSIRO Publishing. pp 159–170.
22. Rousk J, Brookes PC, Bååth E (2009) Contrasting soil pH effects on fungal and bacterial growth suggest functional redundancy in carbon mineralization. Appl Environ Microbiol 75: 1589–1596.
23. Rousk J, Brookes PC, Glanville HC, Jones DL (2011) Lack of correlation between turnover of low-molecular-weight dissolved organic carbon and differences in microbial community composition or growth across a soil pH gradient. Appl Environ Microbiol 77: 2791–2795.
24. Fernández-Calviño D, Rousk J, Brookes PC, Bååth E (2011) Bacterial pH-optima for growth track soil pH, but are higher than expected at low pH. Soil Biol Biochem 43: 1569–1575.
25. Fernández-Calviño D, Pateiro-Moure M, López-Periago E, Arias-Estévez M, Nóvoa-Muñoz JC (2008) Copper distribution and acid-base mobilization in vineyard soils and sediments from Galicia (NW Spain). Eur J Soil Sci 59: 315–326.
26. Broos K, St J Warne M, Heemsbergen DA, Stevens D, Barnes MB, et al. (2007) Soil factors controlling the toxicity of copper and zinc to microbial processes in Australian soils. Environ Toxicol Chem 26: 583–590.
27. Rooney C, Zhai FJ, McGrath SP (2006) Soil factors controlling the expression of Copper toxicity for plants in a wide range of European soils. Environ Toxicol Chem 25: 726–732.
28. Díaz-Raviña M, Bååth E, Frostegård Å (1994) Multiple heavy metal tolerance of soil bacterial communities and its measurement by a thymidine incorporation technique. Appl Environ Microbiol 60: 2238–2247.
29. Lin D, Tian X, Wu F, Xing B (2010) Fate and transport of engineered nanomaterial in the environment. J Environ Qual 39: 1896–1908.
30. Pan B, Zing B (2008) Adsorption mechanisms of organic chemicals on carbon nanotubes. Environ Sci Technol 42: 9005–9013.
31. Keller AA, Wang H, Zhou D, Lenihan HS, Cherr G, et al. (2010) Stability and aggregation of metal oxide nanoparticles in natural aqueous matrices. Environ Sci Technol 44: 1962–1967.
32. Saleh NB, Pferfferle LD, Elimelech M (2010) Influence of biomacromolecules and humic acid on the aggregation kinetics of single walled carbon nanotubes. Environ Sci Technol 44: 2412–2418.
33. Giller KE, Witter E, McGrath SP (2009) Heavy metals and soil microbes. Soil Biol Biochem 41: 2031–2037.
34. Wang Z, Li J, Zhao J, Xing B (2011) Toxicity and internalization of CuO nanoparticles to prokaryotic alga Microcytis aeruginosa as affected by dissolved organic matter. Environ Sci Technol 45: 6032–6040.
35. Lofts S, Spurgeon DJ, Svendsen C, Tipping E (2004) Deriving soil critical limits for Cu, Zn, Cd, and Pb: A method based on free ion concentrations. Environ Sci Technol 38: 3623–3631.
36. Zhao FJ, Rooney CP, Zhang H, McGrath SP (2006) Comparison of soil solution speciation and DGT measurement as an indicator of copper bioavailability to plants. Environ Toxicol Chem 25: 733–742.
37. Li HF, Gray C, Micó C, Zhao FJ, McGrath SP (2009) Phytotoxicity and bioavailability of cobalt to plants in a range of soils. Chemosphere 75: 1066–1077.
38. Speir TW, Kettles HA, Percival HJ, Parshotam A (1999) Is soil acidification the cause of biochemical responses when soils are amended with heavy metal salts? Soil Biol Biochem 31: 1953–1961.
39. Thakali S, Allen HE, Di Toro DM, Ponizovsky AA, Rooney CP, et al. (2006) Terrestrial biotic ligand model. 2. Application to Ni and Cu toxicities to plants, invertebrates, and microbes in soil. Environ Sci Technol 40: 7094–7100.
40. Mertens J, Degryse F, Springael D, Smolders E (2007) Zinc toxicity to nitrification in soil and soilless culture can be predicted with the same biotic ligand model. Environ Sci Techn 41: 2992–2997.
41. Ore S, Mertens J, Brandt KK, Smolders E (2010) Copper toxicity to bioluminescent Nitrosomonas europaea in soil is explained by the free metal ion activity in pore water. Environ Sci Technol 44: 9201–9206.
42. Smolders E, Oorts K, Sprang PV, Schoeters I, Janssen CR, et al. (2009) Toxicity of trace metals in soil as affected by soil type and aging after contamination: using calibrated bioavailability models to set ecological standards. Environ Toxicol Chem 28: 1633–1642.
43. Hanley C, Layne J, Punnoose A, Reddy KM, Coombs I, et al. (2008) Preferential killing of cancer cells and activated human T cells using ZnO nanoparticles. Nanotechnology 19: 29.
44. Miller RJ, Lenihan HS, Muller EB, Tseng N, Hanna SK, et al. (2010) Impacts of metal oxide nanoparticles on marine phytoplankton. Environ Sci Technol 44: 7329–7334.
45. Adams LK, Lyon DY, Alvarez PJ (2006) Comparative ecotoxicity of nanoscale TiO$_2$, SiO$_2$, and ZnO water suspensions. Water Res 40: 3627–3532.
46. Auffan M, Rose J, Wiesner MR, Bottero JY (2009) Chemical stability of metallic nanoparticles: A parameter controlling their potential cellular toxicity in vitro. Environ Pollut 157: 1127–1133.
47. Giller KE, Witter E, McGrath SP (1998) Toxicity of heavy metals to microorganisms and microbial processes in agricultural soils: A review. Soil Biol Biochem 30: 1389–1414.
48. Käkinen A, Bondarenko O, Ivask A, Kahru A (2011) The effect of composition of different ecotoxicological test media on free and bioavailable copper from CuSO$_4$ and CuO nanoparticles: comparative evidence from a cu-selective electrode and a Cu-Biosensor. Sensors 11: 10502–10521.
49. Brandt KK, Holm PE, Nybroe O (2008) Evidence for bioavailable Cu-dissolved organic matter complexes and transiently increased Cu bioavailability in manure-amended soils as determined by bioluminescent bacterial biosensors. Environ Sci Technol 42: 3102–3108.

18

Does *S*-Metolachlor Affect the Performance of *Pseudomonas* sp. Strain ADP as Bioaugmentation Bacterium for Atrazine-Contaminated Soils?

Cristina A. Viegas[1,2]*, Catarina Costa[2], Sandra André[4], Paula Viana[4], Rui Ribeiro[3], Matilde Moreira-Santos[3]

1 Department of Bioengineering, Instituto Superior Técnico (IST), Technical University of Lisbon (TUL), Lisboa, Portugal, 2 Institute for Biotechnology and Bioengineering (IBB), Centre for Biological and Chemical Engineering, Instituto Superior Técnico (IST), Technical University of Lisbon (TUL), Lisboa, Portugal, 3 Instituto do Mar (IMAR), Department of Life Sciences, University of Coimbra, Coimbra, Portugal, 4 Agência Portuguesa do Ambiente (APA), Amadora, Portugal

Abstract

Atrazine (ATZ) and *S*-metolachlor (*S*-MET) are two herbicides widely used, often as mixtures. The present work examined whether the presence of *S*-MET affects the ATZ-biodegradation activity of the bioaugmentation bacterium *Pseudomonas* sp. strain ADP in a crop soil. *S*-MET concentrations were selected for their relevance in worst-case scenarios of soil contamination by a commercial formulation containing both herbicides. At concentrations representative of application of high doses of the formulation (up to 50 µg g^{-1} of soil, corresponding to a dose approximately 50× higher than the recommended field dose (RD)), the presence of pure *S*-MET significantly affected neither bacteria survival (~10^7 initial viable cells g^{-1} of soil) nor its ATZ-mineralization activity. Consistently, biodegradation experiments, in larger soil microcosms spiked with 20× or 50×RD of the double formulation and inoculated with the bacterium, revealed ATZ to be rapidly (in up to 5 days) and extensively (>96%) removed from the soil. During the 5 days, concentration of *S*-MET decreased moderately to about 60% of the initial, both in inoculated and non-inoculated microcosms. Concomitantly, an accumulation of the two metabolites *S*-MET ethanesulfonic acid and *S*-MET oxanilic acid was found. Despite the dissipation of almost all the ATZ from the treated soils, the respective eluates were still highly toxic to an aquatic microalgae species, being as toxic as those from the untreated soil. We suggest that this high toxicity may be due to the *S*-MET and/or its metabolites remaining in the soil.

Editor: Stephen J. Johnson, University of Kansas, United States of America

Funding: This work was supported by Fundo Europeu de Desenvolvimento Regional and Fundação para a Ciência e Tecnologia (http://www.fct.pt/), Portugal (Contracts PTDC/AAC-AMB/111317/2009, PTDC/AMB/64230/2006, PPCDT/AMB/56039/2004, and Ciência 2007-FSE and POPH funds). The funders had no role in study design, data collection and analysis, decision to publish, or preparation of the manuscript.

Competing Interests: The authors have declared that no competing interests exist.

* E-mail: cristina.viegas@ist.utl.pt

Introduction

Diverse pesticides (and their metabolites), fertilizers and organic components used in commercial formulations can be found in agricultural soils due to intensive use in crop cultivation or accidental spills. In particular, agricultural dealership sites and mix-load or disposal sites may represent potential sources of environmental contamination with high levels of diverse compounds [1,2]. For example, 205, 2272, 13, 1829 and 108 µg g^{-1} of soil of the *s*-triazine herbicides atrazine (ATZ), cyanazine, simazine and the chloroacetanilide herbicides metolachlor (MET) and alachlor, respectively, were measured in soil samples from a mix-load site [1]. As a consequence, mixtures of chemicals are likely to contaminate freshwaters via leaching, drainage and runoff events [3–5], and toxicological synergies of multiple chemicals may enhance deleterious effects on ecosystems and human health [6]. Herbicidal formulations containing ATZ, either as the sole active ingredient or in combination with the racemic mixture of MET or, more recently, with the product enriched in the active *S*-enantiomer (*S*-MET), have been used worldwide for broad-spectrum weed control in several crops [7–10].

In soil, ATZ presents moderate persistency (DT$_{50}$ = 28–150 days, field, aerobic) and mobility (soil organic carbon/water partition coefficient, K$_{OC}$ = 100 L kg^{-1}) [8]. These properties and increasing concerns regarding potential impact of ATZ and its toxic chlorinated N-dehalkylated metabolites on human health and ecosystems [11–13], promoted active research on ATZ-degrading microorganisms and on bioremediation strategies aiming to reduce soil contamination to safe levels and to minimize dispersion into surrounding aquatic compartments [1,14–19]. *Pseudomonas* sp. ADP is the best-characterized ATZ-mineralizing bacteria and uses ATZ as sole N source by means of a catabolic pathway encoded in the plasmid pADP-1 [16,18]. It has high potential for the bioaugmentation of ATZ-contaminated soils, but, presumably due to C limitation and low survival in soil, it was found to be less effective at high ATZ concentrations that are relevant in the case of spill or careless disposal scenarios [17,18]. As part of a framework for the rational bioremediation of ATZ-contaminated land, we recently presented evidences that a clean-up strategy, combining soil bioaugmentation with this bacterial strain and biostimulation with citrate [17], was effective at a larger microcosms scale [14,15]. It led to the rapid removal of ATZ from a natural crop soil spiked with a commercial formulation containing ATZ as single active ingredient at doses mimicking worst-case scenarios (e.g. spills and/or concentrated hotspots) [14,15]. More importantly, assessment of the ecotoxicity of soil samples and eluates/leachates from the bioremediated soils proved the effective soil decontamination in less than 10 days, hence contributing to significantly diminish the toxicity impact in the aquatic compartment [14].

S-metolachlor is relatively non-persistent ($DT_{50} = 21$ days, field, aerobic) and moderately mobile ($K_{OC} = 226$ L kg^{-1}) in soil, but very persistent in water [8,9]. Both the racemic mixture and the S-enantiomer have the potential to cause adverse effects on microorganisms [2,20] and on higher terrestrial and aquatic organisms [8,20–22]. Freshwater macrophytes and microalgae are however the most vulnerable ones [6,8,23]. To our knowledge, the few published studies addressing the effect of MET or S-MET on ATZ-degradation in soil have been limited to indigenous soil microorganisms and intrinsic biodegradation. In some cases a negative effect was demonstrated [2] while in others there was apparently no impact [24].

Possible deleterious effects exerted by S-MET [2] could contribute to diminish the efficacy of the bioremediation tool previously shown to be effective for the clean-up of soils contaminated with high concentrations of ATZ [14,15,17]. Therefore, the present work aimed examining whether S-MET could negatively affect the survival and the ATZ-biodegradation performance of the bioaugmentation bacterium Pseudomonas sp. strain ADP in a crop soil. We used a representative crop soil from Central Portugal [14,15] spiked with mixtures of ATZ and S-MET (either as the commercial formulation Primextra S-Gold or as pure active ingredients). Doses mimicked worst-case scenarios of soil contamination, being thus higher than the recommended dose (RD) for weed control in corn plantations. The following issues were addressed: first, the ability of Pseudomonas sp. ADP, either combined or not with citrate amendment, to mineralize [ring-UL-^{14}C]ATZ mixed with increasing doses of Primextra S-Gold (up to 50×RD), in soil at small laboratory scale. Second, the effects

of S-MET in a pure form on the bacteria survival and on its ATZ-mineralizing ability, at concentrations of the pure active ingredients representing up to 50×RD, also in small soil microcosms. Third, the rate and extent of biodegradation of ATZ from 20× or 50×RD Primextra S-Gold when applying the bioremediation tool, in larger and more realistic soil microcosms [14,15]. The efficacy of ATZ-biodegradation in soil microcosms was evaluated mainly by performing microalgae ecotoxicity tests (72-hours Pseudokirchneriella subcapitata growth tests) on the eluates from soil samples. At this stage, eluates from soils contaminated with the herbicidal formulation Atrazerba FL (with ATZ as single active ingredient) and previously reported to be decontaminated after a 10-days treatment with the bioremediation tool [14] were also used for comparison purposes. The extent of ATZ removal was also examined with chemical analysis of ATZ and its metabolites in soil samples. In addition, possible modifications in the concentration of S-MET initially present in the soil microcosms contaminated with Primextra S-Gold as well as the possible formation of its major degradation products S-MET oxanilic acid (OA) and S-MET ethanesulfonic acid (ESA) [8,25,26] were also examined during the biodegradation experiments carried out.

Results

ATZ mineralization by Pseudomonas sp. ADP in soil spiked with Primextra S-Gold

Following inoculation with viable cells of the bioaugmentation bacterium ($1.3 \pm 0.5 \times 10^7$ CFU g^{-1} soil dry weight) of soil spiked with mixtures of [^{14}C]ATZ plus increasing doses of non-labeled ATZ from the double formulation Primextra S-Gold (5×, 20× or 50×RD), ATZ mineralization started rapidly. The percentage of ^{14}CO$_2$ produced from the ^{14}C-labeled ATZ attained maximal values within 3 days (Fig. 1A). Comparatively, ATZ mineralization was negligible (<2%) in the non-inoculated control soil during the same time-period (Fig. 1A). However, a moderate inhibition of rate and extent of ATZ mineralization occurred in the soils contaminated with doses of Primextra S-Gold increasing from 5×

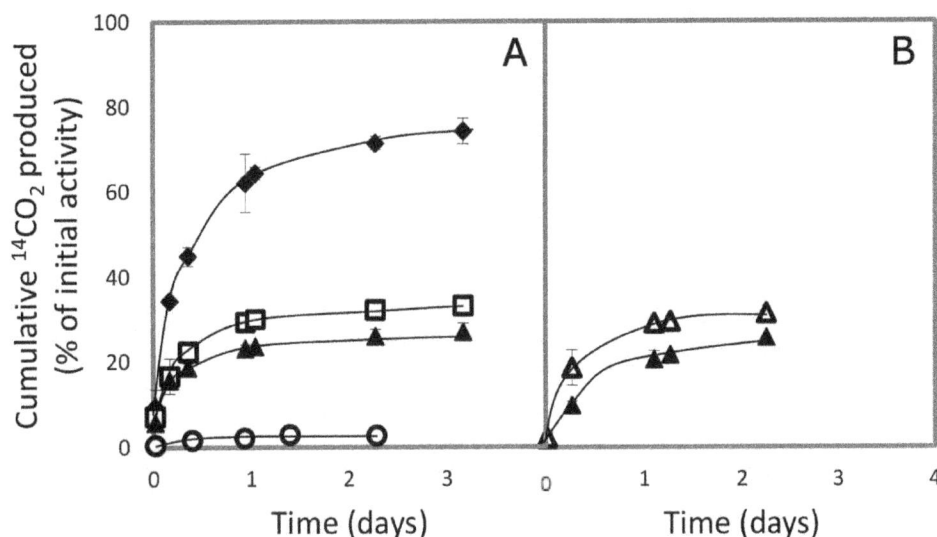

Figure 1. ATZ mineralization by Pseudomonas sp. ADP Rif^R in soil contaminated with Primextra S-Gold. Time-course formation of ^{14}CO$_2$ from [ring-UL-^{14}C]ATZ in (**A**) soil spiked with [^{14}C]ATZ plus increasing doses of Primextra S-Gold as follows: 5× (◆), 20× (□) and 50×RD (▲), without citrate amendment; or (**B**) soil spiked with [^{14}C]ATZ plus 50×RD of Primextra S-Gold and amended with 3.4 mg g^{-1} trisodium citrate (△) or non-amended (▲). The average amount of ^{14}CO$_2$ released from non-inoculated control soil is shown for comparison (○) in (**A**). Data are means±SD of measurements from three replicated samples from at least two independent experiments under each condition.

up to 50×RD (Fig. 1A). For example, the percentage of initial labeled ATZ evolving as $^{14}CO_2$ at day 3 were 74.3±3.0% and 27.0±2.1% in the soil microcosms with 5× and 50×RD of Primextra S-Gold, respectively (Fig. 1A). These values correspond to approximately 37 and 24%, respectively, of the total estimated amount of ATZ mixed into the soil (assuming that labeled and non-labeled ATZ may be mineralized homogeneously). The soil that had been contaminated with 50×RD of Primextra S-Gold showed the lowest percentage of $[^{14}C]$ATZ mineralization following bioaugmentation (Fig. 1A). In this soil, enhancement of the ratio of soluble carbon to nitrogen from atrazine (C_s:N_{atz}) from ∼1 (in the crop soil used) up to ∼50 [15], due to soil amendment with trisodium citrate, led to a slight but significant increase in the rate and extent of $^{14}CO_2$ formation (Fig. 1B).

Effects of S-MET in Pseudomonas sp. ADP survival and ability to mineralize ATZ in soil

Viable populations of Pseudomonas sp. ADP inoculated into the soil were exposed to mixtures of ATZ plus S-MET at concentrations representing applications of approximately 30× or 50×RD of each active ingredient (Fig. 2). For the soil contaminated with 24 μg ATZ g^{-1}, the addition of S-MET at 30 μg g^{-1} (representing ∼30×RD of each active substance) and up to 60 μg g^{-1}, did not significantly affect the bacterial survival (Fig. 2A). Moreover, the rate and extent of $[^{14}C]$ATZ mineralization was essentially the same whether or not S-MET was added to soil previously contaminated with a total of 24 or 40 μg ATZ g^{-1} of soil (Fig. 2B), indicating that presence of S-MET does not significantly affect the ability of Pseudomonas sp. ADP to mineralize ATZ in the worst-case conditions tested herein.

Biodegradation of ATZ from Primextra S-Gold in larger soil microcosms

The performance of the bioaugmentation/biostimulation treatment for ATZ-contaminated soils, consisting on one initial inoculation with Pseudomonas sp. ADP (4.1±1.2×10⁷ CFU g^{-1} soil dry weight) combined with soil amendment with trisodium citrate (C_s:N_{atz}∼50) [15], was examined in larger soil microcosms spiked with 20× or 50×RD of Primextra S-Gold (Fig. 3). For both doses of the commercial formulation and upon soil bioaugmentation, bacterial numbers were always higher (2.5-fold, in average) in the soil amended with citrate compared with non-amended one (Figs. 3A and B). Despite that, whether or not soil was amended with citrate, most of the initial ATZ was rapidly removed from soil in up to 2 days with no lag period required (Figs. 3C and D). This high rate of ATZ biodegradation contrasted with the high levels of ATZ remaining in the untreated soils (Figs. 3C and D). Nevertheless, slight differences on the extent of ATZ biodegradation in the inoculated soils were observed depending on the initial level of soil contamination. For example, in the soil spiked with 20×RD of Primextra S-Gold, the ATZ concentration declined by ∼96%, from 12.8±0.4 to less than 0.5 μg g^{-1}, in only 5 days (Fig. 3C). On the other hand, in the soil with the highest dose of Primextra S-Gold (50×RD, corresponding to an initial measured ATZ of 29.1±12.5 μg g^{-1} soil dry weight), inoculation with the bacterium without citrate amendment led to quite high levels of ATZ still remaining in the soil, namely 2.4 and 1.4±0.8 μg ATZ g^{-1} at days 5 and 8, respectively (Fig. 3D and data not shown). Combination of soil bioaugmentation with citrate amendment apparently allowed a slight but significant improvement in ATZ removal from the soil, with its concentration decreasing to 0.4±0.1 μg g^{-1} in 8 days (Fig. 3D and data not shown).

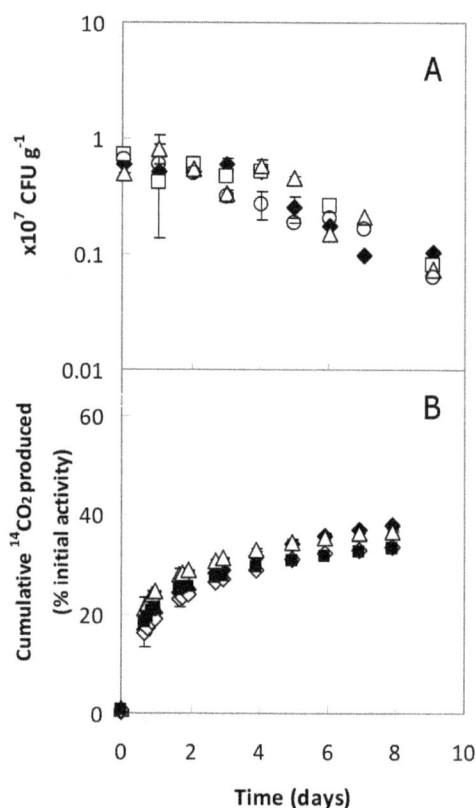

Figure 2. Effects of S-MET in Pseudomonas sp. ADP Rif^R (A) survival and (B) ATZ- mineralization in soil. In (A), soil previously contaminated with 24 μg ATZ g^{-1} of soil was supplemented with increasing concentrations of S-MET, namely 0 (◆), 15 (□), 30 (△) or 60 (○) μg g^{-1} of soil, prior to inoculation. In (B), soil was spiked with a total of 24 μg ATZ g^{-1} of soil (including $[^{14}C]$ATZ) plus 30 μg S-MET g^{-1} (△) or no S-MET (◆), or with a total of 40 μg ATZ g^{-1} plus 50 μg S-MET g^{-1} (■) or no S-MET (◇), prior to inoculation. Data are means±SD of measurements from at least duplicate determinations from two or three independent experiments under identical conditions.

Fate of S-MET and its major degradation products in the soil microcosms contaminated with Primextra S-Gold

Dissipation of S-MET in the soil microscosms was not different in untreated or inoculated soil (Figs. 3E and 3F). Indeed, in all the different conditions tested, its concentration was moderately reduced by around 40% on average during the first 5 days. For example, S-MET concentration declined from the initial 21±5 and 35±19 μg g^{-1} in the 20×RD and the 50×RD-contaminated soils, respectively, to 13±2 and 21±4 μg g^{-1} at day 5 (Figs. 3E and 3F), and maintained identical values after 8 days (data not shown). This decrease in the concentration of S-MET was accompanied by the accumulation of its derivatives S-MET ESA and S-MET OA in the soil microcosms. Namely, S-MET ESA concentration increased from undetectable values (<6 μg kg^{-1} soil dry weight), at time zero, to 27±9 and 46±19 μg kg^{-1} after 5 days of incubation, in the 20×RD and the 50×RD-contaminated soils, respectively. The S-MET OA concentration increased from <6 to 42±22 and 87±33 μg kg^{-1}, respectively, in the same period of time. At day 8, the concentrations of these compounds measured in the soil microcosms were somewhat lower, namely 22±2 and 28±7 μg S-MET ESA kg^{-1}, or 29±5 and 43±35 μg S-MET OA kg^{-1}, in the 20×RD and the 50×RD-contaminated soils, respectively (data not shown).

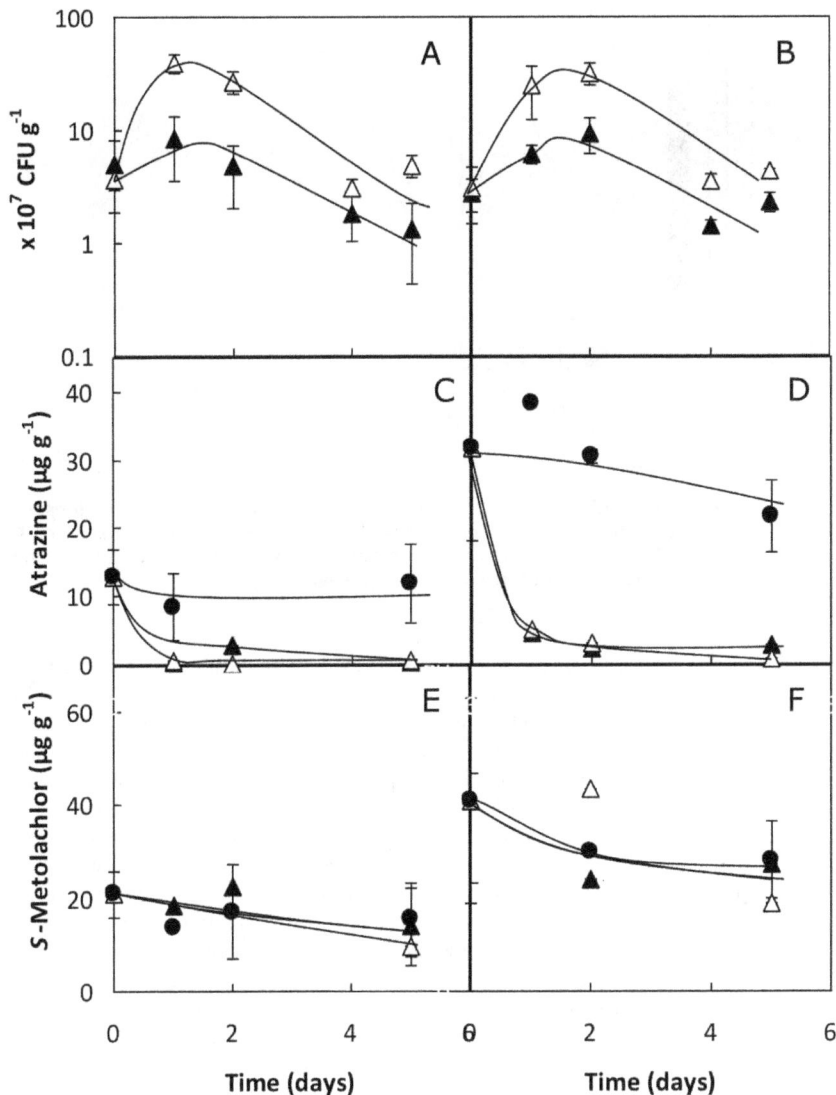

Figure 3. Biodegradation of ATZ from Primextra S-Gold and fate of S-MET in larger soil microcosms. Time-course variation of (**A, B**) the concentration of viable cells of *Pseudomonas* sp. ADP Rif^R, (**C, D**) the average concentration of ATZ and (**E, F**) the average concentration of S-MET, in the soil microcosms contaminated with (**A, C, E**) 20×RD or (**B, D, F**) 50×RD of Primextra S-Gold, and bioaugmented with *Pseudomonas* sp. ADP with (Δ) or without (▲) citrate amendment, during incubation at 25°C. ATZ and S-MET concentrations in the non-inoculated control soil (●) are also shown in (C, D) and (E, F), respectively, for comparison. Data are means±SD of measurements from at least two replicated samples from two independent experiments under each condition.

Ecotoxicity removal efficacy

To estimate potential ecotoxicological effects of addition of the two doses of Primextra S-Gold (20× and 50×RD) to soil, and to assess the efficacy of treatment of these soils with the bioremediation tool, the ecotoxicity to microalgae was assessed in eluates prepared from soil samples collected in the microcosms during the biodegradation experiments with *Pseudomonas* sp. ADP plus citrate (Fig. 3), as examined before for the case of soils contaminated with up to 20×RD of the single ATZ formulation Atrazerba FL [14]. Since there was a similarity between toxicity data obtained with the eluates from soil samples collected after 5 or 8 days upon treatment with the bioremediation tool, only the 5 days toxicity data are presented in Fig. 4.

The eluates from the soil microcosms that were spiked with Primextra S-Gold but not subjected to biaugmentation/biostimu-lation treatment caused serious deleterious effects on microalgae

growth, compared with those from soil not contaminated with the commercial formulation (Fig. 4A). After treatment with the bioremediation tool, the toxicity of the soil eluates was still significantly higher than that of the eluates from the control soil not-contaminated with the herbicide (Fig. 4A). Indeed, microalgae growth rates were similar in the eluates from the treated or untreated soils, for both doses of Primextra S-Gold (Fig. 4A). On the contrary, treatment of Atrazerba-FL-contaminated soils (up to 50×RD) with the same bioaugmentation/biostimulation tool, led to an effective decrease on the ecotoxicity of the respective eluates, compared to the eluates from untreated soil (Fig. 4B). More importantly and corroborating previous observations [14], toxicity values were achieved which were comparable to those obtained for eluates from the soil not contaminated with the herbicide (Fig. 4B).

In addition, we examined whether the S-MET (and/or its degradation products) that remained in the soil during the

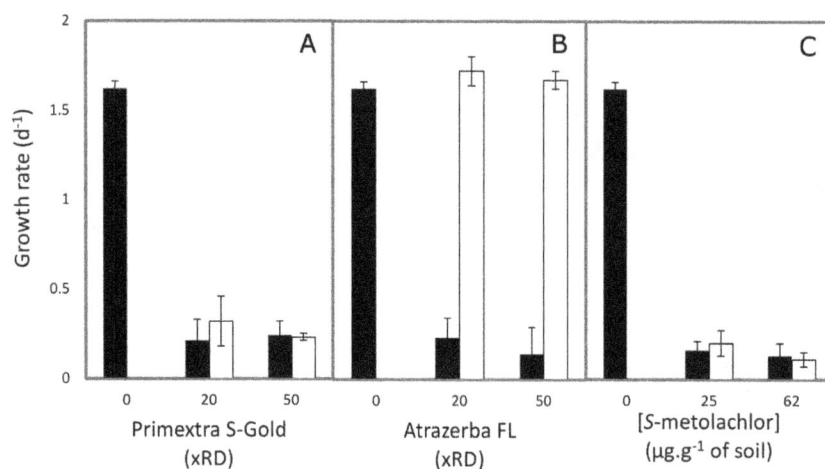

Figure 4. Ecotoxicity of eluates prepared from herbicide-contaminated soil. Mean 72-hours growth rate of *Pseudokirchneriella subcapitata* on eluates prepared with soil samples collected from microcosms spiked with the indicated doses of (**A**) Primextra S-Gold, (**B**) Atrazerba FL, or (**C**) pure S-MET, at day 5 after soil amendment with *Pseudomonas* sp. ADP Rif^R plus citrate (white columns) or without the bioaugmentation/ biostimulation tool (black columns). Data are means±SD from toxicity tests in at least three independent eluate samples.

biodegradation experiments in the microcosms contaminated with Primextra S-Gold (Figs. 3E and F, and data described in previous section) could be responsible, at least partially, for the high toxicity of the respective eluates (Fig. 4A). Consistently, eluates from soils that were contaminated with S-MET in the pure form (25 or 62 μg g^{-1} soil dry weight) were also highly toxic to the microalgae irrespective of the soil having been subjected or not to the *Pseudomonas* sp. ADP plus citrate treatment (Fig. 4C).

Discussion

Effects of S-MET on *Pseudomonas* sp. ADP performance

When considering *in situ* bioremediation strategies, bioaugmentation may fail in the field as a result of the susceptibility of the specialized degrading microorganisms to high concentrations of non-target compounds; other plausible reasons may include microbial competition for a limiting nutrient and/or inhibition of degradation through catabolic repression/competitive inhibition phenomena in the soil [1,2,26,27]. The significance of pesticide interactions with intrinsic xenobiotic degraders or with bioaugmentation bacteria in soil is thus a relevant issue [2,24,26,27]. In the present work, we examined whether the presence of levels of S-MET representing worst-case scenarios of soil contamination with a commercial formulation containing both S-MET and ATZ may affect the performance of the bioaugmentation bacterium *Pseudomonas* sp. ADP for ATZ-biodegradation in soil. Soil amendment with viable cells of this bacterial strain plus an adequate provision of citrate to get a ratio C$_s$:N$_{atz}$~50, has previously been shown to be effective in the cleanup of a crop soil contaminated with high doses (up to 20×RD, mimicking similar worst-case situations) of the formulation Atrazerba FL [14]. It is worth mentioning that intrinsic ATZ-mineralization is absent or shows a very long lag-phase in the soil used ([14,15], and this work). In the present study, we demonstrated that the presence of S-MET in the soil at concentrations between 15 and 50 μg g^{-1} of soil, representative of applications of up to approximately 50×RD of the double herbicidal formulation Primextra S-Gold, affected neither the survival of *Pseudomonas* sp. ADP used to bioaugment the contaminated soil (~10^7 CFU g^{-1} of soil) nor the rate and extent of mineralization of ^{14}C-labeled ATZ by this bacterium. Consistently, its growth curve in liquid PADP medium (initial inoculum around 4×10^7 CFU ml^{-1})

was not affected as a result of medium supplementation with S-MET at concentrations up to 60 μg ml^{-1} (Costa C and Viegas CA, unpublished results). Based on the results obtained herein, we speculate that the observed moderate decrease in the bacterium ability to mineralize [^{14}C]ATZ in the presence of increasing doses of Primextra S-Gold (from 5× up to 50×RD) may mainly reflect inhibitory effects on the bacterium physiology presumably caused by high concentrations of ATZ in soil (as suggested in Silva et al. [17]) and/or by unknown substances that may be present in the commercial formulation (even though not addressed in more detail in the present work), rather than by the active ingredient S-MET. Other authors [24] also reported that the pattern of intrinsic degradation of ATZ (initial concentration 40 μg g^{-1} of soil) in a soil from a pesticide-contaminated site was not significantly affected by the presence of 50 μg g^{-1} of MET or of another herbicide, pendimethalin [24]. On the contrary, Moorman et al. [2] reported that the viability of indigenous ATZ-degrading microorganisms and hence ATZ-mineralization in a contaminated soil from an agricultural chemical dealership area were negatively affected by high concentrations of MET. In the latter study, however, the authors tested the effects of MET at 200 μg g^{-1} of soil [2], which is a concentration quite higher and thus presumably more toxic than the ones tested in the present work.

Consistently, based solely on soil chemical analysis, we present evidences that in soil microcosms contaminated with 20× or 50×RD Primextra S-Gold and subsequently treated with optimized quantities of *Pseudomonas* sp. ADP and citrate, ATZ removal from soil, due to its biodegradation, was extensive (>96% of the initial, to less than 0.5 μg g^{-1} of soil) and rapid (in less than 1 week) as reported before for the soils contaminated with the formulation Atrazerba FL that contains ATZ as the sole active herbicide [14,15]. Moreover, these results together with indications from the chemical data that, in the time-frame of the biodegradation experiments, deethylatrazine (DEA) and deisopropylatrazine (DIA) did not accumulate in the soils (data not shown), point to a significant potential reduction in the dispersion of the herbicide and of its highly toxic N-dealkylated metabolites into the adjacent water compartments, as reported before [14]. In agreement with previous studies, this would indicate an important environmental impact of the treatment of the contaminated soils with the bioremediation tool [14,15,18]. However, contrarily to what is reported for the case of soils contaminated with Atrazerba FL ([14], and present work), the eluates prepared from the soils spiked with Primextra S-Gold and subsequently treated with the ATZ-degrading bacterium plus citrate remained significantly toxic to the microalgae P. subcapitata in the time-frame of the biodegradation experiments. Based on experimental evidences, we propose that the high ecotoxicity of these eluates may be mainly associated with the presence of S-MET that remained in the soil and/or of its degradates, as is further discussed below.

Contribution of S-MET for water extracts ecotoxicity

On one hand, soil bioaugmentation with Pseudomonas sp. ADP apparently had no significant effect on the dissipation of S-MET in the soil microcosms contaminated with Primextra S-Gold, suggesting that this bacterial strain is not able to degrade the chloroacetanilide herbicide, as reported by others [19]. In spite of that, the concentration of S-MET in the soils decreased moderately during the 8 days of the biodegradation experiments even though keeping values always higher than 60% of the initial concentration as described above. We thus suggest that the mobilization of a significant portion of intact S-MET from the soil microcosms to the water extracts may have contributed at least partially for the high ecotoxicity of the eluates towards the microalgae. Even though S-MET levels were not measured in the eluates prepared from the soil samples, using an estimated soil/solution distribution coefficient (K_d value) approximately equal to 2.5 for S-MET in a sandy loam soil with 3.1% organic matter [28], it can be anticipated that ~40% of the S-MET present in the soil may be mobilized into the water. Since a soil:water 1:10 ratio (w/v) was used in the eluate preparation, this points for predictable S-MET concentrations in the water extracts considerably higher (about 50-fold) than the value 8 μg L^{-1} reported as the acute 72 hours median effective concentration for P. subcapitata growth [8], or of similar order of magnitude of the MET concentrations reported recently as causing almost complete inhibition of the growth of this microalgae species [6]. According to published toxicological data, phytoplankton species are important targets for damage caused by S-MET and other chloroacetanilide herbicides, being more susceptible than aquatic organisms from higher trophic levels [6,8,21,23]. In addition to the specific mode of action of these compounds in the target plants (inhibition of fatty acids synthesis) [23], they also have potential to cause non-specific toxicity over diverse soil and aquatic non-target organisms, including microorganisms, related with the lipophilic nature of their molecules [20,21,23]. Since water extracts obtained from the soils contaminated with equivalent quantities of pure S-MET were as highly inhibitors of microalgae growth as those prepared from the soils contaminated with Primextra S-Gold, the high ecotoxicity of the latter may be less likely attributable to the presence of ingredients from the commercial formulation other than S-MET (and/or the S-MET derivatives herein found to accumulate in the soil microcosms, which could also be detrimental for the microalgae). Nevertheless, possible toxic effects of unknown ingredients should not be ruled out as they might have been hidden due to the high toxicity exerted by S-MET (and/or derivatives).

The metabolites S-MET ESA and S-MET OA did build-up in the soil microcosms concomitantly with the moderate dissipation of S-MET down to 60% of its initial value. MET can suffer biodegradation in soil [8,25,26] and in aquatic systems [29] to form these major metabolites, among others [26,30]. The main route of chloroacetanilides degradation in soil has been reported to be microbial [26,29,30]. For example, ESA and OA derivatives of chloroacetanilides were reported to arise primarily from aerobic dechlorination via GST-mediated reactions and further metabolism

of glutathione conjugates by diverse soil bacteria such as pseudomonads and Enterobacteriacea [29,30]. We speculate that under the conditions used in the present work some indigenous active bacteria presumably present in the soil may have contributed for the formation of these two metabolites during the time-course of the biodegradation experiments. Their relative values measured in the soil samples (approximately in a 35:20 ratio OA:ESA) are within the range of published values for other soils [26].

ESA and OA degradates of chloroacetanilide herbicides [8] are frequently detected in ground and surface waters worldwide, often at higher concentrations than the parent compounds [29,31], and are considered of potential concern [31]. Recently, Gadabgui et al. [32] performed a toxicological risk assessment of acetochlor, alachlor and the respective ESA and OA metabolites, and concluded that the toxicity of the degradates for mammals may be lower than that of the parent compounds [32]. To our knowledge scarce information exists about the toxicological properties of ESA and OA metabolites for microorganisms and aquatic organisms. We suggest that the S-MET ESA and OA that accumulated in the soils examined herein (and possibly other metabolites that may have formed but were not analysed [26]) might also contribute for the toxicity of the soil water extracts to the microalgae P. subcapitata. Importantly, the transformation of S-MET into these degradates, whose mobility in soil appear to be greater than that of the parental compound [25,33], suggests that they may have potential to contaminate water compartments and hence impact water quality [31,32].

In conclusion, the results herein presented point to the complexity of bioremediating soils contaminated with mixtures of pesticides, in this particular case of ATZ and S-MET. Both active substances have been frequently used alone or combined in herbicidal formulations [1,8,10,26]. Due to the potential effects of these substances and their metabolites for non-target aquatic organisms, risk mitigation measures are recommended particularly when the herbicidal formulation is applied in regions with vulnerable soil [31]. To date, as far as we are aware of, effective biodegradation and detoxification of MET/S-MET or derivatives ESA and OA in soil have not been well succeeded, apparently because microorganisms do not easily metabolize their aromatic ring [1,19,34]. In the present work, there are evidences of the transformation of 40% initial S-MET in the contaminated soils, presumably performed by intrinsic microorganisms present in the soil, but this is not associated with an effective decontamination as discussed above. Further work is needed for the development of efficient bioremediation strategies for land contaminated with herbicide mixtures such as the one herein examined. It is necessary, in one hand, to focus on the isolation and optimization of performance of soil bacteria (most probably working in consortia) able to concomitantly biodegrade multiple chemicals to less toxic derivatives; and, in the other hand, to address the effects of multiple pesticides on the performance of specific degrading microorganisms [1,35]. The present work is a contribution to enhance the knowledge on the latter issue, particularly with respect to the interaction of S-MET with the well-known ATZ-degrading bacterium Pseudomonas sp. ADP [14–18]. Overall results also highlight the importance of monitoring the efficacy of the soil clean-up processes based on ecotoxicity assessments of soil aqueous extracts before and after the implementation of the bioremediation treatment, besides chemical analysis [14]. Ecotoxicological evaluation provide a more realistic glimpse over the ecological risk assessment of soil remediation, being particularly important, in aquatic ecosystems, for the evaluation of the potential impact of the mobilization of extractable and bounded fractions of ATZ, S-

MET and other pesticides (including their possible metabolites) via water, mainly due to leaching and runoff events.

Materials and Methods

Chemicals

Atrazine (ATZ; Pestanal, purity 99.1%), S-metolachlor (S-MET; Pestanal, purity 98.2%) and 4-morpholinepropanesulfonic acid (MOPS) were purchased from Sigma-Aldrich (Seelze, Germany), trisodium citrate from Merck (Darmstadt, Germany), and [ring-UL-^{14}C]Atrazine (purity 99%, specific activity 1.85 GBq mmol^{-1}) from American Radiolabeled Chemicals (St. Louis, MO, USA). The formulations Primextra S-Gold (320 g of ATZ L^{-1} and 400 g of S-MET L^{-1} as active ingredients; 3 L ha^{-1} as the recommended dose (RD) for weeds in corn plantations) and Atrazerba FL (500 g ATZ L^{-1} as single active ingredient; RD = 2 L ha^{-1}) were purchased from Syngenta Portugal (Lisbon, Portugal) and Sapec (Setúbal, Portugal), respectively. The RD of ATZ and S-MET (0.96 kg ha^{-1} and 1.2 kg ha^{-1}, respectively) were estimated as equivalent to approximately 0.8 and 1.0 μg g^{-1} of soil, respectively, assuming a possible field scenario of herbicide distribution through a 5 cm diameter ×10 cm height soil column and an average soil density of 1.5 g cm^{-3} [15].

Bacterial strain and culture conditions

A spontaneous rifampicin-resistant (Rif^R) mutant of *Pseudomonas* sp. ADP which can mineralize atrazine with equal efficiency than the wild-type [36] was used. The cell suspension used as inoculum was prepared from a late-exponential culture (6.5±0.4×10^8 colony forming units (CFU) ml^{-1}, corresponding to OD$_{640 nm}$~1) grown at 30°C in liquid PADP medium (adapted from [16,17]) following a procedure reported elsewhere [15]. Briefly, the growth medium was buffered using MOPS (0.1 M; pH 6.2) and supplied with trisodium citrate (10 g L^{-1}) as C source [17], and with ATZ (300 mg L^{-1}) from Atrazerba FL as sole N source [15]. Cells were harvested by centrifugation (5 min, 9000 g), and the cell pellet was washed once and resuspended in sterile saline solution (0.9% w/v NaCl). The concentrated cell suspension obtained was used as inoculum in survival, mineralization and biodegradation experiments, as is described below. The inoculum density, expressed as CFU ml^{-1}, was determined by plating culture serial dilutions onto agarized Lennox Broth (LB) medium and further incubation at 30°C during 72 hours.

Soil parameters

A natural sandy loam soil (pH 6.1; organic matter 3.1%; water content 9.8±0.9%; water holding capacity (WHC) 39.4±1.6%; cation exchange capacity 0.013 cmol g^{-1} soil dry weight; soluble carbon 23.5±5.2 μg g^{-1} soil dry weight) was used [14,15]. This soil is representative of a corn production field from Central Portugal (Escola Superior Agrária de Coimbra - ESAC, Coimbra, Portugal) with no history of pesticide applications. The soil was sieved (5 mm mesh) and stored in plastic bags at −20°C. Prior to use in the survival, mineralization or biodegradation experiments, soil was defrosted for at least 4 days at 4°C.

Survival experiments

To examine the effects of S-MET on the survival of *Pseudomonas* sp. ADP in ATZ-contaminated soil, experiments were carried out in sterilized EPA vials (40 ml, gastight PTFE/Silicone septa, Sigma–Aldrich) each containing 5 g soil dry weight freshly spiked with ATZ to give an approximate concentration of 24 μg g^{-1} soil dry weight (equivalent to 30×RD of the active substance). Then, three different concentrations of S-MET (15, 30 or 60 μg g^{-1})

were added to these soils to represent doses equivalent to 15×, 30× and 60× higher than the RD of this active ingredient. Each compound was added from a stock solution in methanol (18 or 20 mg ml^{-1} for ATZ or S-MET, respectively), mixed with pure methanol when needed (to guarantee a similar total volume of methanol in all vials) and with sterile deionized water (to obtain an initial soil moisture of 40% soil WHC). Vials non-supplemented with S-MET were included as controls. Vials were vigorously stirred in a vortex to promote homogeneous distribution into the soil and were left uncapped in the laminar flow chamber for 1 hour to allow most of the methanol to evaporate. Each vial was then inoculated with an adequate amount of a suspension of bacteria viable cells prepared as described above, in order to have an initial inoculum density of approximately 10^7 CFU g^{-1} soil dry weight. Vials were stirred again, capped and incubated at 25.1±0.2°C in the dark. The total volume of liquid added to each vial (298 μl; 40% soil WHC) took in account the volumes of herbicide solutions, inoculum and deionized water. In each experiment, 16 replicates of each condition were prepared. To determine *Pseudomonas* sp. ADP Rif^R viable cells in soil, at each time interval (up to 9 days) one replicated vial was destructively sampled and processed immediately for the determination of CFU concentration as is described below (microbiological analysis). At least duplicate determinations from two or three independent experiments under identical conditions were carried out.

Mineralization experiments

ATZ mineralization assays were carried out in sterilized EPA 40-ml vials containing 5 g soil dry weight, as previously described [15] with minor adaptations. Briefly, a mixture of [ring-UL-^{14}C]ATZ (stock solution in acetonitrile: 467.7 KBq ml^{-1}) plus non-labeled ATZ was incorporated into the soil to give a total activity of 0.65 KBq g^{-1} dry weight of soil and different total concentrations of ATZ in soil. Non-labeled ATZ was supplied from Primextra S-Gold or as the pure substance, depending on the type of experiment to be carried out, as follows: first, to evaluate the effect of increasing doses of Primextra S-Gold in ATZ mineralization by *Pseudomonas* sp. ADP, soil in the vials was freshly spiked with mixtures containing the [ring-UL-^{14}C]ATZ stock solution, aqueous suspensions of Primextra S-Gold (to give approximately 5×, 20× and 50×RD corresponding to 15, 60 and 150 L ha^{-1}, respectively) and sterile deionized water (to obtain an initial soil moisture of 40% soil WHC). Second, to examine the influence of S-MET in ATZ-mineralization bacterial activity, soil in the vials was freshly spiked with mixtures of the [ring-UL-^{14}C]ATZ stock solution plus adequate amounts of stock solutions of herbicides in methanol to have total concentrations of 24 or 40 μg of ATZ g^{-1} soil dry weight plus, respectively, 30 or 50 μg g^{-1} of S-MET (equivalent to approximately 30× or 50×RD of the active ingredients, respectively) or no added S-MET. Pure methanol (to guarantee a similar volume of methanol in all vials) and sterile deionized water (to obtain soil moisture of 40% soil WHC) were also added. Then, soil was vigorously mixed with a vortex apparatus to incorporate the substances, followed by the inoculation with an adequate amount of the bacterial cell suspension prepared as described above (to obtain approximately 10^7 viable cells g^{-1} soil dry weight) [15]. Non-inoculated controls were included in each set of experiments to account for intrinsic mineralization activity in the soil. In addition, in experiments aiming to examine combination of soil bioaugmentation with biostimulation with citrate on ATZ mineralization, sterile concentrated solutions of trisodium citrate were added to soil to obtain a ratio of C$_s$:N$_{atz}$ equal to 50, as described before [15,17]. The total volume of liquid added to each vial (298 μl; 40% soil

WHC) took in account the volumes added of herbicide solutions, inoculum, a citrate solution and deionized water. After all soil amendments, the vials were stirred once more for homogeneity. Triplicate vials were included for each treatment. A test tube containing 1 ml solution of NaOH 1 M was placed inside each vial to trap the released $^{14}CO_2$. All vials were incubated at $25.1 \pm 0.2°C$ in the dark for up to 8 days. At adequate time intervals, the quantity of released $^{14}CO_2$ was quantified using a Beckman LS 5000TD scintillation counter, as described before [17]. Ultima Gold (Perkin Elmer, Waltham, USA) was used as the scintillation cocktail in a 1:4 sample-to-cocktail ratio. At least triplicate determinations from two independent experiments under each condition were carried out.

Biodegradation experiments at a larger scale

Soil microcosms consisting of glass cylinders (10 cm height×4.5 cm interior diameter) containing 160 g dry weight of soil (~7 cm×4.5 cm) over a 2 cm height layer of 2-mm-diameter glass beads supported by a fine Teflon mesh were used as described elsewhere [14,15]. Briefly, the soil was freshly spiked with 5 ml of aqueous suspensions of Primextra S-Gold to obtain approximately $20\times$ or $50\times RD$ (60 or 150 L ha^{-1}, respectively). After homogenization with a glass rod to promote incorporation of the herbicides, soil was inoculated with an adequate quantity of the inoculum bacterial suspension (initial density: 2.5–4.1×10^7 CFU g^{-1} soil dry weight), and either amended or not with trisodium citrate at 1.3 or 3.4 mg g^{-1} soil dry weight in the $20\times$ or $50\times RD$ amended soils, respectively (to give a ratio C_s:N_{atz}~50 [15]). Identical experiments with $20\times$ or $50\times RD$ of Atrazerba FL or with pure S-MET (25 or 62 µg g^{-1} soil dry weight) instead of Primextra S-Gold, were also carried out for comparison purposes. Microcosms non-contaminated with the herbicides or contaminated but not inoculated with the bacterium were also included in each set of experiments as controls. In all experiments, soil moisture was adjusted to 40% soil WHC as described above, taking in account the total volume of liquid (9.5 ml, comprising herbicide suspensions, inoculum, citrate solutions, and deionized water). Amended soils were again mixed, gently packed into the glass cylinders, and incubated at $25.1 \pm 0.2°C$ in the dark for up to 8 days. Soil microcosms were weighted every day to replace the water lost by evaporation with sterile deionized water. Soil samples were collected from the surface, at days 0, 1, 2, 5 and/or 8, and processed immediately for microbiology analysis (1.0 ± 0.3 g dry weight), or stored at $-20°C$ for chemical and ecotoxicological analyses (~20 g), which were performed as described below. At least two determinations with samples from two independent experiments were carried out.

Ecotoxicity analysis

Ecotoxicity tests with the model aquatic microalgae *Pseudokirchneriella subcapitata* (strain Nr.WW 15–2521; Carolina Biological Supply Company, Burlington, NC, USA) were carried out on eluates obtained from soil samples collected in the biodegradation experiments (0, 5 and 8 days) as described above. Microalgae culturing procedures were as previously outlined [37]. Soil eluates were prepared following standard methods as described elsewhere [14]. 72-hours *P. subcapitata* growth tests followed standard guidelines [38,39], on 24-well sterile microplates, at 21 to 23°C and under continuous cool-white fluorescent illumination (100 µE m^{-2} s^{-1}). Three 900 µl sub-replicate cultures per replicated eluate, and a standard control with six replicates, were set up and inoculated with 100 µl of algal inoculum, so that the cell density at the start of the tests would be 10^4 cells ml^{-1}. Both pH and conductivity were measured at the start of each test, and measured levels were comparable across treatments and not expected to have deleterious effects on the test organisms [38,39]. At the end of the 72-hours exposure, algal growth was estimated as the mean specific growth rate (per day). Further details on testing procedures are described elsewhere [37]. For eluates from soil samples taken from non-inoculated microcosms within the same time interval (5 or 8 days), effects of the two doses ($20\times$ or $50\times RD$) of each herbicide application (Primextra S-Gold, Atrazerba FL, or S-MET) on microalgae growth were evaluated through one-way analysis of variance followed by Dunnet's test to compare each dose with the respective non-contaminated control soil. One-tailed Student *t*-tests were employed to compare microalgae growth responses between spiked soils that were treated or not with the bioremediation tool, within the same herbicide dose and time interval. The latter tests were also used to assess decontamination efficacy due to bioaugmentation/biostimulation treatments, by comparing microalgae growth in eluates from non-contaminated (controls) and treated contaminated soils, within the same pesticide dose and time interval.

Microbiological analysis of soil samples

To determine the concentration of CFU of *Pseudomonas* sp. ADP Rif^R, soil samples from survival and biodegradation experiments were used as a basis for 10-fold dilution series in saline solution (0.9% NaCl w/v) in triplicate. Dilutions were spread plated onto agarized selective LB medium supplemented with rifampicin (50 mg L^{-1}) and cycloheximide (100 mg L^{-1}). Petri dishes were incubated at 30°C and colonies counted after 96 hours.

Chemical analysis

Soil samples were thawed at room temperature, dried at 40°C and further processed for analysis of the herbicides and respective metabolites in soil. For analysis of ATZ, DEA and DIA, extracts preparation and analysis by GC-Electron Ionization (EI)-MS (Perkin Elmer-Clarus 500) was carried out as described elsewhere [15]. Recovery range was between 75 and 90%, and the limits of quantification were 25 ng g^{-1} soil dry weight for ATZ, DEA, and DIA. For analysis of S-MET and of its OA and ESA derivatives, soil samples were dried at 40°C, extracted three times with a mixture of methanol and water (1:2) using a Liarre 60 ultrasonic apparatus (15 min) and centrifuged for 15 min at 3000 g. Analysis of the combined extracts was performed by GC-EI-MS (Perkin Elmer-Clarus 500) for S-MET, and by LC-EI-MS (Agilent 1100 Series) for S-MET OA and S-MET ESA. All extracts were injected in scan mode to confirm the presence of each analyte and in SIM (single ion monitoring) for quantification purposes. Recovery ranged from 75 and 90% and the limit of quantification was 16 and 6 ng g^{-1} soil dry weight for S-MET and for S-MET ESA or S-MET OA, respectively.

Acknowledgments

We are greatful to S. Chelinho (IMAR, Univ Coimbra, P), J.P. Sousa (IMAR, Univ Coimbra, P), and R. Guilherme (ESAC, Coimbra, P) for providing the soil used in these studies, and to A.M. Fialho (Instituto Superior Técnico, Lisboa, P) for advice and valuable discussions.

Author Contributions

Conceived and designed the experiments: CAV MMS. Performed the experiments: CC MMS SA PV. Analyzed the data: CC MMS PV CAV. Contributed reagents/materials/analysis tools: MMS RR PV CAV. Wrote the paper: CAV MMS RR.

References

1. Chirnside AEM, Ritter WF, Radosevich M (2009) Biodegradation of aged residues of atrazine and alachlor in a mix-load site soil. Soil Biol Biochem 41: 2484–2492.

2. Moorman TB, Cowan JK, Arthur EL, Coats JR (2001) Organic amendments to enhance herbicide biodegradation in contaminated soils. Biol Fertil Soils 33: 541–545.

3. Hildebrandt A, Guillamon M, Lacorte S, Tauler R, Barcelo D (2008) Impact of pesticides used in agriculture and vineyards to surface and groundwater quality (North Spain). Water Res 42: 3315–3326.

4. Seybold CA, Mersie W (1996) Adsorption and desorption of atrazine, deethylatrazine, deisopropylatrazine, hydroxyatrazine, and metolachlor in two soils from Virginia. J Environ Qual 25: 1179–1185.

5. Southwick LM, Grigg BC, Fouss JL, Kornecki TS (2003) Atrazine and metolachlor in surface runoff under typical rainfall conditions in southern Louisiana. J Agric Food Chem 51: 5355–5361.

6. Pérez J, Domingues I, Soares AMVM, Loureiro S (2011) Growth rate of *Pseudokirchneriella subcapitata* exposed to herbicides found in surface waters in the Alqueva reservoir (Portugal): a bottom-up approach using binary mixtures. Ecotoxicology 20: 1167–1175.

7. Health Canada (2008) Proposed maximum residue limit for S-metolachlor. Available: http://www.hc-sc.gc.ca/cps-spc/pest/part/consultations/_pmrl2008-18/index-eng.php. Accessed 30 June 2010.

8. PPDB The Pesticide Properties Database (PPDB) developed by the Agriculture & Environment Research Unit (AERU), University of Hertfordshire, funded by UK national sources and the EU-funded FOOTPRINT project (FP6-SSP-022704). Available: http://sitem.herts.ac.uk/aeru/footprint/en/index.htm. Accessed 20 April 2012.

9. Shaner DL, Brunk G, Belles D, Westra P, Nissen S (2006) Soil dissipation and biological activity of metolachlor and S-metolachlor in five soils. Pest Manag Sci 62: 617–623.

10. Whaley CM, Armel GR, Wilson HP, Hines TE (2009) Evaluation of S-metolachlor and S-metolachlor plus atrazine mixtures with mesotrione for broadleaf weed control in corn. Weed Technol 23: 193–196.

11. DeLorenzo ME, Scott GI, Ross PE (2001) Toxicity of pesticides to aquatic microorganisms: a review. Environ Toxicol Chem 20: 84–98.

12. Oh SM, Shim SH, Chung, KH (2003) Antiestrogenic action of atrazine and its major metabolites *in vitro*. J Health Sci 49: 65–71.

13. Sanderson JT, Seinen W, Giesy JP, van den Berg M (2000) 2-chloro-s-triazine herbicides induce aromatase (CYP-19) activity in H295R human adrenocortical carcinoma cells: A novel mechanism for estrogenicity. Toxicol Sci 54: 121–127.

14. Chelinho S, Moreira-Santos M, Lima D, Silva C, Viana P, et al. (2010) Cleanup of atrazine-contaminated soils: ecotoxicological study on the efficacy of a bioremediation tool with *Pseudomonas* sp. ADP. J Soils Sediments 10: 568–578.

15. Lima D, Viana P, André S, Chelinho S, Costa C, et al. (2009) Evaluating a bioremediation tool for atrazine contaminated soils in open soil microcosms: The effectiveness of bioaugmentation and biostimulation approaches. Chemosphere 74: 187–192.

16. Mandelbaum RT, Allan DL, Wackett LP (1995) Isolation and characterization of a *Pseudomonas* sp. that mineralizes the s-triazine herbicide atrazine. Appl Environ Microbiol 61: 1451–1457.

17. Silva E, Fialho AM, Sá-Correia I, Burns RG, Shaw LG (2004) Combined bioaugmentation and biostimulation to cleanup soil contaminated with high concentrations of atrazine. Environ Sci Technol 38: 632–637.

18. Wackett LP, Sadowsky MJ, Martinez B, Shapir N (2002) Biodegradation of atrazine and related s-triazine compounds: from enzymes to field studies. Appl Microbiol Biotechnol 58: 39–45.

19. Zhao S, Arthur EL, Moorman TB, Coats JR (2003) Influence of microbial inoculation (*Pseudomonas* sp. strain ADP), the enzyme atrazine chlorohydrolase, and vegetation on the degradation of atrazine and metolachlor in soil. J Agric Food Chem 51: 3043–3048.

20. Papaefthimiou C, Cabral MG, Mixailidou C, Viegas CA, Sá-correia I, et al. (2004) Comparison of two screening bioassays, based on the frog sciatic nerve and yeast cells, for the assessment of herbicide toxicity. Environ Toxicol Chem 23: 1211–1218.

21. Liu H, Ye W, Zhan, Liu W (2006) A comparative study of *rac-* and S-metolachlor toxicity to *Daphnia magna*. Ecotoxicol Environ Saf 63: 451–455.

22. Xu D, Wen Y, Wang K (2010) Effect of chiral differences of metolachlor and its (S)-isomer on their toxicity to earthworms. Ecotoxicol Environ Saf 73: 1925–1931.

23. Junghans M, Backhaus T, Faust M, Scholze M, Grimme LH (2003) Predictability of combined effects of eight chloroacetanilide herbicides on algal reproduction. Pest Manag Sci 59: 1101–1110.

24. Anhalt JC, Arthur EL, Anderson TA, Coats JR (2000) Degradation of atrazine, metolachlor, and pendimethalin in pesticide-contaminated soils: effects of aged residues on soil respiration and plant survival. J Environ Sci Health – Part B 35: 417–438.

25. Aga DS, Thurman EM (2001) Formation and transport of sulfonic acid metabolites of alachlor and metolachlor in soil. Environ Sci Technol 35: 2455–2460.

26. White PM, Thomas LP, Culbreath AK (2010) Fungicide dissipation and impact on metolachlor aerobic soil degradation and soil microbial dynamics. Sci Total Environ 408: 1393–1402.

27. Qiu Y, Pang H, Zhou Z, Zhang P, Feng Y, et al. (2009) Competitive biodegradation of dichlobenil and atrazine coexisting in soil amended with a char and citrate. Environ Pollut 157: 2964–2969.

28. Weber JB, Wilkerson GG, Reinhardt CF (2004) Calculating pesticide sorption coefficients (K_d) using selected soil properties. Chemosphere 55: 157–166.

29. Graham WH, Graham DW, Denoyelles F, Jr., Smith VH, Larive CK, et al. (1999) Metolachlor and alachlor breakdown product formation patterns in aquatic field mesocosms. Environ Sci Technol 33: 4471–4476.

30. Zablotowicz RM, Hoagland RE, Locke MA, Hickey WJ (1995) Glutathione-S-transferase activity and metabolism of glutathione conjugates by rhizosphere bacteria. Appl Environ Microbiol 61: 1054–1060.

31. European Commission (2004) Review report for the active substance S-Metolachlor. Available: http://ec.europa.eu/food/plant/protection/evaluation/newactive/s_metolachlor.pdf. Accessed 30 June 2010.

32. Gadagbui B, Maier A, Dourson M, Parker A, Willis A, et al. (2010) Derived Reference Doses (RfDs) for the environmental degradates of the herbicides alachlor and acetochlor: Results of an independent expert panel deliberation. Regul Toxicol Pharmacol 57: 220–234.

33. Krutz LJ, Senseman SA, McInnes KJ, Hoffman DW, Tierney DP (2004) Adsorption and desorption of metolachlor and metolachlor metabolites in vegetated filter strip and cultivated soil. J Environ Qual 33: 939–945.

34. Saxena A, Renwu Z, Bollag J-M (1987) Microorganisms capable of metabolizing the herbicide metolachlor. Appl Environ Microbiol 53: 390–396.

35. Viegas CA, Chelinho S, Moreira-Santos M, Costa C, Gil FN, et al. (2012) Bioremediation of soils contaminated with atrazine and other s-triazine herbicides: current state and prospects. In: Justin A Daniels, ed. Advances in Environmental Research - Volume 6, Nova Science Publishers Inc., NY. pp 1–49.

36. Garcia-González V, Govantes F, Shaw LJ, Burns R, Santero E (2003) Nitrogen control of atrazine utilization in *Pseudomonas* sp. strain ADP. Appl Environ Microbiol 69: 6987–6993.

37. Rosa R, Moreira-Santos M, Lopes I, Silva L, Rebola J, et al. (2010) Comparison of a test battery for assessing the toxicity of a bleached-kraft pulp mill effluent before and after secondary treatment implementation. Environ Monit Assess 161: 439–451.

38. OECD (1984) Guideline for testing of chemicals 201: Alga growth inhibition test (test guideline 201, updated, adopted 7th June). Organization for Economic Co-Ordination and Development, Paris, France.

39. Environment Canada (1992) Biological test method: growth inhibition test using the freshwater alga *Selenastrum capricornutum*. EC Report EPS 1/RM/25. Environment Canada, Ottawa, Ontario, Canada.

19

Where to Restore Ecological Connectivity? Detecting Barriers and Quantifying Restoration Benefits

Brad H. McRae[1]*, Sonia A. Hall[2], Paul Beier[3], David M. Theobald[4]

1 The Nature Conservancy, North America Region, Seattle, Washington, United States of America, 2 The Nature Conservancy, Washington Chapter, Wenatchee, Washington, United States of America, 3 School of Forestry and Merriam-Powell Center for Environmental Research, Northern Arizona University, Flagstaff, Arizona, United States of America, 4 National Park Service, Inventory and Monitoring Division, Fort Collins, Colorado, United States of America

Abstract

Landscape connectivity is crucial for many ecological processes, including dispersal, gene flow, demographic rescue, and movement in response to climate change. As a result, governmental and non-governmental organizations are focusing efforts to map and conserve areas that facilitate movement to maintain population connectivity and promote climate adaptation. In contrast, little focus has been placed on identifying barriers—landscape features which impede movement between ecologically important areas—where restoration could most improve connectivity. Yet knowing where barriers most strongly reduce connectivity can complement traditional analyses aimed at mapping best movement routes. We introduce a novel method to detect important barriers and provide example applications. Our method uses GIS neighborhood analyses in conjunction with effective distance analyses to detect barriers that, if removed, would significantly improve connectivity. Applicable in least-cost, circuit-theoretic, and simulation modeling frameworks, the method detects both complete (impermeable) barriers and those that impede but do not completely block movement. Barrier mapping complements corridor mapping by broadening the range of connectivity conservation alternatives available to practitioners. The method can help practitioners move beyond maintaining currently important areas to restoring and enhancing connectivity through active barrier removal. It can inform decisions on trade-offs between restoration and protection; for example, purchasing an intact corridor may be substantially more costly than restoring a barrier that blocks an alternative corridor. And it extends the concept of centrality to barriers, highlighting areas that most diminish connectivity across broad networks. Identifying which modeled barriers have the greatest impact can also help prioritize error checking of land cover data and collection of field data to improve connectivity maps. Barrier detection provides a different way to view the landscape, broadening thinking about connectivity and fragmentation while increasing conservation options.

Editor: Adina Maya Merenlender, University of California, Berkeley, United States of America

Funding: This project was funded by the North Pacific Landscape Conservation Cooperative (www.fws.gov/nplcc; grant #13170BG105) and the Great Northern Landscape Conservation Cooperative (greatnorthernlcc.org; grant # 60181AG501). The funders had no role in study design, data collection and analysis, decision to publish, or preparation of the manuscript.

Competing Interests: The authors have declared that no competing interests exist.

* E-mail: bmcrae@tnc.org

Introduction

Landscape connectivity, or "the degree to which the landscape facilitates or impedes movement among resource patches" [1], is crucial for many ecological and evolutionary processes, including dispersal, gene flow, demographic rescue, and movement in response to climate change [2–7]. Many research and conservation planning efforts have focused on mapping areas important for connectivity using GIS models (e.g., [8–12]). The results of these analyses are guiding investments by governmental and non-governmental organizations to promote ecological connectivity across large areas. In the USA and Canada, for example, numerous broad-scale conservation efforts such as the U.S. Department of Interior Landscape Conservation Cooperatives, the Western Governors' Association's Initiative on Wildlife Corridors and Crucial Habitat, and the Yellowstone to Yukon Conservation Initiative are working to integrate and coordinate connectivity conservation actions spanning millions of acres and crossing many political and ecoregional boundaries.

Conservation practitioners employ two primary strategies to promote connectivity. The first focuses on conserving areas that *facilitate* movement; the second focuses on restoring connectivity across areas that *impede* movement (e.g., by removing a fence or building a wildlife-friendly highway underpass). Most connectivity analyses have focused on the former strategy by modeling and mapping areas important for movement under present landscape conditions. A wide array of tools have been developed for this purpose: least-cost corridor modeling [8,13,14], circuit theory [15], individual-based movement models (e.g., [16–18]), graph theory [19,20], and centrality analyses (e.g., [21,22]) have all been used to identify areas important for movement of plants and animals. Outputs from such models are now being used as inputs to reserve selection algorithms (e.g., [23]) to optimize actions to conserve connectivity.

In contrast, there has been little effort by conservation scientists towards identifying candidate areas for the second strategy: that is detecting restoration opportunities by mapping barriers that strongly reduce movement potential. We define a barrier as a landscape feature that impedes movement between ecologically important areas, the removal of which would increase the potential for movements between those areas. Here we are concerned with movements important for access to resources, demographic rescue, gene flow, range shifts, and other ecological and evolutionary processes. In this context, barriers are distinguished from features that are impermeable but not situated such that they block biologically relevant movement routes. Barriers are thus the inverse of corridors, which delineate pathways facilitating movement. Barriers can either be complete (impermeable) or partial (e.g., land cover types that hinder movement relative to ideal conditions, but may still provide some connectivity value). Barriers may be human-made (e.g., roads, fences, or urban areas) or natural (rivers or canyons); they may be linear (e.g., highways) or span large areas (agricultural fields). As with traditional connectivity concepts [1,24], what constitutes a barrier, the impact it has, and whether it reduces connectivity through behavioral inhibition, increased mortality, or other means will differ among species.

Detecting barriers to movement would complement traditional connectivity analyses in several important ways. First, some barriers may be restorable. Knowing where barriers have the greatest impact would help practitioners decide where and how to invest scarce conservation resources to conserve and enhance connectivity. For example, it may be cheaper to restore a barrier that blocks a movement corridor through public land than to establish permanent protection of a functioning corridor that runs through private land [25]. Quantifying such trade-offs would be necessary to integrate connectivity restoration into systematic conservation planning analyses aimed at optimizing conservation investments [26–28], but tools to incorporate connectivity conservation and/or restoration into such efforts remain rare [29–31]. Second, consider that corridor modeling often produces corridors that may not be good enough to realistically support movement [32]. Barrier detection analysis could reveal such cases, allowing practitioners to 'triage' a landscape, focusing efforts on more viable movement routes. Finally, surprising results in a barrier analysis could alert analysts to situations in which poor land cover data or incorrect model parameterization may be causing spurious results.

In this paper, we introduce a new method to identify barriers and rank them by their impact on connectivity. Our method complements existing connectivity modeling approaches, is applicable in least-cost and other connectivity modeling frameworks, and can be extended to centrality analyses. The method can be readily applied across large landscapes, efficiently analyzing barriers among many locations and at different scales corresponding to different sizes of barriers and types of restoration activities. It also quantifies the extent to which restoration can be expected to improve connectivity. We provide example applications of the method, showing that the potential for connectivity conservation is not constrained to narrow corridors, but includes options spanning much more of the landscape when restoration options are considered. We also discuss how our approach can facilitate sensitivity analyses, data quality screening, and prioritization of areas for error checking of GIS base data.

Method for Detecting Barriers and Restoration Opportunities

Our method identifies areas that most reduce connectivity between two locations on a landscape. Making these areas permeable to movement would therefore most increase connectivity between the locations. Thus, these are areas that practition-

ers should consider when implementing restoration to promote connectivity.

To illustrate the method, we use a least-cost corridor modeling framework [13,14,32], which is commonly used to map and prioritize areas important for connectivity conservation (e.g., [8–12]). However, our approach could also be used with other modeling frameworks capable of producing measures of effective distance, such as circuit theory and individual-based movement models (see Discussion).

As with least-cost corridor models, input data include locations to be connected (hereafter, "patches") and a raster resistance surface (Figure 1A). The former may consist of points or polygons, and typically represent natural landscape blocks, protected areas, or core habitat for a particular species or species guild [33]. The resistance surface represents the difficulty, energetic cost, or mortality risk associated with movement through each pixel (see [34] for a review of resistance surface development).

Least-cost methods calculate the cost-weighted distance (CWD) of all pixels to a source location, creating a raster of CWD values (Figures 1B and 1C). Adding together CWD rasters from two locations produces a corridor (Figure 1D), showing the pathways with the lowest cumulative movement cost between the locations [14]. The minimum value of the corridor raster is the least-cost distance (LCD); this represents the cumulative resistance encountered moving along the optimal path from one location to the other, and is a common measure of isolation in spatial ecology (e.g., [35,36]), landscape genetics (e.g., [37,38]), and related fields.

Our method is based on this simple assumption: if a certain area (the size is defined by the user) is restored such that the resistance across it is reduced, then the LCD of the best route connecting the patches through the restoration area will also be reduced. Systematically quantifying the potential reduction across a landscape will allow us to detect those areas where restoration would lead to the greatest reduction in least-cost distance.

The method begins with CWD calculations from two patches (Figures 1B and 1C). However, rather than adding the two CWD surfaces together to produce a corridor, we instead calculate the minimum value of each CWD surface within a localized area around each pixel location (e.g., within a 500 m radius). We then add the minimum values from both CWD surfaces to calculate the cumulative resistance that would be incurred moving between the patches and through the focal pixel assuming the area within the search window is restored:

$$LCD' = CWD1_{MIN} + CWD2_{MIN} + (L * R'), \quad (1)$$

where LCD' is the least-cost distance of the best path between the patches passing through the focal pixel after barrier removal, $CWDX_{MIN}$ is the minimum CWD value from patch X within the search window, L is the length of the longest axis of the search window, and R' is the resistance value of the feature replacing (or cutting through) the barrier. We use a circular moving window to illustrate the method (Figure 2), but consider alternative search window shapes in the Discussion. Note that the longest axis of a circle is its diameter.

For each pixel, this formula yields the cost of the best corridor that would pass through that pixel if the resistance of a strip of land crossing the search window were changed to R'. Including R' and the search window length accounts for the cost of moving across the search window, assuming restoration or removal of the intervening barrier.

If LCD' is less than LCD, then restoration across the moving window (e.g., the circle in Figure 2) would reduce effective distance and increase connectivity between the two patches. When this is

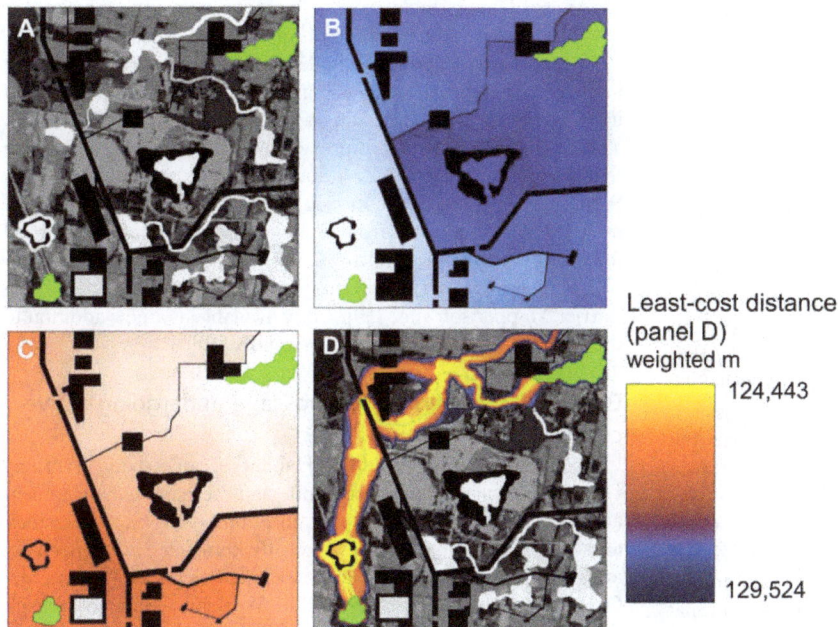

Figure 1. Cost-weighted distance modeling. (A) Example 3 km×3 km landscape with a pixel size of 3 m (from [15]). Two habitat patches (green) are embedded in a matrix of land cover types with differing resistance to movement. Resistances range from 1 (white) to 100 (dark grey); complete barriers with infinite resistance (e.g., linear features representing roads and highways) are shown in black. (B) Cost-weighted distance (*CWD*) from leftmost patch, with darker shades representing higher cumulative resistance from the patch. (C) *CWD* from rightmost patch, with darker shades representing higher cumulative resistance from the patch. (D) Modeled least-cost corridor produced by adding *CWD* surfaces shown in panels B and C (best 20% of study area shown). The least-costly path (traced in green) has a cumulative least-cost distance (*LCD*) of 124,443 weighted meters.

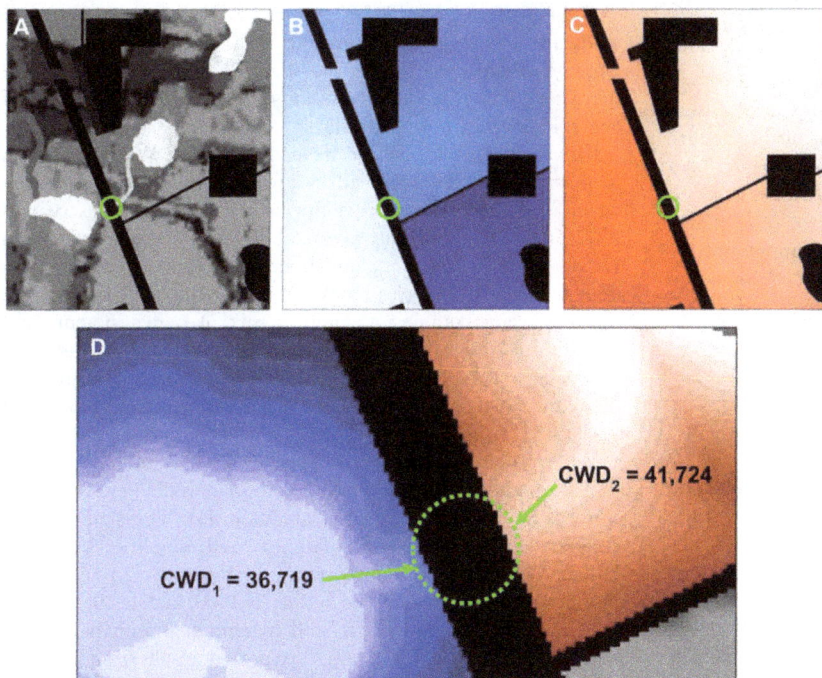

Figure 2. Detail of resistance and *CWD* surfaces with circular moving window. For a window with a diameter of 60 m (20 pixels) centered on the barrier, the arrows show the pixels in the window that have the lowest *CWD* to each patch (values shown are in weighted meters). Because the lowest *CWD* values from each patch will always be found on the edge of a moving window, only pixels on the perimeter need to be examined, increasing processing efficiency.

the case, a simple metric of connectivity improvement that would result from restoration across the moving window is:

$$\Delta LCD = LCD - LCD' \qquad (2)$$

Dividing ΔLCD by the search diameter gives the connectivity benefit per unit distance restored; dividing ΔLCD by LCD gives the proportional improvement relative to unrestored effective distance.

To illustrate the method, we first apply it to the relatively simple landscape described in Figure 1 using a search window with a diameter of 60 m (20 pixels at 3 m resolution; Figure 2). The search window size is chosen to match the size of the barrier that one is interested in detecting: a diameter of 20 pixels will fully incorporate effects of barriers up to 20 pixels across. We assign a resistance of 1 to optimal movement habitat, so that the cumulative cost of movement is identical to the Euclidean distance traversed when no barriers are encountered. For the circular window centered on the highway in Figure 2, the lowest CWD values from the left and right patches are 36,719 and 41,724 weighted meters, respectively. Summing these values and adding 60 (the cost of crossing the circle if it were restored to optimal movement habitat with a resistance of 1), gives the least-cost distance of the path crossing through the restored area (36,719+41,724+60 = 78,503 weighted meters). Since this is considerably lower than the least-cost distance between the patches without restoration (124,443 weighted meters), this location is a potent barrier, and the center pixel is assigned an improvement value of 45,940 weighted meters. This is repeated for every pixel on the landscape using standard GIS neighborhood analyses, resulting in a raster surface of improvement scores (Figure 3A).

The removal of the barrier where the improvement score is maximal – for example, by constructing a wildlife crossing structure – would re-route the best movement path (Figure 3B) and lower the effective distance between the two patches by 37% (45,940/124,443). Once that improvement is carried out, a second barrier analysis with the altered landscape conditions suggests that additional restorations along the highway will not further reduce the LCD at this point (Figure 3C). The next priority would be a road crossing in the upper right of the panel (dark orange in Figure 3C), connecting the rightmost patch to high-quality movement habitat above the road. The method is computationally efficient enough that different restoration scenarios can be tested iteratively: a barrier analysis with a 20- pixel search diameter across a landscape with 1 million pixels takes less than 2 seconds using a 2.7 GHz notebook computer.

Identifying barriers across scales and across large landscapes with multiple patches

The method described above can be extended across scales and across networks of patches, and we explore a few approaches to accomplish this here. By modifying the search diameter, the method can detect barriers of different sizes (Figure 4). Windows the width of a highway will best highlight where highways act as barriers, as in Figures 2 and 3. Larger windows will best detect barriers like agricultural fields, or cases in which narrow barriers run parallel to one another. Summary maps showing barrier effects across search window sizes may be created by first dividing improvement scores by the window size to produce maps of barrier strength per unit width, and then taking the maximum pixel score across scales (Figure 4B). This puts results from different analysis scales in the same units, allowing them to be

summarized in a single map. Alternative summary metrics are possible, and we address some of them in the Discussion.

To summarize across multiple sets of patch pairs, we have implemented a similar approach in which the maximum or sum of improvement scores across all patch pairs is assigned to a pixel. Taking the maximum of improvement scores shows the features that have the greatest effect for any patch pair (Figure 4C). Summing improvement scores highlights those barriers that isolate multiple pairs of patches from one another, extending the method to quantify barrier centrality (Figure 4D).

The methods described in this paper have been implemented in Barrier Mapper software [39], freely available as a new addition to the Linkage Mapper Toolkit for ArcGIS [40].

Example application in a landscape undergoing active conservation planning

The Washington Wildlife Habitat Connectivity Working Group, a collaboration of land and resource management agencies, non-governmental organizations (NGOs), universities, and Washington treaty tribes, recently completed a connectivity analysis across the Columbia Plateau Ecoregion in Washington, Oregon, and Idaho, USA [41]. The Working Group focused on the Columbia Plateau because the ecoregion is home to a large portion of Washington's sensitive plant and animal species but is also highly fragmented by agriculture and other anthropogenic activities. The Group modeled corridors to connect habitat for 11 focal species and also to connect natural landscape blocks scoring highly on an index of landscape integrity (i.e., large areas with relatively low levels human modification). Products from the analysis are being used to inform conservation planning efforts by several state and federal agencies and NGOs. Many of the corridors identified by the analysis pass through human-dominated landscapes, where roads, agricultural fields, and other human uses likely still act as barriers to movement.

We reanalyzed results for a corridor connecting two natural landscape blocks identified by the Working Group in Douglas County, Washington (Figure 5). We chose these blocks because they have been identified as important for many species of concern; for example, the blocks contain important habitat or corridors for 8 of 11 focal species analyzed by the Working Group. Moreover, both are occupied by greater sage-grouse (*Centrocercus urophasianus*, categorized as a Species of Greatest Conservation Need in Washington and a candidate for listing under the US Endangered Species Act), and both fall within a recovery area designated for the species by Washington State [42]. In addition, this landscape contains a complex mix of native systems and agricultural lands – the latter including both annual cropland and perennial vegetation cover – and includes roads, transmission lines, and other human-made features affecting animal movement [41].

To represent species with differing degrees of sensitivity to human modification, the Working Group used different resistance surfaces for landscape integrity analyses [41]. These surfaces all contained resistance values that increased with the degrees of human modification, differing only in the range of resistances assigned. Resistance scores of 1–100, 1–1000, and 1–10,000 were used for minimum, medium, and maximum sensitivity surfaces respectively (see [41] for details). We present results from a barrier analysis using the medium sensitivity resistance surface.

The modeled least-cost corridor connecting the patches dips south from the western patch, runs east to Banks Lake, and then north along a narrow strip of native vegetation and cliffs sitting between the lake and cropland (Figure 6A). A secondary and much longer corridor follows broad swaths of native vegetation through

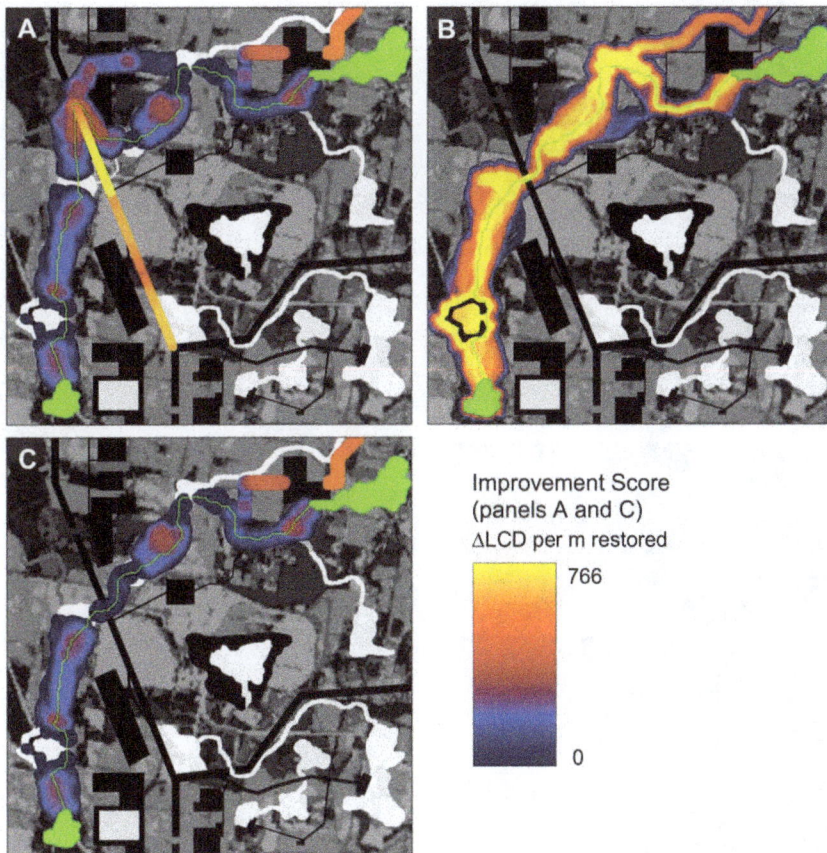

Figure 3. Barrier analysis of landscape. (A) Improvement scores (ΔLCD) for a 60 m search diameter using an enhanced version of Linkage Mapper software ([40]). Only positive values (indicating barriers whose removal would reduce isolation) are shown. To facilitate visualization of the barriers, scores were mapped so that they filled the search window (i.e. the maximum ΔLCD value within the search radius of each pixel is displayed). The greatest improvement potential was detected crossing the highway. Note that a natural corridor is bisected by the highway at the point with the highest improvement potential (see detail in Figure 2). (B) Creating a new gap in the barrier where restoration potential is highest re-routes the modeled least-cost corridor and greatly reduces resistance between the patches ($LCD = 78{,}503$ weighted meters compared with an LCD of 124,443 pre-restoration). Best 20% of study area shown. (C) Barrier detection at 60 m search diameter after restoration.

Moses Coulee and Beezley Hills to the south. A barrier analysis indicates numerous opportunities for improving the least-cost corridor, particularly within its east-west segment (Figure 6B). There are also opportunities outside of the main corridor, occurring along the longer route to the south and to the north as well (Figure 6B). Restoring any of these latter areas would re-route the modeled least-cost corridor, causing it to occur in a different location than it did in the unrestored landscape.

Restoration of any of several barriers identified to the south would improve connectivity as measured by LCD (Figure 6B); however, this would result in a much longer least-cost corridor. Restoration to the north has the potential to both improve LCD and shorten the distance traversed by the corridor. We simulated a restoration by changing a 1 km^2 (500 m×2 km) swath of agricultural land (indicated by the arrow in Figure 6B) to a resistance of 1. We chose 2 km because the greatest improvement was detected at the 2 km scale, and we assumed 500 m was wide enough to accommodate movement. A second corridor analysis following the simulated restoration shows the new corridor to the north (Figure 6C). The corridor has 9.4% less cumulative resistance than the original (1348 weighted km *vs.* 1489 weighted km), and its least-cost path is 44% shorter in un-weighted length. A post-restoration barrier analysis indicates that the highest improvement scores now fall along the new corridor (Figure 6D);

restoring a second 1 km^2 swath in this new corridor at the point indicated by the arrow would further reduce LCD by 50%.

Discussion

Connectivity models have provided valuable guidance to conservation planning efforts, as well as predictions of movement, gene flow, and isolation important to landscape genetics and other fields concerned with movement ecology. Yet they have almost exclusively emphasized identifying features that facilitate, rather than impede, movement; this emphasis gives an incomplete picture of how landscape features affect connectivity, what connectivity management strategies might be appropriate, and the uncertainty underlying model predictions. We see considerable potential for barrier detection analyses to help practitioners overcome these limitations. In particular, the ability to identify restoration opportunities can provide valuable alternatives to traditional conservation efforts focused on existing movement corridors.

Our reanalysis of the Columbia Plateau data (Figure 6) illustrates these points, showing how detecting barriers can increase conservation options available to practitioners, improve understanding of analysis products, and result in more robust conservation plans. Without a barrier analysis, conservation

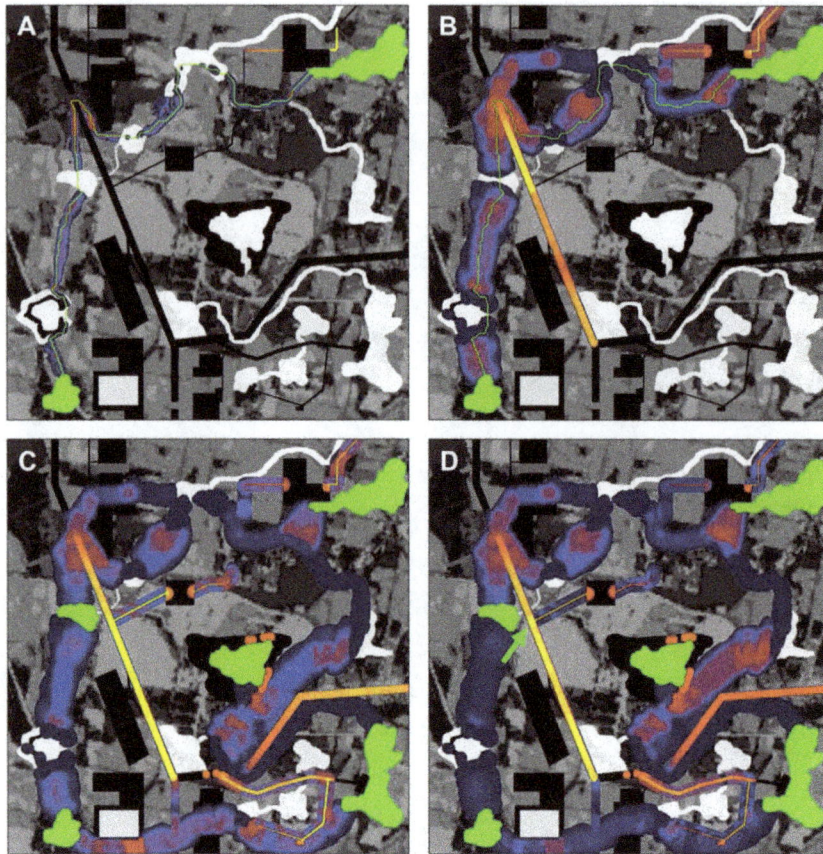

Figure 4. Barrier analyses integrating across multiple scales and patch pairs. (A) Results of barrier analysis with original patch pair at 12 m search diameter, which detects restoration opportunities equal to or less than 12 m across (e.g., local roads). (B) Maximum per-meter improvement value across 10 search window sizes (from 6 m to 60 m, with 6 m steps between search diameters). The map highlights where actions at different scales would have highest impact per meter restored. (C) Maximum per-meter improvement value across same window sizes and 5 patches, showing where greatest improvement could be achieved for any single pair of patches. (D) Sum of improvement scores among 5 patches (green). As in Panels A–C, the maximum per-meter improvement score was calculated for each patch pair at each scale. These were then summed across patch pairs to incorporate cumulative benefit for multiple patch pairs across multiple scales. The area scoring highest (bright yellow) had high improvement scores for multiple patch pairs; we interpret this area as having high 'barrier centrality,' i.e. being an important restoration opportunity for keeping the overall network connected. Note that the area occurs at a road intersection; if practical, placing a wildlife crossing structure here would re-route four corridors connecting the two leftmost patches to both the central and upper-right patch.

planners would likely have focused on conserving land in or adjacent to the original least-cost corridor. Our analysis revealed numerous opportunities to improve this corridor, but also that restoration of a 1 km² swath of cropland would create a new corridor with several desirable characteristics. Specifically, the new corridor has a lower least-cost distance, is shorter in length, and appears to have fewer pinch-points (narrow sections) than the original corridor—all desirable characteristics for corridor design [15,32]. Moreover, if the two original corridors remain in place, the new, northern corridor adds redundancy to connections between the natural landscape blocks. This is important because organisms seldom follow a single optimal path [43], and because redundant connections help to ensure continued connectivity in the face of unpredictable environmental changes [15].

The analysis showed that connectivity conservation options need not be limited to a small portion of the landscape, opening up much more area for actions that could conserve or enhance connectivity and illustrating tradeoffs between different conservation strategies and target locations. Beyond the corridor quality differences cited above, we note that the original corridor runs along a narrow stretch of land bordering Banks Lake, sometimes

traversing cliffs. The cliffs were assigned low resistance because the landscape integrity model used by the Working Group only quantified the degree to which pixels have been converted to human land uses. Practitioners, however, may consider cliffs to be impermeable for some species of conservation concern. The barrier analysis allows the user to quickly focus a more critical examination of corridor characteristics on areas influencing the results, and to identify options for alternative corridors that may better fit specific planning needs.

Similarly, the analysis underscored the potential sensitivity of corridor mapping to errors in GIS base data: our results show how the misclassification of a single agricultural field could have entirely altered the location of the original least-cost corridor shown in Figure 6A. The sensitivity of connectivity analysis results to landscape features at key locations has consequences for disciplines that depend on corridor maps (like conservation planning) and for disciplines that depend on connectivity measures (like landscape genetics). We discuss applicability of barrier detection methods to sensitivity analysis and error checking below.

Following the first barrier analysis and simulated restoration, a subsequent barrier analysis indicated that the restoration would

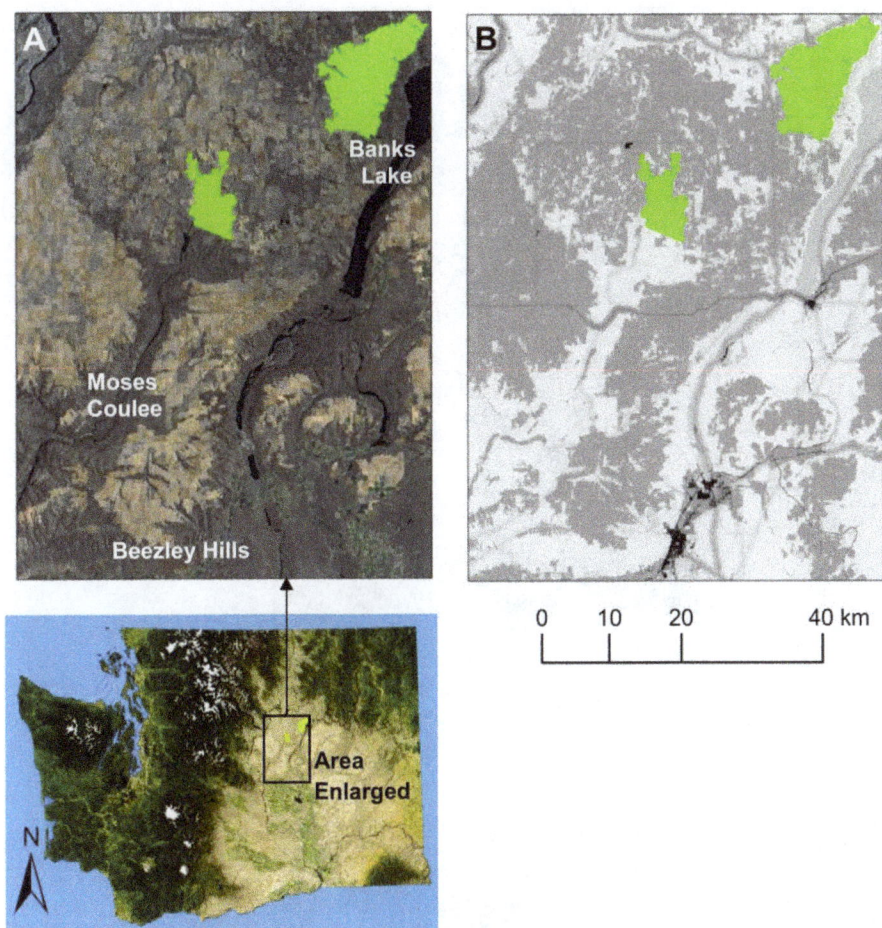

Figure 5. Corridor analysis in a landscape undergoing active conservation planning. (A) 60 km by 80 km study area in eastern Washington, USA, containing two natural landscape blocks to be connected (green). (B) Resistance map used to model corridors in a recent multi-partner connectivity analysis across the Columbia Plateau Ecoregion [41]; values range from 1 (white) to 1000 (black). Low resistance areas include native grassland and shrub-steppe, whereas high resistance areas include roads, developed areas, and agriculture.

open up further restoration opportunities of considerable value, one of which would cut *LCD* values by half. Thus, simulating restorations and re-running corridor and barrier analyses will likely improve final conservation and restoration plans.

Although we are aware of no other efforts to automate identification of terrestrial connectivity restoration opportunities, least-cost corridor analyses have been used to guide placement of crossing structures across roads to restore connectivity for wildlife. For example, Beier et al. [44] assigned a single, finite resistance value to all segments of a highway between two protected areas, regardless of whether a segment contained wildlife crossing structures. The least-cost corridor between the areas crossed the highway at the location where a crossing structure would result in the lowest ecological cost of travel. If highway crossing structures were not located in this corridor, Beier et al. [44] recommended specific structures at particular locations. This approach is useful, but does not quantify the improvement compared to existing conditions, does not identify restoration opportunities outside of least-cost corridors, and cannot be readily applied to barriers more complex than roads.

In addition to overcoming these limitations, our method is also amenable to highlighting barriers that affect multiple corridors, introducing the concept of barrier centrality. As shown in Figure 4, barriers can be mapped across all patch pairs, and the results

summed. This identifies barriers with high network centrality, similar to analyses that identify corridors or pathways with high centrality [21,22,45,46].

Applications for error checking and sensitivity analyses

GIS land cover data used to develop resistance layers for connectivity analyses are typically based on satellite or aerial imagery and often suffer from high levels of classification error [34,47]. Although our method relies on these same base data, it can help to prioritize error checking of the data by highlighting mapped features that strongly influence corridor locations. If a permeable feature is misclassified as impermeable and identified as a barrier, the misclassification could entirely alter a corridor's location. We recommend examining detected barriers, either by manually checking aerial imagery or conducting field surveys. Similarly, impermeable features misclassified as permeable that occur along least-cost paths can change corridor locations as well. Examining features along least-cost paths in tandem with barriers could thus further reduce the effects of classification error in connectivity analysis products.

Barrier detection can also be applied to parameter sensitivity analyses, important because resistances are often assigned based on expert opinion, which can be unreliable [34,47,48]. For example, if a given land cover type fell along a corridor's least-cost

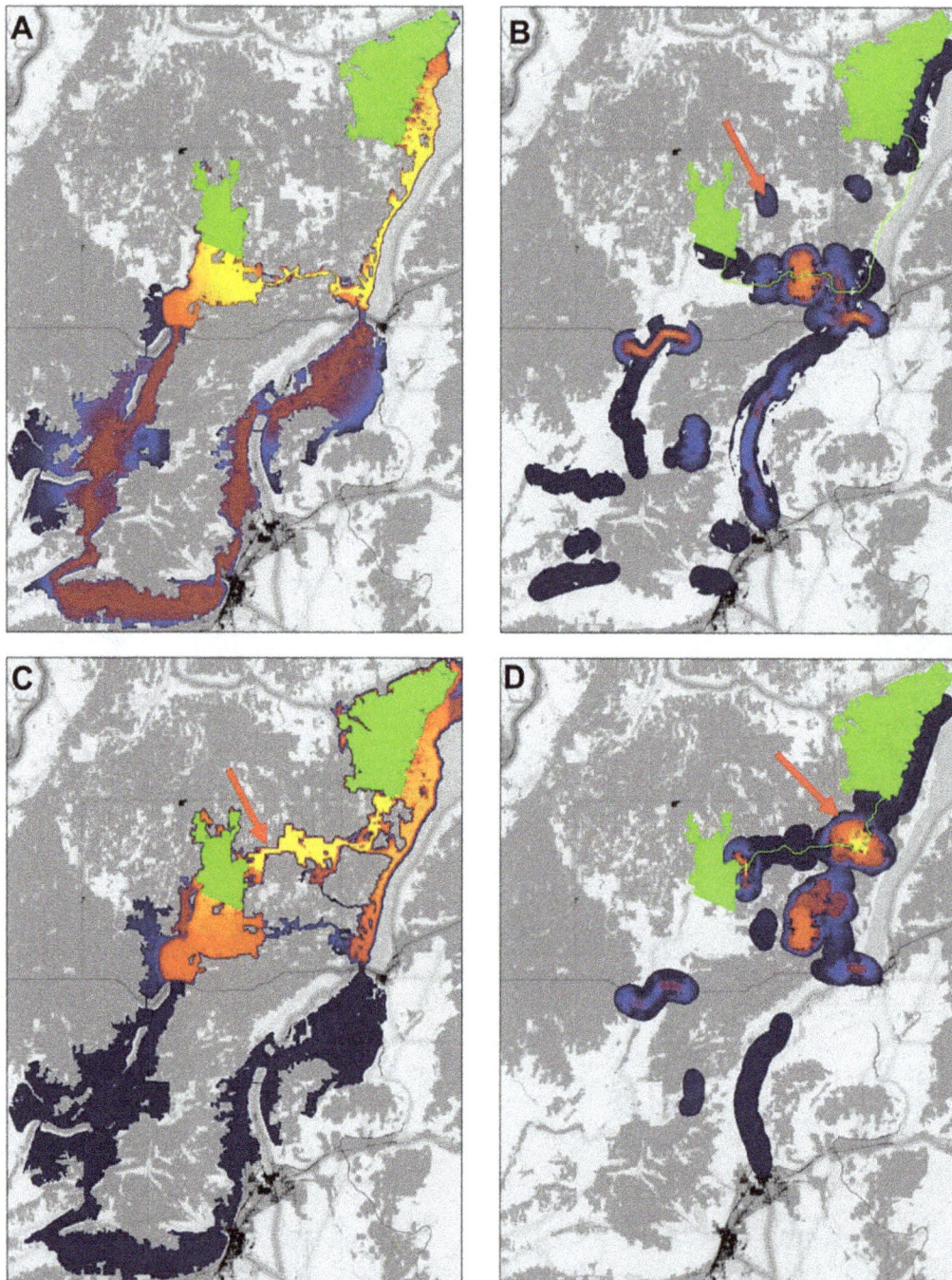

Figure 6. Reanalysis of connectivity modeling results using barrier detection algorithm. (A) Corridor connecting natural landscape blocks, showing least-cost movement routes. Best 20% of study area shown. (B) Barriers detected at diameters from 200 m to 2 km, with original least-cost path shown in green for reference. Mitigating barriers along the least-cost path (i.e., intersecting the green line) would improve the existing corridor without changing its location; mitigating barriers away from the path would re-route the best modeled corridor. (C) Restoring a 1 km^2 (500 m $\times$ 2 km) swath spanning the barrier indicated by the arrow establishes a new least-cost corridor to the North. (D) A barrier analysis incorporating the simulated restoration indicates opportunities to substantially improve the new corridor with additional restorations.

path or encompassed an influential barrier outside of the corridor, the resistance assigned to that land cover type would be known to influence the corridor's location. The sensitivity of the corridor's location to the resistance value assigned the land cover type could then be analyzed using alternative parameterization methods as described by Beier et al. [47]. As with connectivity models, our method will depend on the grain size of the resistance raster; to

adequately resolve features that potentially impede movement, we recommend pixels no larger than ½ the width of barriers one is interested in detecting.

Potential enhancements

Directionality of barrier effects. Our methods could be improved to more precisely pinpoint barriers. For example, elongated moving windows (search polygons) could perform better than circles to identify the best path for an improved corridor design. Measuring ΔLCD along elongate polygons placed at different angles, although more computationally complex than measuring across circles, would allow the attribution of directionality to barrier effects as well as adjustment of improvement scores at large search distances to reflect improvement achievable at smaller (nested) distances. New procedures to select the best orientation and width of such polygons could obviate the need to subjectively orient restoration polygons, like the 500 m×2 km polygon in our simulated restoration (Figure 6).

Restoration cost. We measured barrier strength by conservation improvement per meter restored because it was the simplest way to illustrate our approach. An alternative metric would be conservation improvement per restoration dollar; this would reflect, for example, that the cost per meter of a 10 m road crossing structure exceeds the cost per meter of a 50 m crossing structure, which in turn exceeds the cost per meter of restoring agricultural land. This enhancement would facilitate incorporation of connectivity restoration into return-on-investment analyses [49,50], helping managers to balance improvement potential, corridor importance, costs, and risk of conversion or degradation when deciding which parts of a landscape should be conserved or restored. The disadvantage, or course, is that this metric would require more data to calculate.

Restoration efficacy. Different R' resistance values could be applied to different land cover types to reflect the fact that some barriers would be more permeable to movement following restoration than others. For example, a highway underpass installed to allow animal movement may still have considerable resistance, whereas a restored forest stand may have resistance similar to undisturbed forest.

Other enhancements. Just as areas that cannot be conserved can be removed from reserve selection algorithms [51], unremovable barriers, such as urban areas, could be excluded from barrier analyses. The metrics described in equation (2) could be modified to incorporate restoration costs that vary by feature type, or land prices mapped using parcel data. Metrics of corridor importance (e.g., link centrality) could be integrated by multiplying improvement scores by such metrics, which would highlight opportunities to restore the most potent barriers in the most important corridors. Or, rather than focusing on pairs of patches, the method could be altered to focus on the connectedness of each patch by summing barriers detected between each patch and all others. Lastly, improvement scores may be expressed in terms of absolute improvement or percent improvement relative to unrestored corridor resistance. An advantage of the latter approach is that it would favor restoration in corridors in which LCD values are already low, presumably meaning they are more viable.

Which of these enhancements are most valuable will depend on the objectives of individual users and projects.

Application in other connectivity modeling frameworks

Although least-cost corridor models are by far the most commonly applied connectivity planning tool, they rely on simple assumptions about animal movement and other processes [43,48,52,53]. However, our approach can be applied in any connectivity modeling framework that produces measures of effective distance. For example, circuit-based connectivity analyses can model the relative proximity of each pixel to two patches by setting the voltage of one patch to 1 and the other to ground (see [15] for details on applying circuit modeling to landscapes). The resulting voltage surface gives the probability that a random walker will reach one patch before reaching the other [15,54]. Strong gradients in voltage indicate barriers that separate areas relatively accessible to one patch from areas relatively accessible to the other. If removed, such barriers would reduce effective resistance between the patches, an analog to LCD that takes into account the availability of multiple, parallel connections. A similar approach is widely used in microchip design: simulated voltage levels reveal areas with strong voltage gradients (known as IR drops) where electrical connectivity must be enhanced [55]. Thus barrier analysis using circuit theory can identify opportunities to provide valuable redundant connections even when LCD would not be reduced. In contrast, barrier analysis using least-cost methods will not identify these opportunities.

Individual-based movement models provide a more complex but also more powerful framework for modeling connectivity, capable of incorporating more biological realism and behavioral information than least-cost or circuit analyses [56]. As long as an individual-based model can produce maps of effective distance (e.g., based on the probability of, or energetic expenditure associated with, reaching different locations from a source patch), the approach described here could be applied to the model. Models such as PATH [16] and HexSim [17] can be used to derive such measures.

Potential for integration with systematic conservation planning

Our method is not a substitute for algorithms like Marxan [57] or Zonation [51], which are designed to optimize selection of reserves or sets of conservation actions. Although our method identifies and ranks candidate areas for restoration actions, it does not select optimal *sets* or portfolios of conservation actions to achieve given conservation goals while minimizing cost. The same can be said for algorithms designed to map areas that most facilitate movement and connectivity (e.g., [22,40,46,58–60]); rather than incorporating optimization routines, such algorithms instead produce maps that must be interpreted by practitioners, who then make conservation decisions in light of costs, benefits, and other management objectives.

Although it has long been recognized as important to reserve network design [61], incorporating connectivity directly into optimization algorithms has proven difficult. Most such efforts can be characterized as minimizing local fragmentation by either considering the geographic proximity of candidate areas to other areas (e.g., [62–64]) or maximizing the compactness and contiguity of reserves by favoring selection of adjacent cells or using boundary quality or length penalties (e.g., [29,57,65,66]). Because these algorithms favor conserving or restoring contiguous natural areas, they may neglect areas that, although fragmented, contribute to connectivity between natural areas. Thus, relying solely on maximizing the proximity or contiguity of protected areas could lead to elimination of movement routes that cross human-dominated landscapes.

Progress toward synthesizing connectivity and optimization algorithms has likely been hampered by the 'network' nature of connectivity planning: conservation in one area can affect the function and value of distant areas, contingent upon the conservation status and characteristics of the intervening landscape. Incorporating this complexity into optimization algorithms becomes computationally prohibitive with large numbers of planning units [67]. Still, practitioners are beginning to use outputs of multi-species connectivity models as inputs to optimization algorithms like Zonation [23,68]. Such examples are promising, and should be equally applicable with restoration-oriented algorithms such as ours.

An alternative to our approach that would seek to develop a near-optimal set of conservation actions would be to employ a routine similar to that used by Zonation software, which begins with an intact landscape and iteratively removes grid cells with low conservation value [51,69]. Starting with a landscape in which all restorable barriers have been removed, different sets of barriers could be added back in and connectivity metrics recalculated at each iteration. As with traditional connectivity models, however, this would be computationally prohibitive with large numbers of patches or restoration sites because of the computational time required for recalculating connectivity metrics. A promising hybrid approach could be to use the method described in this paper to identify sets of pre-screened restoration opportunities, which could then be removed from a resistance surface and added back in using an algorithm like Zonation's.

Practical considerations for improving conservation and restoration decisions

Managing for connectivity to facilitate gene flow, climate adaptation, and other processes is challenging without reliable maps to guide practitioners [33]. Connectivity analyses have provided valuable implementation guidance in the past; barrier mapping can increase the rigor of such analyses and the range of conservation options they reveal. It can help practitioners a) decide if connectivity conservation is a worthy investment in a landscape; b) identify opportunities to restore vs. conserve different areas; c) reduce uncertainty due to errors in GIS base data; and d) balance potential improvement against costs so that investments can be prioritized.

The goals of managers and planners can be used to guide applications of barrier detection methods. For example, if a transportation agency is interested in determining which highway segments are likely to have the greatest impact on wildlife movement, the search window should correspond to the width of highways, with outputs clipped to highways and the R' value determined based on the estimated resistance of the kind of crossing structure (or alternative structures) being considered. If a land management agency is prioritizing restoration of degraded native vegetation, the search window should relate to the size of appropriate restoration projects, and outputs should be clipped to the eligible land base (e.g., limited to the type of vegetation the restoration would target). If an NGO is identifying landowners interested in obtaining voluntary incentive payments for wildlife-friendly management, the window should reflect the scale of such management. Summarizing barrier analyses across multiple scales will be desirable for collaborations among organizations with differing goals and mandates. As noted above, iterative application of the model with simulated restorations will likely provide the most informative results and most robust conservation plans.

Similarly, the method may have potential to help adapt results from coarse-filter connectivity assessments, such as landscape integrity/human modification-based connectivity maps, to more fine-filter objectives (see [70] for a review of coarse- and fine-filter conservation planning). Alternative corridors revealed by the method could be assessed for their suitability under different planning constraints (e.g., corridors for species that must avoid cliffs, as in the Columbia Plateau example). While not a replacement for species-specific connectivity analyses, such an approach could help land managers evaluate alternatives if a mapped corridor is deemed unsuitable for their particular needs.

Connectivity maps do not always identify functioning routes that need to be maintained and protected; rather, they frequently map routes that may not be currently viable, but appear to provide the best opportunities for future work toward enhancing connectivity. In this sense connectivity maps often represent visions and goals for desired future conditions [71]. Barrier detection can add insight into the practicality of these goals, and identify specific options for achieving them. It can also help practitioners to 'triage' a connectivity plan, identifying corridors that traverse numerous barriers – and therefore would require significant investment to fully restore – so that efforts may be focused on more viable movement routes.

Perhaps most importantly, the ability to detect options to re-route corridors also opens up a broader suite of potential actions to improve connectivity. It can help managers identify new corridors that add additional movement pathways in areas important to the overall connectivity of a landscape (i.e. linkages with high centrality). Combined with spatially explicit land cost data, the method could help to improve conservation efficacy while reducing costs.

We hope barrier analyses will expand conservation options available to managers, and broaden conversations about restoration of connectivity more generally. By identifying new ways to improve connectivity in a particular area, the method can allow managers to consider different suites of strategies, or engage with new sets of stakeholders with interests in different areas. Both from the perspective of entities mandated to carry out conservation actions, and from the perspective of stakeholders with interests in the lands that are the focus of such actions, broadening the suite of alternatives and tools can only increase the opportunities for finding common ground in pursuit of multiple objectives.

Acknowledgments

We are grateful to the Washington Wildlife Habitat Connectivity Working Group for making data used in this manuscript publicly available. We also thank Theresa Nogeire, Adina Merenlender, and two anonymous reviewers for comments on early drafts.

Author Contributions

Conceived and designed the experiments: BHM. Performed the experiments: BHM. Analyzed the data: BHM. Contributed reagents/materials/analysis tools: BHM. Wrote the paper: BHM SAH PB DMT. Suggested refinements to methods and analyses: SAH, PB, DMT.

References

1. Taylor P, Fahrig L, Henein K, Merriam G (1993) Connectivity is a vital element of landscape structure. Oikos 68: 571–572.

2. Ricketts TH (2001) The matrix matters: effective isolation in fragmented landscapes. The American Naturalist 158: 87–99.

3. Kareiva P, Wennergren U (1995) Connecting landscape patterns to ecosystem and population processes. Nature 373: 299–302.

4. Moilanen A, Nieminen M (2002) Simple connectivity measures in spatial ecology. Ecology 83: 1131–1145.

5. Crooks KR, Sanjayan M (2006) Connectivity Conservation. Cambridge: Cambridge University Press. 712 p.

6. Damschen EI, Haddad NM, Orrock JL, Tewksbury JJ, Levey DJ (2006) Corridors increase plant species richness at large scales. Science 313: 1284–1286.

7. Heller NE, Zavaleta ES (2009) Biodiversity management in the face of climate change: A review of 22 years of recommendations. Biological Conservation 142: 14–32.

8. Singleton P, Gaines W, Lehmkuhl J (2002) Landscape permeability for large carnivores in Washington: a geographic information system weighted distance and least-cost corridor assessment. Research Paper PNW-RP-549. Portland, Oregon. 89 p. Available: http://www.arlis.org/docs/vol1/51864782.pdf. Accessed 2012 Nov 29.

9. South Coast Wildlands (2008) South coast missing linkages: a wildland network for the South Coast Ecoregion. Fair Oaks, CA: South Coast Missing Linkages Initiative. Available: http://www.scwildlands.org/reports/SCMLRegionalReport.pdf. Accessed 2012 Nov 29.

10. Baldwin RF, Perkl RM, Trombulak SC, Burwell WB (2010) Modeling Ecoregional Connectivity. In: Trombulak SC, Baldwin RF, editors. Landscape-scale Conservation Planning. Houten, Netherlands: Springer Netherlands. pp. 349–367.

11. Spencer WD, Beier P, Penrod K, Parisi M, Pettler A, et al. (2010) California Essential Habitat Connectivity Project: a Strategy for Conserving a Connected California. Sacramento, CA: California Department of Transportation and California Department of Fish and Game. Available: http://www.dfg.ca.gov/habcon/connectivity/. Accessed 2012 Nov 29.

12. Washington Wildlife Habitat Connectivity Working Group (2010) Washington Connected Landscapes Project: Statewide Analysis. Olympia, WA: Washington Departments of Fish and Wildlife and Transportation. Available: http://waconnected.org/statewide-analysis/. Accessed 2012 Nov 29.

13. Knaapen JP, Scheffer M, Harms B (1992) Estimating habitat isolation in landscape planning. Landscape and Urban Planning 23: 1–16.

14. Adriaensen F, Chardon JP, De Blust G, Swinnen E, Villalba S, et al. (2003) The application of "least-cost" modelling as a functional landscape model. Landscape and Urban Planning 64: 233–247.

15. McRae BH, Dickson BG, Keitt TH, Shah VB (2008) Using circuit theory to model connectivity in ecology, evolution, and conservation. Ecology 89: 2712–2724.

16. Hargrove WW, Hoffman FM, Efroymson RA (2005) A practical map-analysis tool for detecting potential dispersal corridors. Landscape Ecology 20: 361–373.

17. Schumaker N (2011) HexSim. U.S. Environmental Protection Agency, Environmental Research Laboratory. Available: http://www.epa.gov/hexsim. Accessed 2012 Nov 29.

18. Tracey JA (2006) Individual-based modeling as a tool for conserving connectivity. In: Crooks KR, Sanjayan M, editors. Connectivity Conservation. Cambridge: Cambridge University Press. pp. 343–368.

19. Bunn AG, Urban DL, Keitt TH (2000) Landscape connectivity: A conservation application of graph theory. Journal of Environmental Management 59: 265–278.

20. Urban D, Keitt T (2001) Landscape connectivity: a graph-theoretic perspective. Ecology 82: 1205–1218.

21. Carroll C, McRae BH, Brookes A (2012) Use of linkage mapping and centrality analysis across habitat gradients to conserve connectivity of gray wolf populations in western North America. Conservation Biology 26: 78–87.

22. Theobald DM, Reed SE, Fields K, Soulé M (2012) Connecting natural landscapes using a landscape permeability model to prioritize conservation activities in the United States. Conservation Letters 5: 123–133.

23. Breckheimer I (2012) Mapping Habitat Quality in Conservation's Neglected Geography. M.S. Thesis, Curriculum for the Environment and Ecology, University of North Carolina at Chapel Hill.

24. Tischendorf L, Fahrig L (2000) How should we measure landscape connectivity? Landscape Ecology 15: 633–641.

25. Baldwin R., Reed SE, McRae BH, Theobald DM, Sutherland RW (2012) Connectivity restoration in large landscapes: modeling landscape condition and ecological flows. Ecological Restoration 30(4): 274–279.

26. Margules CR, Pressey RL (2000) Systematic conservation planning. Nature 405: 243–253. doi:10.1038/35012251.

27. Possingham HP, Ball I, Andelman S (2000) Mathematical methods for identifying representative reserve networks. In: Ferson S, Burgman M, editors. Quantitative Methods for Conservation Biology. New York: Springer-Verlag. pp. 291–305.

28. Wilson KA, Underwood EC, Morrison SA, Klausmeyer KR, Murdoch WW, et al. (2007) Conserving biodiversity efficiently: what to do, where, and when. PLOS biology 5: e223. doi:10.1371/journal.pbio.0050223.

29. Thomson JR, Moilanen AJ, Vesk PA, Bennett AF, MacNally R (2009) Where and when to revegetate: a quantitative method for scheduling landscape reconstruction. Ecological Applications 19: 817–828. doi:10.1890/08-0915.1.

30. Lethbridge MR, Westphal MI, Possingham HP, Harper ML, Souter NJ, et al. (2010) Optimal restoration of altered habitats. Environmental Modelling & Software 25: 737–746. doi:10.1016/j.envsoft.2009.11.013.

31. McBride MF, Wilson K, Burger J, Fang Y-C, Lulow M, et al. (2010) Mathematical problem definition for ecological restoration planning. Ecological Modelling 221: 2243–2250. doi:10.1016/j.ecolmodel.2010.04.012.

32. Beier P, Majka DR, Spencer WD (2008) Forks in the road: choices in procedures for designing wildland linkages. Conservation Biology 22: 836–851.

33. Beier P, Spencer W, Baldwin RF, McRae BH (2011) Toward Best Practices for Developing Regional Connectivity Maps. Conservation Biology 25: 879–892.

34. Zeller KA, McGarigal K, Whiteley AR (2012) Estimating landscape resistance to movement: a review. Landscape Ecology 27: 777–797.

35. Graham CH (2001) Factors Influencing Movement Patterns of Keel-Billed Toucans in a Fragmented Tropical Landscape in Southern Mexico. Conservation Biology 15: 1789–1798.

36. Chardon JP, Adriaensen F, Matthysen E (2003) Incorporating landscape elements into a connectivity measure: a case study for the Speckled wood butterfly (Pararge aegeria L.). Landscape Ecology 18: 561–573.

37. Schwartz MK, Copeland JP, Anderson NJ, Squires JR, Inman RM, et al. (2009) Wolverine gene flow across a narrow climatic niche. Ecology 90: 3222–3232.

38. Cushman SA, McKelvey KS, Hayden J, Schwartz MK (2006) Gene flow in complex landscapes: testing multiple hypotheses with causal modeling. The American Naturalist 168: 486–499.

39. McRae BH (2012) Barrier Mapper Connectivity Analysis Software. Seattle, WA: The Nature Conservancy. Available: http://www.circuitscape.org/linkagemapper. Accessed 2012 Nov 29.

40. McRae BH, Kavanagh DM (2011) Linkage Mapper Connectivity Analysis Software. Seattle, WA: The Nature Conservancy. Available: http://www.circuitscape.org/linkagemapper. Accessed 2012 Nov 29.

41. Washington Wildlife Habitat Connectivity Working Group (2012) Washington Connected Landscapes Project: Analysis of the Columbia Plateau Ecoregion. Olympia, WA: Washington Departments of Fish and Wildlife, and Transportation. Available: http://waconnected.org/columbia-plateau-ecoregion/. Accessed 2012 Nov 29.

42. Stinson DW, Hays D, Schroeder M (2004) Washington State recovery plan for the greater sage-grouse. Olympia, WA: Washington State Department of Fish and Wildlife. 121 p. Available: http://wdfw.wa.gov/publications/00395/. Accessed 2012 Nov 29.

43. Pinto N, Keitt TH (2008) Beyond the least-cost path: evaluating corridor redundancy using a graph-theoretic approach. Landscape Ecology 24: 253–266.

44. Beier P, Garding E, Majka D (2008) Arizona Missing Linkages: Linkage Designs for 16 Landscapes. Phoenix, AZ: Arizona Game and Fish Department. Available: http://corridordesign.org/linkages/arizona. Accessed 2012 Nov 29.

45. Estrada E, Bodin O (2008) Using network centrality measures to manage landscape connectivity. Ecological Applications 18: 1810–1825.

46. Landguth EL, Hand BK, Glassy J, Cushmann SA, Sawaya MA (2011) UNICOR: a species connectivity and corridor network simulator. Ecography 35: 9–14.

47. Beier P, Majka DR, Newell SL (2009) Uncertainty analysis of least-cost modeling for designing wildlife linkages. Ecological Applications 19: 2067–2077.

48. Spear SF, Balkenhol N, Fortin M-J, McRae BH, Scribner K (2010) Use of resistance surfaces for landscape genetic studies: considerations for parameterization and analysis. Molecular ecology 19: 3576–3591.

49. Underwood EC, Shaw MR, Wilson KA, Kareiva P, Klausmeyer KR, et al. (2008) Protecting biodiversity when money matters: maximizing return on investment. PLOS ONE 3: e1515. doi:10.1371/journal.pone.0001515.

50. Polasky S, Nelson E, Camm J, Csuti B, Fackler P, et al. (2008) Where to put things? Spatial land management to sustain biodiversity and economic returns. Biological Conservation 141: 1505–1524. doi:10.1016/j.biocon.2008.03.022.

51. Moilanen A, Franco AMA, Early RI, Fox R, Wintle B, et al. (2005) Prioritizing multiple-use landscapes for conservation: methods for large multi-species planning problems. Proceedings Biological sciences/The Royal Society 272: 1885–1891. doi:10.1098/rspb.2005.3164.

52. Rosenberg DK, Noon BR, Meslow EC (1997) Biological corridors: form, function, and efficacy. BioScience 47: 677–687.

53. McRae BH (2006) Isolation by resistance. Evolution 60: 1551–1561.

54. Doyle P, Snell J (1984) Random walks and electric networks. Washington, DC: The Mathematical Association of America. 159 p.

55. Blaauw D, Pant S, Chaudhry R, Panda R (2005) Design and Analysis of Power Supply Networks. In: Lavagno L, Martin G, Sheffer L, editors. Electronic Design Automation for Integrated Circuits Handbook. Boca Raton, FL: CRC Press.

56. Grimm V, Railsback SF (2005) Individual-based modeling and ecology. Princeton, NJ: Princeton University Press. 428 p.

57. Ball IR, Possingham HP (2000) MARXAN (V1.8.2): Marine Reserve Design Using Spatially Explicit Annealing. Available: http://www.marineplanning.org/pdf/marxan_manual_1_8_2.pdf. Accessed 2012 Nov 29.

58. Majka D, Jenness J, Beier P (2007) CorridorDesigner: ArcGIS tools for designing and evaluating corridors. Available: http://corridordesign.org. Accessed 2012 Nov 29.

59. McRae BH, Shah VB (2009) Circuitscape Connectivity Analysis Software. The University of California, Santa Barbara. Available: http://www.circuitscape.org. Accessed 2012 Nov 29.

60. Carroll C (2010) Connectivity Analysis Toolkit. Avilable: http://www.klamathconservation.org/science_blog/software/. Accessed 2012 Nov 29.

61. Possingham HP, Wilson KA, Andelman SJ, Vynne CH (2006) Protected areas: goals, limitations, and design. In: Groom MJ, Meffe GK, Carroll C., editors. Principles of Conservation Biology. Sunderland, MA: Sinauer Associates Inc. pp. 509–533.

62. Briers R (2002) Incorporating connectivity into reserve selection procedures. Biological Conservation 103: 77–83. doi:10.1016/S0006-3207(01)00123-9.

63. Arponen A, Lehtomäki J, Leppänen J, Tomppo E, Moilanen A (2012) Effects of connectivity and spatial resolution of analyses on conservation prioritization across large extents. Conservation biology: the journal of the Society for Conservation Biology 26: 294–304. doi:10.1111/j.1523-1739.2011.01814.x.

64. Cabeza M (2003) Habitat loss and connectivity of reserve networks in probability approaches to reserve design. Ecology Letters 6: 665–672. doi:10.1046/j.1461-0248.2003.00475.x.

65. Moilanen A (2007) Landscape Zonation, benefit functions and target-based planning: Unifying reserve selection strategies. Biological Conservation 134: 571–579. doi:10.1016/j.biocon.2006.09.008.

66. Westphal MI, Field SA, Possingham HP (2007) Optimizing landscape configuration: A case study of woodland birds in the Mount Lofty Ranges,

South Australia. Landscape and Urban Planning 81: 56–66. doi:10.1016/
j.landurbplan.2006.10.015.

67. Moilanen A, Possingham HP, Polasky S (2009) A mathematical classification of
conservation prioritisation problems. In: Moilanen A, Wilson KA, Possingham
HP, editors. Spatial Conservation Prioritization. Oxford: Oxford University
Press. pp. 28–42.

68. Breckheimer I, Milt A (2012) Connect: Landscape Connectivity Modeling
Toolbox. Department of Geography, University of North Carolina. Available:
http://www.unc.edu/depts/geog/lbe/Connect/. Accessed 2012 Nov 29.

69. Moilanen A, Leathwick JR, Quinn JM (2011) Spatial prioritization of
conservation management. Conservation Letters 4: 383–393. doi:10.1111/
j.1755-263X.2011.00190.x.

70. Schwartz MW (1999) Choosing the appropriate scale of reserves for
conservation. Annual Review of Ecology and Systematics 30: 83–108.

71. Hall SA, McRae BH, Gregory A, Krosby MB, Myers W, et al. (2012) Future
work and conclusions. Washington Connected Landscapes Project: Analysis of
the Columbia Plateau Ecoregion. Olympia, WA: Washington Department of
Fish andWildlife and Washington Department of Transportation. Available:
http://www.waconnected.org/wp-content/themes/whcwg/docs/WHCWG_
Columbia_Plateau_Ecoregion_Feb%202012_Ch5.pdf. Accessed 2012 Nov 29.

20

Individual to Community-Level Faunal Responses to Environmental Change from a Marine Fossil Record of Early Miocene Global Warming

Christina L. Belanger*

Department of the Geophysical Sciences, University of Chicago, Chicago, Illinois, United States of America

Abstract

Modern climate change has a strong potential to shift earth systems and biological communities into novel states that have no present-day analog, leaving ecologists with no observational basis to predict the likely biotic effects. Fossil records contain long time-series of past environmental changes outside the range of modern observation, which are vital for predicting future ecological responses, and are capable of (a) providing detailed information on rates of ecological change, (b) illuminating the environmental drivers of those changes, and (c) recording the effects of environmental change on individual physiological rates. Outcrops of Early Miocene Newport Member of the Astoria Formation (Oregon) provide one such time series. This record of benthic foraminiferal and molluscan community change from continental shelf depths spans a past interval environmental change (~20.3-16.7 mya) during which the region warmed 2.1–4.5°C, surface productivity and benthic organic carbon flux increased, and benthic oxygenation decreased, perhaps driven by intensified upwelling as on the modern Oregon coast. The Newport Member record shows that (a) ecological responses to natural environmental change can be abrupt, (b) productivity can be the primary driver of faunal change during global warming, (c) molluscs had a threshold response to productivity change while foraminifera changed gradually, and (d) changes in bivalve body size and growth rates parallel changes in taxonomic composition at the community level, indicating that, either directly or indirectly through some other biological parameter, the physiological tolerances of species do influence community change. Ecological studies in modern and fossil records that consider multiple ecological levels, environmental parameters, and taxonomic groups can provide critical information for predicting future ecological change and evaluating species vulnerability.

Editor: Anna Stepanova, Texas A&M University, United States of America

Funding: This material is based upon work supported by the National Science Foundation under Grant No. 0910026, the William V. Sliter award from the Cushman Foundation, Conchologists of America Research Award, and Sigma Xi Grants-in-Aid of Research. CLB was also supported by the National Science Foundation Graduate Research Fellowship and the Frank H. and Eva B. Buck Foundation. The funders had no role in study design, data collection and analysis, decision to publish, or preparation of the manuscript.

Competing Interests: The author has declared that no competing interests exist.

* E-mail: belanger@uchicago.edu

Introduction

The fossil record is a key source of data on biotic responses to past and ongoing climate change and its importance to establishing natural baselines of environmental and ecological change are increasingly realized [1–3]. Fossil records also permit long-term studies of ecological responses to environmental changes over millennia to millions of years – time scales unobtainable by modern ecological studies. Fossil records that pre-date human influences are especially valuable because the effects of climate change can be assessed without confounding anthropogenic factors such as pollution, eutrophication, fishing, and human-facilitated invasive species. In addition, future environmental conditions and ecosystem states may have no present-day analog [4–5]. Investigating past conditions can broaden the range of observed ecosystem states and increase our ability to predict the behavior of systems outside modern observation.

Coastal marine environments need critical study because they are highly productive systems rich in biodiversity and important human resources. Coastal environments are also especially

vulnerable to climate change and have experienced greater environmental changes with modern global warming than open ocean settings [6–7]. Marine benthic communities from subtidal soft-sediment environments contain taphonomically durable taxa, like benthic foraminifera and mollusks, which provide excellent fossil time-series of community structure and composition with high fidelity to the original, living, assemblages [8–10]. These marine benthic groups are thus valuable for examining past ecological change in coastal systems.

In addition to illuminating ecosystem states with no present-day analog, fossil records can reveal the tempo of past faunal responses, which are important to predicting modern ecological changes. Temporal changes in ecological communities may occur gradually or have stepped, pulsed shifts that are rapid compared to ecological dynamics before and after the shift [11]. Large, sudden shifts across an ecological threshold may be difficult to reverse if the new community represents an alternative stable state or ecological regime [11–13]. Much of environmental management, however, is built upon a model of gradual ecosystem changes [14]. Furthermore, documented shifts to alternative stable states in the

Table 1. Faunal variables measured in the Newport Member of the Astoria Formation; source indicates the materials measured for each variable.

Faunal Variable	Ecological Level	Source	Ecological Metric
Species Richness	Community	Molluscan and Foraminiferal bulk samples	sample size standardized number of species
Evenness	Community	Molluscan and Foraminiferal bulk samples	probability of intraspecific encounter (PIE)
Relative abundance of ecologically distinct groups	Community	Molluscan, Foraminiferal bulk samples	proportional abundance of subsurface deposit-feeding bivalves and organic-loving benthic foraminifera
Taxonomic Composition	Community	Molluscan and Foraminiferal bulk samples	Ordination score (DCA, NMDS)
Average Body Size 1	Community	Bivalves	geometric mean size of valve for all bivalve taxa
Average Body Size 2	Population	Bivalves	geometric mean size of valve for bivalve species
Shell growth rate	Individual	Bivalves	distance between external growth bands, internal growth bands, or $\delta^{18}O$ maxima
Age at first reproduction*	Individual	Bivalves	reproductive growth checks; mismatch between "annual" growth bands and $\delta^{18}O$

* = Variables not measured in the present study, which have the potential to yield additional information after methodological improvements.

marine realm often result from, or at least occur in the context of, direct human influence (ie. eutrophication or over-fishing) rather than anthropogenic climate change alone [15]. Fossil records that pre-date human impacts are thus essential testing grounds for determining if abrupt, threshold, shifts are a general mode of biological response to climate change.

Many studies of biotic responses to climate change focus on temperature, but other stressors such as changes in productivity and benthic oxygenation that coincide with temperature change can play a major role in marine benthic community structure [16–19]. Changes in productivity and oxygenation can be linked to temperature change, but these environmental properties can be variably coupled or decoupled so that temperature itself may not be the direct driver of biotic responses [20–22]. Proxies, such as $\delta^{18}O$ (temperature and salinity) and $\delta^{13}C$ (productivity) derived from benthic foraminiferal tests and information on bottom water oxygenation and water energy from sedimentology, can provide climate data for fossil records and allow us to identify the proximal drivers of marine benthic faunal change.

Further complicating the analysis of biotic responses, both in modern and ancient settings, is that different taxonomic groups may respond to the same environmental change but in different ways, may respond most strongly to different environmental parameters, or may respond at different thresholds of change. For example, molluscs and benthic foraminifera commonly co-occur and have fossil records preserved in the same sediments, but interact with different aspects of the environment. Examining multiple taxonomic groups from the same fossil record provides a test of the generality of faunal responses, and the environmental factors that drive those faunal changes, despite biological disparity.

Paleoecological studies rarely examine multiple ecological levels, but studying the community, population, and individual levels in a single fauna can lead to an understanding of the biological mechanisms underlying faunal changes. For example, physiological limits are often used to explain the geographic range shifts of marine organisms that are associated with climate change [23–25], but this is rarely tested in paleoecological studies and few studies consider both processes at multiple levels. However, in organisms that grow by accretion, we can use growth banding and stable isotope sclerochronology in fossil material to reconstruct growth

rates, which are a proxy for physiological rates, and thus directly test whether physiological limits underlie faunal change. If body size and growth rates decline in species whose abundances decline during an environmental change, we can infer that community-level change is driven, at least in part, by mechanisms at the level of individual physiological tolerances. Alternative hypotheses for climate-linked biotic change within the marine realm include changes in settlement or recruitment potentials [16,26] and changes in ecological interactions [27–28].

To extract biotic responses and their environmental drivers at the individual, population, and community levels from past intervals of long-term climate change, we need fossil records that (a) were deposited across past climate change events, (b) can be analyzed at high enough resolution to evaluate trends in regional faunal and environmental changes, and (c) have fossils well enough preserved for geochemical and growth line analyses. Here I present an example of the utility of fossil records that fit these criteria using a rich record of co-occurring benthic foraminifera and molluscs from the Early Miocene Newport Member Astoria Formation exposed in coastal cliffs along central Oregon (see Table 1 for faunal variables measured). This record shows that (a) ecological responses to natural environmental change can be abrupt, (b) productivity can be the primary driver of faunal change during global warming, (c) molluscs and foraminifera respond differently to the same environmental changes, and (d) physiological mechanisms underlie community-level change.

Geological and Environmental Background

The Early Miocene was an interval of increasing temperature leading into the Middle Miocene Climate Optimum, the warmest time in the past 20 million years [29]. Earlier work on the paleoenvironmental changes in the Newport Member using the same samples described herein documents regional warming on the order of 2.1 to 4.5°C between 19.5 and 18.5 Ma; the total ~80 m thick stratigraphic record spans ~20.3-16.7 mya [30]. In addition to temperature change, surface water productivity and organic carbon flux to the seafloor both increased at ~19 mya, resulting in dysoxic benthic conditions [30].

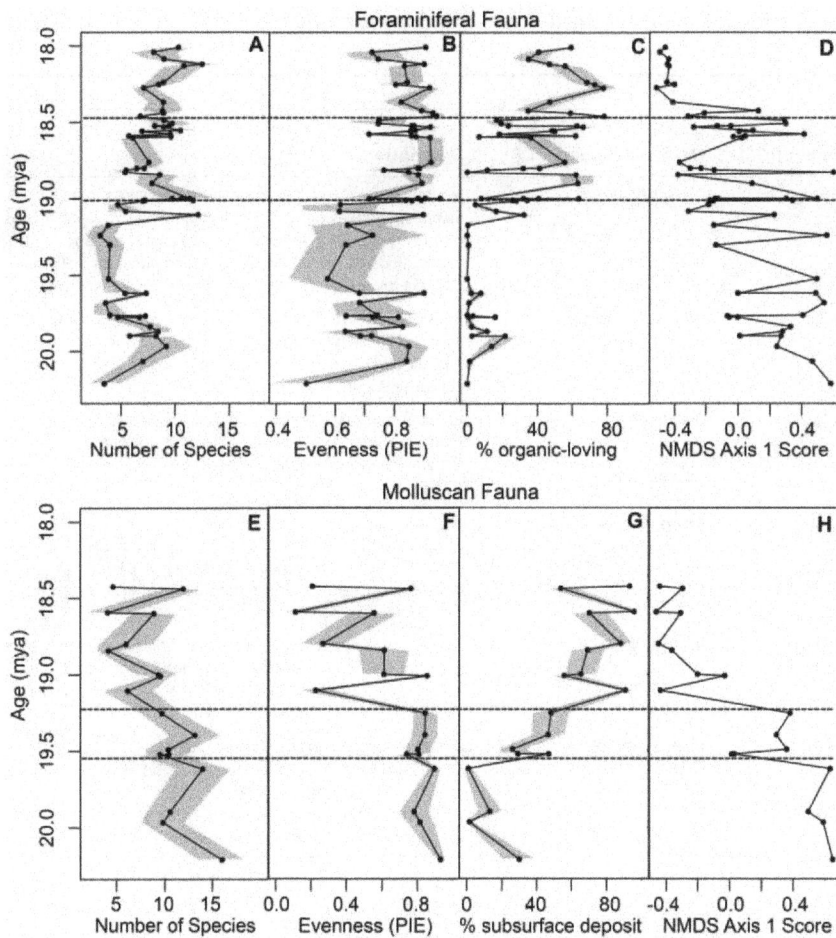

Figure 1. Changes in community structure over time. (A) Foraminiferal species richness, (B) foraminiferal community evenness, (C) the proportion of organic-loving foraminifers, (D) NMDS axis 1 ordination scores from foraminiferal species abundances, (E) molluscan species richness, (F) molluscan community evenness, (G) the proportion of subsurface deposit-feeding bivalves, and (H) NMDS axis 1 ordination scores from molluscan species abundances. Gray shading indicates the 95% confidence intervals of the ecological metrics following sample size standardization. Dashed lines indicate the boundaries between communities (Table 2).

The stratigraphic section was deposited in a structural embayment on the Miocene Oregon coast [31] and is dominated by middle to outer shelf siliciclastic sandy-mud facies. Most of the section is composed of mixed-grain populations of fine sand and mud, but moderately well-sorted fine to coarse sands and laminated muds are common in the upper 50 m of section (after ~18.3 Ma). In the lower 30 m of the section (before ~18.9 Ma), the medium gray strata are pervasively burrow-mottled with meter-scale bedding, but the upper 50 m is comprised of centimeter- to decimeter-scale bedding with dark gray silt and clay laminations in many beds. Sedimentary accumulation rates are ~16.5 m/myr from 20.3 to 19 Ma then increase to ~29 m/ myr between 18.8 and 18 Ma [30].

Estimated absolute water depths are ~25 m based on the molluscan fauna [32–34] and <150 m based on the benthic foraminifera fauna [34]. Hummocky cross-stratification preserved in tuff beds and in mud-capped sandy beds are found at stratigraphic levels throughout the section, further supporting deposition in shelf water depths, specifically between the lower shoreface and storm wave base (e.g., [35]). This restricted depth of deposition throughout the section ensures that faunal assemblages in the time series are drawn from similar environments.

Fossil collections are from a ~5 km stretch of coastline from Beverly Beach State Park to Newport, Or (see [30] for location map and stratigraphic log). Fossil densities are highest in the lower 54 m of the section (~20.3-18 mya) and only one molluscan and one foraminiferal faunal count were recovered from sediments younger than 18 Ma (Table S1). This younger sample (~16.7 Ma) is included in analyses, but is excluded from Figure 1 for clarity. Fossiliferous beds are typically poorly sorted very-fine to fine sandstones; specimens from coarser sediments (intervals 31–35 m and 55–62 m above the base of the section) are excluded from analysis because the sedimentology and faunal composition suggest a sandy shoreface environment distinct from the shelf depths inferred for the rest of the section.

The molluscan fauna is preserved as original aragonite and calcite shells and are taphonomically comparable throughout the section. Bivalves are preserved as whole to large fragments (umbo included) and are articulated and closed in ~50% of specimens. Occasionally, valves are preserved articulated but open and many specimens are disarticulated but found in associated valve pairs. This suggests minimal transport and disruption of fossil material. Preservation quality in the total foraminiferal assemblage varies among samples, but differences in sample quality are largely due to

Table 2. Community states recognized in benthic foraminiferal and molluscan assemblages based on shifts in species richness, evenness and taxonomic composition.

Foraminiferal Community	Age Range (mya)	Median Richness (IQR)	Median Evenness (IQR)	Description
C	18.48-16.71	9.00 (7.74–9.50)	0.77 (0.72–0.81)	Semi-infaunal *Bolivina astoriensis* and fully-infaunal *Buliminella bassendorfensis* and *Nonionella* spp. each >20% of individuals. Semi-infaunal *Buccella mansfieldi* and fully-infaunal *Bolivina ovata* and *Fursenkoina punctata* >5% each. Organic-loving taxa together comprise 60% of individuals.
B	19.03-18.48	8.98 (7.10–10.56)	0.77 (0.72–0.79)	*B. astoriensis* and *Pseudononion costiferum* (fully infaunal) each >20% of individuals. *B. mansfieldi* and *B. bassendorfensis* each >10%. Organic-loving taxa together comprise 33%.
A	20.26-19.03	5.72 (4.16–8.05)	0.56 (0.49–0.7)	Characterized by *B. astoriensis* (44% of individuals). *B. mansfiedli* and *P. costiferum* each >10% of individuals. Organic-loving taxa together comprise 7%.

Molluscan Community	Age Range (mya)	Median Richness (IQR)	Median Evenness (IQR)	Description
3	19.25-16.71	6.70 (5.12–10.30)	0.38 (0.19–0.55)	Subsurface deposit-feeder *Saccella* spp. >78% of individuals. Filter-feeder *Anadara devincta* and surface deposit-feeder *Macoma albaria* each >3%.
2	19.6-19.25	11.35 (11.10–11.51)	0.74 (0.74–0.77)	Surface deposit-feeders *M. albaria* and *Macoma arctata*, and subsurface deposit-feeders *Acila conradi* and *Saccella* spp. each >10% of individuals. Filter-feeders *A. devincta* and *Katherinella augustifrons* each >3%.
1	20.26-19.6	14.21 (12.00–16.45)	0.80 (0.74–0.86)	Predatory scaphopods >20% of individuals. Filter-feeder *Chione ensifera* and surface-deposit feeder *Macoma arctata* each >15%. Subsurface deposit feeders *A. conradi* and *Litorhadia astoriana* and predatory gastropods *Cryptonatica oregonensis* and *Sinum scopulosum* each >3%.

IQR = interquartile range.

changing abundances of thin-walled species and do not affect paleoecological interpretations [36].

Results and Discussion

Changes in community structure and composition are abrupt

For both benthic foraminiferal (n = 58) and molluscan (n = 19) time series, faunal composition changes in steps that are abrupt with respect to the duration of each distinct assemblage or community state (Figure 1). Among foraminiferal assemblages, species richness has two statistically supported steps (a decrease at 19.60 and an increase at 19.17 Ma; see Methods and Table S2), community evenness has one supported increase (at 19.03 Ma), and two increases in the proportional abundances of dysoxic taxa are supported (at 19.03 and 18.48 Ma). Among molluscan assemblages, species diversity and community evenness decrease once at 19.25 Ma, average body size decreases in two steps (at 19.50 and 19.25 Ma), and the proportional abundance of subsurface deposit feeders increases in two steps (at 19.60 and 19.25 Ma). Abrupt shifts in species richness, evenness, body size, and the proportion of ecologically distinctive taxa appear to be coordinated within each taxonomic group allowing each faunal time series to be divided into three distinct community states (Table 2). Each step is instantaneous at the resolution of sampling (<0.1 Ma for the foraminifera and <0.3 Ma for the molluscs); changes between communities may in fact be gradual on finer timescales.

Offsets in the timing of steps between the foraminiferal and molluscan data sets (Figure 1) likely result from slight differences in sample ages and differences in sampling resolution (beds rich in molluscs are not always rich in benthic foraminifera and vice versa). The general coordination of steps suggests that both faunas are responding to an environmental change occurring at the same time. However, not all faunal shifts are coordinated between the two faunas. For example, the shift between molluscan Communities 1 and 2 is not reflected in the foraminiferal community nor is the shift between foraminiferal Communities B and C reflected in the molluscan fauna. Such offsets suggests that the two faunas are responding to different environmental parameters for at least some faunal changes or are responding at different threshold values of the same factor.

Abrupt changes are characterized both by threshold responses and linear tracking of productivity

Despite some similarity in the timing of community steps, the directionality of changes in ecological metrics differs between the molluscan and foraminiferal time series. In the molluscs, species richness and evenness decline, whereas in the foraminifera species richness and evenness increase (Figure 1). On the other hand, species that favor organic rich environments increase in proportional abundance in both groups (Figure 1) suggesting that productivity is a key driver of ecological change in both groups. Increased upwelling during Miocene warming has been observed in other coastal settings [37–39] and could also have driven increased productivity on the Miocene Oregon Coast [30]. In addition, intensification of wind-driven upwelling is thought to drive seasonally dysoxic conditions on the modern Oregon Coast [40–41]; dysoxia is recognized in the Newport Member via a stratigraphic decrease in bioturbation and by the presence of well-

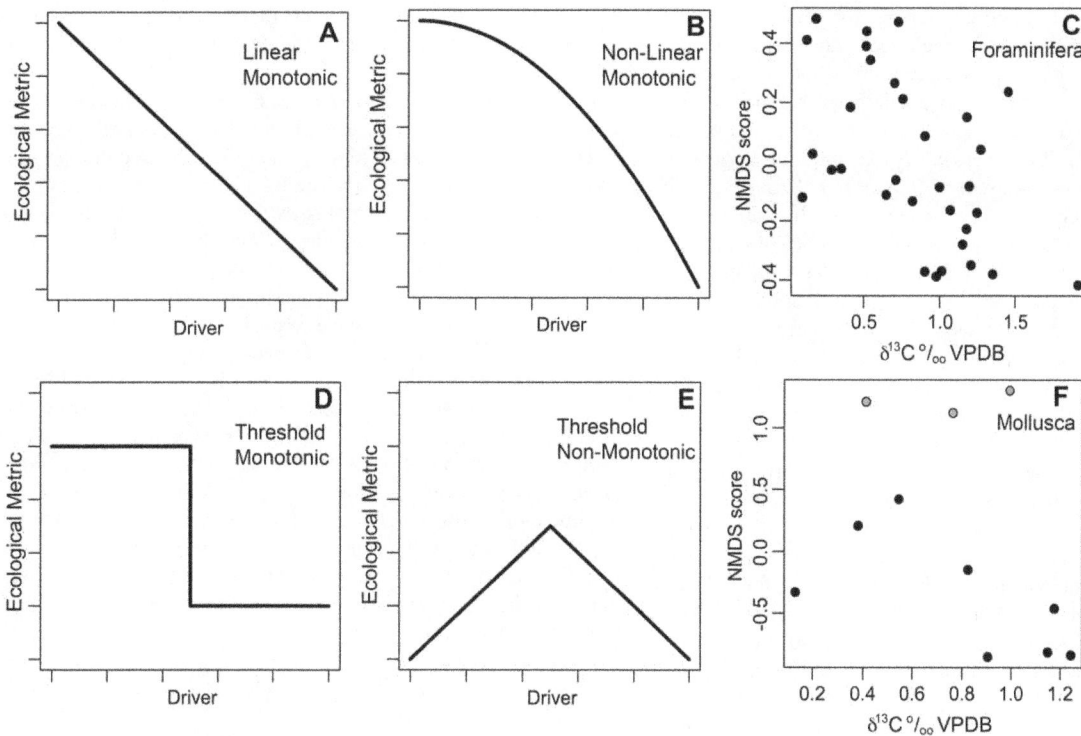

Figure 2. Relationship between ecological variables and environmental drivers. Models of relationships between ecological variables and environmental drivers (A–D). Observed relationship between NMDS ordination score and productivity (δ^{13}C) in (E) foraminifera and (F) molluscs. Gray points in (F) depict Molluscan Community 1 and black depict Molluscan Communities 2 and 3. The foraminiferal relationship most closely resembles model A. The molluscan relationship most closely resembles model D.

laminated sediments after the shift to a high-productivity environment, as indicated by δ^{13}C [30].

The relationship between ecological metrics and environmental factors reveals whether step changes in ecological metrics are due to abrupt changes in underlying environmental changes or due to crossing ecological thresholds (Figure 2) [13]. Foraminiferal ecological metrics have approximately linear relationships with geochemical proxies for surface productivity and organic carbon flux (δ^{13}C and $\Delta\delta^{13}$C respectively) suggesting that foraminiferal community structure is closely tracking productivity changes and that the abrupt ecological changes occur due to abrupt changes in the environment (Figure 2). This relationship to productivity is reflected in strong negative rank order correlations between measures of foraminiferal community composition and the environmental proxies δ^{13}C and $\Delta\delta^{13}$C (Table 3). Species richness and community evenness do not have a strong relationship with any of the measured environmental variables.

Molluscan ecological metrics in Communities 2 and 3, on the other hand, have non-linear and non-monotonic relationships to surface productivity (Figure 2). Multivariate ordination axis 1 scores of taxonomic composition increase with increasing δ^{13}C until a value of ~0.8‰ at which point axis scores decrease sharply as δ^{13}C value continue to rise. PIE also increases as δ^{13}C increases from ~0.1 to 0.8‰, but then decreases strongly as δ^{13}C increases above ~0.8‰. Low species diversity is only observed at the highest δ^{13}C values. This pattern indicates a threshold level of productivity, or a related environmental factor such as benthic oxygenation, that triggers a major reorganization in the molluscan community. In contrast, the average community-level body size has an approximately linear relationship with δ^{13}C, reflected in a strong negative correlation between body size and δ^{13}C (body size

decreases as productivity increases; Table 4). The relationship between productivity and body size evident in Molluscan Communities 2 and 3 is paralleled by Community 1, but differs due to differences in taxonomic composition (e.g. the dominate deposit-feeder is *Macoma arctata* and the dominate suspension-feeder is *Chione ensifera* in Community 1, but in Communities 2 and 3 the dominate deposit-feeder is *Saccella* spp. and the dominate suspension feeders are *Anadara devincta* and *Katherinella augustifrons*; Table 2).

Molluscan community evenness and the detrended correspondence analysis axis 1 scores of taxonomic composition are significantly positively rank ordered with δ^{18}O after accounting for mutual correlations with sample age (Table 4). This suggests

Table 3. Partial correlations based on Spearman rank order between foraminiferal ecological and environmental variables.

	Richness	Evenness	NMDS 1	DCA 1	% Organic-loving
δ^{18}O	−0.18	−0.255	0.174	0.317	−0.361*
δ^{13}C	0.264	0.174	−0.325	−0.447**	0.537**
$\Delta\delta^{13}$C	−0.029	−0.006	−0.787***	−0.458**	0.383*
% mud	0.058	0.315	−0.132	−0.19	0.267

*p<0.05,
**p<0.01,
***p<0.001.

either a temperature or a salinity driver for changes in the molluscan fauna. However, the molluscan and foraminiferal assemblages represent fully marine communities and the latitudinal ranges of modern congeners do not indicate an increase in equatorward (warm water) taxa between successive communities; the median modern latitude over which the majority of genera overlap is similar among communities (Figure S1). Thus neither a temperature or salinity driver is compelling. Instead, these correlations with $\delta^{18}O$ may be a spurious consequence of sampling resolution: when the foraminiferal data set is culled to match the resolution of the molluscan data set, $\delta^{18}O$ also appears more important despite the strong relationship to carbon cycle processes evident in the full foraminiferal data set (Table S3). The correlations between $\delta^{18}O$ and molluscan ecological metrics are also less significant than environmental correlates with the foraminiferal metrics and could be type 1 errors.

Differences in the timing, tempo, and directionality of faunal changes between the molluscan and foraminiferal communities could be explained by differences in the temporal scale of community turnover. Foraminiferal lifespans are much shorter than most molluscs. For example, *Nonion depresulus* lives for only 3–4 months during which time it reproduces once [42]. Larger benthic foraminifera have longer life spans, but even for these taxa longevity is only between 4 months to two years [43]. These short life spans allow populations to respond rapidly to favorable conditions and complete their life cycles before those conditions vanish [44]. In addition, foraminiferal propagules can remain dormant in sediments for at least 2 years until favorable conditions return, increasing the rapidity with which foraminifera can respond to environmental change [45–46]. For example, *Nonion depresulus* is abundant in spring and fall coinciding with diatom blooms in coastal waters, but has very low abundances in the winter and summer when food resources are lower [42]. Standing crops also vary in parallel with upwelling in northern California in *Gabratella ornatissima* [47], and vary with phytoplankton blooms in many Southern California margin taxa [48]. Seasonal fluctuations in the standing crops of shallow benthic foraminifer species appear to be a general rule [44]. When time averaged in the rock record, these fluctuations between populations that are favored at different times of year, or in different years, would yield a more diverse and more even assemblage [49–50]. Thus, the increase in community evenness among the benthic foraminifera may arise from natural time-averaging of strong seasonality in environmental conditions and intra-annual community turnover rather than reflect the diversity that would be captured in a single sample of standing populations. Increased time-averaging up-section could also increase richness and evenness in the foraminiferal fauna,

however, if variation in time-averaging was a major factor molluscan richness and evenness should co-vary with foraminiferal metrics.

In contrast to foraminifera, molluscs typically live multiple years with a median maximum lifespan of 8–9 years and may not reproduce in their first year of life [51]. Seasonality in the composition of molluscan communities is common among juvenile individuals due to the timing of reproductive events, but this is not reflected in shell assemblages drawn only from adult specimens >2 mm [52]. Only shells >4 mm are used in the present study.

With eutrophication, marine benthic species dominance tends to shift from K to r reproductive strategies [53–54] suggesting that foraminifera should generally fare better than molluscs during a rise in productivity. Low-oxygen tied to high productivity could also ultimately explain the observed differences between the foraminiferal and molluscan communities in their relationship to productivity. Metazoans that live in well-oxygenated conditions exhibit oxygen stress at higher oxygen concentrations than do foraminifera; most metazoans begin to respond negatively at ≤ 1.42 ml/l whereas foraminifera do not respond until oxygen levels are ≤ 0.5 ml/l [55–57]. In addition, species that reproduce only once in a lifetime tend to be better at habitat tracking than those that reproduce in multiple years when environmental changes occur faster than evolutionary responses [58], thus molluscs may be more negatively affected than foraminifera if conditions vary seasonally.

Body size changes at the population level parallel community-level change

Body size at the community-level (bivalves only) declines significantly between subsequent molluscan community states (Mann-Whitney, $p < 0.001$; Figure 3). This corresponds with an increase in the proportional abundance of subsurface deposit-feeding bivalves while both suspension feeders and surface deposit feeders decline (Figure 3). In Community 3, subsurface deposit feeders comprise ~83% of the community and a single taxon, *Saccella* spp., alone comprises ~76% of individuals. The decrease in community evenness, increase in dominance of a single taxon, decrease in species richness, and decrease in average body size between molluscan Community 2 and 3 are consistent with an increasingly stressed community [59–60]. In environments of high stress, including those with high organic loading or eutrophication, small-bodied opportunistic taxa often bloom and species with r reproductive strategies dominate [53–54,59,61–62]. In the Newport Member, *Saccella* spp. appears to be an "explosive opportunist" as recognized by its aggregation in pods and clusters in thin horizons and its overwhelming dominance in the youngest samples [61]. Much of the body-size change at the community-level can thus be explained by changes in the proportional abundances of species.

In general, the body size at the population level is also predicted to decrease with increased temperature (temperature size rule) or increased environmental stress [59,63–66]. However, species with different ecological requirements are not all stressed by the same factors and can respond individualistically to environmental change sensu [67]. Under conditions of high productivity and organic carbon flux as seen in the Newport Member, subsurface deposit-feeders would find greater food resources and thus may be able to maintain or even increase their body size despite an increase in temperature or accompanying decreases in oxygenation. For example, the deposit-feeding bivalve *Saccella* spp. is significantly larger in Community 3 (n = 247) than in Community 2 (n = 22) (Mann-Whitney, $p < 0.05$; Figure 3). The average body size of subsamples of 22 individuals from Community 3 are

Table 4. Partial correlations based on Spearman rank order between molluscan ecological and environmental variables.

	Richness	Evenness	NMDS 1	DCA 1	% Deposit-feeding	Body Size
$\delta^{18}O$	0.152	0.712*	0.419	0.616*	−0.406	0.287
$\delta^{13}C$	−0.063	−0.28	−0.148	−0.299	0.293	−0.816**
$\Delta\delta^{13}C$	−0.152	−0.372	−0.271	−0.423	0.302	−0.393
% mud	−0.37	−0.107	−0.452	−0.329	0.382	−0.482

*p<0.05,
**p<0.01,
***p<0.001.

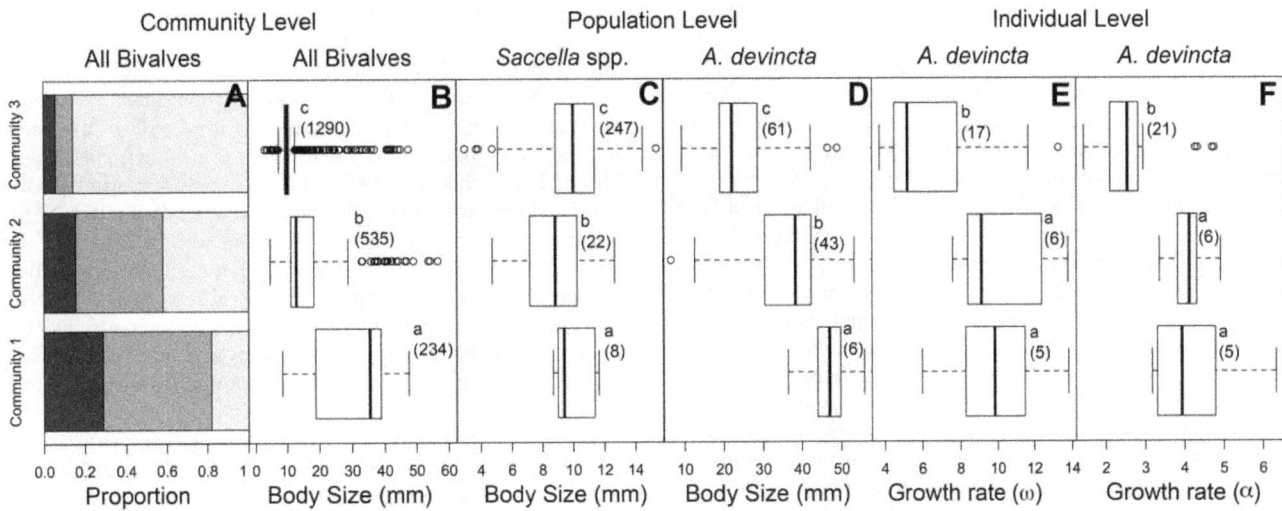

Figure 3. Bivalve body size and growth rate changes over time. (A) Proportion of bivalves that are suspension feeders (dark gray), surface deposit-feeders (medium gray), and subsurface deposit-feeders (light gray), (B) community-level body size, (C) *Saccella* spp. body size, (D) *Anadara devincta* body size, and (E) average growth rate (ω) of *A. devincta* based on external growth breaks and (F) average growth rate (α) of *A. devincta* based on external growth breaks in each of the three molluscan community states. Lower case letters to the upper right of each box plot denotes significant differences (Mann-Whitney, p<0.05). Homogenous groups within each panel are indicated by the same letter. Numbers in parenthesis are the number of individuals represented in each box plot.

significantly greater than Community 2 at the 0.05 level in 87% of cases, indicating that the difference in body size is not due to sample size differences. The shift to larger median body size in *Saccella* spp. from Community 2 to 3 could, however, be due to an increase in the variance of body sizes because minimum sizes decrease while maximum sizes increase (Figure 3). In either case, the body size changes in *Saccella* spp. are not consistent with the hypothesis that body size declines with increasing temperature or stress.

In contrast, body size in the suspension-feeding bivalve *Anadara devincta* is significantly smaller in Community 3 where suspension feeders are rare than in Communities 1 and 2 (Figure 3). Sample sizes are low in Community 1 (6 individuals compared to 43 and 49 in Communities 2 and 3 respectively), but when individuals from Communities 1 and 2 are pooled *A. devincta* from Community 3 remains significantly smaller (Mann-Whitney test, p<0.01). Both maximum and minimum body sizes of *A. devincta* decrease from Community 1 to Community 3, indicating a directional decrease in intraspecific body size (Figure 3). Changes in the size frequency distribution of fossil species can occur if taphonomic loss is size selective; most often it is the smaller individuals that are lost [68–69]. However, small-bodied *A. devincta* in Community 1 were probably not lost due to poor preservation or to wave transport because smaller-bodied, thinner-shelled *Saccella* spp. are found in Community 1. Instead, the size shift likely reflects a true change in size-at-death.

The persistence of *A. devincta* in younger strata despite the sharp decline of other suspension feeding taxa is similar to the occurrence of *A. montereyana* from the Miocene Monterey Formation, which is found on bedding planes preserving the dysoxic/oxic boundary [70]. Like the Monterey congener, *A. devincta* may also be adapted to tolerate low oxygen conditions or be able to take advantage of seasonally high oxygen conditions. Living *Anadara* spp., such as *A. ovalis*, have two hemoglobin components, which are thought to be an adaptation to environments with varying oxygen concentrations [71]. Nonetheless, despite this potential adaptive advantage, *A. devincta* body size

still declines as productivity and deoxygenation increase through the Newport section, suggesting that individuals are either dying younger or growing slower under stressed conditions. Low-oxygen conditions can reduce growth and feeding in many marine animals and this lowered functional ability is often more pronounced in larger individuals [72–73].

In sum, body-size reduction at the community level is driven by both an increase in the abundance of smaller-bodied species and a reduction in size of larger-bodied species. Body size changes also parallel changes in proportional abundance: subsurface deposit-feeders increase in abundance and body size, while suspension feeders decrease in abundance and body size.

Decreased body size in *Anadara* is driven by decreased growth rates

Measuring changes in growth rates provides a direct test for the influence of individual physiological processes on community and population-level change. Median growth rate is significantly lower in *A. devincta* from Community 3 regardless of the age marker or growth rate metric used (Mann-Whitney, p<0.05; Figure 3; Table S4). Internal and external growth breaks often correspond with $\delta^{18}O$ maxima (winter low temperatures) suggesting growth breaks are annual, consistent with the extant *A. senilis* in which growth breaks occur annually during the winter [74]. One exceptional fossil specimen of *A. devincta* had an "extra" break between the third and fourth $\delta^{18}O$ maxima and other specimens had more $\delta^{18}O$ maxima than growth breaks. Ages estimated from growth break age-markers sometimes differ from ages estimated from geochemical age-markers (e.g., [75]), but in the present study these minor disagreements do not affect the consistent pattern of declining growth rates over the time series.

Both the maximum and minimum growth rates observed in *A. devincta* specimens decrease from Community 1 to Community 3 indicating that the decrease in median growth rate across communities is driven by individuals growing slower (Figure 3). Some individuals from the oldest Community 3 beds have growth rates comparable to Communities 1 and 2, suggesting that the environmental changes that drive the shift in community composition from Community 2 to Community 3 did not negatively affect *A. devincta* as immediately as they did the suspension and surface deposit-feeders that went locally extinct. This lag in responses by *A. devincta* supports the hypothesis that it was fundamentally better adapted to low-oxygen and high-productivity conditions than were other suspension feeders.

Paleoecological studies of mass extinctions find reduced body sizes in surviving taxa and have termed this phenomenon the "Lilliput Effect" [76–77]. For example, the body size reduction following the late Permian mass extinction appears to have been driven by reduced overall growth rates and more frequent interruptions to growth, which have been attributed to environmental factors such as dysoxic episodes, extreme temperature events, or low productivity [78]. In modern systems, reduced growth rates or reduced "size-at-age" in ectotherms are associated with exposure to increased temperature and concurrent reductions in oxygen availability [66,79]. Often it is this combined thermal and oxygen stress that demands physiological adaptation, making predictions of faunal responses based on temperature alone difficult [80]. The Newport Member provides an example of reduced growth rates in a bivalve over a period of global warming that may also have been driven by low-oxygen stress as its proximal cause. Other environmental variables that could retard growth rate include low food quality and predator density but these are not testable in the Newport Member.

Parallel changes between physiological rates and community properties are generally taken as signifying that changes at the community-level emerge from processes operating at the individual-level [81]. In the case of the Newport Member, declines in the proportional abundance of suspension feeders and in community-level body size are paralleled by decreased body size and decreased growth rates in the suspension feeder *A. devincta*, a numerically important component of the community. Thus, the physiological tolerance of individuals provides a mechanistic link to community-level change in the Newport Member, and changes in community composition and structure did not arise simply from changes in propagule supply.

The decline in *A. devincta*'s physiological health also implies that this species did not respond evolutionarily to the new environmental conditions despite having ~1 Ma to adapt. This suggests that the environmental stress was imposed on only part of its geographic range, such that gene flow among stressed and unstressed populations prevented local adaptation; *A. devincta* is known to range from Alaska to Southern California during the Early Miocene [24,82]. Alternatively, the decrease in body size could be a locally adaptive paedomorphic response to environmental stress, which would be accompanied by an earlier onset of reproductive maturity marked by an earlier ontogentic decline in growth rates [83,84]. In contrast, the age at which growth rates decline most sharply in *A. devincta* is consistent across the three communities (age-marker 4), indicating that the paedomorphic hypothesis can be rejected. Additional sclerochronological examination of shells for reproductive growth checks might reveal the onset of reproduction. However, in the six shells for which both geochemical age-markers and growth line age-markers are available, reproductive breaks are rarely preserved and there is no evidence for an ontogenically younger onset of reproduction to suggest that paedomorphosis played a role.

Implications for predicting future ecosystem change

Abrupt changes across an environmental threshold are common in modern ecology and receive much attention because they affect our ability to predict how and when biota will respond to

environmental drivers. Abrupt changes can also indicate shifts to new, persistent, ecological regimes and are often caused by a combination of climate changes and anthropogenic impacts [13,85]. Such "threshold shifts" with eutrophication are recognized in modern, anthropogenically-influenced ecological systems and in large marine ecosystems with overfishing [14,85–86]. Threshold shifts have also been observed in pelagic marine systems associated with anthropogenic warming and ice melt in the North Atlantic, which may or may not be influenced by fishing in the region [87–88]. Threshold shifts with respect to environmental variables in a pre-anthropogenic ecological record over intervals <0.1–0.3 Ma suggest that rapid changes due human influences are not necessary to push ecosystems over a threshold: climate changes occurring at natural rates and within the range of natural variability are sufficient.

The Newport Member also demonstrates that the proximal driver of ecological changes during a global warming event can be related to changes in productivity and oxygenation in some settings, as in areas of upwelling, instead of temperature itself. Identifying the particular environmental variables that drive ecological change is integral to anticipating which taxa or functional groups will most likely be affected. In addition, if changes in productivity typically drive ecological responses to global warming in coastal marine environments, then we may be late in detecting changes in many modern systems that are already eutrophied, either naturally or from human activities in the watershed. The life history characteristics of taxa also influence how they respond to environmental changes and, thus, the predictions derived from studying one group cannot necessarily be extrapolated to others. This study highlights the power of examining multiple environmental variables and multiple taxonomic groups on multiple biological levels in order to develop better predictive frameworks for biotic response to climate change.

Although fossil records are usually limited to skeletal hard parts preserved in time-averaged assemblages, information on physiological rates can be extracted from growth rate analyses and, in that way, mechanistic links can be drawn between the environmental tolerances of individuals and changes at the community level. These links strengthen confidence in identifying the particular environmental factors driving change and also provide a physiological basis for predicting the most vulnerable taxa. Bivalve growth rates in living populations are already used to detect physiological stress during anthropogenically driven environmental deterioration (e.g., [89–90]). The present study shows that bivalve growth rates can also be instrumental in detecting individual responses to fully natural environmental changes on a range of time scales. Growth rate analysis is thus a potential tool for detecting the early stages ecological change in modern systems, before change is manifested at the community level.

Fossil records with good temporal resolution that span past climate change events, like that provided by the Newport Member, can yield detailed information on (a) the tempo of paleoecological changes, (b) the environmental drivers of those changes, and (c) and the effects on individual physiology. This high level of detail is invaluable to forecasts of modern ecosystem change as the climate system moves outside the realm observed by modern ecology, and perhaps into no-analog regimes known only from the deeper geological record. Systematic exploration of fossil records that span histories of environmental change and ecosystem response will illuminate the conditions that induce regime shifts and the rates at which such abrupt shifts between community states can occur. In addition, targeting environments, like coastal settings, that are especially vulnerable to modern global warming will heighten the value that paleoecological analysis brings to present ecological challenges.

Materials and Methods

Sediment samples were collected from fresh outcrop surfaces for benthic foraminifera sampling at a resolution of ~1 sample/m of stratigraphic section. Approximately 100 g of sediment from each sample were disaggregated in ultrapure water before wet-sieving through a 150 mm nylon sieve. The total sample was picked if it contained <300 individuals >150 mm. Richer samples were split using a microsplitter to obtain ~300 individuals each. The minimum sample size included in analyses was 100 individuals. Bulk samples for molluscan sampling were extracted from nineteen stratigraphic beds exposed in outcrop or from fallen blocks whose original stratigraphic position was clear for a resolution of ~1 sample/10 m. The minimum molluscan sample size included in analyses was 85 individuals

Molluscan samples are drawn from the total volume of the bed, regardless of bed thickness. The smallest molluscan individual identified was 4 mm. For thicker units, multiple samples are taken from different vertical levels within the bed for foraminiferal faunal analysis so that within bed changes are recorded. All necessary permits were obtained for the described field studies from the Oregon Parks and Recreation Department.

For each faunal sample, species richness (sample-size standardized to the minimum number of specimens), evenness (PIE following Hurlbert, 1971) [91], and the proportional abundances of ecological groups are calculated. Detrended correspondence analysis (DCA) is performed on the proportional abundances of species using the decorana function in the vegan package in the R programming language [92–93]. Non-metric multidimensional scaling (NMDS) is performed on the Bray-Curtis distances among samples based on the proportional abundances of species using the isoMDS function in the MASS package in the R programming language [92]. Correlations between ordination score and environmental variables are preformed on the axis that summarizes the greatest proportion of variance. Axis 1 summarizes 63% of the variance in DCA and 73% in NMDS (stress = 5.86; 3 dimensions) for the molluscan analysis and 65% and 56% (stress = 9.04; 3 dimensions) respectively for the foraminiferal analyses. The proportion of variance in the data set summarized by each ordination axis is computed via a Mantel Test on the Euclidean distances between points and the Bray-Curtis distances between faunal assemblages in the unreduced space (following [94]).

For the bivalves, species are categorized as suspension-feeders, surface deposit-feeders, or sub-surface deposit-feeders based on the feeding preferences of living congeners. For the foraminifera, the proportional abundance of taxa tolerant of low oxygen and high organic sediments (buliminids, *Fursenkoina* spp. and *Nonionella* spp.) is calculated. Molluscan body size is also measured at the community level and is calculated as the square root of the product of valve length and height (geometric mean size) e.g. [95]. The tempo of changes in the above ecological metrics are tested using maximum likelihood model selection fitting to four models of temporal change: no change, change without directional bias, directional change, and stepped change. Procedures follow Hunt [96–97] and use the paleoTS package in the R programming environment [92].

At the population level, body size is examined using 110 *Anadara devincta* and 277 *Saccella* spp. shells from bulk samples and from opportunistic outcrop collections. Growth rates were only examined in *A. devincta*: this species has a thick shell with well-preserved growth banding unlike *Saccella* whose valves <1 mm thick. Size-at-age was estimated using (1) external growth lines, (2) internal growth lines, and (3) $\delta^{18}O$ maxima along the growth axis as annual markers of winter temperatures. Collectively, these are referred to here as "age markers."

Prominent external growth lines were measured on 32 well-preserved individuals of *A. devincta* with dial calipers to the nearest 0.1 mm. All measurements were made from the umbo along the maximum growth axis of the shell. Bivalve shells were then coated in epoxy and cross-sectioned along the maximum growth axis using a low-speed saw. Shell cross-sections were ground on glass plates (320 and 600 grit powder), polished with 1200 grit pat gel, and sampled for stable isotope analysis with a 300 mm drill bit. Cross-sections with visible internal growth lines (14) were digitally scanned and internal growth lines were measured from a fixed point on the umbo to the intersection of the growth line with shell edge using the software Image J (NIH) to the nearest 0.01 mm. From six shells, 12–34 microsamples were taken along the maximum growth axis for a resolution of ~1 sample every 0.75 mm for stable isotope analysis. Shell powders were analyzed at the University of Michigan Stable Isotope Laboratory on a Finnigan MAT 251 mass spectrometer for $\delta^{18}O$ and $\delta^{13}C$. Analytical precision is 0.06‰ for $\delta^{18}O$. Prior to stable-isotope sampling, preservation of original aragonite was confirmed using x-ray diffraction for mineral composition and scanning-electron microscopy for mineral structure. The $\delta^{18}O$ maxima (cool temperatures) were chosen as age markers instead of $\delta^{18}O$ minima because the winter values form sharp cusps in plots of the $\delta^{18}O$ values in the direction of growth whereas summer values form broader plateaus and thus yield less precise size-at-age estimates.

Growth rates for individual shells were calculated from size-at-age for each age-marker method in two ways: (a) using the slope of the linear regression (α) of the first 5 age markers and (b) fitting the von Bertalanffy growth function to all age markers. The distance from the umbo for each age marker were fitted to a growth curve using the von Bertalanffy growth function and growth rate (ω) is defined from the model parameters following Jones et al. [98]. Omega (ω) is representative of the growth rate near t_0. Differences in α and ω between communities are tested using a Kruskal-Wallis test and post-hoc Mann-Whitney tests.

The position and number of internal growth increments and $\delta^{18}O$ maxima are also compared. If internal growth increments occur more frequently than $\delta^{18}O$ maxima, these additional increments may be reproductive growth checks and thus indicate reproductive maturity.

Strontium isotope ages were obtained from molluscan carbonate from 12 locations and have a precision of ±0.23 Ma (Table S1) [30]. Constant sedimentation rates are assumed between Sr-isotope ages to assign intervening faunal samples to an age. Sr-isotope ages from each stratigraphic section along the coast combined with lithologic marker beds allow faulted blocks to be placed in correct temporal order and for repeated sections to be identified. All stratigraphic sections have portions of overlap, which allow them to be joined in a composite section [30].

Four environmental proxies from the Newport Member were compared to faunal patterns to identify potential environmental drivers of faunal changes including temperature ($\delta^{18}O$), water column productivity ($\delta^{13}C$), organic carbon flux ($\Delta\delta^{13}C$), and percent mud (see Table 5 for summary of all paleoenvironmental data types considered in this study). The $\delta^{18}O$ and $\delta^{13}C$ measurements were made on *Buccella mansfieldi*, a semi-infaunal benthic foraminifer. Temporal trends in $\delta^{18}O$ are consistent among the three foraminiferal taxa and among fault repeated stratigraphic sections, but there are clear offsets in $\delta^{13}C$ among the

Table 5. Environmental variables measured in Newport Member of the Astoria Formation.

Environmental Variable	Source	Environmental Proxy
Temperature	Foraminifera	$\delta^{18}O$, Mg/Ca*
Productivity	Foraminifera	$\delta^{13}C$, Cd/Ca*
Organic Matter Flux 1	Foraminifera	$\Delta\delta^{13}C$
Organic Matter Flux 2	Foraminifera, Molluscs	habitat of modern congenerics
Sediment Grain Size	Sediments	% mud
Seasonality*	Bivalve	$\delta^{18}O$ along maximum growth axis
Freshwater flux	Foraminifera, Sediments	$\delta^{18}O$, max grain size
Water depth	Foraminifera, Molluscs, Sedimentology	habitat of modern congenerics, grain size, sedimentary structures
Oxygenation	Foraminifera, Molluscs, Sedimentology	sedimentary laminations, habitat of modern congenerics

Source indicates the materials measured for each variable.
* = Variables or proxies not measured in the present study, which have the potential to yield additional information after methodological improvements.

taxa consistent with life habit differences indicating that original calcite is preserved [30]. The $\Delta\delta^{13}C$ was calculated as the difference between the $\delta^{13}C$ values of *B. mansfieldi* and *Pseudononion costiferum*, an infaunal foraminifer. Mud content was measured as the weight percent of sediment grains <63 mm [30]. Trace element analyses were also performed on a limited number of samples to reconstruct paleotemperature (Mg/Ca) and productivity (Cd/Ca), but samples were judged to be contaminated by siliciclastic material.

Supporting Information

Figure S1   Boxplots of horizontal distributional mean calculations of modern equivalent latitude for molluscan communities 1–3 showing similar estimated latitudes for each community. For each community pool, the horizontal distributional mean characteristic curve (HDM) of the modern latitudinal ranges of constituent genera is calculated following [99]. The median of the curve is taken as the modern equivalent latitude of the sample. Species abundances from the pooled communities are bootstrap resampled to 300 individuals for each calculation and resampling is repeated 1000 times to test for sampling effects in the result. Heavy horizontal bar is the median latitude from the resampling procedure, the box encompasses the interquartile range, whiskers extend to the most extreme data point within 1.5 times the interquartile range, and open circles denote data outside 1.5 times the interquartile range.

Table S1   Sampling coverage with meters above base of composite stratigraphic section and ages for each sample. F = foraminiferal faunal sample; M = molluscan faunal sample; S = sediment grains size (% mud); SI = stable isotope ($\delta^{18}O$ and $\delta^{13}C$); Sr = Strontium isotope, G = bivalve growth. The first 2–3 letters in the sample code indicate the stratigraphic section and the number indicates the bed from which the sample was obtained. Letters following the bed number differentiated samples that were taken at different levels within the same bed. YH = Yaquina Head, NYF = North Yaquina, DM = Moolach Beach, SE = Section E, SP = Schooner, RR = Reef Rocks. These names correspond to section labels in the stratigraphic log published as Figure 1 in [30]. The Schooner-Yaquina column contains, in order from oldest to youngest, SP, RR, NYF, and YH.

Table S2   AIC values, AICc values, and Akaike weights based on AICc values for five models of changes in species diversity, PIE, the proportion of ecologically important taxa: URW (unbiased random walk), GRW (general random walk), stasis, 1-Shift, and 2-Shifts. "n" is the number of samples in the time series. "Shift start" indicates the age of the youngest sample prior to shift.

Table S3   Partial correlations based on Spearman rank order between foraminiferal ecological and environmental variables where the resolution of the foraminiferal data set has been reduced to match the resolution of the molluscan data set. None of the partial correlations are significant.

Table S4   Median growth rates for *Anadara devincta* in each community calculated as the slope of a linear regression and as the parameter ω derived from the von Bertalanffy growth equation for external growth increments, internal growth increments, and $\delta^{18}O$ maxima. Krustal-Wallis (K-W) and Mann-Whittney (M-W) tests report which communities have significantly different growth rates. Homogeneous groups (p>0.05) in pairwise tests are indicated by the same letter (A, B). Communities 1 and 2 are also combined due to small sample sizes in these two communities. IQR = interquartile range, N = number of individuals.

Acknowledgments

I thank J. Belanger, N. Bian, L. Chang, S. Kidwell, G. Patton and S. Ladd for fieldwork assistance, S. Kidwell, D. Jablonski, C. K. Boyce, and S. Lidgard for helpful discussions and input on this work, and P. Martin for laboratory support at U Chicago. I also thank L. Wingate at U Michigan and D. Winter at UC Davis for assistance with geochemical analyses.

Author Contributions

Conceived and designed the experiments: CLB. Performed the experiments: CLB. Analyzed the data: CLB. Contributed reagents/materials/analysis tools: CLB. Wrote the paper: CLB.

References

1. National Research Council (2005) The Geological Record of Ecological Dynamics: Understanding the Biotic Effects of Future Environmental Change. Washington D.C.: National Academies Press. 216 p.
2. Willis KJ, Bailey RM, Bhagwat SA, Birks HJB (2010) Biodiversity baselines, thresholds and resilience: testing predictions and assumptions using palaeoecological data. Trends in Ecology and Evolution 25: 583–591.
3. Dietl GP, Flessa KW (2011) Conservation paleobiology: putting the dead to work. Trends in Ecology and Evolution 26(1): 30–37.
4. Williams JW, Jackson ST (2007) Novel climates, no-analog communities, and ecological surprises. Frontiers in Ecology and the Environment 6(9): 475–482.
5. Jackson ST, Betancout JL, Booth RK, Gray ST (2009) Ecology and the ratchet of events: climate variability, niche dimensions, and species distributions. Proceedings of the National Academy of Sciences 106: 19685–19692.
6. Rabalais NN, Turner RE, Diaz RJ, Justic D (2009) Global change and eutrophication of coastal waters. ICES Journal of Marine Science 66(7): 1528–1537.
7. Gilbert D, Rabalais NN, Diaz RJ, Zhang J (2010) Evidence for greater oxygen decline rates in the coastal ocean than in the open ocean. Biogeosciences 7: 2283–2296.
8. Kidwell SM (2001) Preservation of species abundance in marine death assemblages. Science 294: 1091–1094.
9. Kowalewski M, Carroll M, Casazza L, Gupta NS, Hannisdal B, et al. (2003) Quantitative fidelity of brachiopod-mollusk assemblages from modern subtidal environments of San Juan Islands, USA. Journal of Taphonomy 1(1): 43–65.
10. Murray JW (2006) Ecology and Applications of Benthic Foraminifera. Cambridge: Cambridge University Press. 426 p.
11. Scheffer M, Carpenter SR (2003) Catastrophic regime shifts in ecosystems: linking theory to observation. Trends in Ecology and Evolution 18(12): 648–656.
12. Collie JS, Richardson K, Steele JH (2004) Regime shifts: can ecological theory illuminate the mechanisms? Progress in Oceanography 60: 281–302.
13. Andersen TF, Carstensen J, Hernandez-Garcia E, Duarte CM (2009) Ecological thresholds and regime shifts: approaches to identification. Trends in Ecology and Evolution 24(1): 49–57.
14. Petersen JK, Hansen JW, Laursen MB, Clausen P, Carstensen J, et al. (2008) Regime shift in a coastal marine ecosystem. Ecological Applications 18(2): 497–510.
15. Knowlton N (2004) Multiple "stable" states and the conservation of marine ecosystems. Progress in Oceanography 60: 387–396.
16. Clarke A (1993) Temperature and extinction in the sea: a physiologist's view. Paleobiology 19(4): 499–518.
17. Allmon WD (2001) Nutrients, temperature, disturbance, and evolution: a model for the late Cenozoic marine record of the western Atlantic. Palaeogeography, Palaeoclimatology, Palaeoecology 166: 9–26.
18. Walther G (2002) Ecological responses to recent climate change. Nature 416: 389–395.
19. Smith JR, Fong P, Ambrose RF (2006) Dramatic declines in mussel bed community diversity: response to climate change. Ecology 87(8): 1153–1161.
20. Baukun A, Field DB, Redondo-Rodriguez R, Weeks SJ (2010) Greenhouse gas, upwelling-favorable winds, and the future of coastal ocean upwelling ecosystems. Global Change Biology 16: 1213–1228.
21. Keeling RF, Kortzinger A, Gruber N (2010) Ocean deoxygenation in a warming world. Annual Review of Marine Science 2: 199–229.
22. Miles EL (2009) On the increasing vulnerability of the world ocean to multiple stresses. Annual Review of Environmental Resources 34: 17–41.
23. Valentine JW (1961) Paleoecologic Molluscan Geography of the Californian Pleistocene. University of California Publications in Geological Sciences 34(7): 309–442.
24. Hall CAJ (2002) Nearshore marine paleoclimatic regions, increasing zoogeographic proviniciality, molluscan extinctions, and paleoshorelines, California: Late Oligocene (27 Ma) to Late Pliocene (2.5 Ma). Geological Society of America Special Paper 357: 1–489.
25. Helmuth B, Miezkowska N, Moore P, Hawkins SJ (2006) Living on the edge of two changing worlds: forecasting the responses of rocky intertidal ecosystems to climate change. Annual Review of Ecology and Systematics 37: 373–404.
26. Gaylord B, Gaines SD (2000) Temperature or Transport?: Range limits in marine species mediated solely by flow. The American Naturalist 155(6): 769–789.
27. Case TJ, Holt RD, McPeek MA, Keitt TH (2005) The community context of species' borders ecological and evolutionary perspectives. Oikos 108: 28–46.
28. Thomas CD (2005) Recent evolutionary effects of climate change Yale University. pp 75–88.
29. Zachos J, Pagani M, Sloan L, Thomas E, Billups K (2001) Trends, rhythms, and aberrations in global climate 65 Ma to Present. Science 292: 686–693.
30. Belanger CB (2011) Continental shelf dysoxia accompanies Early Miocene warming based on benthic foraminifera and sediments from Oregon. Marine Micropaleontology 80: 101–113.
31. Niem WA, Niem AR, Snavely PD (1992) Late Cenozoic continental margin of the Pacific Northwest - Sedimentary Embayments of the Washington-Oregon Coast. In: Burchfiel BC, Lipman PW, Zoback ML, eds. The Cordilleran Orogen: Conterminous U.S.: The Geology of North America G-3: 314–319.
32. Moore EJ (1963) Miocene marine mollusks from the Astoria Formation in Oregon. United States Geological Survey Professional Paper 419: 1–109, 132 plates.
33. Addicott WO (1970) Latitudinal gradients in Tertiary molluscan faunas of the Pacific Coast. Palaeogeography, Palaeoclimatology, Palaeoecology 8: 287–312.
34. Snavely PD, Rau WW, Wagner HC (1964) Miocene stratigraphy of the Yaquina Bay area, Newport, Oregon. The ORE BIN 26(8): 133–150.
35. Dumas S, Arnott RWC (2006) Origin of hummocky and swaley cross-stratification - the controlling influence of unidirectional current strength and aggradation rate. Geology 34(12): 1073–1076.
36. Belanger CB (2011) Evaluating taphonomic bias of paleoecological data in fossil benthic foraminiferal assemblages. Palaios 26(12): 767–778.
37. Li Q, McGowran B (1994) Miocene upwelling events: Neritic foraminiferal evidence from southern Australia. Australian Journal of Earth Sciences 41: 593–603.
38. Kender S, Peck VL, Jones RW, Kaminski MA (2009) Middle Miocene oxygen minimum zone expansion offshore West Africa: evidence for global cooling precursor events. Geology 37(8): 699–702.
39. Grunert P, Soliman A, Harzhauser M, Mullegger S, Piller WE, et al. (2010) Upwelling conditions in the Early Miocene Central Paratethys Sea. Geologica Carpathica 61(2): 129–145.
40. Grantham BA, Chan F, Nielsen KJ, Fox DS, Barth JA, et al. (2004) Upwelling-driven nearshore hypoxia signals ecosystem and oceanographic changes in the northeast Pacific. Nature 429(749–754.
41. Chan F, Barth JA, Lubchenco J, Kirinchich A, Weeks H, et al. (2008) Emergence of anoxia in the California Current large marine ecosystem. Science 319: 920.
42. Murray JW (1983) Population dynamics of benthic foraminifera: results from the Exe Estuary, England. Journal of Foraminiferal Research 13: 1–12.
43. Hallock P (1985) Why are larger foraminifera large? Paleobiology 11: 195–208.
44. Murray JW (1991) Ecology and Paleoecology of Benthic Foraminifera. New York: Longman Scientific & Technical. 397 p.
45. Alve E, Goldstein ST (2010) Dispersal, survival and delayed growth of benthic foraminiferal propagules. Journal of Sea Research 63: 36–51.
46. Goldstein ST, Alve E (2011) Experimental assembly of foraminiferal communities from coastal propagule banks. Marine Ecology Progress Series 437: 1–11.
47. Erskian MG, Lipps JH (1987) Population dynamics of the foraminiferan Glabratella ornatissima (Cushman) in northern California. Journal of Foraminiferal Research 17: 240–256.
48. Sheperd AS, Rathbun AE, Perez ME (2007) Living foraminiferal assemblages from the southern California margin: a comparison of the >150, 63–150, and >63 um fractions. Marine Micropaleontology 65: 54–77.
49. Fursich FT, Aberhan M (1990) Significance of time-averaging for paleocommunity analysis. Lethaia 23: 134–152.
50. Tomasovych A, Kidwell SM (2011) Accounting for the effects of biological variability and temporal autocorrelation in assessing of species abundance. Paleobiology 37: 332–354.
51. Kidwell SM, Rothfus TA (2010) The live, the dead, and the expected dead: variation in life spans yields little bias of proportional abundances in bivalve death assemblages. Paleobiology 36: 615–640.
52. Kidwell SM (2002) Mesh-size effects on the ecological fidelity of death assemblages: a meta-analysis of molluscan live-dead studies. Geobios Memoire special no 24.
53. Lillebo AI, Pardal MA, Marques JC (1999) Population structure, dynamics and production of Hydrobia ulvae (Pennant) (Mollusca: Prosobranchia) along an eutrophication gradient in the Mondego estuary (Portugal). Acta Oecologia 20: 289–304.
54. Grall J, Chauvaud L (2002) Marine eutrophication and benthos: the need for new approaches and concepts. Global Change Biology 8: 818–830.
55. Levin LA, Gage JD, Martin C, Lamont PA (2000) Macrobenthic community structure within and beneath the oxygen minimum, NW Arabian Sea. Deep-Sea Research II 47: 189–226.
56. Murray JW (2001) The niche of benthic foraminifera, critical thresholds and proxies. Marine Micropaleontology 41: 1–7.
57. Levin LA, Ekau W, Gooday AJ, Jorissen F, Middelburg JJ, et al. (2009) Effects of natural and human-induced hypoxia on coastal benthos. Biogeosciences 6: 2063–2098.
58. Zeineddine M, Jansen VAA (2009) To age, to die: parity, evolutionary tracking and Cole's Paradox. Evolution 63: 1498–1507.
59. Odum EP (1985) Trends expected in stressed ecosystems. Bio Science 35(7): 419–422.
60. Levin LA (2003) Oxygen minimum zone benthos: adaptation and community response to hypoxia. Oceanography and Marine Biology: an Annual Review 41: 1–45.
61. Levinton JS (1970) The paleoecological significance of opportunistic species. Lethaia 3: 69–78.
62. Pearson TH, Rosenberg G (1978) Macrobenthic succession in relation to organic enrichment and pollution of the marine environment. Oceanography and Marine Biology: an Annual Review 16: 229–311.

63. Moore M, Folt C (1993) Zooplankton body size and community structure: effects of thermal and toxicant stress. Trends in Ecology and Evolution 8: 187–200.

64. Atkinson D (1994) Temperature and organism size - a biological law for ectotherms. Advances in Ecological Research 25: 1–58.

65. Millien V, Lyons SK, Olson L, Smith F, Wilson AB, et al. (2006) Ecotypic variation in the context of global climate change: revisiting the rules. Ecology Letters 9: 853–869.

66. Daufresne M, Lengfellner K, Sommer U (2009) Global warming benefits the small in aquatic ecosystems. Proceedings of the National Academy of Sciences 106(31): 12788–12793.

67. Gleason HA (1926) The individualistic concept of the plant association. Bulletin of the Torrey Botanical Club 53: 7–26.

68. Hallam A (1967) The interpretation of size frequency distribution in molluscan death assemblages. Palaeontology 10: 25–42.

69. Cummins H, Powell EN, Stanton RJJ, Staff G (1986) The size frequency distribution in paleoecology: effects of taphonomic processes during formation of molluscan death assemblages in Texas Bays. Palaeontology 29: 495–518.

70. Savrda CE, Bottjer DJ (1987) The exaerobic zone, a new oxygen-deficient marine biofacies. Nature 327: 54–56.

71. Borgese TA, Harrington JP, Hoffman D, San George RC, Nagel RL (1987) *Anadara ovalis* hemoglobins: distinct dissociation and ligand binding characteristics. Comparative Biochemistry and Physiology 86B: 155–165.

72. Wu RSS (2002) Hypoxia: from molecular responses to ecosystem responses. Marine Pollution Bulletin 45: 35–45.

73. Peck LS, Morley SA, Pörtner HO, Clark MS (2007) Thermal limits of burrowing capacity are linked to oxygen availability and size in the Antarctic clam *Laternula elliptica*. Oecologia 154: 479–484.

74. Debenay JP, Leung Tack D, Ba M, Sy I (1994) Environmental conditions, growth and production of *Anadara senilis* (Linnaeus, 1758) in a Senegal lagoon. Journal of Molluscan Studies 60: 113–121.

75. Goewert AE, Surge D (2008) Seasonality and growth patterns using isotope sclerochronology in shells of the Pliocene scallop *Chesapecten madisonius*. Geo-Marine Letters 28: 327–338.

76. Urbanek A (1993) Biotic crises in the history of Upper Silurian graptoloids: a palaeobiological model. Historical Biology 7: 29–50.

77. Twitchett RJ (2007) The Lilliput effect in the aftermath of the end-Permian extinction event. Palaeogeography, Palaeoclimatology, Palaeoecology 252(1–2): 132–144.

78. Metcalfe B, Twitchett RJ, Price-Lloyd N (2011) Changes in size and growth rate of 'Lilliput' animals in the earliest Triassic. Palaeogeography, Palaeoclimatology, Palaeoecology 308: 171–180.

79. Pörtner HO, Knust R (2007) Climate change affects marine fishes through oxygen limitation of thermal tolerance. Science 315: 95–97.

80. Pörtner HO (2001) Climate change and temperature-dependent biogeography: oxygen limitation of thermal tolerance in animals. Naturwissenschaften 88: 137–146.

81. Harley CDG, Hughes AR, Hultgren KM, Miner BG, Sorte CJB, et al. (2006) The impacts of climate change in coastal marine systems. Ecology Letters 9: 228–241.

82. Marincovich LJ, Moriya S (1992) Early middle Miocene mollusks and benthic foraminifers from Kodiak Island, Alaska. United States Geological Survey Bulletin 199: 163–169.

83. McNamara KJ (2001) Importance of heterochrony. In: Briggs DEG, Crowther PR, eds. Palaeobiology II. Malden, MA: Wiley-Blackwell. pp 180–188.

84. Hallam A (1965) Environmental causes of stunting in living and fossil marine benthonic invertebrates. Palaeontology 8(1): 132–155.

85. deYoung B, Barange M, Beaugrand G, Harris R, Perry RI, et al. (2008) Regime shifts in marine ecosystems: detection, prediction and management. Trends in Ecology and Evolution 23(7): 402–409.

86. Carpenter SR, Ludwig D, Brock WA (1999) Management of eutrophication for lakes subject to potentially irreversible change. Ecological Applications 9: 751–771.

87. Frank KT, Petrie B, Choi JS, Leggett WC (2005) Trophic cascades in a formerly cod-dominated ecosystem. Science 308: 1621–1623.

88. Greene CH, Pershing AJ, Cronin TM, Ceci N (2008) Arctic climate change and its implications on the ecology of the North Atlantic. Ecology 89(1): 524–538.

89. Martel P, Kovacs T, Voss R, Megraw S (2003) Evaluation of caged freshwater mussels as an alternative method for environmental effects monitoring (EEM) studies. Environmental Pollution 124: 471–483.

90. Kirby MX, Miller HM (2005) Response of a benthic suspension feeder (*Crassostrea virginica* Gmelin) to three centuries of anthropogenic eutrophication in Chesapeake Bay. Estuarine, Coastal and Shelf Science 62: 679–689.

91. Hurlbert SH (1971) The nonconcept of species diversity: a critique and alternative parameters. Ecology 52(4): 577–586.

92. R Development Core Team (2010) R: a language and environment for statistical computing. R Foundation for Statistical Computer, Vienna, Austria. Available: http://www.R-project.org. Accessed 2012 Jan 1.

93. Oksanen J, Blanchet FG, Kindt R, Legendre P, Minchin PR, et al. (2011) vegan: Community Ecology Package. R package version 2.0-2. Available: http://CRAN.R-project.org/package=vegan. Accessed 2012 Jan 1.

94. McCune B, Grace JB (2002) Analysis of Ecological Communities. Mjm Software Design, Glenden Beach, OR, US. 300 p.

95. Kosnik MA, Jablonski D, Lockwood R, Novack-Gottshall PM (2006) Quantifying molluscan body size in evolutionary and ecological analyses: maximizing the return on data-collection efforts. Palaios 21: 588–597.

96. Hunt G, Roy K (2006) Climate change, body size evolution, and Cope's Rule in deep-sea ostracodes. Proceedings of the National Academy of Sciences 103(5): 1347–1352.

97. Hunt G (2008) Gradual or pulsed evolution: when should punctuational explanations be preferred? Paleobiology 34(3): 360–377.

98. Jones DS, Arthur MA, Allard DJ (1989) Sclerochronological records of temperature and growth from shells of *Mercenaria mercenaria* from Narragansett Bay, Rhode Island. Marine Biology 102: 225–234.

99. Amano K (1994) An attempt to estimate the surface temperature of the Japan Sea in the Early Pleistocene by using a molluscan assemblage. Palaeogeography, Palaeoclimatology, Palaeoecology 108: 369–378.

21

The Invisible Prevalence of Citizen Science in Global Research: Migratory Birds and Climate Change

Caren B. Cooper[1]*, Jennifer Shirk[1], Benjamin Zuckerberg[2]

1 Cornell Lab of Ornithology, Ithaca, New York, United States of America, **2** University of Wisconsin, Madison, Wisconsin, United States of America

Abstract

Citizen science is a research practice that relies on public contributions of data. The strong recognition of its educational value combined with the need for novel methods to handle subsequent large and complex data sets raises the question: Is citizen science effective at science? A quantitative assessment of the contributions of citizen science for its core purpose – *scientific research* – is lacking. We examined the contribution of citizen science to a review paper by ornithologists in which they formulated ten central claims about the impact of climate change on avian migration. Citizen science was never explicitly mentioned in the review article. For each of the claims, these ornithologists scored their opinions about the amount of research effort invested in each claim and how strongly the claim was supported by evidence. This allowed us to also determine whether their trust in claims was, unwittingly or not, related to the degree to which the claims relied primarily on data generated by citizen scientists. We found that papers based on citizen science constituted between 24 and 77% of the references backing each claim, with no evidence of a mistrust of claims that relied heavily on citizen-science data. We reveal that many of these papers may not easily be recognized as drawing upon volunteer contributions, as the search terms "citizen science" and "volunteer" would have overlooked the majority of the studies that back the ten claims about birds and climate change. Our results suggest that the significance of citizen science to global research, an endeavor that is reliant on long-term information at large spatial scales, might be far greater than is readily perceived. To better understand and track the contributions of citizen science in the future, we urge researchers to use the keyword "citizen science" in papers that draw on efforts of non-professionals.

Editor: Robert Guralnick, University of Colorado, United States of America

Funding: These authors have no support or funding to report.

Competing Interests: The authors have declared that no competing interests exist.

* Email: caren.cooper@cornell.edu

Introduction

Citizen science, the practice of involving the public in scientific research, is undergoing a period of rapid growth across numerous disciplines. Effective use of citizen science as a method of data collection relies on techniques from many disciplines, such as science communication, informal science education, and informatics. As an interdisciplinary field of practice, citizen science is experiencing a process of professionalization, which is evident from newly organized societies such as the Citizen Science Association (http://citizenscienceassociation.org), the European Citizen Science Association (http://ecsa.biodiv.naturkundemuseum-berlin.de/), and the Citizen Science Network Australia (http://csna.gaiaresources.com.au). Other hallmarks of professionalization include new educational- and cyber-infrastructure initiatives, international conferences, and numerous advisory boards [1], [2].

Research has documented that citizen science can engage the public through hobbies and games [3], [4], support learning in science, technology, engineering, and math (STEM) [5] provide sources of large and complex datasets that catalyze advances in data visualization [6], [7], advance cyber-infrastructure and new analysis techniques [8], [9], [10], [11], and influence management and policy [12], [13], [14], [15]. But is citizen science effective at science?

Research focused on the scientific value of citizen science has yet to directly quantify the impact of citizen science to any specific area of scientific research, but has instead focused on concerns regarding data quality e.g., [16], [17]. Data from citizen-science efforts can indeed have biases, the largest source of which may originate from the extent to which people self-select to participate, which affect the effort they expend, their level of skill, and the spatiotemporal distribution of the data [18]. Yet, the quality of data collected by volunteers, on a project-by-project basis, has generally been found as reliable as the data collected by professionals in community-based research [19] and contributory projects across a wide variety of subjects, including lady beetles [20], moths [21], wolves [22], trees [23], air pollution [24], light pollution [25], plants [26], pikas [27], invasive plants [28], and bees [29]. Researchers address the biases and variable data quality of volunteered observations through the use of novel online data filters and workflows for detecting erroneous submissions [9], [30], [31], and modified project design and analysis [11], [16], [32]. These advances have been central in the development, vetting, and dissemination of citizen-science datasets that have allowed for new areas of scientific research that would not otherwise be possible; particularly questions requiring data collected over broad spatiotemporal scales.

Many citizen-science programs have operated for decades (some, for more than a century) and frequently span multiple countries and continents [3], [33]. To investigate unanticipated questions arising from global changes, researchers have begun relying on diverse sources of data including the re-purposing of data contributed through citizen science [33]. We expect that research related to climate change may be particularly well suited to draw on citizen science because: (i) the influence of climate change spans broad spatial scales, possibly affecting species throughout their entire ranges, (ii) ecological responses are highly variable across space, that is, not all populations are exposed to similar trends in climate, and (iii) climate-induced impacts occur over long periods of time, from decades to centuries, generally longer than the duration of any scientists' careers or a typical funding cycle.

Given this presumption, we focused on examining citizen science related to research on birds and climate change. Our objectives were to i) quantify the scientific contribution of citizen science to this area of active inquiry, ii) assess whether professionals held volunteer-based research in equal regard to research by professionals, and iii) evaluate the extent to which citizen science was readily visible or noted in the focal studies. Taken together, our goal was to evaluate the use and confidence of citizen science in advancing understanding in an important area of ecology and global change research.

Materials and Methods

One of the most widely cited lines of evidence that species are responding to modern climate change relates to shifts in phenology. Changing spring phenology in migratory birds is a rapidly developing field of study that has been identified as a critical bellwether for assessing the ecological impacts of climate change. We based our study on a review paper by Knudsen et al. [34] to evaluate the influence of citizen science on the study of climate change research. The Knudsen et al. [34] review represents a critical synthesis of the existing scientific support for the patterns, mechanisms, and consequences of phenological changes in bird migration. The authors (all active researchers in the field of migration phenology) reviewed literature to formulate 10 specific scientific claims about migratory birds and global climate change (Table 1). Also, 18 of the 27 authors scored their opinion regarding the amount of research effort so far invested in each claim (hereafter referred to as "knowledge basis") and whether each claim held in general (hereafter referred to as "support"), on a continuous scale from 0 (least) to 10 (most) (Table 1). They report the mean for each value and the associated standard deviation.

There were 205 papers referenced in the 10 claims, 15–36 references per claim, in Knudsen et al. [34]. We excluded 42 references that were reviews (including books, chapters, and meta-analyses) from our analysis. We were able to classify 171 of the 173 research papers according to the sources of data: either as including observations collected by volunteers (citizen science) or not (professionals). We noted sources of museum collections (n = 4) and found two of these papers also used data from a citizen science source; we classified the remaining two papers using museum collections as professional. Some (n = 45) papers were referenced in more than one claim, thirteen of which appear in 2–5 claims. We computed the Pearson's correlation between the mean knowledge basis and percent citizen science per claim, and between the mean support and proportion of citizen per

claim. We repeated these analyses using the reported standard deviation associated with knowledge basis and support in relation to proportion of citizen science per claim.

For papers that used citizen-science data, we classified the type of project as: large-scale coordinated scheme (dispersed network of volunteers following a protocol sharing centralized objectives and data management), local volunteers following a protocol but with multiple sampling objectives (e.g., ringers at an observatory), local groups (e.g., local bird clubs and societies), and other (e.g., journals and diaries of amateur naturalists). We also noted the terminology we relied on to classify each citizen-science paper, including explicit mention of volunteers, and/or a specific program, other term, or through contact with the author.

Results

We examined 173 original research papers that were used by Knudsen and colleagues to formulate 10 central claims about the impacts of climate change on avian migration [34]. We found that 85 of the 171 papers that we could classify were based on citizen science, constituting 5 to 20 papers per claim (Appendix S1). Citizen science heavily informed claims related to ecological patterns and consequences and was less frequently cited for claims about mechanisms (Table 1).

Data from a wide range of citizen-science efforts were included in these papers, including observations from large-scale coordinated programs (n = 35 papers), local volunteers following protocols (n = 40), ad hoc counts from local bird clubs (n = 12 papers), other volunteer sources (n = 4), and some papers (n = 6) drew on several of these citizen-science efforts.

Of the 84 papers that were based on citizen science, 74 were published in, or prior to, 1995, which was when "citizen science" was coined in the context of bird research [35]. Despite the importance of citizen science in substantiating the claims in Knudsen et al. [34], the term "citizen science" *never* appeared in any referenced publications (Appendix S1). The term "volunteer" was used in 37 of the citizen-science papers, typically only in the acknowledgements (Appendix S1). We viewed 37 citizen-science papers as "invisible" because our ability to identify these as citizen science was based on i) the name of specific programs (n = 35; e.g., BTO Common Bird Census), ii) the mention of general programs or efforts (n = 7; e.g., local ornithological societies), iii) terms "bird watchers," "ringers," "bander," "public," "naturalist," and "people" (n = 42), iv) by contacting authors and data-providing organizations for confirmation (n = 11 authors contacted and 9 responded), or v) a combination of these identifiers (n = 22). The papers that were difficult to classify were evenly distributed across claims, comprising roughly half of the citizen-science papers per claim.

Knudsen et al. [34] provided the mean values of the opinions of 18 of their authors on the strength of the knowledge basis and support for each of the claims they reviewed. We found that the mean values of the expert opinions were not correlated to the proportion of citizen science supporting each claim (r = -0.02, $p = 0.9$ for knowledge basis, Figure 1; r = 0.07, $p = 0.8$ for support). Similarly, the standard deviation associated with the mean values of the expert opinions were not correlated to the proportion of citizen science supporting each claim (r = -0.38, $p = 0.3$ for knowledge basis; r = 0.17, $p = 0.6$ for support).

Discussion

Our findings are strongly indicative of the usefulness and credibility of citizen science in the field of global change research. We found that more than half the central claims about the impacts

Table 1. Ten claims with mean values for support and knowledge basis from Knudsen et al. [34].

	#	Statement	Expert Opinion		# papers*	Citizen Science	
			Support	Knowledge basis		#	%
Pattern	1	Spring migration advances due to global climate change	8.52	8.28	32	17	53
	2	Phenological response depends on migration distance	5.68	5.99	22	11	50
	3	Climate change affects migratory distance and routes	4.86	3.83	20	15	77
Mechanism	4	Controlling mechanisms are hardwired	4.06	3.79	21	5	24
	5	Changes are mainly due to phenotypic plasticity	4.84	3.79	15	5	33
	6	Phenotypic variability is mainly due to weather en route	4.87	4.42	29*	14	48
	7	Responses are constrained by the annual cycle	5.54	3.27	24	11	46
Consequence	8	Increased trophic mismatch on breeding grounds	4.97	4.00	21	5	24
	9	Climate change causes population declines	4.53	3.69	35	20	57
	10	Climate change affects community composition	4.10	2.97	21*	14	67

*excluded papers that we were unable to classify (1 in claim 6 and 1 in claim 10).
The number of papers used for assessment of each claim, and the number and percent of papers that used citizen science.

of climate change on avian migration were based on studies that depended on data from citizen scientists. In addition, we did not find any statistical relationship between the knowledge basis or scientific support for each claim and the proportion of citizen science. The use of citizen science data in an active field of ecological research, such as migration phenology, is strong evidence that any stigma associated with the use of data collected by volunteers is unwarranted. Yet, the contributions of citizen science were not readily detectable in most cases. Thus, the stigma may persist unless researchers begin to draw attention to the citizen-science elements in their research papers.

Citizen science was more critical in supporting claims related to ecological patterns and consequences when compared to claims about mechanisms (Table 1). The reason for this finding highlights a simultaneous strength and limitation of citizen science. One of the primary motivations of developing and deploying a citizen science project is to collect data over spatiotemporal scales that would be difficult (if not impossible) with professional scientists. As a result, these programs have been particularly useful for documenting broad-scale patterns (e.g., macroecology) or long-term consequences (e.g., population trends). Many of these programs, in many ecological disciplines in addition to ornithology, were initiated for purposes other than documenting the ecological responses to climate change, but as our findings emphasize, the scale of data collection has proven essential for analyzing patterns and consequences. Cost savings are another advantage of re-purposing existing data. Climate change studies focused on mechanisms tend to involve experimental investigation (e.g., active or passive warming devices) or more intensive field studies (e.g., use of geolocators). That being said, data generated from long-term citizen science programs have also been repurposed to focus on mechanistic hypotheses and objectives. As an example, several of the papers classified as supporting mechanisms of migration focused on explicit hypotheses such as the geographical variation in phenological mismatch [36], [37] and buffer effects in population dynamics [38]; these hypotheses could not be tested without data from coordinated banding studies involving volunteers in different regions.

Given the invisible prevalence of citizen science in advancing this one area of global change research, we suspect it also common in many other areas of inquiry such as studies of land-use change, invasive species, and environmental pollutants, to name a few. We urge future use of consistent terminology and acknowledgement to facilitate tracking the impact of citizen science across numerous disciplines. Specifically, we urge use of the keyword phrase "citizen science" in papers that rely on scientific contributions from the public. Continued assessment of the value of citizen science in other areas of research could help increase overall public participation as well as identify new frontiers in multiple research fields and improve the interdisciplinary practice of citizen science.

An additional consequence of the invisibility of the scientific impact of citizen science is that projects may miss the broader social impacts of their work. Unique positive societal impacts, such as increased scientific literacy, depend on participants being not merely engaged as instruments or human sensors, but upon being informed and engaged with research progress and outcomes [39]. Yet, many long-running volunteer efforts did not originate with the specific purpose of understanding the consequences of global climate change, and as a result, most of these projects were not designed to foster communication of scientific findings back to project participants; this is particularly true for studies using data from online repositories. Explicit recognition of citizen science in published papers could promote the communication linkages necessary for broader impacts by helping shift public discourse

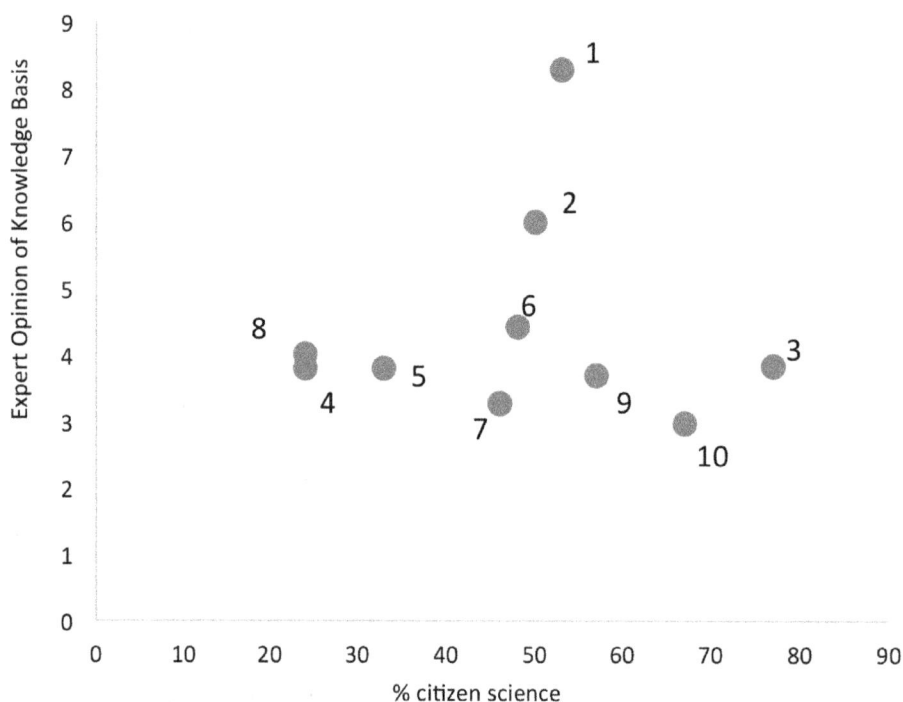

Figure 1. No relationship between the mean values of the expert opinions on the knowledge base of each claim in Knudsen et al. [34] were not correlated to the proportion of citizen science supporting each claim.

associated with modern climate change from controversy to acceptance. Our findings demonstrate the exceptional value of the efforts of thousands of participants whose data informed the 10 claims, and point to the potential of the millions of global participants whose "invisible" efforts may be contributing to new discoveries.

Supporting Information

Appendix S1 Research papers in Knudsen et al. [34], classified as involving data from citizen science (yes = 1, no = 0).

Acknowledgments

We thank R Bonney, A Shwartz, and an anonymous reviewer for comments and insights that improved this manuscript.

Author Contributions

Conceived and designed the experiments: CC BZ. Analyzed the data: CC. Contributed reagents/materials/analysis tools: CC JS. Contributed to the writing of the manuscript: CC JS BZ.

References

1. Miller-Rushing A, Benz S (2013) Workshop 1: Conference on Public Participation in Scientific Research 2012: An International, Interdisciplinary Conference. Bull Ecol Soc Am 94: 112–117.
2. Silvertown JA (2009) A new dawn for Citizen Science. Trends Ecol Evol 24: 467–471.
3. Roy HE, Pocock MJ, Preston CD, Roy DB, Savage J (2012)Understanding Citizen Science and Environmental Monitoring. Final Report on behalf of UK-EOF. NERC Centre for Ecology & Hydrology and Natural History Museum. 179 p.
4. Dickinson JL, Zuckerberg B, Bonter DN (2010) Citizen Science as an Ecological Research Tool: Challenges and Benefits. Annual Rev Ecol Evol Syst 41: 149–172.
5. Bonney R, Cooper CB, Dickinson J, Kelling S, Phillips T, et al. (2009) Citizen Science: a developing tool for expanding science knowledge and scientific literacy. BioSci 59: 977–984.
6. Hochachka WM, Fink D, Hutchinson RA, Sheldon D, Wong W, et al. (2012) Data-intensive science applied to broad-scale citizen science. Trends Ecol Evol 27: 130–137.
7. Kelling S, Hochachka WM, Fink D, Riedewald M, Caruana R, et al. (2009) Data-intensive Science: A New Paradigm for Biodiversity Studies. BioSci 59: 613–620.
8. Newman G, Graham J, Crall AW, Laituri M (2011) The art and science of multi-scale citizen science support. Ecol Infor 6: 217–227.
9. Sullivan BL, Wood CL, Illiff MJ, Bonney RE, Fink D, et al. (2014) The eBird enterprise: An integrated approach to development and application of citizen science. Biol Cons 169: 31–40.
10. Switzer A, Schwille K, Russell E, Edelson D (2012) National Geographic FieldScope: a platform for community geography. Front Ecol Envir 10: 334–335.
11. Fink D, Hochachka WM, Zuckerberg B, Winkler DW, Shaby B, et al. (2010) Spatiotemporal exploratory models for broad-scale survey data. Front Ecol Envir 20: 2131–2147.
12. Danielsen F, Burgess ND, Jensen PM, Pirhofer-Walzl K (2010) Environmental monitoring: the scale and speed of implementation varies according to the degree of people's involvement. J Appl Ecol 47: 1166–1168.
13. McCormick S (2012) After the Cap: Risk Assessment, Citizen Science and Disaster Recovery. Ecol Soc 17: 31.
14. Cornwell ML, Campbell LM (2011) Co-producing conservation and knowledge: Citizen-based sea turtle monitoring in North Carolina, USA. Soc Stud Sci 42: 101–120.
15. Bonilla NO, Scholl J, Armstrong M, Pieri D, Otero B, et al. (2012) Ecological Science and Public Policy: an Intersection of Action Ecology. Front Ecol Envir 93: 340–345.
16. Engel SR, Voshell JR Jr (2002) Volunteer biological monitoring: can it accurately assess the ecological condition of streams? Am Entomol 48: 164–177.
17. Genet KS, Sargent LG (2003) Evaluation of methods and data quality from a volunteer-based amphibian call survey. Wildl Soc Bull 31: 703–714.
18. Cooper CB, Hochachka WH, Dhondt AA (2012) The opportunities and challenges of Citizen Science as a tool for ecological research. In: Dickinson JL, Bonney B, editors.Citizen Science: Public Collaboration in Environmental Research.Ithaca: Cornell University Press. pp. 99–113.

19. Danielsen F, Jensen PM, Burgess ND, Altamirano R, Alviola PA, et al. (2014) A multicountry assessment of tropical resource monitoring by local communities. BioSci 64: 236–251.
20. Gardiner MM, Allee LL, Brown PMJ, Losey JE, Roy HE, et al. (2012) Lessons from lady beetles: accuracy of monitoring data from US and UK citizen-science programs. Front Ecol Envir 10: 471–476.
21. Bates AJ, Sadler JP, Everett G, Grundy D, Lowe N, et al. (2013) Assessing the value of the Garden Moth Scheme citizen science dataset: how does light trap type affect catch? Entom Expt Appl 146: 386–397.
22. Miller DAW, Weir LA, McClintock BT, Grant EHC, Bailey LL, et al. (2013) Experimental investigation of false positive errors in auditory species occurrence surveys. Eco Appl 22: 1665–1674.
23. Galloway AWE, Tudor MT, Vander Haegen WM (2006) The reliability of citizen science: A case study of Oregon white oak stand surveys. Wildl Soc Bull 34: 1425–1429.
24. Tregidgo DJ, West SE, Ashmore MR (2013) Can citizen science produce good science? Testing the OPAL Air Survey methodology, using lichens as indicators of nitrogenous pollution. Envir Pollut 182: 448–451.
25. Kyba CCM, Wagner JM, Kuechly HU, Walker CE, Elvidge CD, et al. (2013) Citizen Science Provides Valuable Data for Monitoring Global Night Sky Luminance. Sci Rep 3, 10.1038/srep01835
26. Gollan J, de Bruyn LL, Reid N, Wilkie L (2012) Can Volunteers Collect Data that are Comparable to Professional Scientists? A Study of Variables Used in Monitoring the Outcomes of Ecosystem Rehabilitation. Envir Manage 50: 969–978.
27. Moyer-Horner L, Smith MM, Belt J (2012) Citizen science and observer variability during American pika surveys. J Wildl Manage 76: 1472–1479.
28. Jordan RC, Brooks WR, Howe DV, Ehrenfeld JG (2012) Evaluating the Performance of Volunteers in Mapping Invasive Plants in Public Conservation Lands. Envir Manage 49: 425–434.
29. Kremen C, Ullmann KS, Thorp RW (2011) Evaluating the Quality of Citizen-Scientist Data on Pollinator Communities. Con Bio 25: 607–617.
30. Bonter DN, Cooper CB (2012). A process for improving data quality and a strategy for ensuring sustainability in a citizen science project. Front Ecol Envir 10: 305–307.
31. Parsons J, Lukyanenko R, Wiersma Y (2011). Easier citizen science is better. Nature 471: 37–37.
32. Cooper CB (2014) Is there weekend bias in clutch-initiation dates from citizen science? Implications for studies of avian breeding phenology. Intl J Biometeor in press.
33. Devictor V, Whittaker RJ, Beltrame C (2010) Beyond scarcity: citizen science programmes as useful tools for conservation biogeography. Div Distrib 16: 354–362.
34. Knudsen E, Lindén A, Both C, Jonzén N, Pulido F, et al. (2011) Challenging claims in the study of migratory birds and climate change. Biol Rev 86: 928–946.
35. Bonney R, Ballard H, Jordan R, McCallie E, Phillips T, et al. (2009) Public Participation in Scientific Research: Defining the Field and Assessing Its Potential for Informal Science Education. A CAISE Inquiry Group Report. Washington, D.C.: Center for Advancement of Informal Science Education (CAISE).
36. Both C (2010) Flexibility of timing of avian migration to climate change masked by environmental constraints en route. Current Biol 20: 243–248.
37. Both C, te Marvelde L (2007) Climate change and timing of avian breeding and migration throughout Europe. Climate Res 35: 93–105.
38. Gill JA, Norris K, Potts PM, Gunnarsson TG, Atkinson PW, et al. (2001) The buffer effect and large scale regulation in migratory birds. Nature 412: 436–438.
39. Lawrence A (2009) The first cuckoo in winter: phenology, recording, credibility, and meaning in Britain. Global Envir Change 19: 173–179.

22

Why Do Some People Do "More" to Mitigate Climate Change than Others? Exploring Heterogeneity in Psycho-Social Associations

José Manuel Ortega-Egea*, Nieves García-de-Frutos, Raquel Antolín-López

Department of Economics and Business, University of Almería (ceiA3), Almería, Spain

Abstract

The urgency of climate change mitigation calls for a profound shift in personal behavior. This paper investigates psycho-social correlates of *extra* mitigation behavior in response to climate change, while also testing for potential (unobserved) heterogeneity in European citizens' decision-making. A person's extra mitigation behavior in response to climate change is conceptualized—and differentiated from common mitigation behavior—as some people's broader and greater levels of behavioral engagement (compared to others) across specific self-reported mitigation actions and behavioral domains. Regression analyses highlight the importance of environmental psychographics (i.e., attitudes, motivations, and knowledge about climate change) and socio-demographics (especially country-level variables) in understanding extra mitigation behavior. By looking at the data through the lens of segmentation, significant heterogeneity is uncovered in the associations of attitudes and knowledge about climate change—but not in motivational or socio-demographic links—with extra mitigation behavior in response to climate change, across two groups of environmentally active respondents. The study has implications for promoting more ambitious behavioral responses to climate change, both at the individual level and across countries.

Editor: Bayden D. Russell, The University of Adelaide, Australia

Funding: The authors acknowledge financial support from the Spanish Ministry of Economy and Science (National R&D Project ECO2011-24921 and predoctoral grant program) and the European Regional Development Fund (ERDF/FEDER), from the University of Almería's (UAL, ceiA3) predoctoral grant program, and CySOC. The funders had no role in study design, data collection and analysis, decision to publish, or preparation of the manuscript.

Competing Interests: The authors have declared that no competing interests exist.

* Email: jmortega@ual.es

Introduction

Climate change is recognized as a major, anthropogenically-induced environmental threat, with potentially severe and far-reaching consequences for human and natural systems [1–2]. It is now beyond dispute that, in industrialized countries, people contribute to climate change through unsustainable high-carbon lifestyles [3]. To illustrate, in the European Union (EU), private households are directly responsible for as much as 20% of total greenhouse gas (GHG) emissions and over one-fourth of the final energy consumed [4–5]; substantially larger shares of EU energy use and GHG emissions are attributable in an indirect way—via consumption expenditures—to household/consumer activities [5–6].

Clearly, then, individuals have a central role to play in addressing climate change risks [7]. People can respond to climate change through adaptation—to potential and unavoidable climate impacts—and mitigation efforts—focused on reducing GHG emissions to prevent (or delay) further damages [3,8]. With climate change being primarily rooted in excessive energy consumption, the public's engagement in mitigation activities is recognized as critical to achieving a sustainable future—that is, shifting towards a new, low-carbon paradigm [1,3,7]. Over the past two decades, increased media coverage—coupled with economic incentives, subsidies, and related interventions—has substantially raised citizens' awareness and concern about climate change, but has typically failed to induce persistent behavioral changes [1,9]. In Europe, estimates of (some form of) personal action to mitigate climate change have ranged from 53% to 63% of the population, according to four surveys conducted between 2008 and 2011 in all EU-27 countries [10–11]; however, as shown elsewhere [12], there is limited engagement of most EU citizens in mitigation efforts beyond recycling [10–11].

Given the urgency of climate change mitigation, a profound shift is needed in personal behavior—from inaction or limited action levels—towards broader and greater levels of behavioral engagement [12–14]. Such *extra* behavioral responses—comprising additional mitigation actions and specific behavioral levels that go beyond what most people do—hold promise for a further incremental impact in addressing climate change. Hence, an important question arises as to what makes some people make an extra commitment (i.e., do "more") to mitigate climate change through personal action, as compared to others.

Considerable attention has been directed at the correlates of individual pro-environmental behavior, in general, or at specific types (or subsets) of personal action—both in private and public spheres [15–16]. For example, past research has investigated recycling [17–18], reducing car use and choosing an environmentally friendly mode of transport [19–20], engaging in environmental citizenship [21], and many other forms of personal pro-environmental behavior. Such green behaviors have been studied through a variety of social-psychological variables and factors, including psychographics—such as environmental attitudes [15,22], concern [23], values and motivations [24], or knowledge [25]—and socio-demographics—such as gender [26], age [27], education [28], income [29], or nationality [30–31].

Corresponding advances are being rapidly made in the field of climate change research, in relation to the correlates of personal mitigation (and adaptation) behavior [1–3,8]. The present article aims to enhance understanding of the psycho-social correlates of extra (i.e., broader and greater) mitigation behavior in response to climate change, among the EU public. It contributes to ongoing climate change research in the following three ways. First, the authors investigate factors that relate to a person's extra behavioral engagement to address climate change, as opposed to more common (or typical) types and levels of mitigation behavior. This is important because, as advocated in previous research [14], "we must do a lot" to effectively respond to climate change. For example, if altruism is the most important motivational process associated with extra mitigation behavior, then this suggests that reinforcing altruistic values—rather than egoistic ones—is a more powerful way to support this increasingly critical action pattern. Second, the study tests for potential unobserved heterogeneity—across respondents—in the relations of psychographics and socio-demographics to extra mitigation behavior in response to climate change. For instance, by looking at the data through the lens of segmentation (i.e., through latent class regression), the analyses will clarify whether specific types of knowledge/information about climate change issues (i.e., the causes, consequences, and ways of fighting this threat) relate to similarly or differently to the breath and level of self-reported behavioral engagement (i.e., extra vs. common) in climate change mitigation. Third, through the analysis of a large-scale survey dataset of European citizens, the study increases current understanding of cross-national variations in personal mitigation behavior in response to climate change. Overall, the findings reported here should help researchers and policy-makers to promote broader and greater levels of personal response to climate change, both at the individual level and across countries.

Extra Mitigation Behavior (in Response to Climate Change)

Pro-environmental behavior—and, more specifically, mitigation (and adaptation) responses to climate change—can be operationalized at multiple levels of analysis, such as individual, group, organizational, or regional/national levels. The focus here is on individual-level, personal mitigation behavior that, according to previous literature, can be broadly described as comprising voluntary and future-oriented behavioral responses to climate change (e.g., consumer reduction in energy consumption with mid- to long-term positive impacts on climate change) [3,32]. Given the multi-faceted nature of personal mitigation behavior, it potentially encompasses a broad range of actions in private and public spheres of life, one-off and regular decisions, simple and more difficult steps, as well as low and high impact actions—as regards their effectiveness in mitigating climate change [1,12,14].

Extra mitigation behavior—the focal variable of interest in this study—is viewed as comprising additional mitigation actions (i.e., broader) and enactment levels of specific mitigation behaviors (i.e., greater) that go beyond what most people do to address climate change. A person's extra behavioral engagement to mitigate climate change is captured—and differentiated from common mitigation behavior—on a set of self-reported past/present mitigation actions spanning in-home and out-of-home settings

(e.g., saving energy at home vs. reducing car use), high- and low-impact mitigation practices (e.g., energy conservation vs. recycling), one-off and frequent choices (e.g., installing solar panels in the home vs. buying seasonal, locally produced food), relating to four specific, environmentally significant domains of personal action (i.e., domestic energy/water conservation, waste reduction, eco-friendly transportation, and eco-shopping) [12]. Thus, several different types of personal action are covered so as to profile (and measure) the behavior of people who report extra behavioral engagement to mitigate climate change, as compared to others; this approach allows for sufficient variability in the difficulty, impact, frequency, and context of the specific mitigation actions being examined.

Proposed Models of Extra Mitigation Behavior

The models tested in this study propose that environmental psychographics and socio-demographics help to explain why some people make an extra commitment to mitigate climate change through personal action, as compared to others, while testing for potential (unobserved) heterogeneity among respondent groups.

Psychographics and extra mitigation behavior

To date, numerous studies have assessed how psychological traits relate to ecological behavior (e.g., [33]). The importance of psychographics (for pro-environmental behavior) has remained fairly stable over time [15,34], suggesting their association with different types of climate change-motivated behavior. This study focuses first on three psychographic processes—i.e., attitudes, motivations, and knowledge about climate change (model set 1)—that likely relate to an individual's self-reported extra behavioral engagement to mitigate climate change, as compared to others.

Attitude is defined as the positive or negative feeling that an individual holds about a psychological object [25,35]. The potential targets of attitudes cover a broad spectrum of discriminable aspects of the physical world, such as a physical entity, a person or group of people, an abstract concept or issue, or a behavior [36]. In the environmental literature, attitude is acknowledged as a major proximal factor for ecological intention and behavior [15,33]. Two meta-analyses have confirmed a significant, moderate association between attitude and pro-environmental behavior, with estimated mean correlations of approximately 0.4 [15,34]. Nonetheless, the empirical evidence has been mixed for attitudinal associations with behavior—in line with the widely reported attitude–action gap [1]. Attitude–behavior links across environmental studies are contingent on various factors, including the consideration or omission of intention as a possible mediator of this relationship [15,35], attitude strength [37], attitude certainty and ambivalence [13], situational constraints to pro-environmental action [33,38], and importantly, congruence in the level of specificity/generality (i.e., the scope) and time interval (e.g., simultaneous or lagged) between attitude and behavior measures [33,39]. On the latter point, expectancy-value models recommend that attitudes be measured at the level of behavior—e.g., by measuring attitudes toward pro-environmental action, rather than attitudes toward the environment—, lagged but close in time to the measure of behavior, in order to optimize congruence in attitude–behavior measurement and to reinforce attitude associations with behavior [25,40–41]. Importantly also, the analysis of attitude relations to behavior is likely to be affected by the focus on self-reports or objective outcome measures of actual behavior [41].

It is worth stressing here that investigating the attitude–action gap in climate change-motivated behavior is beyond the scope of this study, given the focus here on self-reported (instead of actual or objectively measured) mitigation behaviors, the simultaneous measurement of attitudes and behavior, and not measuring important factors such as intentions, attitudinal strength, certainty, or ambivalence. Nonetheless, attitude variables should be significantly related to EU citizens' breath and level of self-reported behavioral engagement (i.e., extra vs. common) in response to climate change, owing to: (1) the matching level of specificity/generality in attitudes (towards the threat of climate change and the role of mitigation efforts) and mitigation behavior (in response to climate change)—in a way that measurement congruence exists at the level of climate change as an environmental problem and, at least partly, at the level of behavior; (2) the close time correspondence between (present) attitudes and (past/present) self-reported behaviors; and, (3) the high degree of volitional control over performance of the behavior, which is modeled as an aggregate of specific (and highly voluntary) mitigation actions [40,42].

Respondents with more positive/desirable attitudes, understood and measured here as positive (desirable) feelings that climate change is harmful and that mitigation actions are important, should be more likely to display extra behavioral engagement to mitigate climate change than others [12,14–15]. Likewise, respondents with negative (undesirable) attitudes towards the threat of climate change (i.e., that climate change is not harmful) and the role of mitigation efforts (i.e., that mitigation behavior is not important or ineffective) should not be disposed towards extra mitigation behavior in response to climate change, as compared to others [12,14].

H1a. *Positive/desirable attitudes* (towards the threat of climate change and the role of mitigation efforts) *will be positively associated with extra mitigation behavior.*

H1b. *Negative/undesirable attitudes* (towards the threat of climate change and the role of mitigation efforts) *will be negatively associated with extra mitigation behavior.*

Motivation is usually described as the driving force of behavior [25] or the "reason why a given behavior occurs" [43]. Motives can be both overt and hidden, depending on people's awareness (or not) of their motives for behavior [43]. Researchers also distinguish between primary (general) motives for a whole class of behaviors—e.g., acting in environmentally responsible ways—and selective (domain-specific) motives for particular actions, such as recycling or reducing car use [25,43]. The present study examines the role of primary/general environmental motivations (for mitigating climate change) because of the assessment of aggregate, self-reported pro-environmental actions—i.e., people's breath and level of behavioral engagement in climate change mitigation—at a comparable level of generality [42,44].

People may be concerned about environmental issues for several reasons [45–46]. Thus, previous research has explored the different types of value orientations underlying motivations for environmentally significant behavior (see [47]). Owing to the prominence of Schwartz's norm-activation model [48], most studies have differentiated between self-transcendent (altruistic) and self-enhancement (egoistic) values [47–49]. Stern et al. [24] further subdivided altruistic values into social-altruistic and biospheric value orientations. Similarly, Gagnon Thompson and Barton [45] drew a distinction between ecocentric and anthropocentric motives and values. Ecocentric individuals attach importance to the environment for itself and will engage in pro-environmental behavior, even if it involves some sort of sacrifice on their part [45]; this behavior pattern is largely rooted in biospheric values [24]. Anthropocentrics' actions are more deeply

grounded in social-altruistic and egoistic values [24,47]; that is, these individuals will engage in pro-environmental behavior, such as climate change mitigation behavior, only if it has positive consequences for mankind and does not diminish their quality of life or wealth [45].

Previous research suggests that, in general, pro-environmental behavior is more closely linked to biospheric values than to social-altruistic or egoistic ones [24,29,45–46]. However, not only ecocentric (biospheric) motivations, but also anthropocentric (social or egoistic) ones can relate to environmentally significant behavior [24,29]; as shown in previous research [8,12], non-altruistic motives—such as financial motivations—often underpin mitigation actions (e.g., energy conservation practices). There is in fact evidence of multiple motivations—i.e., altruistic and egoistic ones—for mitigation behavior [8], suggesting that both self-transcendent (altruistic) and self-enhancement (egoistic) motives may independently relate to, conflict, or converge for extra (i.e., broader and greater) behavioral engagement to mitigate climate change (among the EU public) [7,49–50].

Given the mixed evidence from previous research regarding the link between self-transcendent and self-enhancement motives for and personal mitigation behavior, the issue is posed here as an explorative research question rather than as a hypothesis:

RQ1. *How do self-transcendent (altruistic) and self-enhancement (egoistic) motives relate to extra mitigation behavior?*

Knowledge of environmental issues and problems has been a significant correlate of pro-environmental awareness, moral norms, attitude, intention, and behavior [15,25,33]. In particular, recent meta-analytic evidence suggests the close association of environmental knowledge with pro-environmental behavior [15]. Regardless of the assumed importance of environmental knowledge (and information) as a major, but not sufficient, rational precondition for ecological action [51], its specific role in pro-environmental decision-making has long been debated [15,34]. Information deficit approaches to behavior change—depicted largely as linear-sequential models [environmental knowledge → awareness and concern (environmental attitude) → pro-environmental behavior]—have been criticized as being too simplistic or ineffective [25]. Consistent with this, informational efforts to encourage voluntary, public engagement in climate change mitigation actions—mostly through the provision of scientifically sound information—appear to have had little success [1,9,13].

Multiple factors may work to strengthen knowledge/information associations with climate change mitigation action (and general pro-environmental behavior), or to cause the widely reported knowledge–action gap. First, the distinction between objective and subjective (self-reported) environmental knowledge is important; most past research implicitly assumes or explicitly states that self-assessments—used here to measure subjective knowledge about three climate change issues—serve as valid proxy measures of objective environmental knowledge (e.g., [52–53]), although each of these two knowledge types (objective and subjective) can be differently associated with specific pro-environmental behaviors [54]. Second, besides structural and situational constraints (see, e.g., [7]), the level and type of environmental knowledge and information have been shown to affect the strength of knowledge–behavior links [25,40,55]. Basic information provision is necessary for people to recognize environmental problems—e.g., to overcome the public's lack of knowledge about climate change—and consciously engage in mitigation behavior [1,3,25]. In contrast, an excessive amount of environmental information or very detailed technical data, concerning complex and far-reaching environmental issues such as climate change and global warming, can lead to public confusion and frustration [25,56]. Following the distinction between declarative, procedural, and effectiveness environmental knowledge [55], knowledge about the nature and causes of environmental problems (declarative knowledge) and knowledge about ecological action strategies or "how-to" knowledge (procedural knowledge) have been especially linked to individual pro-environmental behavior [33–34,55] and—importantly here—to mitigation behavior in response to climate change [8,57]. Arguably, extra (i.e., broader and greater) personal engagement in mitigation actions should depend on the public's knowledge about the causes of climate change (and global warming)—e.g., knowledge about human contributions to climate change—and knowledge about available courses of action [8,12]. The role of effectiveness environmental knowledge—which can be approached from two related, but distinct, angles—is somewhat more controversial. First, effectiveness environmental knowledge, when understood as knowledge about the relative ecological consequences (i.e., effectiveness) of different behavioral alternatives [55], has been shown to relate to individual pro-environmental behavior [52] and thus could be expected to be associated with extra mitigation behavior in response to climate change [8]. A second view of effectiveness environmental knowledge, referring to knowledge about the consequences of environmental problems [52]—and examined in this study in relation to self-reported climate change-motivated behavior—, has been criticized for eliciting feelings of frustration, owing to people's increased awareness of the limited impact of their actions on environmental protection [25]; thus, improving knowledge about the serious consequences of climate change—e.g., emphasizing the consequences of not engaging in mitigation behavior [52], or using fear appeals in climate change communication [9]—is not likely to result in an extra (i.e., broader and greater) level of mitigation behavior in response to climate change.

H2. *Both knowledge about the causes (declarative knowledge) and knowledge about the ways of fighting climate change (procedural knowledge) will be positively associated with extra mitigation behavior.*

Socio-demographics and extra mitigation behavior

Socio-demographic variables have generally shown modest or equivocal associations with pro-environmental behavior [16,26,51]. In fact, some authors have argued that socio-demographics (e.g., age, sex, race, or political orientation) will be less associated with environmental concern and ecological behavior over time, due to widespread green concerns across many demographic groups [58–59]—particularly in Western countries [26]. This contention contrasts with recent climate change studies showing that socio-demographic variables can be significant correlates of general or specific types of personal action to mitigate climate change [3,8,12,60].

Part of the between-study variation in socio-demographic associations with ecological behavior can be ascribed to methodological problems and differences across studies [61], analysis of direct vs. indirect relationships [29,62], country-specific factors [26], or the type of behavior studied [63]. An important argument here is that some socio-demographic (background) variables may be proxies for personal capabilities—that is, the knowledge and skills necessary for particular behaviors [16,36]. Thus, demographic variables like age, education, and income should be related to climate change mitigation efforts that depend strongly on personal capabilities [16]—i.e., mitigation actions potentially influenced by objective or subjective constraints [1,64]. This is particularly the case of high-impact mitigation actions—i.e., energy conservation practices—which appear to be significantly associated with an individual's age (see [8]). This study evaluates

the association of external socio-demographic factors with self-reported extra mitigation behavior, both at the person level—i.e., gender, age, education, and political ideology—and at the country level—i.e., country values and country wealth.

Gender. The evidence from prior environmental and climate change research—though far from conclusive (see [61])—suggests that women typically report greater environmental concern and involvement in environmentally significant behaviors, relative to men [3,12,26,36,65]. Specifically, women appear more likely than men to engage in private-sphere and regular pro-environmental activities in response to climate change, such as reducing waste (e.g., recycling) [12,66] and conserving energy in the course of daily routines [12,66].

Three theoretical explanations have been offered for gender distinctions in general environmentalism and climate change behavior. The first rationale is that traditional gender roles and socialization patterns largely underlie women's greater environmental involvement [65]. Traditional female socialization has been linked to pro-environmental behavior, owing to women's other and ecocentric value orientations [65] and caretaker role [66]. Women tend to be more attentive to the interconnections between the natural environment and things they value—e.g., other people [24]; as a result, women will be more sensitive than men to the environmental consequences of their actions [51]. The second rationale lies in the fact that, overall, women tend to judge the world as more risky [67], perceive higher levels of environmental risk [68], and thus are likely to take more pro-environmental actions than men [62]. Finally, women appear to perceive fewer (subjective and objective) constraints on personal engagement with climate change mitigation, relative to men [64].

H3. *Female gender will be positively associated with extra mitigation behavior.*

Age. There is much controversy surrounding age relations to environmental knowledge, attitudes, and behavior [26,61]. Most studies have reported negative associations of age with environmental attitudes and concern, indicating that younger people tend to be more concerned with environmental problems such as climate change [2,38,61]. Less clear is the relationship between age and environmental knowledge, with age showing either non-significant or weak negative associations with knowledge about various environmental issues [26]. Also, researchers have studied the linkage between age and pro-environmental behavior with differing results—that is, age has been reported to be negatively, positively, or non-significantly related to environmentally-significant behavior (see [26]). This mixed evidence is also reflected in the study of personal mitigation behavior in response to climate change [3,8]. In fact, asymmetric age associations with different types of action have been observed [8,12]; to illustrate, as regards energy conservation actions, older individuals appear more likely to engage in less painful or simple energy conservation activities (e.g., buying energy-saving light bulbs or turning off unused lights), but less likely than younger people to engage in more difficult transport-related energy conservation [8,12].

Consistent with the common negative links between age and environmental concern (i.e., attitudinal measures), Diamantopoulos et al. [26] argued that age tends to negatively correlate with *intended* ecological behavior—i.e., with intentional measures of behavior [12]; conversely, positive age–behavior linkages are typically found in studies using measures of *current* pro-environmental behavior [8,12,26]. Environmental attitudes and intentions may not translate into climate change mitigation behavior in younger people, partly because of their lack of necessary resources (e.g., financial means) for environmentally significant actions [26,64]. Life-cycle and cohort effects may account as well for age differences in pro-environmental decisions [38,61]. The life-cycle age effect points to a non-linear (inverted U-shaped) relationship between age and climate change concern; that is, highest levels of environmental concern during middle-age [38,61]. In addition, researchers generally agree on a cohort effect, resulting from greater exposure of birth cohorts from the 1950s (or 1960s) to public discussion and concern about environmental problems—such as climate change and global warming—, compared to previous cohorts [38,63]. Finally, "differences in time horizons in relation to climate change" [13] would be suggestive of negative age associations with personal mitigation behavior.

Taking together the available evidence—particularly the life-cycle and cohort rationales—, middle-aged European citizens are more likely to report extra mitigation behavior in response to climate change, compared to their younger and older counterparts.

H4. *There will be a curvilinear (inverted U-shaped) association between age and extra mitigation behavior.*

Education. Several studies have examined the potential role of education level as an indirect correlate—e.g., through environmental knowledge, attitudes, or concern—and direct correlate of pro-environmental behavior [26,69]. With few exceptions (see [12,69]), findings have been fairly consistent across studies: better-educated individuals tend to be more knowledgeable, concerned, and involved in pro-environmental activities [26,61,63]—including climate change mitigation actions [2–3,12,57]. Much like age and income, educational attainment may be a good proxy for personal capabilities involved in environmentally significant behavior [16,62]. In this regard, people with more years of formal education have shown greater concern and behavioral commitment to environmental protection [26]; such individuals have access to more sources and types of information [70], and can be expected to understand highly-complex environmental issues, such as climate change, more fully than less educated citizens [26,57,61]. As a result, the following hypothesis is proposed:

H5. *Education level will be positively associated with extra mitigation behavior.*

Political ideology has often been employed, along with other psychological and demographic variables, to gain a deeper understanding of individual green behaviors (e.g., [8,47,69]). Studies including political ideology have reported very consistent results; on a left to right (liberalism–conservatism) continuum, people with left-of-center political views tend to show higher levels of concern, verbal commitment, attitudes, and environmentally significant behavior, compared to conservatives [22,51,63,69]; moreover, conservative political values appear to be strongly associated with skepticism about climate change [8]. Only a few studies have not found significant associations of political orientation with pro-environmental behavior [60,71] or climate change mitigation behavior [12]. Thus, political ideology is considered one of the most robust and stable socio-demographic correlates of environmental concern and behavior [51,61].

H6. *Left-of-center political ideology will be positively associated with extra mitigation behavior.*

Country variations. In accordance with calls for more international research [25], cross-cultural analyses of pro-environmental and climate change-motivated behavior have garnered increased attention over the past decade [21,31,66,72–79]. The far-reaching consequences of environmental degradation, coupled with increased (societal and governmental) environmental activism in wealthy and developing countries [80–81], led some authors to conclude the emergence of global environmentalism [30,79,82].

However, the *globalization hypothesis* has been disputed by several authors (e.g., [31,83]); in this respect, substantial empirical evidence has accumulated in support of cross-national variations in public environmental concern and protection [46,66,74–75,81]—including climate change mitigation efforts [78]—, both within Europe and across continents.

Post-materialism hypothesis. It is widely believed that international variations in pro-environmental attitudes and behavior are a consequence of different values and primary goals held across cultures [46,75]. Particularly, Inglehart's theory of post-materialism [80] provides a prevalent value priorities approach to understanding country differences in public environmental concern (see [81]). Post-materialist theory posits that environmental concern emerges only once basic individual needs are fulfilled [80]. According to this view, people from countries with a predominant post-materialist orientation tend to be more concerned about the environment and climate change and, consequently, can be expected to make and report an extra commitment to pro-environmental and climate change mitigation behaviors [31,80], compared to people from non-post-materialist countries. Although challenged on important points [30,79,82], the post-materialism hypothesis has received strong support from recent cross-cultural environmental studies [21,31,73,81,83].

H7a. *Post-materialist EU-27 countries will be positively associated with extra mitigation behavior.*

Wealth hypothesis. Cross-national variations in pro-environmental attitudes can also be explained by national differences in wealth [76,81,83]. Environmental concern may be an indirect consequence of wealth—i.e., mediated through post-materialist values, as asserted by Inglehart [80]—; in contrast, the prosperity/affluence hypothesis posits a direct link from wealth to environmental concern [31,76,83]. Regardless of small differences between the direct and indirect influence paths of wealth [31], sufficient evidence exists to suggest that citizens of wealthier nations tend to give higher priority to global environmental protection goals, compared to individuals in poorer nations [76,80–81,83]. Strong correlations have been obtained between wealth (GDP per capita) and priority/global indexes of environmental concern in the works of Franzen and colleagues—i.e., correlations of approximately 0.8, accounting for more than 50% of the cross-national variance in environmental concern (see [81]).The opposite relationship—i.e., negative correlations—has also been observed between wealth and measures of local environmental concern [30,83]. In poorer nations, lower environmental quality and pressing ecological problems are more likely to be sources of public concern and support for local environmental protection than in rich countries [31,81,83]. In the present study, given the global scope of climate change, respondents from wealthier EU-27 countries are expected to report more pro-environmental attitudes and extra behavioral engagement in climate change mitigation, compared to citizens of less-wealthy nations.

H7b. *Wealthier EU-27 countries will be positively associated with extra mitigation behavior.*

Unobserved heterogeneity in psycho-social associations

The implicit assumption in most environmental and climate change studies that data are collected from a single homogeneous population is, in general, unrealistic [84]. Individuals are often heterogeneous with regard to environmental psychographics—i.e., people hold different views and have different information levels of environmental problems, such as climate change—and relevant socio-demographic characteristics that can positively (or negatively) relate to pro-environmental action [85]. Such heterogeneity in public conceptualizations and preferences for mitigating climate change impacts is acknowledged as a central question in the climate change research literature [1–2,9,60]. Imposing the assumption of homogeneity—when, in fact, there is substantial (psychographic or socio-demographic) heterogeneity within the population—is likely to produce misleading inferences and biased results [84,86]. In particular, if heterogeneity across individuals is present but ignored in regression-based studies, researchers run the risk of obtaining inconsistent model parameters and probability estimates [86].

It is important to clarify that the term heterogeneity—as used in this article—refers to both *distinct subpopulations* and *variation across individuals* [87]. In general, two forms of heterogeneity are present in data sampled from a heterogeneous population: observed and unobserved to an analyst [86–87]. Observed heterogeneity has been frequently dealt with—in the study of pro-environmental and climate change behavior—by the use of observed socio-economic variables (e.g., demographics like gender) that define a priori subgroups [85,87–88]. However, few studies have incorporated—or attempted to uncover—unobserved heterogeneity in models of individual pro-environmental behavior or (even less so) of climate change mitigation behavior [85–86,88]. With long tradition in the marketing and management literature, unobserved heterogeneity is commonly given precedence over observed heterogeneity in uncovering subpopulations (i.e., for segmentation purposes) [84,87], probably for two main reasons. First, in uncovering unobserved heterogeneity researchers "let the data speak for itself" [87]; that is, subpopulations are unobserved by the analyst (not predefined) and have to be inferred from the data [87]. Second, unobserved heterogeneity is arguably the preferred approach for uncovering subpopulations on psychographic constructs, such as environmental attitudes and motivations [88].

Put simply, the issue of unobserved heterogeneity revolves around uncovering subgroups or segments with distinctive path model estimates [84]. In this study, potential unobserved heterogeneity is accounted for in both psychographic and socio-demographic correlates of extra personal mitigation behavior in response to climate change. By looking at the data through the lens of segmentation (i.e., response-based segmentation) [89], the findings will clarify if psycho-social associations with people's breath and level of self-reported behavioral engagement (i.e., extra vs. common) in climate change mitigation is affected by heterogeneity, or the extent to which heterogeneity exists [84].

Methods

Data source

The empirical analyses are performed on the cross-national dataset "Eurobarometer 69.2—Europeans' attitudes towards climate change". A primary goal of this EU-wide survey was to investigate European citizens' climate change-related attitudes and behavior. Data were collected between March 25th and May 4th 2008 by TNS Opinion & Social, at the request of the European Commission, Directorate-General for Communication, Research and Political Analysis Unit. The Eurobarometer survey covers the population—aged 15 and over—of the 27 EU member states, three candidate countries (Croatia, Turkey, and the Former Yugoslav Republic of Macedonia), and the Turkish Cypriot Community. In each country, a stratified, multistage probability sampling design was used to guarantee the reliability of national and European estimates. A total of 30,170 individuals were interviewed face-to-face in their homes, and in the appropriate national language. The questionnaire addressed European citi-

zens' self-reported attitudes, motivations, level of knowledge, and personal enactment of specific mitigation activities in response to climate change; other relevant measures were available, including materialist/post-materialist values and socio-demographic indicators, such as gender, age, education, political ideology, and country. Access to the Eurobarometer data was provided by the GESIS Data Archive for the Social Sciences (Cologne, Germany). A detailed description of the Eurobarometer dataset used here is available as electronic supplementary information (Appendix S1).

Measurement items

Indicators of mixed scale types (i.e., categorical and continuous items) were used to measure the outcome and independent variables of the study—detailed in Tables 1 to 4.

Measures of personal mitigation behavior. Respondents already engaged in some form of climate change-motivated activity were asked to report, on a binary nominal scale (1 = *yes*; 0 = *no*), whether they had undertaken each of 11 types of actions aimed at fighting climate change (see Table 2 for the list of behaviors); these activities entail different levels of mitigation difficulty, impact, and frequency in the behavioral domains of domestic energy/water conservation, waste reduction, eco-friendly transportation, and eco-shopping. An important behavioral domain not covered in this study—and reflective of high-impact mitigation behavior—is that of public sphere (public/political) environmental activism and citizenship. The focal outcome of interest, a categorical aggregate score of people's behavioral engagement (i.e., extra vs. common) in climate change mitigation, was created through segmentation analysis on all self-reported mitigation activities.

Measures of attitudes, motivations, and knowledge (see Table 3). Environmental attitudes (five items) and self-reported knowledge about climate change issues (three items) were both rated on four-point scales from 1 to 4; in the attitude measures, 1 indicates *totally* disagree and 4 indicates *totally agree*; in the self-reported knowledge items, 1 denotes *not at all informed* and 4 denotes *very well informed*. Ecological motivations were measured through five possible reasons for fighting climate change; all motivation items were rated on a binary scale (1 = *yes*; 0 = *no*).

Individual-level demographics (see Tables 1 and 4). Gender, measured as biological sex, was coded with 1 designating *male* and 2 designating *female*. Age, initially measured as a continuous variable, was divided into six age categories (coded from 1 to 6): *15–24, 25–34, 35–44, 45–54, 55–64,* and *65 years and over*. Education was measured by the age at which respondents stopped full-time education, and recoded into a five-category variable ranging from 0 to 4: *no full-time education, up to 15 years of age, 16–19 years old, 20 years and over,* and *still studying*. Political ideology was assessed through respondents' self-placement on a 10-point, left-to-right continuum; scores 1–4 were combined into 1 = *left/liberal*; categories 5–6 into 2 = *moderate*; and scores 7–10 into 3 = *right/conservative*.

Country-level variables (see Table 4). Participants' country is a nominal variable, with categories ranging from 1 to 33. Post-materialism, the first hypothesized explanation for country variations in citizens' self-reported behavioral engagement (i.e., extra vs. common) in climate change mitigation, was measured at the individual level through Inglehart's four-item materialist/post-materialist value battery [80]; latent class segmentation performed on these value priorities, and profiled by country, led to the classification of European countries into three materialist vs. post-materialist groups; the grouping variable was coded 1 = *materialist countries*, 2 = *countries with mixed values*, and 3 = *post-materialist*

Table 1. Demographic profile of respondents.

Demographic variables	Total sample (n = 30,170)	Subsample 1: environmentally active citizens (n = 17,233)	Subsample 2: environmentally inactive citizens (n = 12,937)	χ^2 tests: subsample 1 vs. 2			
				χ^2	d.f.	p	Phi/Cramer's V
Gender				7.72	1	0.005	0.016
Male	45.6%	44.9%	46.5%				
Female	54.4%	55.1%	53.5%				
Age				150.39	5	<0.001	0.071
15–24	12.6%	10.8%	15.1%				
25–34	15.1%	15.0%	15.2%				
35–44	17.2%	18.0%	16.2%				
45–54	17.3%	17.9%	16.5%				
55–64	16.9%	17.7%	15.8%				
65+ years	20.9%	20.7%	21.2%				
Education (age at which respondents stopped full-time education)				528.37	5	<0.001	0.132
No full-time education	0.5%	0.2%	0.8%				
15– years	21.9%	19.7%	24.8%				
16–19	41.7%	41.3%	42.3%				
20+ years	25.3%	29.5%	19.7%				
Still studying	8.4%	7.7%	9.3%				
Refusal/DK	2.2%	1.6%	3.1%				
Political ideology				659.16	3	<0.001	0.148
Left/liberal	24.1%	27.1%	20.3%				
Moderate	30.2%	31.7%	28.3%				
Right/conservative	24.6%	25.3%	23.8%				
Refusal/DK	21.0%	16.0%	27.7%				

Table 2. Behavioral characterization of "extra vs. common" mitigation behavior.

Relative sizes/Mitigation actions		Segment membership probabilities [a]	
		Segment 1: "common" mitigation behavior	**Segment 2: "extra"** mitigation behavior
Relative size of segments		0.7706	0.2294
qe6	Which of the following actions aimed at fighting climate change have you personally taken?		
qe6.1	You have purchased a car that consumes less fuel, or is more environmentally friendly	0.1345	0.2973
qe6.2	You are reducing the use of your car, for example by car-sharing or using your car more efficiently	0.1651	0.4316
qe6.3	You have chosen an environmentally friendly way of transportation (by foot, bicycle, public transport)	0.2654	0.4769
qe6.4	You are reducing your consumption of energy at home (for example by turning down air conditioning or heating, not leaving appliances on stand-by buying energy efficient products such as low-energy bulbs or appliances)	0.5842	0.9254
qe6.5	You are reducing your consumption of water at home (for example not leaving water running when washing the dishes, etc.)	0.5116	0.7926
qe6.6	Where possible you avoid taking short-haul flights	0.0570	0.2958
qe6.7	You have switched to an energy supplier or tariff supplying a greater share of energy from renewable sources than your previous one	0.0537	0.1558
qe6.8	You are separating most of your waste for recycling	0.6284	0.9305
qe6.9	You are reducing the consumption of disposable items (for example plastic bags, certain kind of packaging, etc.)	0.2618	0.8318
qe6.10	You buy seasonal and local products to avoid products that come from far away, and thus contribute to CO_2 emissions (because of the transport)	0.1500	0.6398
qe6.11	You have installed equipment in your own home that generates renewable energy (for example, a wind turbine, solar panels)	0.0411	0.1051

[a]Conditional (marginal) probabilities clarifying how segment-membership relates to each climate change mitigation action.

countries, which closely resembles the three-class classification used in Inglehart's short post-materialism index [80,90]. Eurostat data on GDP per capita (year 2008) was used as a proxy for EU countries' wealth; based on terciles of GDP per capita, countries were divided into three wealth groups (coded from 1 to 3): countries with *low*, *intermediate*, and *high* wealth levels.

Statistical methodology

This paper applies latent class models (LC cluster and regression), as implemented in the Latent Gold v4.5 software, to synthesize the outcome variable and test the hypotheses linking psychographic and socio-demographic variables to the breath and level of personal mitigation behavior (i.e., extra vs. common) in response to climate change. Latent class analysis provides a powerful probabilistic approach for capturing *unobserved* heterogeneity in survey responses, and is especially useful for modeling (dependent and independent) categorical variables with varying numbers of categories [91–92], as in the present study. Alternative methods such as multi-group SEM allow researchers to account for *observed* heterogeneity—instead of unobserved heterogeneity—, where both the source of variation and subpopulations are known and defined a priori by the analyst. A more detailed description of statistical analysis and procedures is available as electronic supplementary information (Appendix S2).

Results

Descriptive profile of respondents

The sample is well-balanced in terms of gender, age, education, and political ideology (see Table 1 for results); yet, there was greater participation of female (54.4%), middle-aged (mean age = 47.6 years), and moderately educated individuals (41.7%), with a center political orientation (30.2%). All participants were asked to report whether they had "personally taken actions aimed at helping to fight climate change"; more than half of the sample (57.1%; n = 17,233) *totally agreed* or *tended to agree* with this statement. Demographically, the subsamples of environmentally active and inactive EU citizens differed significantly (based on χ^2 tests), but weakly (based on association measures such as Phi and Cramer's V), in gender, age, education, and political ideology. As expected, the subsample of EU citizens already engaged in some form of climate change-motivated activity is an older, better-educated, leftist/liberal, female group, compared to environmentally inactive respondents. Subsequent analyses focused only on environmentally (i.e., climate change) active EU citizens in the year 2008 (subsample 1 in Table 1). The decision to restrict the analyses to the subsample of climate change-active citizens is consistent with this study's investigation of the correlates of "extra vs. common" personal engagement in mitigation behavior—i.e., why some people go beyond what most other environmentally active people do to mitigate climate change through extra personal action. However, it is worth raising a cautionary note about

Table 3. Psychographic associations with extra mitigation behavior (model set 1).

Independent variables	Parameter estimates (z-values)[a]		Wald	p-value	Wald (=)	p-value
	Class 1	Class 2				
Model 1a: Environmental attitudes						
qe5... *Agreement with the following statements:*					*class-independent*	
qe5.1 Climate change is an unstoppable process, we cannot do anything about it	−0.04 (−2.56)	−0.04 (−2.56)	6.5	0.011		
qe5.2 The seriousness of climate change has been exaggerated	n.s.	−0.30 (−2.52)	6.4	0.012	-	-
qe5.3 Emission of CO_2 (carbon dioxide) has only a marginal impact on climate change	−0.10 (−5.51)	n.s.	30.4	3.5e-08	-	-
qe5.4 Fighting climate change can have a positive impact on the European economy	n.s.	0.26 (1.80)	3.3	0.072	-	-
qe5.5 Alternative fuels, such as 'bio fuels', should be used to reduce GHG emissions	0.19 (2.55)	−1.07 (−3.24)	71.2	3.5e-16	20.5	6.1e-06
Model 1b: Ecological motivations						
qe7... *Reasons for taking mitigation actions:*						
qe7.1 You think that if everybody changed their behavior, it will have a real impact on CC	0.14 (12.3)		152.0	6.4e-35		
qe7.2 You think that it is your duty as a citizen to protect the environment	0.08 (7.36)		54.2	1.8e-13		
qe7.3 You are very concerned about the world that you will leave for the future and young generations	0.18 (16.9)		286.6	2.8e-64		
qe7.4 You think that taking these actions will save your money	0.09 (7.78)		60.5	7.5e-15		
qe7.5 You have been directly exposed to the consequences	n.s.		0.9	0.34		
Model 1c: Environmental knowledge						
qe3... *Well informed about:*						
qe3.1 The different causes of climate change	−0.20 (−1.36)	0.67 (2.35)	9.8	0.007	9.5	0.002
qe3.2 The different consequences of climate change	0.40 (3.06)	−0.40 (−1.33)	17.4	0.000	8.4	0.004
qe3.3 Ways in which we can fight climate change	−0.16 (−1.28)	0.41 (2.27)	7.2	0.027	7.2	0.007

[a]Parameter estimates represent class-specific associations, of each independent variable, with extra mitigation behavior; z-values in brackets.

Table 4. Socio-demographic associations with extra mitigation behavior (model set 2).

Independent variables	Model 2a			Model 2b		
	Parameter estimates (z-values)[a]	Wald	p-value	Parameter estimates (z-values)[a]	Wald	p-value
Individual-level demographics:						
Gender		57.2	3.9e-14		61.1	5.3e-15
Male	-0.083 (-7.566)			-0.086 (-7.819)		
Female	0.083 (7.566)			0.086 (7.819)		
Age		82.4	2.6e-16		87.3	2.5e-17
15–24	-0.370 (-6.924)			-0.380 (-7.132)		
25–34	-0.073 (-2.635)			-0.084 (-3.013)		
35–44	0.095 (3.848)			0.097 (3.906)		
45–54	0.163 (6.688)			0.157 (6.427)		
55–64	0.120 (4.804)			0.132 (5.276)		
65+ years	0.066 (2.595)			0.078 (3.079)		
Education		147.7	8.2e-32		157.8	5.6e-34
15– years	-0.285 (-9.638)			-0.300 (-10.097)		
16–19	-0.038 (-1.682)			-0.051 (-2.276)		
20+ years	0.125 (5.469)			0.119 (5.182)		
Still studying	0.198 (3.827)			0.232 (4.490)		
Political ideology		19.7	5.4e-05		20.3	3.9e-05
Left/liberal	0.058 (3.774)			0.064 (4.138)		
Moderate	0.008 (0.539)			-0.002 (-0.108)		
Right/conservative	-0.066 (-4.030)			-0.062 (-3.781)		
Country-level variables:						
Materialism/post-materialism (country groups)		542.8	1.3e-118			
Materialist countries	-0.395 (-21.925)					
Countries with mixed values	0.102 (6.610)					
Post-materialist countries	0.294 (19.231)					
Wealth (country groups)					684.6	2.1e-149
Low				-0.296 (-12.530)		
Intermediate				-0.125 (-6.311)		
High				0.421 (26.050)		

Note: Models 2a and 2b separately include each of the two country-level variables, along with the four individual-level demographics.
[a]Parameter estimates represent category-specific associations, of each independent variable, with extra mitigation behavior; z-values in brackets.

generalizing the main study's findings to environmentally inactive people; that is, what drives extra mitigation behavior may differ greatly between environmentally active vs. inactive population segments.

Capturing extra mitigation behavior

As suggested in previous sections, LC cluster analysis was conducted on a set of 11 self-reported climate change mitigation actions (see Table 2), so as to capture and define the levels of the focal outcome variable of the study; thus, respondents were classified as showing different breath and level of personal engagement in mitigation behavior in response to climate change. The values of three segment retention criteria (BIC, AIC3, CAIC) and their percent reductions were examined for each of 10 potential cluster-solutions. These reached a minimum value for either 8 or 9 clusters; however, cluster-solutions with more than two segments were not regarded as appropriate, owing to practically insignificant percent reductions in all segment retention criteria (less than 1%), and excessive classification errors (over 10%) for the use of segment-membership as the outcome variable in subsequent regression analyses (see supplementary information—Appendix S3). The 2-cluster solution was also preferred over more complex models in terms of interpretability of the segment profiles, thus providing a higher theoretical and practical value. In sum, the results yielded an optimal solution of two segments (i.e., two differentiated behavior patterns) of EU citizens currently engaged in some form of climate change-motivated activity. This finding confirmed the distinction of two levels of "extra vs. common" personal engagement in mitigation behaviors.

Latent class sizes and levels of engagement in climate change mitigation behavior were substantially unbalanced between the two differentiated segments; *segment 1* (extra mitigation behavior) and *segment 2* (common mitigation behavior) respectively accounted for 77.1% and 22.9% of environmentally active respondents. The observations were reweighted to correct for potential biases in the latent class (segment) sizes (see Appendix S1 for details); nonetheless, the comparison of reweighted and unweighted LC cluster analyses revealed very similar results. As detailed in Table 2, respondents in segment 2 reported greater participation in mitigation efforts, compared to EU citizens showing more common types and levels of mitigation behavior (and classified into segment 1), in areas such as separating garbage for recycling (93.1% vs. 62.8%), and reducing the consumption of energy (92.5% vs. 58.4%) and water (79.3% vs. 51.2%) at home. Segment profiles differed even more in anti-shopping actions; a large majority of respondents in segment 2 claimed to be reducing their consumption of disposable items (83.2%), and avoiding products that come from far away-places (64.0%), compared to much lower shares in segment 1. The less common pro-environmental behaviors, both in the segments of people reporting extra and common mitigation behavior, refer to installing renewable energy systems in the household, switching to a greener energy supplier or tariff, and (where possible) avoiding taking short-haul flights. Overall, the profile of segment 2—and its comparison with that of segment 1—is largely in accordance with the definition of extra mitigation behavior in response to climate change given here: broader and greater engagement levels across all the specific mitigation actions and behavioral domains being examined (i.e., domestic energy/water conservation, waste reduction, eco-friendly transportation, and eco-shopping), compared to people showing more common mitigation behavior to address climate change.

Regression results

Following LC cluster (segmentation) analysis, respondents' segment membership—i.e., a two-level categorical measure differentiating extra from common behavioral engagement in climate change mitigation—was treated as the outcome variable in two sets of LC regression models. Respectively, model sets 1 and 2 examine how environmental psychographics and socio-demographics relate to the breath and level of self-reported personal mitigation behavior in response to climate change, while exploring the extent of unobserved heterogeneity.

Psychographic correlates (model set 1). The psychographic variables assessed in *model set 1* include respondents' self-reported knowledge (about the causes, ways of fighting, and consequences of climate change), positive/desirable and negative/undesirable attitudes (towards the threat of climate change and the role of mitigation efforts), and altruistic and egoistic ecological motivations (for mitigating climate change). Knowledge, attitude, and motivation associations were first examined in separate models, so as to ascertain the level of heterogeneity (if any) in the relationship of each subset of psychographic variables to the breath and level of self-reported personal mitigation behavior. Within the range of 1 to 3 latent classes, *2-class* solutions were deemed optimal in "attitude" and "knowledge" models (models 1a and 1c, respectively)—thus suggesting that unobserved heterogeneity exists in how knowledge and attitude variables relate to extra mitigation behavior, across two subgroups of environmentally active EU citizens; conversely, the *1-class* solution of homogeneity was preferable in the "motivational" model (model 1b). For a detailed description of the criteria supporting the model comparison and selection process, please see supplementary information (Appendix S3).

As an important matter of clarification at this point, the following should be noted: (1) *1-class* models entail the existence of a similar (perfectly homogeneous) pattern of association between the examined correlates with extra mitigation behavior for all environmentally active respondents; (2) *2-class* models imply that the correlates show a different pattern of association with extra mitigation behavior between two subgroups of environmentally active respondents.

All variables analyzed in the "attitude model" (model 1a) had significant links to the breath and level of self-reported personal mitigation behavior (see Table 3). For the most part, attitude associations were class-dependent—that is, four of the five attitude variables under investigation (items qe5.2 to qe5.5) behaved differently across two subgroups of environmentally active respondents. In the largest subgroup (class 1), as expected, positive/desirable attitude items (qe5.5) linked positively, and negative/undesirable attitude items (qe5.1 and qe5.3) negatively, to extra mitigation behavior. In the smallest subgroup (class 2), negative/undesirable attitude items (qe5.1 and qe5.2) were also, as hypothesized, negatively linked to extra mitigation behavior; the findings were mixed for the positive/desirable attitude items in class 2—that is, showing positive and negative associations (respectively, items qe5.4 and qe5.5) with extra mitigation behavior. In summary, the trends observed here for attitude items—despite heterogeneity across two groups environmentally active EU citizens—provide substantial support for hypotheses H1a and H1b positing that positive/desirable attitudes (towards the threat of climate change and the role of mitigation efforts) would be associated positively, and negative/undesirable attitudes negatively, with some people's extra mitigation behavior in response to climate change, as compared to others.

In the 1-*class* "motivational model" (model 1b), almost all tested variables were significantly associated with the breath and level of self-reported personal mitigation behavior—except for one motivation item: qe7.5, *you have been directly exposed to the consequences of climate change* (Wald $= 0.92$; $p = 0.34$)—a finding that relates to the literature on personal experience of and behavioral responses to climate change (see the Discussion section). The other four altruistic and egoistic motivational variables (items qe7.1 to qe7.4) were significant positive correlates of extra mitigation behavior. The findings stressed the greater importance of respondents' social-altruistic motivations in understanding why some people engage in extra mitigation behavior to address climate change, as compared to others; two such social-altruistic items (qe7.3 and qe7.1) ranked first and second, respectively, in order of statistical significance (see Table 3). These results shed light on RQ1, by showing the existence of both self-transcendent (altruistic) and self-enhancement (egoistic) motives for some people's extra mitigation behavior, as compared to others, and that self-transcendent (altruistic) motives can be expected to show greater positive (desirable) associations than egoistic ones.

The three knowledge variables (about climate change issues) tested in the "knowledge model" (model 1c) were significant, class-dependent correlates of the breath and level of self-reported personal mitigation behavior (see Table 3). Interestingly, the associations of environmental knowledge items were all in the opposite direction across two classes of environmentally active respondents. In the largest subgroup (class 1), only item qe3.2, *knowledge about the consequences of climate change*, was positively associated with extra behavioral engagement to mitigate climate change; conversely, better *knowledge about the causes* (item qe3.1) and *ways of fighting climate change* (item qe3.3) was linked to extra mitigation behavior in the smallest segment (class 2). These findings are considered to partially support H2, in that both knowledge about the causes and ways of fighting climate change positively relate to extra mitigation behavior in response to climate change, but only in the comparatively small segment of more environmentally engaged EU citizens. Contrary to the authors' expectations, *knowledge about the consequences* of climate change may relate to extra mitigation efforts (in response to climate change) for the majority of environmentally active EU citizens.

Overall, the findings of the three related psychographic models (model set 1) provide evidence of unobserved heterogeneity in attitude and knowledge correlates—across two subgroups of environmentally active citizens—, and homogeneity in motivational correlates of people's breath and level of behavioral engagement (i.e., extra vs. common) in mitigation behavior to address climate change. Because of different number of *latent classes* in the "motivational model" (1-*class* model)—compared to the "attitude model" and the "knowledge model" (2-*class* models), these three types of psychographic variables were not entered into a single model simultaneously.

Socio-demographic correlates (model set 2). Preliminary inspection of separate socio-demographic associations with the breath and level of personal mitigation behavior suggested the consideration of *1-class* models of homogeneity as optimal solutions across the variables under investigation. Thus—unlike in the regression model set 1—, individual and country-level socio-demographics could be entered jointly into LC regression models. Two different specifications of *model set 2* (models 2a and 2b) were then tested to examine the relationships of four individual-level demographics (gender, age, education, and political ideology) and two country-level variables (materialism/post-materialism and wealth) with EU citizens' extra (vs. common) behavioral engagement in climate change mitigation. In models 2a and 2b, each country-level variable was separately investigated, along with individual-level demographics. Within the range of 1 to 3 latent

classes, 1-*class* solutions were deemed optimal in the socio-demographic models 2a and 2b, thus revealing homogeneity in socio-demographic associations with extra mitigation behavior (see Appendix S3 for details).

All tested socio-demographic variables were significantly associated with the breath and level (extra vs. common) of climate change mitigation behavior in the final, 1-*class* versions of models 2a and 2b (see Table 4). Country-level variables were most significantly associated with extra mitigation behavior, with wealth ranking first and materialism/post-materialism second in statistical significance. The post-materialism hypothesis for explaining country variations in EU citizens' extra (vs. common) mitigation behavior was tested and strongly supported in model 2a; accordingly, the post-materialist country group showed the strongest positive association with extra personal engagement in mitigation behavior. Further, the results of model 2b confirmed the importance of the wealth hypothesis; only the wealthiest country group showed a positive association with extra behavioral engagement to mitigate climate change. In summary, the previous findings yielded full support for the two hypotheses involving country-level variables (support for H7a and H7b).

The parameter estimates for individual-level demographics were almost identical in the alternative model specifications 2a and 2b (all significant at $p < 0.001$). As expected, female gender was positively, although somewhat weakly, related to extra mitigation behavior to address climate change (support for H3). As suggested by the visual inspection of findings in Table 4, a curvilinear-like or "plateau" relationship appeared to exist between age and the breath and level of personal mitigation behavior, with the 45–56 age group showing the highest association with extra mitigation action. A complementary test of non-linearity was performed to clarify the precise form of age associations; for this purpose, the age variable—with linear ranges and of equal span—was converted into a quasi-interval scale and the authors tested whether a linear, curvilinear, or quadrilateral relationship was the best way to describe age relations to extra mitigation behavior. As expected, and detailed in the supplementary information (Appendix S4), a curvilinear function provided the optimal fit to the age data. Thus, there was considerable support for hypothesis H4 of a curvilinear (inverted U-shaped) relationship between age and EU citizens' extra behavioral engagement in mitigation behavior—mostly grounded in the interplay of cohort and life-cycle age effects on pro-environmental (climate change) concern and behavior. Regarding education, a positive association was identified between the age at which respondents stopped full-time education and extra mitigation behavior—which is fully supportive of H5. Finally, the results supported the hypothesis (H6) that people with a leftist/liberal political orientation would display extra (i.e., broader and greater) behavioral engagement in climate change mitigation, compared to right-wing/conservative environmentally active respondents.

Discussion

A fundamental shift is needed towards broader and greater levels of behavioral engagement in the population that provide further incremental benefits in addressing climate change. The cross-national research reported here aimed to enhance understanding of what makes some people make an extra commitment (i.e., do "more") to mitigate climate change through personal action, as compared to others. The authors assessed the role of psychographics, individual and country-level socio-demographics, and the extent of (unobserved) heterogeneity in an individual's extra behavioral engagement to address climate change.

In line with past environmental research (e.g., [93–94]), segmentation analysis revealed two heterogeneous (and differentiated) action patterns in response to climate change, among environmentally active EU citizens. These findings validated the a priori, intuitive two-level distinction of "extra vs. common" mitigation behavior in the focal variable of interest. As expected, engaging extra to mitigate climate change through personal behavior, compared to more common (or typical) mitigation behavior, was reflected in broader and greater self-reported engagement in all specific mitigation endeavors—entailing different levels of difficulty, impact, and frequency—and behavioral domains being examined [8,14].

The definition and measurement in aggregate of extra mitigation behavior allowed the analysis of factors that could influence broader and greater levels of mitigation behavior in a variety of settings. Overall, the findings reinforce earlier evidence that environmental psychographics (i.e., attitudes, motivations, and knowledge about climate change) are more associated with personal mitigation efforts in response to climate change than socio-demographics [12].

A central question in this study was: Is there heterogeneity in psycho-social correlates of extra mitigation behavior? The findings revealed that unobserved heterogeneity significantly affects how attitude and knowledge variables relate to extra (vs. common) personal behavior to mitigate climate change, but does not affect the links with motivations or socio-demographics. These results warn of the risk of ignoring the potential presence of unobserved heterogeneity in the analysis of attitudinal and knowledge associations with climate change-motivated behavior—i.e., researchers run the risk of obtaining biased or inconsistent results.

Positive/desirable and negative/undesirable attitude variables were significantly associated with extra mitigation behavior. Unlike most expectancy-value models [15,34,40], attitudes were not measured here only at the level of behavior; however, the congruent level of specificity/generality in attitudes (towards the impact of climate change and mitigation efforts) and self-reported mitigation behavior (in response to climate change), the close time correspondence between attitudes and behaviors, coupled with the assessment of aggregate mitigation behavior—i.e., offsetting differences in behavioral control across 11 types of mitigation actions—, is likely to have reinforced the significance attitude relations to behavior [33,39–40]. Overall, the findings emphasize the importance of building citizens' positive/desirable attitudes toward climate change issues (i.e., feelings that climate change is harmful and that mitigation actions are important), and reducing negative/undesirable ones (i.e., feelings that climate change is not harmful and that mitigation behavior is not important or ineffective)—but differently for two subgroups of environmentally active EU citizens—, so as to effectively promote broader and greater levels of engagement in mitigation action in response to climate change.

An important finding concerned the role of self-reported knowledge about climate change issues. Respondents' level of knowledge about the causes and ways of fighting climate change—on the one side—, and knowledge about the consequences of climate change—on the other side—were all significantly, but inversely, related to extra mitigation behavior, across two (environmentally active) respondent subgroups. In contrast with most previous studies [8,33,55], informing the public about the consequences of climate change appears to be more useful, than informing about the causes and ways of fighting climate change, in promoting extra mitigation behavior for the majority of environmentally active citizens—i.e., people that currently display limited action levels in response to climate change. Thus, not all types of

climate change messages and information can be assumed to be equally effective in promoting extra mitigation efforts in the population.

In line with previous studies (e.g., [50]), both altruistic and egoistic motivations were positively associated with extra (vs. common) personal behavior to mitigate climate change for all environmentally active respondents. Thus, interventions aimed at encouraging citizens to undertake more ambitious mitigation efforts should appeal to both altruistic and egoistic ecological motives and values. Nonetheless, as suggested in earlier work (e.g., [95]), the results warn that motivational influences on pro-environmental behavior tend to vary in strength (and importance). The only motivational variable not significantly associated with the extra mitigation behavior concerned respondents' direct exposure to the consequences of climate change. Such a finding (as noted earlier) adds to the inconclusive evidence in the literature on the association of first-hand experience of climate change consequences and personal engagement in mitigation behavior [96]. There is evidence to suggest that personal experience of different types of climate change consequences (e.g., flooding vs. air pollution) are likely to elicit different personal behavioral responses [97]—an issue that could not be addressed here and requires further investigation.

This study further suggests that—despite equivocal evidence from previous environmental studies (e.g., [16,26,51])—, a variety of individual and country-level socio-demographics are significant correlates of extra mitigation behavior in response to climate change. Country-level variables were most significantly related to respondents' level of climate change mitigation behavior. In line with Franzen's work (e.g., [31,81]), strong support was obtained for the post-materialism and wealth hypotheses—i.e., broader and greater levels of behavioral engagement to mitigate climate change in EU countries with predominant post-materialist values and in wealthier nations. These findings exemplify the cultural and economic underpinnings of personal mitigation behavior in response to climate change. Accordingly, climate change campaigns encouraging people to do "more" or "a lot" through environmentally responsible behaviors should be tailored to each cultural and wealth EU country group.

The results for individual-level demographics also support the contention that demographic variables continue to be significant correlates of pro-environmental behavior [22,57,60,63]. As hypothesized, the findings confirm the need to encourage broader and greater mitigation responses to climate change, especially among men, conservatives, less educated individuals, and both the youngest and oldest population groups. The curvilinear (inverted U-shaped) link between age and extra mitigation behavior warn environmental researchers of potential non-linear associations of socio-demographics with pro-environmental (climate change) behavior. Thus, researchers are advised to use statistical methods which can detect the presence of potential non-linearities (and heterogeneity), such as latent class analysis or neural networks.

Limitations and Recommendations for Further Study

The present study has a number of limitations that should be addressed in future work. First, the use of secondary survey data—i.e., the Eurobarometer dataset—limited the scope of the research questions and the operationalization of the study variables. Despite the advantages of Eurobarometer data, such as providing a rich, cross-national source of information—with minimum time and financial investment—, its use inhibited the analysis of other relevant correlates of extra mitigation behavior, such as behavioral intentions, subjective norm, or perceived behavioral control [40]. Thus, future studies should consider including unmodeled factors such as citizens' perceptions of self-relevance or involvement with climate change and global warming issues, using alternative and refined measures of psychographics (e.g., attitudes toward the behavior) and behavior (e.g., lagged or future mitigation action), and assessing situational deterrents (and enhancers) of broader and greater levels of behavioral engagement to mitigate climate change. In addition, the important behavioral domain of public sphere (public/political) environmental activism and citizenship, and reflective of high-impact mitigation behavior—that could not be covered in this study—should be accounted for in the future.

Second, this study is based on self-reported (instead of actual or objectively measured) mitigation behavior and the correlates of interest. On the one hand, the focus on self-reported mitigation behavior, coupled to the simultaneous measurement of attitudes and behavior, did not allow the analysis of attitude–action gap in climate change-motivated behavior—an important issue that warrants further investigation through comparison of self-reports and objective measures of actual behavior. On the other hand, it is possible that social desirability biases may have played a role, leading some respondents to overstate their knowledge about climate change issues and their engagement in mitigation actions. Past research has shown the correlation and redundancy between measures of perceived and objective environmental knowledge, but also their potentially differential associations with pro-environmental behaviors [54]. The disparity between perceived and objective knowledge (and their relationships) deserves future attention in studies on climate change-motivated behavior. As regards mitigation behavior, social desirability arguably represents a minor problem in light of the analysis of a representative, non-student sample of environmentally active EU citizens, and the satisfactory correspondence of the self-report measures used in this study to *past/present* behavior—measured as *yes/no* present enactment of specific mitigation actions. Non-student samples, people with high scores of ecological behavior, and measures of past/present (rather than *intended/future*) pro-environmental behavior are likely to be less affected by social desirability biases [52].

Third, the analyses reported here were restricted only to participants already engaged in some form of climate change-motivated activity (i.e., environmentally active citizens). This decision was consistent with this study's investigation of the correlates of "extra vs. common" to mitigate climate change through personal behavior. In pursuing this objective, respondents not reporting any action on climate change did not provide relevant information on the focal variable (extra mitigation behavior) and its underlying motivations. Yet, the exclusion from the analyses of environmentally inactive respondents should be acknowledged as a significant methodological and practical limitation that cautions against generalizing the present study's findings to environmentally inactive people—around 40% of EU-27 citizens [10–11]; this is important because environmentally active vs. inactive population segments are likely to differ in what drives extra (i.e., broader and greater levels of) mitigation behavior. The shortcomings of the current approach could be overcome in future studies that extend understanding of what makes people do "a lot" to mitigate climate change beyond the more receptive (and arguably less challenging) population segment of environmentally active citizens. Such broader investigations would make additional progress toward ambitious goals in the public's behavioral engagement to mitigate climate change.

Fourth, this study examined only direct (psychographic and socio-demographic) correlates of extra climate change mitigation behavior. However, full understanding of the complex and dynamic mechanisms involved in such environmentally significant behavior requires that direct, indirect, and moderating influences be considered. Statistical methodologies such as structural equation modeling (SEM) or partial least squares (PLS) are most appropriate to unravel the interplay between internal and external correlates of broader and greater levels of personal response to climate change [69]—for instance, by placing the variables under the nomological structure of expectancy-value models [40]. If possible, future research should explore lagged effects of knowledge, motivations, and attitudes on intentions and future behavior.

Future research should continue to address the important topic of unobserved heterogeneity in pro-environmental decisions [93–94]. Particularly, additional empirical evidence is needed to verify the heterogeneous (class-dependent), psychographic links to personal mitigation behavior in response to climate change, observed in this study. The significance and strength of country-level associations warrant additional international analyses of pro-environmental behavior in response to climate change. Also, the interaction between individual-level psychographics (e.g., knowledge and attitudes) and countries' wealth and post-materialism levels deserves closer attention in relation to personal mitigation behavior. Preliminary analyses (not reported here but available upon request) show that, amongst post-materialist and wealthier EU countries, the findings tend to be more consistent with previous literature [8,33,55]—e.g., there is greater presence of the segment of environmentally active people for whom knowledge about the causes and ways of fighting climate change relates to extra behavioral engagement to mitigate climate change, and positive/desirable and negative/undesirable attitudes are significantly associated (positively and negative, respectively) with extra mitigation behavior. Certainly, wealth and post-materialism are not the only valid approaches to explaining country differences in climate change mitigation behavior. Other potentially relevant cultural dimensions for cross-national environmental studies include harmony [21]; individualism, long-term orientation, and locus of control [77]; and traditional and altruistic values [72]. Country and regional heterogeneity in environmental policy and legislations are also likely to account for national differences in public mitigation behavior. Multilevel analysis seems to be most appropriate for environmental research questions involving variables from different levels—e.g., individual and country-level influences on climate change-motivated behavior.

Concluding Remarks

As a result of the variety of variables analyzed in relation to extra (vs. common) personal behavior to mitigate climate change, the following can be concluded:

1. A profound shift is needed in personal behavior—from inaction or limited action levels—toward extra (i.e., broader and greater levels of) behavioral engagement to mitigate climate change.

2. The population segments of environmentally active and inactive EU citizens differed significantly, but weakly, in demographic terms.

3. Environmentally active citizens—the population segment under study—can be differentiated in two intuitive categories: people reporting "extra" vs. "common" behavioral engagement to mitigate climate change.

4. Both psychographics and (individual and country-level) socio-demographics help to explain why some people make an extra commitment to mitigate climate change through personal action, as compared to others.

5. Psychographics tend to show greater association than socio-demographics with extra mitigation behavior in response to climate change.

6. There is heterogeneity in the associations involving attitude and knowledge variables, whereas homogeneity exists in the links of motivations and socio-demographics with extra mitigation behavior.

7. The findings draw attention to the importance of potential non-linearities in socio-demographic correlates.

8. The study has implications for promoting more ambitious behavioral responses to climate change, both at the individual level and across countries.

Supporting Information

Appendix S1   Detailed description of Eurobarometer 69.2.

Appendix S2   Supplementary description of statistical analysis.

Appendix S3   Criteria supporting the model comparison and selection process: LC cluster and regression models.

Appendix S4   Complementary test of non-linear age associations.

Acknowledgments

The authors gratefully thank the GESIS Data Archive for the Social Sciences (Cologne, Germany) for their continued support and providing access to the dataset used in this study. We are also grateful to Dr. Bayden D. Russell and two anonymous reviewers for their insightful and constructive feedback on earlier versions of this manuscript.

Author Contributions

Conceived and designed the experiments: JMO NGF RAL. Performed the experiments: JMO NGF RAL. Analyzed the data: JMO NGF RAL. Contributed reagents/materials/analysis tools: JMO NGF RAL. Wrote the paper: JMO NGF RAL.

References

1. Lorenzoni I, Nicholson-Cole S, Whitmarsh L (2007) Barriers perceived to engaging with climate change among the UK public and their policy implications. Global Environ Change 17: 445–459. doi: 10.1016/j.gloenvcha.2007.01.004

2. Whitmarsh L (2011) Scepticism and uncertainty about climate change: Dimensions, determinants and change over time. Global Environ Change 21: 690–700. doi: 10.1016/j.gloenvcha.2011.01.016

3. Semenza JC, Hall DE, Wilson DJ, Bontempo BD, Sailor DJ, George LA (2008) Public perception of climate change: Voluntary mitigation and barriers to behavior change. Am J Prev Med 35: 479–487. doi: 10.1016/j.amepre.2008.08.020

4. Junta de AndalucûUa (2010) Andalucía Innova. Especial cambio climático. Available: http://www.andaluciainvestiga.com/revista/pdf/100PreguntasMedioAmbiente/100medioambiente.pdf. Accessed 2013 Jun 3.

5. European Environment Agency (2013) Final energy consumption by sector. Available: http://www.eea.europa.eu/data-and-maps/indicators/final-energy-consumption-by-sector-5/assessment#toc-1. Accessed 2013 Jun 3.

6. Bin S, Dowlatabadi H (2005) Consumer lifestyle approach to US energy use and the related CO₂ emissions. Energy Policy 33: 197–208. doi: 10.1016/S0301-4215(03)00210-6

7. Whitmarsh L, Seyfang G, O'Neill S (2011) Public engagement with carbon and climate change: to what extent is the public 'carbon capable'? Global Environ Change 21: 56–65. doi: 10.1016/j.gloenvcha.2010.07.011

8. Whitmarsh L (2009) Behavioural responses to climate change: Asymmetry of intentions and impacts. J Environ Psychol 29: 13–23. doi: 10.1016/j.jenvp.2008.05.003

9. Howell RA (2011) Lights, camera . . . action? Altered attitudes and behaviour in response to the climate change film *The Age of Stupid*. Global Environ Change 21: 177–187. doi: 10.1016/j.gloenvcha.2010.09.004

10. European Commission (2008) Europeans' attitudes towards climate change. Directorate-General for Communication. Available: http://ec.europa.eu/public_opinion/archives/ebs/ebs_300_full_en.pdf. Accessed 2009 Nov 30.

11. European Commission (2011) Climate change. Directorate-General for Communication. Available: http://ec.europa.eu/public_opinion/archives/ebs/ebs_372_en.pdf. Accessed 2013 Jun 3.

12. Whitmarsh L, O'Neill S (2010) Green identity, green living? The role of pro-environmental self-identity in determining consistency across diverse pro-environmental behaviours. J Environ Psychol 30: 305–314. doi: 10.1016/j.jenvp.2010.01.003

13. Poortinga W, Spence A, Whitmarsh L, Capstick S, Pidgeon NF (2011) Uncertain climate: An investigation into public scepticism about anthropogenic climate change. Global Environ Change 21: 1015–1024. doi: 10.1016/j.gloenvcha.2011.03.001

14. Thøgersen J, Crompton T (2009) Simple and painless? The limitations of spillover in environmental campaigning. J Consum Pol 32: 141–163. doi: 10.1007/s10603-009-9101-1

15. Bamberg S, Moser G (2007). Twenty years after Hines, Hungerford and Tomera: A new meta-analysis of psychosocial determinants of pro-environmental behavior. J Environ Psychol 27: 14–25. doi: 10.1016/j.jenvp.2006.12.002

16. Stern PC (2000) Toward a coherent theory of environmentally significant behavior. J Soc Issues 56: 407–424. doi: 10.1111/0022-4537.00175

17. Cheung SF, Chan DK-S, Wong ZS-Y (1999) Reexamining the theory of planned behavior in understanding wastepaper recycling. Environ Behav 31: 587–612. doi: 10.1177/00139169921972254

18. Hopper JR, Nielsen JM (1991) Recycling as altruistic behavior: Normative and behavioral strategies to expand participation in a community recycling program. Environ Behav 23: 195–220. doi: 10.1177/0013916591232004

19. Abrahamse A, Steg L, Gifford R, Vlek C (2009) Factors influencing car use for commuting and the intention to reduce it: A question of self-interest or morality? Transport Res F 12: 317–324. doi: 10.1016/j.trf.2009.04.004

20. Bamberg S, Schmidt P (2003) Incentives, morality or habit? Predicting students' car use for university routes with the models of Ajzen, Schwartz, and Triandis. Environ Behav 35: 264–285. doi: 10.1177/0013916502250134

21. Oreg S, Katz-Gerro T (2006) Predicting proenvironmental behavior cross-nationally: values, the theory of planned behavior, and value-belief-norm theory. Environ Behav 38: 462–483. doi: 10.1177/0013916505286012

22. Scott D, Willits FK (1994) Environmental attitudes and behavior: A Pennsylvania survey. Environ Behav 26: 239–260. doi: 10.1177/001391659426002006

23. Stern PC, Dietz T, Guagnano GA (1995) The new ecological paradigm in social-psychological context. Environ Behav 27: 723–743. doi: 10.1177/0013916595276001

24. Stern PC, Dietz T, Kalof L (1993) Value orientations, gender and environmental concern. Environ Behav 25: 322–348. doi: 10.1177/0013916593255002

25. Kollmuss A, Agyeman J (2002) Mind the gap: Why do people act environmentally and what are the barriers to pro-environmental behavior? Environ Educ Res 8: 239–260. doi: 10.1080/13504620220145401

26. Diamantopoulos A, Schlegelmilch BB, Sinkovics RR, Bohlen GM (2003) Can socio-demographics still play a role in profiling green consumers? A review of the evidence and an empirical investigation. J Bus Res 56: 465–480. doi: 10.1016/S0148-2963(01)00241-7

27. Mohai P, Twight BW (1987) Age and environmentalism: An elaboration of the Buttel model using national survey evidence. Soc Sci Q 68: 798–815.

28. Iyer ES, Kashyap RK (2007) Consumer recycling: Role of incentives, information, and social class. J Consum Behav 6: 32–47. doi: 10.1002/cb.206

29. Clark CF, Kotchen MJ, Moore RM (2003) Internal and external influences on pro-environmental behavior: Participation in a green electricity program. J Environ Psychol 23: 237–246. doi: 10.1016/S0272-4944(02)00105-6

30. Dunlap RE, Mertig AG (1995) Global concern for the environment: Is affluence a prerequisite? J Soc Issues 51: 121–137. doi: 10.1111/j.1540-4560.1995.tb01351.x

31. Franzen A (2003) Environmental attitudes in international comparison: An analysis of the ISSP surveys 1993 and 2000. Soc Sci Q 84: 297–308. doi: 10.1111/1540-6237.8402005

32. Parker SK, Williams HM, Turner N (2006) Modeling the antecedents of proactive behavior at work. J Appl Psychol 91: 636–652. doi: 10.1037/0021-9010.91.3.636

33. Kaiser FG, Wölfing S, Fuhrer U (1999) Environmental attitude and ecological behaviour. J Environ Psychol 19: 1–19. doi: 10.1006/jevp.1998.0107

34. Hines JM, Hungerford HP, Tomera AN (1986/87) Analysis and synthesis of research on responsible environmental behavior: A meta-analysis. J Environ Educ 18: 1–8.

35. Fishbein M, Ajzen I (1975) Belief, attitude, intention, and behavior: An introduction to theory and research. Reading, MA: Addison-Wesley.

36. Ajzen I, Gilbert N (2008) Attitudes and the prediction of behavior. In: Crano WD, Prislin R, editors. Attitudes and attitude change. New York, NY: Psychology Press. pp. 289–311.

37. Smith SM, Haugtvedt CP, Petty RE (1994) Attitudes and recycling: Does the measurement of affect enhance behavioral prediction? Psychol Market 11: 359–374. doi: 10.1002/mar.4220110405

38. Kaiser FG, Keller C (2001) Disclosing situational constraints to ecological behavior: a confirmatory application of the mixed Rasch model. Eur J Psychol Assess 17: 212–221. doi: 10.1027//1015-5759.17.3.212

39. Fazio RH (1990) Multiple processes by which attitudes guide behavior: The MODE model as an integrative framework. In: Zanna MP, editor. Advances in experimental social psychology, Vol. 23. New York, NY: Academic Press. pp. 75–109.

40. Ajzen I (1991) The theory of planned behavior. Organ Behav Hum Dec 50: 179–211. doi: 10.1016/0749-5978(91)90020-T

41. Webb TL, Sheeran P (2006) Does changing behavioral intentions engender behavior change? A meta-analysis of the experimental evidence. Psychol Bull 132: 249–268. doi: 10.1037/0033-2909.132.2.249

42. Epstein S (1983) Aggregation and beyond: Some basic issues on the prediction of behavior. J Pers 51: 360–392. doi: 10.1111/j.1467-6494.1983.tb00338.x

43. Moisander J (2007) Motivational complexity of green consumerism. Int J Consum Stud 31: 404–409. doi: 10.1111/j.1470-6431.2007.00586.x

44. Booth-Kewley S, Vickers RRJr (1994) Associations between major domains of personality and health behavior. J Pers 62: 281–298. doi: 10.1111/j.1467-6494.1994.tb00298.x

45. Gagnon Thompson S, Barton M (1994) Ecocentric and anthropocentric attitudes toward the environment. J Environ Psychol 14: 149–157. doi: 10.1016/S0272-4944(05)80168-9

46. Schultz PW, Zelezny L (1999) Values as predictors of environmental attitudes: Evidence for consistency across 14 countries. J Environ Psychol 19: 255–265. doi: 10.1006/jevp.1999.0129

47. De Groot JIM, Steg L (2008) Value orientations to explain beliefs related to environmental significant behavior: How to measure egoistic, altruistic, and biospheric value orientations. Environ Behav 40: 330–354. doi: 10.1177/0013916506297831

48. Schwartz SH (1977) Normative influences on altruism. In: Berkowitz L, editor. Advances in experimental social psychology: Vol. 10. New York, NY: Academic Press. pp.221–279. doi: 10.1016/S0065-2601(08)60358-5

49. Nordlund AM, Garvill J (2002). Value structures behind proenvironmental behavior. Environ Behav 34: 740–756. doi: 10.1177/001391602237244

50. De Young R (2000) Expanding and evaluating motives for environmentally responsible behavior. J Soc Issues 56: 509–526. doi: 10.1111/0022-4537.00181

51. Mobley C, Vagias WM, DeWard SL (2010) Exploring additional determinants of environmentally responsible behavior: The influence of environmental literature and environmental attitudes. Environ Behav 42: 420–447. doi: 10.1177/0013916508325002

52. Frick J, Kaiser FG, Wilson M (2004) Environmental knowledge and conservation behavior: Exploring prevalence and structure in a representative sample. Pers Indiv Differ 37: 1597–1613. doi: 10.1016/j.paid.2004.02.015

53. McCright AM, Dunlap RE (2011) Cool dudes: The denial of climate change among conservative white males in the United States. Global Environ Change 21: 1163–1172. doi: 10.1016/j.gloenvcha.2011.06.003

54. Ellen PS (1994) Do we know what we need to know? Objective and subjective knowledge effects on pro-ecological behaviors. J Bus Res 30: 43–52. doi: 10.1016/0148-2963(94)90067-1

55. Kaiser FG, Fuhrer U (2003) Ecological behavior's dependency on different forms of knowledge. Appl Psychol-Int Rev 52: 598–613. doi: 10.1111/1464-0597.00153

56. Immerwahr J (1999) Waiting for a signal: Public attitudes towards global warming, the environment and geophysical research. Public Agenda Foundation. Available: http://www.policyarchive.org/handle/10207/bitstreams/5662.pdf. Accessed 2011 Feb 15.

57. O'Connor RE, Bord RJ, Fisher A (1999) Risk perceptions, general environmental beliefs, and willingness to address climate change. Risk Anal 19: 461–471. doi: 10.1023/A:1007004813446

58. Jones RE, Dunlap RE (1992) The social bases of environmental concern: Have they changed over time? Rural Sociol 57: 28–47. doi: 10.1111/j.1549-0831.1992.tb00455.x

59. Roberts JA (1996) Will the real socially responsible consumer please step forward? Bus Horizons 39: 79–83. doi: 10.1016/S0007-6813(96)90087-7

60. Leiserowitz A (2006) Climate change risk perception and policy preferences: The role of affect, imagery, and values. Climatic Change 77: 45–72. doi: 10.1007/s10584-006-9059-9

61. Xiao C, McCright A (2007) Environmental concern and socio-demographic variables: A case study of statistical models. J Environ Educ 38: 3–13. doi: 10.3200/JOEE.38.1.3-14

62. Slimak MW, Dietz T (2006) Personal values, beliefs, and ecological risk perception. Risk Anal 26: 1689–1705. doi: 10.1111/j.1539-6924.2006.00832.x

63. Dietz T, Stern PC, Guagnano GA (1998) Social structural and social psychological bases of environmental concern. Environ Behav 30: 450–471. doi: 10.1177/001391659803000402

64. Sutton SG, Tobin RC (2011) Constraints on community engagement with Great Barrier Reef climate change reduction and mitigation. Global Environ Change 21: 894–905. doi: 10.1016/j.gloenvcha.2011.05.006

65. Zelezny LC, Chua P, Aldrich C (2000) Elaborating on gender differences in environmentalism. J Soc Issues 56: 443–457. doi: 10.1111/0022-4537.00177

66. Hunter LM, Hatch A, Johnson A (2004) Cross-national gender variation in environmental behaviors. Soc Sci Q 85: 677–694. doi: 10.1111/j.0038-4941.2004.00239.x

67. Slovic P (1999) Trust, emotion, sex, politics, and science: Surveying the risk-assessment battlefield. Risk Anal 19: 689–701. doi: 10.1111/j.1539-6924.1999.tb00439.x

68. Davidson DJ, Freudenburg WR (1996) Gender and environmental risk concerns: A review and analysis of available research. Environ Behav 28: 302–339. doi: 10.1177/0013916596283003

69. Cottrell SP (2003) Influence of sociodemographics and environmental attitudes on general responsible environmental behavior among recreational boaters. Environ Behav 35: 347–375. doi: 10.1177/0013916503035003003

70. Vining J, Ebreo A (1990) What makes a recycler? A comparison of recyclers and nonrecyclers. Environ Behav 22: 55–73. doi: 10.1177/0013916590221003

71. Newman K (1986) Personal values and commitment to energy conservation. Environ Behav 18: 53–74. doi: 10.1177/0013916586181003

72. Aoyagi-Usui M, Vinken H, Kuriyabashi A (2003) Pro-environmental attitudes and behaviors: An international comparison. Hum Ecol Rev 10: 23–31.

73. Gelissen J (2007) Explaining popular support for environmental protection: A multilevel analysis of 50 nations. Environ Behav 39: 392–415. doi: 10.1177/0013916506292014

74. Guerin D, Crete J, Mercier J (2001) A multilevel analysis of the determinants of recycling behavior in the European countries. Soc Sci Res 30: 195–218. doi: 10.1006/ssre.2000.0694

75. Gifford R, Scannell L, Kormos C, Smolova L, Biel A, et al. (2009) Temporal pessimism and spatial optimism in environmental assessments: An 18-nation study. J Environ Psychol 29: 1–12. doi: 10.1016/j.jenvp.2008.06.001

76. Kemmelmeier M, Król G, Kim YH (2002) Values, economics and proenvironmental attitudes in 22 societies. Cross-Cult Res 36: 256–285. doi: 10.1177/10697102036003004

77. Sarigöllü E (2009) A cross-country exploration of environmental attitudes. Environ Behav 41: 365–386. doi: 10.1177/0013916507313920

78. Bord RJ, Fisher A, O'Connor RE (1998) Public perceptions of global warming: United States and international perspectives. Climate Res 11: 75–84. doi: 10.3354/cr011075

79. Brechin SR (2003) Comparative public opinion and knowledge on global climatic change and the Kyoto protocol: The U.S. against the world? Int J Sociol Soc Pol 23: 106–134. doi: 10.1108/01443330310790318

80. Inglehart R (1997) Modernization and postmodernization: Cultural, economic, and political change in 43 societies. Princeton, NJ: Princeton University Press.

81. Franzen A, Meyer R (2010) Environmental attitudes in cross-national perspective: A multilevel analysis of the ISSP 1993 and 2000. Eur Sociol Rev 26: 219–234. doi: 10.1093/esr/jcp018

82. Brechin SR, Kempton W (1994) Global environmentalism: A challenge to the postmaterialism thesis? Soc Sci Q 75: 245–269.

83. Diekmann A, Franzen A (1999) The wealth of nations and environmental concern. Environ Behav 31: 540–549. doi: 10.1177/00139169921972227

84. Vinzi VE, Ringle CM, Squillacciotti S, Trinchera L (2007) Capturing and treating unobserved heterogeneity by response based segmentation in PLS path modeling. A comparison of alternative methods by computational experiments. ESSEC Research Center Working Paper 07019. Paris, France: ESSEC Business School.

85. Ojea E, Loureiro ML (2007) Altruistic, egoistic and biospheric values in willingness to pay (WTP) for wildlife. Ecol Econ 63: 807–814. doi: 10.1016/j.ecolecon.2007.02.003

86. Bhat CR (2000) Incorporating observed and unobserved heterogeneity in urban work travel mode choice modeling. Transport Sci 34: 228–238. doi: 10.1287/trsc.34.2.228.12306

87. Tay L, Diener E, Drasgow F, Vermunt JK (2011) Multilevel mixed-measurement IRT analysis: An explication and application to self-reported emotions across the world. Organ Res Methods 14: 177–207. doi: 10.1177/1094428110372674

88. Aldrich GA, Grimsrud KM, Thacher JA, Kotchen MJ (2007) Relating environmental attitudes and contingent values: How robust are methods for identifying preference heterogeneity? Environ Resour Econ 37: 757–775. doi: 10.1007/s10640-006-9054-7

89. Barr S, Shaw G, Coles T (2011) Times for (Un)sustainability? Challenges and opportunities for developing behaviour change policy. A case-study of consumers at home and away. Global Environ Change 21: 1234–1244. doi: 10.1016/j.gloenvcha.2011.07.011

90. Moors G (2007) Testing the internal validity of the Inglehart thesis by means of a latent class choice model. Acta Sociol 50: 147–160. doi: 10.1177/0001699307077656

91. Magidson J, Vermunt JK (2004) Latent class models. In: Kaplan D, editor. The Sage handbook of quantitative methodology for the social sciences. Thousand Oaks, CA: Sage Publications. pp. 175–198.

92. Vermunt JK, Magidson J (2002) Latent class cluster analysis. In: Hagenaars JA, McCutcheon AL, editors. Applied latent class analysis. Cambridge, UK: Cambridge University Press. pp. 89–106.

93. Bodur M, Sarigöllü E (2005) Environmental sensitivity in a developing country: Consumer classification and implications. Environ Behav 37: 487–510. doi: 10.1177/0013916504269666

94. Dolnicar S, Grün B (2009) Environmentally friendly behavior: Can heterogeneity among individuals and contexts/environments be harvested for improved

sustainable management? Environ Behav 41: 693–714. doi: 10.1177/0013916508319448

95. Thøgersen J (2009) The motivational roots of norms for environmentally responsible behavior. Basic Appl Soc Psychol 31: 348–362. doi: 10.1080/01973530903317144

96. Spence A, Poortinga W, Butler C, Pidgeon NF (2011) Perceptions of climate change and willingness to save energy related to flood experience. Nature Clim Change 1: 46–49. doi: 10.1038/nclimate1059

97. Whitmarsh L (2008) Are flood victims more concerned about climate change than other people? The role of direct experience in risk perception and behavioural response. J Risk Res 11: 351–374. doi: 10.1080/13669870701552235

Permissions

The contributors of this book come from diverse backgrounds, making this book a truly international effort. This book will bring forth new frontiers with its revolutionizing research information and detailed analysis of the nascent developments around the world.

We would like to thank all the contributing authors for lending their expertise to make the book truly unique. They have played a crucial role in the development of this book. Without their invaluable contributions this book wouldn't have been possible. They have made vital efforts to compile up to date information on the varied aspects of this subject to make this book a valuable addition to the collection of many professionals and students.

This book was conceptualized with the vision of imparting up-to-date information and advanced data in this field. To ensure the same, a matchless editorial board was set up. Every individual on the board went through rigorous rounds of assessment to prove their worth. After which they invested a large part of their time researching and compiling the most relevant data for our readers.

The editorial board has been involved in producing this book since its inception. They have spent rigorous hours researching and exploring the diverse topics which have resulted in the successful publishing of this book. They have passed on their knowledge of decades through this book. To expedite this challenging task, the publisher supported the team at every step. A small team of assistant editors was also appointed to further simplify the editing procedure and attain best results for the readers.

Apart from the editorial board, the designing team has also invested a significant amount of their time in understanding the subject and creating the most relevant covers. They scrutinized every image to scout for the most suitable representation of the subject and create an appropriate cover for the book.

The publishing team has been an ardent support to the editorial, designing and production team. Their endless efforts to recruit the best for this project, has resulted in the accomplishment of this book. They are a veteran in the field of academics and their pool of knowledge is as vast as their experience in printing. Their expertise and guidance has proved useful at every step. Their uncompromising quality standards have made this book an exceptional effort. Their encouragement from time to time has been an inspiration for everyone.

The publisher and the editorial board hope that this book will prove to be a valuable piece of knowledge for researchers, students, practitioners and scholars across the globe.

List of Contributors

Matjaž Kuntner
Institute of Biology, Scientific Research Centre, Slovenian Academy of Sciences and Arts, Ljubljana, Slovenia
Centre for Behavioural Ecology and Evolution, College of Life Sciences, Hubei University, Wuhan, Hubei, China
National Museum of Natural History, Smithsonian Institution, Washington, D. C., United States of America

Magdalena Năpăruş
Centre of Landscape–Territory–Information Systems - CeLTIS, University of Bucharest, Bucharest, Romania
Tular Cave Laboratory, Kranj, Slovenia

Daiqin Li
Centre for Behavioural Ecology and Evolution, College of Life Sciences, Hubei University, Wuhan, Hubei, China
Department of Biological Sciences, National University of Singapore, Singapore, Singapore

Jonathan A. Coddington
National Museum of Natural History, Smithsonian Institution, Washington, D. C., United States of America

Matthew Gordon and Jamie Seymour
School of Marine and Tropical Biology, James Cook University, Cairns, Queensland, Australia

Mirco Bundschuh, Frank Seitz, Ricki R. Rosenfeldt and Ralf Schulz
Institute for Environmental Sciences, University of Koblenz-Landau, Landau, Germany

Ernesto Azzurro
Institut de Cie`ncies del Mar (ICM-CSIC), Barcelona, Spain
High Institute for Environmental Protection and Research (ISPRA), Laboratory of Milazzo, Milazzo, Italy

Paula Moschella
Commission Internationale pour l'Exploration Scientifique de la Mer Me´diterrane´e (CIESM), Monaco, Monaco

Francesc Maynou
Institut de Cie`ncies del Mar (ICM-CSIC), Barcelona, Spain

Joshua E. Cinner, Cindy Huchery, Nicholas A. J. Graham and Christina C. Hicks
Australian Research Council Centre of Excellence for Coral Reef Studies, James Cook University, Townsville, Queensland, Australia

Emily S. Darling
Earth to Ocean Research Group, Simon Fraser University, Burnaby, British Columbia, Canada

Austin T. Humphries
Coastal Research Group, Rhodes University, Grahamstown, South Africa
Coral Reef Conservation Project, Wildlife Conservation Society, Mombasa, Kenya

Nadine Marshall
Ecosystem Sciences, Commonwealth Scientific and Industrial Research Organisation, Townsville, Queensland, Australia

Tim R. McClanahan
Marine Programs, Wildlife Conservation Society, Bronx, New York, United States of America

Chris Walzer and Christine Kowalczyk
Research Institute of Wildlife Ecology, Department of Integrative Biology and Evolution, University of Veterinary Medicine, Vienna, Austria

Jake M. Alexander and Christoph Kueffer
Institute of Integrative Biology, Swiss Federal Institute of Technology Zurich, Switzerland

Bruno Baur
Department of Environmental Sciences, University of Basel, Basel, Switzerland

Giuseppe Bogliani
Department of Earth and Environmental Sciences, University of Pavia, Pavia, Italy

Jean-Jacques Brun
Unité Ecostémes Montagnards, National Research Institute of Science and Technology for Environment and Agriculture, Saint Martin d'Hères, France

Leopold Füreder
Institute of Ecology, University of Innsbruck, Innsbruck, Austria

Marie-Odile Guth
Ministry of Ecology, Sustainable Development and Energy, Paris, France

Ruedi Haller
Research and Geoinformation, Swiss National Park, Zernez, Switzerland

Rolf Holderegger
Research Unit Biodiversity and Conservation Biology, Swiss Federal Institute for Forest, Snow and Landscape Research, Birmensdorf, Switzerland

Yann Kohler
Task Force Protected Areas - Permanent Secretariat of the Alpine Convention, Chambery, France

Antonio Righetti
Partner-innen in Umweltfragen, Liebefeld, Switzerland

Reto Spaar
Swiss Ornithological Institute, Sempach, Switzerland

William J. Sutherland
Conservation Science Group, Department of Zoology, University of Cambridge, Cambridge, United Kingdom

Aurelia Ullrich-Schneider
International Commission for the Protection of the Alps-CIPRA International, Schaan, Liechtenstein

Sylvie N. Vanpeene-Bruhier
É cosystèmes Méditerranéens et Risques, National Research Institute of Science and Technology for Environment and Agriculture, Aix-en-Provence, France

Thomas Scheurer
Swiss Academy of Sciences, Berne, Switzerland

Astrid Layton
George W. Woodruff School of Mechanical Engineering, Sustainable Design and Manufacturing, Georgia Institute of Technology, Atlanta, Georgia, United States of America

John Reap
School of Business and Engineering, Quinnipiac University, Hamden, Connecticut, United States of America

Bert Bras
George W. Woodruff School of Mechanical Engineering, Sustainable Design and Manufacturing, Georgia Institute of Technology, Atlanta, Georgia, United States of America
Center for Biologically Inspired Design, Georgia Institute of Technology, Atlanta, Georgia, United States of America

Marc Weissburg
School of Biology, Georgia Institute of Technology, Atlanta, Georgia, United States of America
Center for Biologically Inspired Design, Georgia Institute of Technology, Atlanta, Georgia, United States of America

Sebastián Bustos
Center for International Development, Harvard University, Cambridge, Massachusetts, United States of America
Harvard Kennedy School, Harvard University, Cambridge, Massachusetts, United States of America

Charles Gomez
Graduate School of Education, Stanford University, Stanford, California, United States of America

Ricardo Hausmann
Harvard Kennedy School, Harvard University, Cambridge, Massachusetts, United States of America
Santa Fe Institute, Santa Fe, New Mexico, United States of America

César A. Hidalgo
Center for International Development, Harvard University, Cambridge, Massachusetts, United States of America
The MIT Media Lab, Massachusetts Institute of Technology, Cambridge, Massachusetts, United States of America
Instituto de Sistemas Complejos de Valparaíso, Valparaíso, Chile

Martina G. Vijver
Institute of Environmental Sciences (CML), Leiden University, Leiden, The Netherlands

Paul J. van den Brink
Alterra, Wageningen University and Research centre, Wageningen, The Netherlands
Wageningen University, Wageningen University and Research centre, Wageningen, The Netherlands

Jocelyn L. Aycrigg and Anne Davidson
National Gap Analysis Program, Department of Fish and Wildlife Sciences, University of Idaho, Moscow, Idaho, United States of America

Leona K. Svancara
Idaho Department of Fish and Game, Moscow, Idaho, United States of America

Kevin J. Gergely
United States Geological Survey Gap Analysis Program, Boise, Idaho, United States of America

Alexa McKerrow
United States Geological Survey Gap Analysis Program, Raleigh, North Carolina, United States of America

J. Michael Scott
Department of Fish and Wildlife Sciences, University of Idaho, Moscow, Idaho, United States of America

Pascal Sornom, Eric Gismondi, Céline Vellinger, Simon Devin, Jean-François Férard, Jean-Nicolas Beisel
Laboratoire des Interactions, Ecotoxicologie, Biodiversité, Ecosyste`mes, CNRS UMR 7146, Université de Lorraine, Metz, France

Lluís Brotons
CEMFOR-CTFC, Forest Sciences Center of Catalonia, Solsona, Catalonia, Spain
CREAF, Centre for Ecological Research and Forestry Applications, Cerdanyola del Vallè s, Catalonia, Spain

Ana Luisa Caetano, Catarina R. Marques and Fernando Gonçalves
Department of Biology, University of Aveiro, Campus Universitário de Santiago, Aveiro, Portugal
CESAM, University of Aveiro, Campus Universitário de Santiago, Aveiro, Portugal

Ana Gavina
Interdisciplinary Centre of Marine and Environmental Research (CIIMAR/CIMAR), University of Porto Porto, Portugal
CESAM, University of Aveiro, Campus Universitário de Santiago, Aveiro, Portugal

Fernando Carvalho
Nuclear and Technological Institute (ITN) Department of Radiological Protection and Nuclear Safety, Sacavém, Portugal

Eduardo Ferreira da Silva
Department of Geosciences, University of Aveiro, GeoBioTec Research Center, Campus Universitário de Santiago, Aveiro, Portugal

Ruth Pereira
Interdisciplinary Centre of Marine and Environmental Research (CIIMAR/CIMAR), University of Porto Porto, Portugal
Department of Biology, Faculty of Sciences of the University of Porto, Porto, Portugal

Ameena H. El-Bibany, Andrea G. Bodnar and Helena C. Reinardy
Molecular Discovery Laboratory, Bermuda Institute of Ocean Sciences, St. George's, Bermuda

Hector Galbraith
Manomet Center for Conservation Sciences, Manomet, Massachusetts, United States of America
National Wildlife Federation, Springfield, Massachusetts, United States of America

David W. DesRochers
Department of Natural Sciences, Dalton State College, Dalton, Georgia, United States of America

Stephen Brown
Manomet Center for Conservation Sciences, Manomet, Massachusetts, United States of America

J. Michael Reed
Department of Biology, Tufts University, Medford, Massachusetts, United States of America

Jennifer L. Neuwald and Nicole Valenzuela
Department of Ecology, Evolution and Organismal Biology, Iowa State University, Iowa, United States of America

Johannes Rousk
Environment Centre Wales, Bangor University, Gwynedd, United Kingdom
Section of Microbial Ecology, Department of Biology, Lund University, Lund, Sweden,

Kathrin Ackermann and Davey L. Jones
Environment Centre Wales, Bangor University, Gwynedd, United Kingdom

Simon F. Curling
BioComposites Centre, Bangor University, Gwynedd, United Kingdom

Cristina A. Viegas
Department of Bioengineering, Instituto Superior Técnico (IST), Technical University of Lisbon (TUL), Lisboa, Portugal
Institute for Biotechnology and Bioengineering (IBB), Centre for Biological and Chemical Engineering, Instituto Superior Técnico (IST), Technical University of Lisbon (TUL), Lisboa, Portugal

Catarina Costa
Institute for Biotechnology and Bioengineering (IBB), Centre for Biological and Chemical Engineering, Instituto Superior Técnico (IST), Technical University of Lisbon (TUL), Lisboa, Portugal

Sandra André and Paula Viana
Agência Portuguesa do Ambiente (APA), Amadora, Portugal

Rui Ribeiro and Matilde Moreira-Santos
Instituto do Mar (IMAR), Department of Life Sciences, University of Coimbra, Coimbra, Portugal

Brad H. McRae
The Nature Conservancy, North America Region, Seattle, Washington, United States of America

Sonia A. Hall
The Nature Conservancy, Washington Chapter, Wenatchee, Washington, United States of America

Paul Beier
School of Forestry and Merriam-Powell Center for Environmental Research, Northern Arizona University, Flagstaff, Arizona, United States of America

David M. Theobald
National Park Service, Inventory and Monitoring Division, Fort Collins, Colorado, United States of America

Christina L. Belanger
Department of the Geophysical Sciences, University of Chicago, Chicago, Illinois, United States of America

Caren B. Cooper and Jennifer Shirk
Cornell Lab of Ornithology, Ithaca, New York, United States of America

Benjamin Zuckerberg
University of Wisconsin, Madison, Wisconsin, United States of America

José Manuel Ortega-Egea, Nieves García-de-Frutos and Raquel Antolín-López
Department of Economics and Business, University of Almería (ceiA3), Almería, Spain

Index

www.ingramcontent.com/pod-product-compliance
Lightning Source LLC
Chambersburg PA
CBHW080533200326
41458CB00012B/4421

9781632398345